Digital TV over Broadband

Digital TV over Broadband:
HARVESTING BANDWIDTH

Joan Van Tassel

**Focal
Press**

Boston Oxford Auckland Johannesburg Melbourne New Delhi

Focal Press is an imprint of Butterworth–Heinemann.

Copyright © 2001 by Butterworth–Heinemann and Joan Van Tassel

 A member of the Reed Elsevier group

All rights reserved.

Recognizing the importance of preserving what has been written, Butterworth–Heinemann prints its books on acid-free paper whenever possible.

 Butterworth–Heinemann supports the efforts of American Forests and the Global ReLeaf program in its campaign for the betterment of trees, forests, and our environment.

Library of Congress Cataloging-in-Publication Data

Van Tassel, Joan M.
    Digital TV over broadband : harvesting bandwidth / Joan Van Tassel.
        p. cm.
    Includes bibliographical references and index.
    ISBN 0-240-80357-4 (pbk. : alk. paper)
    1. Digital television. 2. Broadband communication systems. I. Title.

TK6678 V36 2000
384.55'4—dc21   00-048432

British Library Cataloguing-in-Publication Data

A catalogue record for this book is available from the British Library.

The publisher offers special discounts on bulk orders of this book.
For information, please contact:
    Manager of Special Sales
    Butterworth–Heinemann
    225 Wildwood Avenue
    Woburn, MA 01801-2041
    Tel: 781-904-2500
    Fax: 781-904-2620

For information on all Focal Press publications available, contact our World Wide Web home page at: http://www.focalpress.com

10 9 8 7 6 5 4 3 2 1

Printed in the United States of America

This book is dedicated to
Bailey Michael and Emmy Joan Van Tassel.
Citizens of the twenty-first century.

Table of Contents

Acknowledgments

First and foremost, I am deeply grateful to Focal Press publisher Marie Lee for all her support, patience, and kindness over the two and a half years it has taken me to complete this book. Editor Terri Jadick has been a source of guidance and help, contributing a great deal throughout the process. And, of course, thanks to everyone at Focal Press for all your hard work.

Over the years, I have had the good fortune to learn from people who know far more than I ever will about their respective areas of expertise. Steve Rose spent many hours patiently coaching me in the intricacies of cable networks, servers, and video-on-demand service. James Bromley, a friend since high school, taught me about ham radio and TV in the eleventh grade. In the last few years, the courses included the basics of bandwidth, computer architecture, digital signal processing, and RF transmission.

Mark Bunzel, now at Intel, gave me a top-level view of the effects of digital technology on entertainment. He and I were both video producers earlier in our careers who recognized that computers would transform the communications industries. This shared background allowed us to engage in some of the most intense and interesting conversations in which I've ever been privileged to participate.

Then there's Peggy Miles. Founder of the International Webcasting Association, president of Intervox, and author of four books, she knows everything worth knowing about webcasting—and she's a team player who gives great introduction. Mariana Danilovich understands the business side of digital content better than anyone I've ever interviewed, and has generously invested her time in explaining the funding, incubation, and development of business models for creating and producing new media content.

Gary Arlen is a dear friend and knows everybody in the interactive arena. He's a fine teacher as well, with a domain of knowledge that extends from the Big Picture to just about every nook and cranny of the new digital world. Finally, although everyone now knows that Mark

Cuban is brilliant, I was lucky enough to have wonderful extended discussions with him before the world found out.

I am fortunate to cover the topics of this book in my work as a journalist. Editors nurture me and make it possible for me to explore the new dimensions of the digital landscape. Dr. Larry Vanston at Technology Futures, Inc. (www.tfi.com) is very supportive. At PricewaterhouseCoopers, Jennifer Walsh and Irene Mantikas keep me on the cutting edge of infrastructure and telecom with my TeleVistas column on the Telecom Direct website (www.telecomdirect.pwcglobal.com). Natalie Dickter lets me write the challenging articles on the changing world of entertainment for the Entertainment & Media Direct website (www.emdirect.pwcglobal.com). Special acknowledgments must go to Bennett McClellan and Saul Berman for giving me a chance to work with them on projects that let us all peer into the future. Both are gifted writers who kept me busy and humble.

Scott McKim, editor of new media and the website at *The Hollywood Reporter* (www.hollywoodreporter.com) is a pleasure to work for and has given me so much opportunity. Publisher Robert Dowling is deeply knowledgeable about the new landscape and has guided the publication to a pre-eminent position on the digital side, while maintaining its traditional status in the creative community as Hollywood's hometown trade paper. Paula Parisi, now editor of special issues, gave me my first assignment at *THR* and taught me so much about reporting and writing.

Alan Waldman has been an enthusiastic mentor, editor, agent, and friend. He is a tremendous and tireless supporter of writers and a talented journalist in his own right. He introduced me to the *Extra Extra* crew that gets out trade show daily papers with such skill and verve. Editors Kathy Haley and Antonette Goroch are the best—they really run a tight newsroom that manages to be exciting and fun as well as efficient.

I also teach at UCLA Extension School, which provides a venue for organizing and passing on my ideas. Director Ronnie Rubin is a dynamo who knows how to help people make the most of themselves. And co-instructor Deborah Hudson made my first teaching experience such a pleasure.

For work on this book, I thank Kathy Sadlier for helping to round up and organize the graphics. Debbie Reed did her usual wonderful job on the end notes and bibliography. Joyce St. Claire is a wonderful writer and a splendid copy editor as well.

I have a very special mentor, Mary Murphy. She's the wisest woman I know and contributes immeasurably to my life. When I'm in trouble, I turn to her and she's always there, with the right word at the right time. Another wonderful source of inspiration and humor are Uncle Robert and Aunt Irene Newton, who have learned to use the PC in their 60s. Kudos!

My family has been a source of love and support all my life. My mother, Lucille Newton-Van Tassel, always knew how to see life as a great adventure and a creative endeavor, and I thank her for that gift. My sibs are just too amazing and are my valued friends. Nancy is deeply creative and bright, and fun to be with to boot. I share so much with Elaine, who understands the human condition—and lives right down the street from me. Her husband Jim is nothing short of a gem. Gordon is a caring physician who is a genius with computers and helps me build my own. He got me started by suggesting I change out a hard drive. When I protested, he laughed and said: "What are you worried about? Just take it apart and figure it out. After all, no one is going to die!" His wife Karen is a joyous addition to the family, and their son Bailey brings together the best of them both. We are eagerly awaiting their new child, Emmy Joan.

To all of you—my deepest heartfelt thanks.

Introduction

I hope that people for whom the subjects covered in this book are new will take away a greater understanding of the basics of digital technologies, networks, and the industries that are growing up around them. For those who have been digital for awhile, I hope they are able to look at their world with new insights and an appreciation of the totality of the ongoing transformation.

When I wrote my first book, *Advanced Television Systems: Brave New TV,* I had a clear vision of what I wanted to write about. I saw it as an explication of a "high-definition, digital, connected, interactive, global network." Each of these words was a different seed in the digital garden. The first five chapters addressed each one, and the rest of the book considered them all together.

The goal of that book was to introduce people to the future that those five words entailed and to help them thread their way through the conceptual, technical, and verbal tangle. I used to write a weekly column for *The Malibu Times.* When I stopped writing it to concentrate on the book, I went in to let the publisher and editor know. Arnold York asked what the book was about, and I spit out the five-word descriptive phrase. He nodded and told me: "Thanks to two years of your columns, I understand every one of those words." He wished me luck and went on to become quite the digit-head himself. Thanks to Arnold's encouragement, I thought I might be able to achieve my goal.

Now it is nearly five years later and each of the seeds I wrote about in the first book has grown into a huge plant. Some are even gardens unto themselves. I thought that this second book would be a quick update, involving little more than replacing a few new figures, statistics, and research reports here and there. But the maturing garden had become infinitely more complex, and the journalistic work I had done in the intervening years had widened my own horizons.

So this book evolved into something rather different than I had planned. High-definition turned into variable definition. Digital theory became reality. Interactive came to mean the Internet. And a single network grew into multiple networks—cable, telephone, computer, power, and the many flavors of wireless—just before morphing into a ubiquitous broadband network. This book is certainly bigger than the first one and, I hope, better. I'm grateful to the dozens of people I've interviewed who have shared their knowledge and experience with me. The mistakes are mine.

Joan Van Tassel
Malibu, California
June 26, 2000

On the Companion Website at www.focalpress.com/companions:

- Special chapter: "A Matter of Standards"
- Glossary

The Digital Destiny

Brave New TV

Television isn't what it used to be. Nor is it yet what it is going to be. The digital revolution is taking over one of the most formidable bastions of the old analog empire: the television industry and its electronic video programming. Numeric data, text, audio, graphics, and photos have all been co-opted into the digital domain. After TV goes digital, only motion picture film will stand apart, and the advent of the digital distribution of cinema will arrive within a decade.

Television has been evolving for the past twenty years. In the 1970s, TV came into the home almost exclusively over the air. Only three broadcast networks and a handful of local independent stations offered programs. In the early 1980s, cable became available in many markets, and the number of channels people could choose to watch expanded to 15 or so, then to 30 or so, then to 50 or so. Today, most cable systems offer at least 50 channels, and many operators provide 100 or more choices. Satellite television delivers even more programming, bringing from 150 to 300 channels to its subscribers.

The transition now affecting television—the technology, the industry, and the cultural phenomenon—are no longer evolutionary. Although there will be some continuities between the present and the future, the differences between the old and the new will be far greater.

In the next decade, there will be a proliferation of delivery systems into the home. Some residential services will bring TV over wired systems, like the existing cable and telephone company lines. Other systems will bring in moving images over one of several types of wireless networks. The audience will also be watching TV over new displays, ranging from conventional television sets to desktop computer screens, handheld and other portable computers, personal digital assistants, phones with screens, and embedded screens in walls, cabinets, refrigerators, shopping carts, and other as yet unknown venues.

In just a few short years, the new delivery systems and reception devices will blur any distinctions between television and video. The similarities between them are obvious: Both are sequences of realistic, colorful moving images, viewed on screens, that arrive via over-the-air transmission or some kind of cable.

But there are some important differences as well. Television means "seeing far," and programming comes from some distant transmitter to a few locations in the home; video means, "I see," and it is anywhere and everywhere. Television is the professional product of an industry, expensive to acquire, edit, and distribute; video is a commonplace data type that can be relatively cheap to acquire, manipulate, and distribute. Television is programmed by specialists for the public; video is accessed by the public directly. TV is a "lean backward" technology; viewers watch it on a TV set, sitting from four to twenty feet away. Computers are a "lean forward" technology when they are on a desktop; but programs can be ported from the PC to the TV so that users can watch video on any display device—from any distance they want to.

Most important, as traditional television falls under the rule of the digital regime, it becomes entirely overshadowed by its video content, a situation exacerbated by the proliferation of producers, formats, means of distribution, and reception. So thorough is the absorption of television into the video space, there are many instances where the phrase digital video (DV) will replace the term digital television (DTV) altogether. As we will see, one of the most revolutionary aspects of digital technology is its ability to separate the message from the messenger, the content from the conduit.

This book covers all forms of digital video and digital television (DTV), technologies that are replacing the current analog NTSC television system. (NTSC stands for the National Television Standards Committee, an industry group of professional engineers who established the original standards for U.S. television in the 1940s and 1950s.) The eleven varieties of digital television include:

1. Television signal technologies: HDTV, SDTV, eTV, PConTV, and TVonPC
    *   HDTV: high-definition television
    *   SDTV: standard-definition television
    *   eTV: enhanced television
    *   PConTV: Internet and other data-based signals are received by the TV or set-top box and are seen on the TV screen
    *   Datacasting/TVonPC: Television signals sent to the PC and seen on the computer monitor
2. Standalone technologies: DVD and CD-ROM
    *   DVD: digital video or digital versatile disc sometimes called digital video disc
    *   CD-ROM: compact disc, read-only memory

3. Medium-capacity networked technologies: BBTV, or broadband digital TV
4. Narrow-capacity networked technologies:
    - ITV: Internet television
    - TTV: telephone television

Each type of video is tied to an end-to-end system of production, distribution, and reception. However, all types of digital video are much closer to one another than any of them are to traditional analog video.

There are three important aspects to think about when considering how to classify digital television. Table 1.1 shows how they are related to delivery platforms.

- Quality—Ranges from very low, with jerky motion and nearly unrecognizable images, to very high, such as satellite-delivered video
- Connectivity—None or standalone, one-way (linear), and two-way (interactive)
- Distribution platform—A variety of technologies and infrastructures that bring video to viewers and users

Quality, connectivity, and distribution platforms—these three variables are related because the distribution platform defines the channel that conveys the video or other information and the channel determines the levels of quality and the connectivity. Some distribution technologies, like cable, will carry beautifully detailed colorful images at full speed; others, such as existing telephone lines, will carry only small, blurry pictures that are really sequential snapshots taken from the stream of video images. Some, like traditional TV, provide quality pictures but not connectivity.

In the future, probably over a period of some years, the capacity of communication channels will become more or less equal and the infrastructure will be integrated. When this happens, the distribution platform will become largely irrelevant, at least to the final consumer. The quality of the video will be determined by the manner of its origination rather than its means of transport, and the people who receive the material will probably take the ability to interact with it for granted.

Members of the DTV Television Family: HDTV, SDTV, eTV, PConTV, and TVonPC

All these varieties of digital television are carried within the television signal. One result is that all of them are commercially developed and transmitted services created by industry professionals, which assures a

Table 1.1 Levels of Quality of Digital Video

Quality	Connectivity		
	Stand-Alone (Not Connected)	*One-Way (Linear)*	*Two-Way (Interactive)*
LOW— A few kilobits per second: 1.5 Mbps (narrowband)	CD-ROM	Linear video on the Internet, via dialup connection Some eTV signals that accompany and complement the TV picture	Some BBTV over telephone lines: ISDN, slow DSL, T-1 Some eTV signals TVonPC: Data sent to PCs as part of a TV signal, but do not appear on the TV PConTV: Internet on the TV services TTV: Videotelephony and videoconferencing ITV: Computer TV over dialup modem, 28.8–56k
MEDIUM— 1.5Mb/s– 2.5Mb/s (midband and broadband)	VCR	Some digital tier and DSL programming	Some PConTV service: like satellite WebTV Slower-speed interactive cable digital tier and DSL programming
HIGH— 2.5Mb/s + (high bandwidth)	DVD	Broadcast TV Cable TV Satellite TV Digital VCR	Some interactive high-speed cable digital tier and DSL telephone lines such as HDSL and VDSL

Note: If you are not familiar with the meaning of the numbers and abbreviations listed under the different levels of quality, the next chapter will explain them thoroughly. In the meantime, the visually apparent difference in quality between Internet video, a VCR, and broadcast TV are known to enough people to make clear what low-, medium-, and high-quality video images look like.

relatively high quality of production and technical proficiency. Image quality is a product both of the professional people who gather and process the video, as well as the expensive equipment they use. At every point, much effort is expended to create the best-looking, most error-free material possible and to avoid anything that would degrade the signal. For the most part, amateurs and average individuals do not have the technical expertise or the equipment to format content for carriage over satellite, cable, and local television station signals.

Regardless of which platform carries it, any one of these formats of professional DTV might simply be called digital television. The most commonly discussed formats are SDTV and HDTV, although there are several others. Standard-definition television (SDTV) is a digital image that is roughly equivalent to today's analog TV, while HDTV, high-definition television, offers a much more detailed, vibrant image and CD-quality audio. In their initial implementations, HD, SD, and TVonPC are one-way media platforms, whether the images are uncompressed in the studio or compressed for delivery via satellite, terrestrial broadcast, or cable.

Chapter 4, "High-Definition Television: Or How DTV Came Into Being Because HBO Needed Cash," will cover the quarter century of development that it took to reach a deployable DTV system and the plans for implementing it in the United States and Europe. For now, it suffices to say that DBS satellite systems broadcast SDTV nationally and are using its higher quality to take customers away from analog cable systems. Within a few years, every TV station in the United States will transmit it and every cable system will carry it. Both satellite and terrestrial TV systems will also deliver HDTV, although not for the majority of their programming.

ETV, or enhanced TV, refers to data that accompanies the television picture and usually complements it. WebTV and the Wink system are both examples of this kind of programming. However, the additional information may arrive on the consumer TV set in a variety of ways but all of them involve putting it in unused portions of the analog TV signal. Thus, eTV is a way for broadcasters to add data and interaction capabilities to analog TV right now, permitting them to experiment and acquire experience with ideas and formats in the years before a majority of consumers will have these services in their homes.

PConTV services put Internet content on the TV set. Almost everyone has a TV set—about 99% of households. But only about 50% of U.S. homes have computers, and about 35% of all homes are online. Many companies who have products or programs that depend on the Internet see expanding their reach to the TV set as an important way to reach a bigger audience. WebTV is the best-known example, and AOL TV will soon launch as well. In Europe, NetGem offers its NetBox service in France, Germany, Austria, and Sweden, and other countries.

TVonPC is often called datacasting. It is similar to eTV in that it is sent as part of the TV signal, but it is received by the PC rather than the TV. WavePhore is an example of a TVonPC service, where, through an agreement with PBS local affiliate stations, the company sends digital signals to the WaveTop device. The company offers stock quotes and news services that subscribers see on their PC screen. Another TVonPC service is provided by DIRECTV, called DIRECPC, which allows people to get data on their PCs from the same dish they use to get satellite TV service. All these companies require the customer to purchase some kind of home equipment and to pay a monthly subscription fee as well.

Disc-Based Digital Video: DVD and CD-ROM

Both these kinds of digital television are played on standalone platforms. Most often, the player is a peripheral in a game player (like Sony Playstation), or an internal disc player on a personal computer. Of course, the material on DVD and CD-ROM discs can be sent out over the air or over a wired network.

Both DVD and CD-ROM are recorded on a metallic disc. The DVD is variously referred to as a "digital versatile disc" or a "digital video disc." It provides very high-quality video with interactive activities designed right into the disc; programmers embed data that allows users to access the disc's information in different and unique ways. Moreover, when a user accesses a DVD that is played on a DVD player computer peripheral, there may be additional programming external to the DVD provided by someone else, a third party or the user. The added material will allow for entirely new and different applications that were never foreseen by the original disc designers. (See Figure 1.1.) If that computer is connected to a network, then users in remote locations may be able to access the material on the DVD player and use the images in conjunction with data and applications on the remote machine.

The DVD is replacing the CD-ROM; however, DVD players will also play the older discs. (The commonly used term for the ability of newer technologies to continue to function, even with content or processes from the previous model, is "backward compatibility.")

By contrast with the DVD, the images on CD-ROMs are of rather low quality. The video usually runs in a small window of the computer screen, which improves its appearance, although it sometimes appears in full-screen format. In both cases, images lack rich detail and the action is not very smooth.

The New Kid: Broadband Television (BBTV)

BBTV is digital television designed for a medium-capacity network—the digital cable tier or telephone company digital subscriber line (DSL) services. Its quality is about the same or a little lower than a VHS tape that has been out for rental quite a few times. BBTV doesn't have the crisp resolution and brilliant color saturation of satellite-delivered images; it's not even as

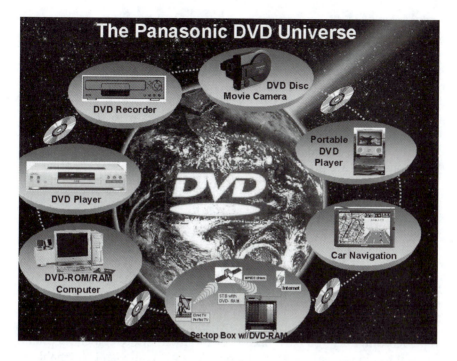

Figure 1.1 The DVD will be central to the entertainment industry. Source: Panasonic.

good as broadcast TV. But it is much better than video over the Internet. The resolution is adequate, the colors are bright, and the motion is smooth; most viewers find it quite watchable.

This is a new content category. There are dozens of broadband projects underway, some of them already playing on cable digital services such as RoadRunner and @Home. Many others are in the works. The programming ranges from interactive, data-laden sports applications from Quokka and BSkyB to interactive Domino pizza ads, to the offerings of Atom Films and other entertainment producers.

The development environment is very fluid. In the digital arena, where companies are creating material for the new technology platforms like the cable digital tier and telephone services like Asymmetrical Digital Subscriber Line (ADSL), enhanced TV often merges with broadband TV.

Narrowband Digital Television: Internet TV (ITV) and TV over the Telephone (TTV)

From the outset, both of these digital television formats were designed to run over narrowband computer and telephone networks. Internet TV

(ITV), called webcasting or netcasting, is streamed from a computer server across the Internet, or other computer network, to the user's PC desktop. Frequently, it isn't referred to as television at all; rather, it is called DV or just video, because on a computer network, video is simply a data type.

When video is webcast over the Internet, it usually looks pretty terrible. Since most people access the Internet using dialup modems over low bit rate telephone wires (at 28.8 kHz, 33.6 kHz, or 56 kHz), streamed material must be highly compressed. The most often-used adjectives used to describe the poor quality of streamed audiovisual content over regular telephone lines are postage-stamp-sized, fuzzy, and jerky. Computer TV (CTV) over larger capacity networks, like organizational Intranets, merges into mid-band BBTV.

It is important to remember that webcasting is new—it didn't even exist before 1995, when it took many minutes to download archived video to play back from the local hard drive. However, it is far better this year than it was last year, and it was better last year than the year before. In short, there has been a significant improvement in the quality as better compression techniques and faster modems have appeared.

TTV is telephone television, which includes videotelephony and videoconferencing. Like other narrowband video, when videotelephony or videoconferencing takes place over the Internet or regular telephone lines, it too looks really bad. Some people say it really isn't watchable at all. For the most part, when large corporations choose to videoconference, they use good quality cameras and send the video over high-capacity telephone lines to carry the higher resolution video, again approaching BBTV service. (See Figure 1.2.)

Digital Television: Part of the Bigger Picture

In and of itself, a new television system doesn't mean much—a few more or less channels, sports specials, and sitcoms are hardly earthshaking. But that's not all there is. The transition from television to digital video is part of a transformation that is significant for nearly every person on the planet: The child born today will not inhabit the same social world as the people now reading this book.

This claim is not exaggerated. We are on the brink of a profound change that may well require several hundred years to unfold in its entirety. Before it is finished, the world will be a very different place than humans have ever experienced before. In one sense, the new environment will be invisible—the sun will still shine and the oceans will ebb and flow, and people will still grow up, get married, have children, and dream of a better life for them.

The transformation taking place is occurring in the means of communication, from nationally bounded, government regulated and often

Figure 1.2 A high-quality videoconferencing system, as used by organizations. Source: VTEL.

operated communication networks, to an interconnected, more open global broadband infrastructure. Ultimately, this new worldwide network will have the ability to link almost every person on earth with every other person, not just by voice as the telephone already does, but by visual "telepresence"—allowing interaction between people that comes very close to traditional face-to-face communication.

Can this development be so important? It will certainly change communication. The explanation that follows is a long answer to a short question, but it is the real reason why the advent of a rich media like DTV, that can be produced and distributed inexpensively, is important.

Some scholars believe that communication isn't just essential to relationships—it *is* the relationship. They argue that a relationship is no more or less than the sum total of all the communications that have ever occurred between the parties engaged in the relationship. As a result, changes in the communication system must be considered in light of their potential effects on individual relationships and the larger social fabric comprised of them.

Throughout the eons of human existence, interpersonal communication has been of necessity limited to the co-incidence of physical presence. Factoring in the time it takes to travel and the ritualistic elements of even relatively impersonal transactions, the most sociable individual's network cannot easily exceed several hundred others. The number of close relationships is typically much smaller, and intimate friends, family, and associates are fewer yet. Indeed, the development and maintenance of close, intimate ties nearly always depend on frequent contact, much facilitated by physical proximity.

Of course, letters, telephone calls, E-mail, and fax serve fairly well. And these technologies have proven to be powerful media, as anyone who uses them regularly can attest. Yet no other form of communication approaches face-to-face communication for establishing close relationships.

The next few decades will see many people of the world linked so that, with the touch of a few buttons, they will be able to conduct real-time exchanges that deliver much of the same impact that in-person contacts do. It may seem unlikely that the ability to see others as well as hear them should matter so much. However, the commonly heard expressions "seeing is believing" and "the eyes are the mirror of the soul" show that there is a common-sense belief that the effect of visual communication is important.[1] Further, the processing of visual perceptions takes up a substantial portion of the brain's power, making vision a primary sensory pathway that listeners use to evaluate the meaning, intention, emotional state, truthfulness, and sincerity of the utterances of others.

The unique ability of human beings to communicate symbolically and abstractly perhaps characterizes us more than any other aspect of our behavior. All organisms can communicate chemically; many can do so with gestures and other instinctive means; there are even instances of apparent symbolic communication between one or more dolphins and between monkeys and humans. However, only people can choose from such a rich palette of possibilities, from smoke signals to Morse code, images, and complex verbal, mathematical, and logical languages.

In the twentieth century, vast numbers of people began to use technologies to interact with each other, a practice usually referred to as "mediated communication," to distinguish it from face-to-face encounters. The telephone is the most familiar and often-used technology for two-way interaction, while radio and television are nearly universal one-way media.

The introduction of each of these technologies has been associated with significant changes in the societies where they are introduced. For example, in 1910 one observer wrote:

> *It is not easy for us to realize to-day how young and primitive was the United States of 1876. Yet the fact is that we have twice the population that we had when the telephone was invented. We have twice the wheat crop and twice as much money in circulation. We have three times the railways, banks, libraries, newspapers, exports, farm values, and national wealth. We have ten million farmers who make four times as much money as seven million farmers made in 1876. We spend four times as much on our public schools, and we put four times as much in the savings bank. We have five times as many students in the colleges. And we have so revolutionized our methods of production that we now produce seven times as much coal, fourteen times as much oil and pig-iron, twenty-two times as much copper, and forty-three times as much steel.[2]*

It is not clear that communication technologies actually cause such changes, but they do seem to occur at the same time as significant transformations in how people live, work, and play. Indeed, while there is little agreement about the actual effects of mass media like radio and television, most people believe that their influence must be considerable.

Until now, many of the most successful technologies of communication have entered the social domain as fresh alternatives, and the systems that support them evolved as more and more people organized their activities around them. The telephone, television, and radio are all good examples of how new technologies become established in the social environment. They do not necessarily replace the existing means of communication; rather, they create new additions and choices.

It is important to recognize that communication technologies never exist by themselves—they are always part of a system. Communication itself is a process that involves more than one entity. At minimum, even two tin cans linked by a string must have a sender, a message, a receiver, and time for the process to take place. Some formulations include the effect of the communications on the receiver, or a feedback loop back to the sender, or both. Others examine the greater social system in which any communication system is nested.

In practice, communication systems require expensive, complex infrastructure. No matter how easy they appear, it takes decades of development for them to become simple, efficient, and reliable. The effort and capital it requires to build such systems are themselves economic and political decisions that arise from the social fabric in which they are made.

So the importance of the coming change in communication infrastructure is simply that it alters how people interrelate. Their activities in many areas of their lives become enmeshed in the various ways they interact with one another. When the technology is used by many people, the overall society changes too. Global virtual telepresence—the ability to experience another person with nearly the same fidelity as face-to-face encounters—is a transformative technology that will have far-reaching effects for decades to come.

Going Digital, Going Broadband

For the last two decades, telephone companies have been installing digital equipment throughout their networks. For the most part, only the last mile, the distance between the central office and the home, is analog. Similarly, cable companies have been upgrading their systems so that they will carry information in two directions—both into and out of the home.

Now the government has mandated that over-the-air broadcasters begin transmitting a digital signal. In practice, this means replacing all of the television equipment that has existed since the late 1940s. In this sense, digital TV really does revolutionize the existing analog television system. Virtually every process and every piece of equipment must change, including program creation, editing, transmission and transport, cameras, and reception devices like TV sets, as shown in Figures 1.3 and 1.4.

The fact that the new TV system is digital has important consequences for the industry and audience alike, because it means that computers, telephones, and now television are all digital. Instead of being separate networks, systems, and data types, they will all be compatible and interchangeable with one another.

Digitization brings information handling to a new level of ease and efficiency. It brings awesome speed to the manipulation of masses of information, including computationally-intensive procedures that simplify and minimize the data. Digital storage and retrieval permit unlimited, virtually error-free recreation of the original material, as distinct from the successive, error-full copying procedures that nondigital media require for duplication.

Digital technology allows the transformation of diverse original materials, into a universal, compact, transportable, storable, processible,

Figure 1.3 Where are these people? This appropriately named "limbo" set is matte blue and can be digitally transformed so that they look like they are just about anywhere.

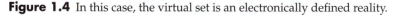

Figure 1.4 In this case, the virtual set is an electronically defined reality.

and retrievable format. Most importantly, one effect of the universality of digital content promotes the interconnection of networks and encourages the elimination of transport problems that may result from differences in their architectures because there are no barriers in the content itself.

The integration of previously separate technologies, such as computers, with telephones, radios, fax machines, and televisions, is often called "convergence." Convergence is a powerful idea that appeared in the 1960s when the power of digital technology became apparent. It means "coming together." When people use the term convergence referring to communication, it can refer to any or all of the following:

- Data—The dictum among information engineers, "bits are bits," means that data, whether it is voice, graphics, moving images, or text files, are all equal before a digital processor;

- Devices—Functions are grouped into single devices in new configurations such as computers that include a CD-ROM, game machine, telephone, and fax devices. Or customers put their TVs and PCs in the same room and use them simultaneously, as research indicates that about 20% of TV viewers do;
- Networks—Telephone, cable, computer, and wireless networks become a single infrastructure;
- Organizations—Activities are pulled under a single departmental umbrella, such as reorganizations to combine telecommunications and information technology services.

There is great theoretical attractiveness in merging the various strengths of computers, television, and telecommunications. It is a grand sweep of imagination to consider uniting the analytical power of computers with the reach and high production quality of TV, the ubiquity and ease of use of TVs and phones, and the interactivity of computers and phones.

However, actual convergence is proving quite difficult to achieve. Among many people in the computer industry, the prevailing view is that the PC will inevitably become the chief communications device in the home and the workplace. So digital companies are anxious to embrace—and encompass—the television, as well as the telephone, radio, fax, answering machine, game player, intercom, home security system, and electrical power and programmable appliance manager.

Not surprisingly, executives in these various industries, especially telecommunications and broadcasting, find the prospect of being dominated by computers less than enticing. Telephone companies look at the prospect of Internet telephony with concern. Broadcasters are not thrilled with the idea of sharing screen real estate with Internet content. But they detest the idea of TV signals coming into the home via the PC even more, and for several years they have fought the adoption of standards that would make it possible.

In some corners, the term convergence and the process it stands for is a hotly debated issue. As writer George Gilder puts it: "Television will converge with computers the way the horse and buggy converged with the automobile, or the way Carthage converged with Rome." Another view of the process is that digital technology and networks "absorb" other media into their sphere.[3]

However the change is characterized, it is evident that digitization has already revolutionized numerical computation, text processing, and image creation and processing, and the ability to transport these forms of data on networks has changed many features of organizational procedures, structures, and work flow.[4] As digital technology penetrates the visual world of television it brings important changes in its wake. In the past, the equipment used by television networks and stations changed every few years.

The larger pieces might remain in service for a decade or longer. Now that the end-to-end infrastructure of companies in the television industry will be digital, this leisurely approach to equipment replacement is over.

Moore's Law will come into play in the lives of these organizations. Gordon Moore was the cofounder of Intel, a leading manufacturer of computer processing chips. He stated that the capacity of such chips would grow exponentially, that it would double every eighteen months. So far, his prediction has proved fairly accurate, as constantly upgrading PC owners are aware.

Now these economics will devolve on the TV industry. Just to get into the digital game, every broadcasting station will spend $1 to 5 million to transmit digital signals; they will spend even more if the figure includes all the professional equipment for the generation, manipulation, and editing of full-motion video. The price tag for the transition to digital television for the 1600 local stations will run between $5 and $12 million dollars, bringing industry costs of between $8 and $19 billion.[5] And Moore's Law guarantees that within three or four years, many of the products they have purchased will be two generations behind state-of-the-art technology.

In addition to the digital television (DTV) of broadcasters and distributors, many companies are working to allow the delivery of DTV across telephone and cable networks for such uses as traditional TV delivery, video-on-demand, videoconferencing, webcasting, immersive entertainment, and multiplayer video games.

When communication devices are networked, another law comes into play—Metcalfe's Law. Robert Metcalfe achieved digital immortality as the individual most responsible for the creation of Ethernet technologies, a method used to move data over computer networks. Metcalfe's Law states that as more nodes (seats, or end-users) are added to a network, the value of the network goes up, and the cost of each node (seat) goes down.

Think about the telephone system. If everyone or nearly everyone is on it, it is much more valuable than a phone system where only 10% of people use it. In a society with thirty incompatible and unconnected phone systems, where about the same number of people used each one, consider how nearly useless a telephone would be—and how much time you would spend coordinating with friends to figure out how to stay in contact.

If everyone is on the same system, it's not only more convenient and useful. Metcalfe's point is that it becomes increasingly less expensive. As costs drop, it becomes more and more possible to add people (nodes, seats) to the network, and the network becomes increasingly useful. And Moore's Law doesn't stop; rather, it works in conjunction with Metcalfe's law, further lowering costs.

So in the brave new digital world we are entering, not only will communication be digital. It will take place over networks that link almost everybody into a unified system of communication that accepts all kinds of digital data, whether voice, video, or information. It's a world where everybody will be able to communicate in any mode they desire. There is a great deal of progress in reasonably-priced technologies that allow people to create, manipulate, and edit both still and moving images on their own computers. (See Figure 1.5.) Analysts predict that each of these product categories will bring in millions, perhaps billions, of dollars to the companies that can bring to market hardware and software to fulfill the demand for these applications, so there is no shortage of elbow grease in trying to develop them.

The digital revolution has happened so quickly that demand outstrips supply in many categories, including some professional equipment. For example, three manufacturers make telecine machines that transfer film to HDTV video. Nevertheless, throughout 1999 there will be such a severe shortage of them that many studios and production companies with large libraries of films are unable to transfer as much of their material to HDTV as they would like. Another kind of shortcoming of current DTV products in the consumer arena is that not all of them provide the quality that today's sophisticated consumers are used to—real-time streamed audio and video still fall far short of established professional standards.

However, as Moore's Law predicts, the processing power of computers continues to grow exponentially, quickly advancing digital video

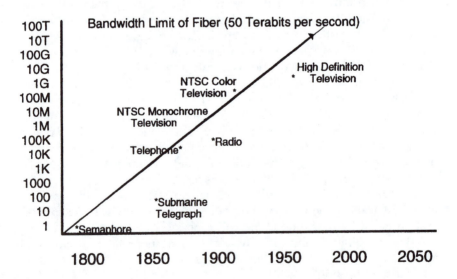

Figure 1.5 People communicate as much as the infrastructure allows them to—the demand for information bandwidth just keeps going up.

technology and infrastructure. New chips that perform ever more giga-flops (billions of operations per second), internal improvements like the broadband bus that accelerates the computer's speed, specialized cards with dedicated processing and additional memory, expanded storage, and more efficient compression techniques all come to market with dizzying rapidity.

The Long View: Freeing the Message from the Messenger

In the 1970s, digital visionary Jonathan Seybold published a newsletter to printing concerns that covered progress in computerized typesetting. Seybold described the sense of wonder he felt when he realized that digital computers had set a historical force in motion, not only because of computer technology itself, but because of the data structure they brought into being. Seybold argued that for the first time in history, translating messages into digital data frees them from the media that has heretofore served as their messengers.[6]

Messages and media have always been communicational Siamese twins. Television was tied to the TV set; telephone calls went with the phone; compact discs played on CD players; and audio-cassettes on tape players. Now, through digitization, paintings become independent of canvas, paint, or even ink on paper; music no longer necessarily resides on a vinyl record or a compact disc; and moving images play, unfettered by celluloid and magnetic tape. All these disparate messages—text, sound, and images—can be expressed by an infinite number of 0s and 1s, tossed into the great bit-bucket of digital processors.

As digital data, messages are infinitely plastic, transformable into other, new forms that allow new interpretations and expressions. Music can become visual, translated into colors, notes, graphs, bars, or dots. Manipulating images is commonplace. Using a system costing less than $5,000, a competent computer user can process and retouch digitized images in ways only experts with hundreds of thousands of dollars of equipment could have done just a decade ago.

The unparalleled plasticity of digital data cannot be overemphasized. It permits changes and amendments to messages, usually accomplished more easily in the digital domain than in the primary format. Once stored, digital data is extremely compact, almost always smaller than the initial data. For example, a novel, or the complete score for a concerto, can be stored on a 3 1/2-inch diskette. From storage, random access to any part of the message is possible: with little difficulty, any page of text, any frame of a series of moving images, and any musical sequence can be called up instantly.

The transformability of digital data allows easy, instant, virtually error-free replication of images, regardless of the complexity of the original material. This near-perfect replication is due to the fact that the image is not copied or reproduced; rather, it is re-created, reconstituted each time from the stored 0s and 1s. Data that describes the original, not the original itself, is copied. Error-free re-creation means that images can be processed and reprocessed in postproduction with no loss of quality, a characteristic termed multigenerational performance.

The problem of a severe loss of picture quality when copying analog images has always been regarded as a shortcoming in the entertainment industry. Ironically, it also acted to protect intellectual property because only a few copies could be made from any given original and even fewer copies could be made from the copy of a copy, because in the analog domain each duplication brings the signal a "generation" further away from the original. With each successive generation, the image becomes less and less clear and more marred by unwanted artifacts.

All these characteristics of digital media are magnified in a networked environment. Messages can be transmitted from one to many others (potentially millions of people), with little more than a few keystrokes. If these users can access the stored material, they can dissect, alter, and disseminate all or part of the original document with startling ease. Moreover, the time is not far away when the network will be able to tap the processing power of multiple computers, enabling users to run applications beyond the power of any single desktop computer.

While these applications are apparently admirable and no doubt will offer substantial benefits, they also entail some negative consequences. The distinctions between different types of public works is confusing and may weaken intellectual property rights.[7] The ease of changing messages may result in unauthorized alterations to important works. Even more serious, documentary evidence may be lost, destroyed, or fabricated.

This chapter now turns to a comparison of analog television to digital video and its highest resolution form, High-Definition Television, or HDTV.

Comparing Analog Television and Digital Video

The differences between analog and digital video begin with the camera. It scans a scene very rapidly and translates its color and brightness into electrical impulses, called a video signal. An analog camera stores the signal as electrical impulses; a digital camera translates them into 0s and 1s and stores them as data.

To show the image to viewers, the analog signals are retranslated back into the color and brightness characteristics of the original scene and displayed on a screen. Similarly, a digital set-top box (STB) or display unit uses the digital data to reconstruct the original scene as it was captured by the camera. In both types of video, the quality of the picture that the viewer sees depends on how much data the camera is capable of picking up, how many times the scene is scanned in a given period of time, and how accurately the display device reproduces the scanned scene.

Analog TV employs an interlaced scanning system, meaning that both the camera and the receiving TV set "read" the 240 even lines by scanning them from the top of the screen to the bottom, 30 times each second. Then the 240 odd lines are scanned from top to bottom. Just as the camera scans the original scene, the resulting image is also scanned onto the TV screen. In the TV set, the scanning starts at the top left corner and moves horizontally across the screen to the right, emitting light as it goes to create a single "scanning line," Line 1. At the end of the line, the scanner shuts off and snaps back to the left hand side of the screen. It skips a line, where Line 2 would be. Then the scanner turns on again and repeats the process, emitting light along Line 3, then Line 5, Line 7, and so on, through line 479 or so. The time during which the scanner turns off and snaps back to the right is called the "horizontal blanking interval."

Each scan is called a field; a double-scan, including both the even and odd lines, is called a frame. Today's analog system transmits about 480 active lines at a rate of 60 fields, or 30 frames per second, often written as 30 fps. The updating of the picture is called the refresh rate, which for NTSC television occurs 60 times per second. The refresh rate is expressed in hertz, so it is written as 60 Hz. Interlaced scanning offers an advantage when the picture changes rapidly, as it does in sports, for example, because the motion is updated on half the lines every 1/60th of a second.

Surprisingly, even though half the picture is shown at a time, the viewer doesn't see a sparse picture. One reason is "persistence of vision," which occurs because the human eye continues to see an image for a fraction of a second after it has disappeared. In addition, phosphor (the incandescent material in the screen) continues to glow after the electrical impulse that excited it decays, long enough for viewers to receive the impression of a whole image.

While analog TV cameras and sets incorporate interlaced scanning, computer cameras and monitors use progressive scanning. Progressive scanning starts at the top left of the screen and scans each line in order, from top to bottom, then starts over at the top. The camera converts what it "sees" into 0s and 1s that represent the brightness, contrast, and color of the original scene.

Other differences between analog and digital video come up in their display, which have been the source of longstanding and fierce disagreements between the broadcast and consumer electronics industries on one side and the computer industry on the other. For example, most computer monitors refresh at the rate of 72 times per second or greater, rather than the 60 Hz refresh rate of analog sets. Finally, computer screens use square pixels (a contraction of "picture elements," which are tiny points of light) and digital HDTV pixels are round.

Characteristics of Color TV Sets and Computer Monitors (SVGA)

	TV Sets	*Computer Monitors*
Scanning	Interlaced	Progressive
Color Coding	Luminance, Hue, Intensity (YUV)	Red, Green, Blue (RGB)
Pixels	640 x 480 (equivalent)	800 x 600
Frames/second	29.97	72

Analog sets do not really have pixels at all. Each scanned line is a continuous stream of light, actually a wave. Along the line, there is an ever-changing mosaic of light and dark, equivalent to about 700 points of information. This continuous wave varies exactly with the information it carries—it is analogous to the actual picture, which is the reason it is called an analog picture.

Precise standards govern the display of analog television and, at first, it seemed that digital video would be the same. But DTV formats are more flexible, demonstrating the plasticity of their technological blood-line. Producers can process and distribute digital video in any resolution lower than the one in which it was originated.

In some cases, decisions about resolution are in the hands of consumers, not just producers.

Many PC users have installed streaming media players like Real-Player, Windows Media Player, and QuickTime. Users can set up the software to choose the screen size and resolution they want to display, or specify the best quality that their system can play. TV tuner cards also let users control what they are watching, and Hauppage, a leading maker of video tuner cards for PCs, plans to market an HDTV-capable card for under $200 in 2000.

HDTV is the highest resolution variation of DTV, and like all digital standards, defining them has proved to be rather slippery. The Federal Communications Commission standards (often referred to as the ATSC standards after the Advanced Television Standards Committee that submitted them for approval) provide for eighteen different resolutions. For the purposes of this discussion, I will use the highest level of those standards to describe HDTV.

The HD scanning line consists of 1,920 pixels, providing much greater horizontal resolution than the equivalent 704 points of NTSC television. Unlike the lines in the analog picture, the digital TV scanning line is not a wave; rather, it is made up of many tiny points of light that are packed so closely together that the human eye cannot tell they are discrete points. HDTV and other forms of DTV are similar to a color comic strip in the Sunday paper: a magnifying glass reveals the individual dots of color that create the picture. (Later in this chapter, I will cover in more detail how the points of a digital image compare to the analog wave.)

There is a big difference in the vertical resolution between today's analog TV and HD digital images as well in the horizontal resolution, as shown in Figure 1.6. TV screens have 525 lines from top to bottom, with about 480 of them used for the picture. Engineers call the 45 lines that aren't used to carry video signals the "vertical blanking interval." HDTV increases the vertical resolution to 1080 lines, with 90 lines that are not used for video signals. The overall result of the greater horizontal and vertical resolution is that an HDTV picture contains about five times as much information as the current analog picture. Table 1.2 summarizes the characteristics of analog TV, digital video, and high-definition television

Table 1.2 Properties of Analog TV, Digital Video, and Digital HDTV

Characteristics	Analog TV	Computer Digital Video	Digital High-Definition TV
Display unit	TV set	Monitor	Display
Horizontal form	Wave	Square pixels	Round pixels
Horizontal resolution	704 equivalent	Variable, to maximum of 1,920 pixels	1,920 pixels
Vertical resolution	525 lines	Variable, to maximum of 1080 lines	1,080; 720; or 480 lines
Scanning	Interlaced	Progressive	Interlaced or progressive
Frames per second	30	Variable, to maximum of 60 fps	60, 30, or 24 fps
Refresh rate	60 Hz	72 Hz to 80 Hz	72 Hz

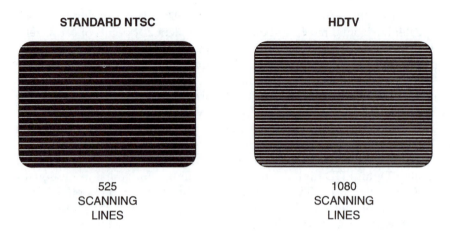

STANDARD NTSC

HDTV

525
SCANNING
LINES

1080
SCANNING
LINES

Figure 1.6 Comparing the resolution of NTSC and HDTV images.

Digital Democracy: Moving Images for Everybody

The ease of producing and distributing streamed AV content is an important part of the transition to digital television—it makes almost everyone a potential producer of video, as shown in Figure 1.7. The cost of DV cameras have already gone down and are going to plummet over the next few years, so that relatively high-quality camcorders will be quite reasonably priced. Low-quality equipment will be disposable—as many still cameras are today. Cameras will come in many different form factors as well. They will be ever smaller, until they will probably fit onto a tie tack or an earring.

Professional-level video is available to any company and to individuals who are serious hobbyists. Today, it is possible to put together a DV kit for $5,000, carried in two small cases that will deliver video with higher quality than 1500 pounds of broadcast-quality equipment that cost $150,000 fifteen years ago. The force behind the lowering price/performance curve is Moore's Law—and we are just in the beginning of the curve in the video production arena.

At the highest professional level, computer processing has surpassed the requirements for commercial production and reception. Digital images at 2k x 2k (2000 x 2000) and 4k (4000 x 4000) images exceed the resolution of film, and they can be created and processed in many high-end postproduction and special effects facilities. In a few years, these daunting resolutions will be available at affordable prices. The postproduction sphere is experiencing the same phenomenon. In network news organizations, the computer-automated newsroom has become common and by 2003 will be the norm in TV stations across the United States. Avid, Matrox, and the Fast Machine are all examples of products that are used, although Avid has captured the lion's share of this market. The cost of a

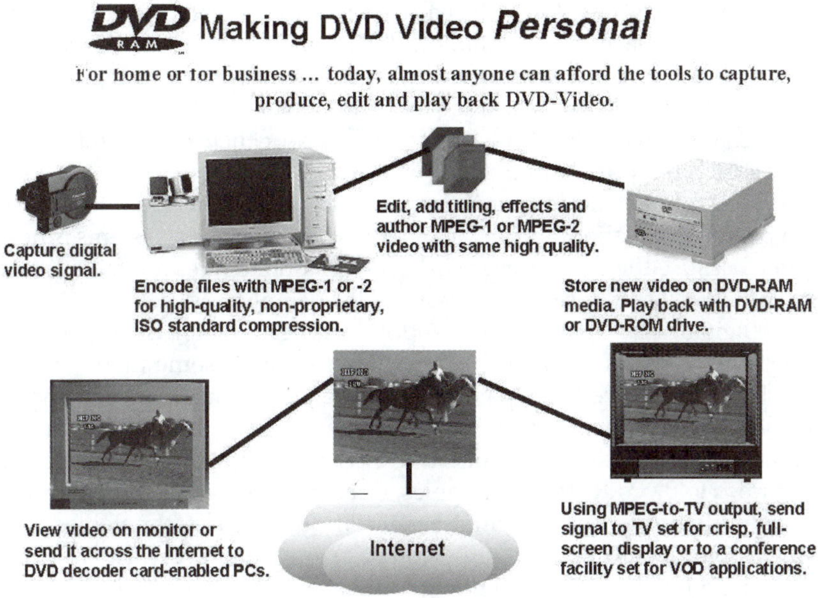

DVD Making DVD Video *Personal*

For home or for business ... today, almost anyone can afford the tools to capture, produce, edit and play back DVD-Video.

Capture digital video signal.

Encode files with MPEG-1 or -2 for high-quality, non-proprietary, ISO standard compression.

Edit, add titling, effects and author MPEG-1 or MPEG-2 video with same high quality.

Store new video on DVD-RAM media. Play back with DVD-RAM or DVD-ROM drive.

View video on monitor or send it across the Internet to DVD decoder card-enabled PCs.

Internet

Using MPEG-to-TV output, send signal to TV set for crisp, full-screen display or to a conference facility set for VOD applications.

Figure 1.7 People who want to will be able to make their own movies too. Source: Panasonic.

professional system exceeds $15,000, so it is out of range of all but the most serious consumers.

However, there are several good-quality consumer video editing systems—a PC card and software—that allow desktop editing. The new PCs can easily handle the processing requirements for professional video and audio. The cost of such a PC in late 1999 was about $2,000, and additional hardware and software might run another $1,000 to $2,000. The package would include a wide range of special effects, previously only available at multimillion dollar postproduction facilities.

Within most medium- to large-sized organizations, video capability has become part of the client-server technology in the workplace. The previous generation of file servers on local area networks carried text and low-resolution graphics; few transported audio, high-resolution graphics, or full-motion video. In just the last decade, there has been a profound transition to higher bandwidth networks. Most companies have Intranets, private versions of the Internet running over their local area and wide area communication networks. For the most part, the higher-capacity networks, servers, and storage devices they have deployed easily handle industrial-strength digital video.

The desktop revolution has also come to videoconferencing. In the 1980s, it took place in multimillion dollar conference rooms, especially designed for the purpose. In the early 1990s, videoconferencing became available for small groups, dubbed "huddleware." These systems ran about $50,000. Now, multiple individuals stay at their desks and everybody's picture appears in a small window and the rest of the screen is used as a "whiteboard" for presenters to make their points and for one or two people to collaborate. The incremental cost to existing systems is negligible because the networks are in place, telephone rates are falling, and the software cost is spread over thousands of employees.

Digital democracy has tremendous implications for the media and entertainment industries. People who wish to invest some money and a lot of time will have the capability to create audiovisual material and to distribute it over the Internet. Some observers believe this presages a cultural renaissance, a flowering of creative energy that will lessen the power of studios and production companies. Certainly the media and entertainment industry has been the first beneficiary of digital technologies. The next chapter looks at how digitization has penetrated every aspect of the creation and distribution of moving images.

Notes

1. W. J. Mitchell, "When is seeing believing," *Scientific American* (February 1994): 68–73.
2. H. N. Casson, *The History of the Telephone*, Chicago: AC McClurg (1910):224–5.
3. G. Gilder, *Life after Television*, Knoxville, TN: Whittle Direct Books (1990).
4. P. Samuelson, "Digital media and the law," *Communication of the ACM* 34:10 (October 1991):23–29.
5. D. I. Sheer, "Cost-effective DTV: Transmitter companies offer digital transition strategies," *Digitaltelevision.com* (December 1998). Article at: http://www.digitaltelevision.com/business1298bp.shtml.
6. A. Harmon, "A digital visionary scans the info horizon," *Los Angeles Times* (June 1, 1994):D6, D8. Also, J. Seybold, personal interview, October 1994.
7. Samuelson, op. cit.

2

From Show Biz to Show Bits: Digital Technology in the Media and Entertainment Industry

Traditional media and entertainment companies may be profitable but they aren't happy these days. The advent of the Internet, MP3, broadcast.com, DVD and CD recordable drives—all these and more are turning the industry on its collective ear, and forcing it to reexamine and redesign its processes and procedures, its audience, content, and business and revenue models.

A new technology can ruin a company's whole bottom line in less time than it takes to type in a URL. At the same time, things aren't as bad as they might seem because new technologies don't really displace old ones. They establish a new balance among themselves to accommodate the altered array of communication choices. The contents, audiences, and uses of each medium become more sharply defined as it assumes the position most appropriate to its unique features and benefits. For companies involved in media and entertainment, such a rebalancing act can be tough to follow—or even survive.

The telephone heralded an explosion of companies that could operate nationally, even internationally, and put people into immediate, intimate contact with one another. Newspapers and film brought a mass audience into being that was solidified by the experience of radio information and entertainment. Television established a set of cultural expectations and societal mores that transcended the country's regional heritage.

As each new medium came online, it was greeted by premature announcements of the death of the old. Now digital communication and the Internet have arrived, and broadband connection to a critical mass of homes is on the horizon. This fundamental transformation of communica-

Figure 2.1 Virtual egos at work. Source: Ultimatte.

tions infrastructure is merely the forerunner of considerable changes in media usage and audience behavior, presaging a new and unknown environment for entertainment enterprises. Companies' views of the marketplace, their products, processes, and business models—all are based on a set of conditions that may not hold sway into the future.

Consider the effect that the Internet has on the traditional mass audience. In a phrase, forget controlling the consumer. Get used to doing anything you can to reach them, please them, cosset them, cater to them, and make them feel good.

To the extent that people are online, they are not the same isolated audience members towards which entertainment enterprises have traditionally aimed their wares. Previously, people's relationships were limited to others within rather small interpersonal networks. Now media giants face a significant challenge from digitally empowered consumers, who have powerful processing at their fingertips and are connected easily, instantaneously, and inexpensively to an infinite number of others around the world.

Anyone can produce and reproduce digital material at will and distribute it to an unlimited number of recipients with just a few keystrokes. Right now, this new reality is affecting companies whose products require only narrow bandwidth, such as music, newspapers, and magazines. However, as consumer access and network capacity increase, the same challenges will arise for filmed entertainment sectors of the industry.

Moreover, the Internet's pace of adoption is more rapid than that of any other technology. Radios existed for 38 years before 50 million people owned them; TV required 13 years to reach the same number; PCs took 16 years. The Internet required only 4 years. Internet traffic doubles every 100 days. In 1996, there were 40 million users around the world; in 1997, there were more than 100 million; by the end of 1999, there will be 130 million. In December 1996, there were 627,000 domain names; in January of 1999, there were 43,498,699—how awesome is that?!

The Internet brings a new kind of free-form "everyone-to-everyone" connectivity between millions of people called peer-to-peer or P2P communication. It poses a new set of challenges and opportunities to just about everybody. But since many entertainment products are information-based from product creation through distribution and fulfillment, the networked environment holds more far-reaching implications for entertainment companies than it does for those in many other industries.

Digital Redefinition

The giant media and entertainment companies are in for a decade of massive change, driven by the new communications infrastructures and technologies. Whether they like it or not, the writing is on the screen for everyone to see. Here are some of the new business conditions and means of adaptation:

- A decisive movement away from traditional analog to digital modes of creation, marketing, distribution, fulfillment, reception, and consumption. Digitally created content such as CGI (computer-generated imagery) in films, television programs, animation, and music consumes an ever-growing portion of project budgets. Just in the last few years, about one-third of the images in the average film have been digitally created or enhanced.

- Replacement of inefficient isolated silos of digital technologies by end-to-end production and supply chain systems—that cost a lot of money.
- Entertainment product customization based on data-mining consumer behaviors, requests, preferences, and end-user access technologies and devices.
- Branding works, sort of—if the company does something for me today, not yesterday.
- Forget intellectual property and copy protection. It's probably impossible.
- Emphasis on timeliness as a major appeal to customers. The effects of real-time communications haven't even begun to be felt.
- Finally, it *is* a small world. The United States and Canada make up less than 5% of the world's population. This represents an enormous marketing opportunity.
- But there's a catch: The appeal of most entertainment products is inherently local. Take yourself . . . what matters most to you?
- Industry consolidation isn't a passing fancy; consolidation is how business is done.
- Too bad bigger isn't smarter; innovation will continue to occur within small, nimble organizations.
- Business says it wants competition. Be careful of what you wish for—media companies can count on dog-eat-dog competition for the foreseeable future.

As the power of digital technology becomes apparent, the next five years will also bring recognition of the limits of digitization and a corresponding resurgence of the real, where the value of entertainment can more easily be controlled and monetized. Media and entertainment giants will increasingly create new experiences for people to participate in actively, rather than merely fashioning new products for them to receive passively.[1] Fortunately, many producers have extraordinary skills developing such products.

Media companies are strong and healthy. But they are not immortal. They are forced to respond to significant challenges from within and without their empires. Nevertheless, digital technologies also place powerful tools at their disposal to ensure their survival. The next sections will explore these issues in more detail.

Digital Technology in the Creation of Content

The digitization of Hollywood has already brought about new methods of production and entirely new types of programming. In the next few years, the digital regime will extend its domination in every area, from the creation of film and television, across high-speed digital networks, all the way to the digital reception and playing devices in consumers' homes, like the PC, DVD, and HDTV.

In the late 1980s, several high-end postproduction companies and in-house operations began constructing studio-quality digital facilities. The Home Box Office cable network was a pioneer, opening their facility in 1989. In mid-1994, Sony built Image Works, an in-house, state of the art, digital postproduction studio, and Disney created its own facility as well. At about the same time, computer giant IBM teamed with the digital special effects wizard Stan Winston (*Star Wars, Terminator*) to open Digital Domain, specializing in state of the art digital postproduction.[2]

During the past decade, the increased contributions of digital technology became clear. On average, nearly 30% of a film produced today contains digital effects, as reflected in the escalating cost of special effects, as shown in Table 2.1. Computer-generated imagery (CGI) accounts for 30% to 50%, compared to 10% just a decade ago.

Table 2.1 The Price of Novelty: Expenditures in Millions for Special Effects for Feature Films

Year	*Expenditures (millions)*
1998	$345
1997	$210
1996	$160
1995	$100
1994	$ 75
1993	$ 60

Source: Industrial Light and Magic.

From preproduction storyboarding and previsualization to postproduction editing and compositing, digital processing is deeply embedded in today's movie and television production. Previsualization is a great improvement on traditional storyboarding, which relied on hand-drawn sketches. Previs, as it is often called, calls for a digital simulation of the scene to be filmed, allowing the director, producers, and executives to see the material with nearly perfect fidelity to how it will appear on screen.

One of the effects of digital technology, especially special effects, is the gradual shift of activities from the production phase of creation to the postproduction phase. (See Figure 2.2.) Digital effects were essential to the success of *Titanic* and were actually the star of the show in *Twister*. Morphing and other high-impact effects added unique entertainment value to *Terminator 2*.

This change has many consequences for both creative artists and filmmakers and the film industry as a whole. At the management level, it transfers powers to executives who control the budgets to purchase access to capital-intensive technology. Overall, the production process is becoming even more capital intensive than it has been previously, and requires much more technical skill. These developments will impact the organizational structure of studios and production companies in unknown and profound ways.

Digital technologies give filmmakers new ways to realize their visions that were never before possible. Until the development of software that made it possible to populate scenes with hundreds of thousands of moving animated creatures, the Dreamworks film *Antz* could not have been made

Figure 2.2 A little video . . . a lot of special effects.

at all. Similarly, the richly produced scenes and perspective effects in that studio's animated *Prince of Egypt* feature required computer assistance.

One of the first instances of sophisticated digital matting that made it possible to replicate extras occurred with the thriller *In the Line of Fire*, and cut that portion of the film's budget by as much as 80%. In 1998, this technology was also used extensively in the blockbuster *Titanic*. By 1999, off-the-shelf software packages made it possible to create crowds of unique individuals as a matter of course. Similarly, producers David and Heather Nicksay were pioneers in the use of computerized storyboarding and scheduling in preproduction, saving considerable time in a very tight production schedule for *Addams Family Values*. A mere five years later, packages that handle everything from script development through editing and release were commonplace.

In the entertainment industry, the pros invoke an old, cynical adage, "Fix it in post!" whenever there are problems in production. The infusion of digital technology is changing that aphorism to, "Create it in post!" In the early 1990s, George Lucas digitally created matte backgrounds for the television show *Young Indiana Jones* and edited the entire program on digital equipment. Now he has taken that even further. According to Adobe Systems cofounder John Warnock, "With the stuff that Pixar and George Lucas do, there are huge cost savings to be achieved. Most movies depend very, very heavily on computers to deal with almost all of the special effects. In the last *Star Wars*, they hardly had any sets at all."[3]

The dinosaurs in Steven Spielberg's *Jurassic Park* demonstrated how digital image generation could create "characters" and effects not otherwise possible at all. And the double waterfall that heightened the drama of Meryl Streep's role in *The River Wild* was created by digitally duplicating the one waterfall that was actually on location.

Broadcasters increasingly rely on digital technology in their less costly productions as well. The Turner Broadcasting Service produced the submarine film *Huntley* using a digitally created underwater environment. Plans originally called for a live-action shoot, but when the producers realized that they could capture many more kinds of scenes that would appear amazingly realistic, to say nothing of the budget savings, they cancelled the live shoot.[4]

All the networks are converting their broadcast facilities to digital. ABC invested $153 million in digital equipment in 1994 and 1995 and replaced analog videotape recorders with digital D-2 VTRs. The company's news footage is stored on a BTS Media Pool tapeless server and routed in and out of the network's digital editing rooms by a BTS routing switcher. ABC built digital control rooms, including digital monitors, switchers, special effects units, and audio equipment. NBC has been constructing its Genesis broadcasting center for the past two years. And CBS has installed a robotic, electronic IBM asset management library to house its collection of news footage.

Networking: In Hollywood, You've Got to Have Connections

In the past, this phrase meant knowing the right people at the right studios; but today, it also means the right communication networks. All the major studios have either already constructed, or plan to construct, sophisticated networks linking the buildings on their own facilities. One company, Wam!Net, offers sophisticated animation services right on the Net, as shown in Figure 2.3.

Overseas, production centers are also wiring up. The historic Babelsberg Studios, located on the outskirts of Berlin, saw its first production in 1911 and since then more than 2,500 films have been shot there. Today it's called Media City Babelsberg and it the home to more than sixty companies, including the Bertelsmann Group's Ufa Film and Television Productions, ORB, and ZDF. It's also a city within the city. In addition to production facilities and office space, residential space is available.

In addition to sound stages and postproduction services, Babelsberg features a fully-stocked Media Communication center for digital

Figure 2.3 Wam!Net's Render-On-Demand (ROD) service lets animators create and store their work online where others can access it. Source: Wam!Net.

production. It has up-to-date workstations and software for creating a full range of digital effects for motion pictures or completely digital projects. The center can be used by all resident companies and is linked into the lot's network.

Babelsberg has been outfitted with the latest in communications gear. It's connected by a fiber optic ATM (asynchronous transfer mode) network capacious enough to move uncompressed video around the lot at 270 Mbps—and probably faster if a company requests it. Communication to the outside world is fully enabled as well, using Germany's state of the art digital telephone system. It is more advanced than U.S. phone networks, and many people have ISDN to the home.

Hollywood studios, like nearly all U.S. corporations, have Ethernet LANs (local area networks) on the main campuses—some even have state-of-the-art high-speed Ethernet or ludicrously expensive ATM LANs (asynchronous transfer mode local area networks). Most have ATM WANs (wide area networks) to connect distributed facilities around the city and the world. They have satellite dishes and uplink transmitters to bring information in, and microwave capability, ISDN (Integrated Services Digital Network), T-1, and T-3 lines to carry it both in and out.

But networking in Hollywood stops at the studio gates. For local transport within the city of Los Angeles or New York, high-speed networks are much more expensive than strapping a package to a motorcycle and sending it by courier for less than $100. Squirting it over a network would cost at least two "kilobucks" ($2,000) a month for the service. It is easier to send high-quality footage over a high-speed network from Burbank to Tokyo than it is to send it from Burbank to nearby Santa Monica. From a work process standpoint, this means that it is faster to get approvals, to exchange ideas, and to rework material with people who are 6,000 miles away than with those 30 miles away.

Networking and Program Creation

Nevertheless, it is clear that networking eliminates the effects of distance. This has profound implications for the creation of entertainment. The increasing globalization of film development and production are likely to reduce the role of motion picture studios in the United States in the next few years. A great deal of shooting is occurring in Canada, where governmental tax breaks and a weak currency favor production.

Figures released by the Entertainment Industry Development Corporation, a Los Angeles area agency that issues film permits, indicate that production began declining in the area in 1998. According to their data, production reached a plateau in mid-1998; then studios and producers significantly tightened activity and lowered costs. Feature film shooting dropped 13% from the previous year and TV commercials fell 7.5%. These numbers reflect the first decline since 1994 and a sharp departure from 1996, when production rose 30%, and 1997, when it went up 27%.

Distributed work groups, where people work from wherever they happen to be, are also coming. The explosion of production keeps the most knowledgeable and skillful people working constantly, all over the world. One of the first instances of distance collaboration was Steven Spielberg's approval of editing on an earlier film while in Poland filming scenes for *Schindler's List*.

This kind of communication is quickly becoming commonplace in the entertainment industry and is expected to continue as producers hire the "best of breed" regardless of location. Animation artists working in the United States will be competing with people doing similar work all over the world. However, it also gives creative people the opportunity to live and raise their families where they want to be, rather than clustering in Los Angeles, New York, or London.

Just about everyone agrees that there is demand for broadband networks from studios, production companies, postproduction and special effects facilities, advertising agencies, TV networks, casting agencies, location services, and prop houses, among others. And, as we will learn in later chapters, such networks are just now being built all around the world. The enthusiasm of film professionals who have worked in a networked environment is contagious. (See Figure 2.4.)

For example, Threshold Entertainment produced *Beowulf in 1998*, an imaginative realization of the Old English eighth-century poem. Larry Kasanoff, CEO of Threshold Entertainment, and Alison Savitch, president of the company's Digital Research Lab, bubble over as one person when they talk about how a partnership with Sprint DRUMS network facilitated the production.[5]

> *Larry: We're setting it 1,000 years in the future in a techno-feudal environment, marrying Romania's ruined castles with a futuristic high tech look—*
>
> *Alison: —a techno future built on a feudal past.*
>
> *Larry: But when I saw the digitally-altered castle, I told Alison I didn't think the interiors in Romania would match.*
>
> *Alison: So we took it to the next level, shooting blank walls, stone floors, and dressing them digitally.*
>
> *Larry: We had high-speed connectivity between Romania and L.A. and we sent digital set-dressing over the telephone line, incorporating the production design, the digital effects, into the live-action shoot, in real time. The director of photography could see both the picture coming into the camera and the digital sets and effects coming over the phone line.*

Alison: We couldn't have done this movie without it.

Larry: Certainly not for the cost we budgeted.

In addition to actually producing the film over digital telephone lines, the company also used special effects and set designers from around the world. Since they were turning in their work over a network, it didn't matter where they lived.

Consider how a production company usually hires an artist for their production design team. A call goes out for artists. Each one drops off a portfolio at the studio and gets a receipt for it. The portfolio goes to a conference room where executives drop in over a two-week period to examine the five or twenty or thirty portfolios kept there. At some point, a meeting is held to select members of the team. Afterwards, the artists reclaim their portfolios at the studio gate.

With networking, all this is done on the desktop. The artists digitize their work (if it doesn't already exist in digital form) and send it by E-mail to the production company. It is distributed to all the relevant executives

Figure 2.4 Silicon Graphics Indy workstation lets people in distant locations work on the same material. They are able to talk and see one another at the same time. Source: Silicon Graphics, Inc.

who examine it at their own convenience. No drop-offs, no pick-ups, no courier charges, no geographical "center" where material has to change hands.

Former Universal executive Thom Mount says it all depends on reasonably-priced broadband connectivity.[6] He's cochairman of Cast-Net.com, an online casting service. It is based on ISDN (Integrated Services Digital Network) and frame-relay telephone technology that works like the Internet, but it's independent—faster and more secure. The company says it can shorten the administrative aspects of the casting process to a day or two, instead of the traditional week or two.

The company has global plans, with expansion to New York and Europe on the drawing board for 1999. CastNet also hopes to add new services in the future, such as CrewNet and LocationNet.

"Our business is overdue for a rapid consolidation of planning and preproduction through the application of digital technologies," believes Mount. "We are at the beginning of a historic five-year transition, and by 2004 this kind of interactivity will be as ubiquitous as a fax machine."

The current lack of cost-effective broadband connectivity is a frustrating problem for CastNet. The quality of audio at ISDN and T-1 rates is very high but video is only marginal—not really good enough for the most serious casting purposes that would involve transporting high-quality video of performers' work. Budgeting for a broadband line between Hollywood and Europe or even New York may make sense, but if it is only used for a few transfers within the L.A. basin, the cost is unacceptable.

One company that does enough transfers to make a dedicated connection is Novocom, a postproduction and special effects facility that handles several Paramount Domestic Television shows, including *Entertainment Tonight, Hard Copy, Leeza, Real TV,* and *Wild Things.* Novocom exchanges enough material to fully utilize two-way 270 Mbps fiber optic cable.[7]

Over the fiber, Hollywood personnel run high-end Henrys (equipment that generates high-quality digital effects) that are actually located in Playa Vista—as if they were in different rooms in the same building. On the Paramount lot, there's a Henry room with a tablet, monitor, and a D-1 tape machine. At Novocom, technicians route the signal to any one of three Henrys in Playa Vista, but give control to the editors at the Paramount studio in Hollywood.

Novocom's director of postproduction, Paul Mitchel, says the service has run perfectly "24/7" for over a year and a half. But it's been a long time coming. Mitchel's memories of the not-so-distant past are a reminder of just how much gerrymandering and innovation is required to get needed service: "My science project a few years ago was smuggling GTE Alcatel encoders onto the Pac Bell system. It took them a year to get a cease-and-desist order," said Mitchel.

The situation is changing, and entertainment facilities in London, Vancouver, and Los Angeles are getting wired. In London, SohoNet moves footage between postproduction houses that are located within a fairly small area of the city. In Vancouver, MediaNet performs the same service.

In Los Angeles, local telephone company Pacific Bell started up a new service called the Advance Video Service or AVS-270, allowing real-time transfer of uncompressed video, as well as compressed video, regardless of the format. The service rolled out in September, 1998, and it also enables real-time collaboration between creatives at remote sites. The incumbent carrier has been moving to address the needs of the region's media industry for several years and has brought fiber to thousands of buildings in the L.A. area. Pac Bell also established a Hollywood Broadcast Operations Center to coordinate its broadband service offerings to the entertainment industry.

Another local incumbent telephone company, GTE, established its own media-oriented 270 Mbps service at the end of 1998. The company signed an agreement with Canada-based Jazz Media Network to provide software for their product. This agreement pointed up one of the barriers to very high-speed broadband connections: the lack of off-the-shelf application software. This kind of software is extremely important to telephone companies and other carriers because it allows them to offer a value-added service, rather than just raw bandwidth at commodity prices. Ready-to-use software is also important to smaller companies that would use such connections because they can't afford to commission proprietary software; they need to be able to buy it or have such software available with the service.

No one has more experience with broadband infrastructure and software for entertainment use than Jim Fancher, Chief Technical Officer at Pacific Ocean Post in Santa Monica.[8] He's a legendary industry guru because of his participation in so many high-speed network trials for entertainment applications: Media Park, Clicklink, EDNet, DRUMS, GTE ATM trial, Studio of the Future, Independent Fiber Network, USC-Hollynet, Century/Continental Cable, MediaOne, and Jazz Media Network. Fancher explains:

> *The software issues have proved to be much more difficult than infrastructure development. If you're a big studio, you might want to be able to send footage between the studio and locations as easily as E-mail. So you hire summer interns to sit at a computer and do it. But then there's nothing to keep them from E-mailing it to their friends, so you have to put in some elaborate security. That means you need someone to write applications in each facility, which then develops its own rules and procedures and won't interoperate with anybody else's software.*

Other companies beside the telephone giants are entering this potentially lucrative market. Enron Corporation and Qwest have both started building fiber networks in Los Angeles. Concentric Networks and Teligent are offering middle-band service to business customers. And London's SohoNet is looking for local partners to wire up entertainment-industry support facilities.

Entrepreneur Scott Tolleson, a dynamic mover and shaker behind cable company MediaOne, established a media industry fiber service, but the cabler ultimately abandoned the project.[9] Tolleson plans to recreate a broadband network over a cable infrastructure:

> *It takes a lot of money, but there's a wide open market with people screaming for the service. It is like America Online for the production environment. When the industry is spending $22 million just for film production, it is screaming. So we want to get into the market as soon as we can and build majority penetration.*

Another project comes from Lightway, which wants to establish an entertainment broadband fiber optic network for the Pacific Coast. It would carry streams from Vancouver, Canada (a production center) to Los Angeles and points between (Seattle, Portland, San Francisco). It would maintain a satellite uplink at each point to redirect the signal to any KU- or C-band satellite from the mid-Pacific to the mid-Atlantic. This network design would enable seamless delivery from L.A. to Vancouver to anywhere in the world.

An unusual development in the growing networking of the entertainment companies of Los Angeles is the role of municipalities. The city of Burbank has forged agreements with nearby cities and is in the process of extending its fiber network into Glendale, Pasadena, and ultimately, L.A.'s Television Academy area that adjoins Burbank.

Visionary Fred Fletcher is the assistant general manager of Burbank's Public Service Department, and he believes that the city's fiber, used in electrical operations, could serve the entertainment industry:

> *I could see there would be a need for very, very high-speed service, and I thought the right role for a municipality would be to provide dark fiber, fiber optic cable that is simply buried in the ground. That lets private companies provide the smart devices and services, and permits users to do what they need to do, yet takes advantage of the city's rights-of-way to get the fiber in the streets.[10]*

The city of Los Angles has been slow to get the picture, so that a MAN—a metropolitan area network of high-speed that could serve the media industry—doesn't really exist as it does in Burbank. While there's enough dark fiber to hold L.A. together during the next earthquake, there are only three clusters of dense interconnection—Santa Monica, Hollywood, and Burbank. And these primetime media centers are not linked in any convenient, inexpensive, or straightforward way.

Networking in the Television Industry

The last few years has seen a significant emphasis on networking throughout the television industry. FOX Broadcasting lit up its Network Center in late 1998 and plays out all its programming, commercials, promotions, and long-form programs from networked video servers. In 1998 and 1999, NBC redesigned its facilities and implemented the Genesis digital server system. Cable networks, which often distribute multiple TV programming services, called "screens," digitize their material and play it out through servers, such as Viacom channels Showtime and MTV2, CNN, and Discovery.

Consider as well how digital technology has created and networked an entire virtual national television network, the WeB.[11] The WeB came into being because the Warner Brothers WB television network was unable to get carriage in most of the smaller 100+ television markets, despite its popular programs such as *Dawson's Creek*. That's because in the largest 1-to-99 markets, there are an average of 5.5 stations, while the 100+ markets average only 3.5 stations—not enough to carry all the networks.

To solve the problem, WBT president Jamie Kellner came up with the idea to create virtual television stations in as many 100+ markets as possible. These WeB affiliates are a partnership between the WBT network, a local station, and a local cable operator. WBT supplies original and syndicated programming and signal, the local station (which might be a CBS, NBC, ABC, or FOX affiliate) provides a sales force and over-the-air promotion, and the cable operator provides carriage.

The network began beaming its programming to new eighty affiliates from the California Video Center facility on September 22, at 7:00 P.M., Eastern time. "We launched in eighty markets at the same time, and that's never been done," noted WB Senior Vice President Russ Myerson.

The process of virtual network operation begins when the local station sends station IDs, promos, and local commercials to the WB for digitization into a compressed format and insertion into the signal. IBM designed an addressable Station-in-a-Box that sits in the cable headend. When WB transmits an MPEG-2 digital signal from the Hewlett-Packard servers in the video center, the local commercials have an address header that the box reads; it accepts only the commercials that are addressed to it.

Each local bricks-and-mortar station handles its own ad trafficking. They send a log file via the Internet to the WBT, where trafficking and local insertion is handled by a Novar system. An enterprise traffic system traffics the national program feed and national programming, while commercials are run out of a Hewlett-Packard server. WBT inserts the schedule and any new commercials into the digital feed. The box holds ninety minutes of material, runs the commercials and promos, and generates an "as-run" log of the spots that actually aired.

The entire virtual station chain is digital, although WBT still sends an analog signal to its network of over-the-air affiliates. The only analog process occurring in the cable headend is when the box converts the digital signal to run over the cable system.

Cable systems have also wired up. A few years ago, cable operators looked at ad revenue and saw that while viewers watched cable 20% of the time, systems were not receiving that percentage of advertising dollars. The reason? Cable systems were fragmented and advertisers couldn't track how many viewers saw their ads.

Cable interconnects were built in every major market, so that today, an advertiser can buy a national spot on CNN and expect that it will run on that network at the same time in every market. Interconnects also allow for consolidated regional buys, and advertisers know their spot will be seen by consumers throughout the interconnected area at the same time on the same networks. These moves have helped the cable industry increase its share of overall TV advertising dollars.

At the local TV station level, broadcasters are busy digitizing their newsrooms as well as their transmission facilities. One of the first stations to do so was KHNL in Honolulu. The newsroom may shoot material with digital cameras, or they digitize their analog footage, and place it on a server. Reporters and editors access the material from their desktops to edit and produce their packages. Local broadcast executives also see that the server-based station is only a few years away. Avid has captured a significant share of the market. Matrox Studio and the Fast Video Machine are competitors that also manufacture editing systems for newsrooms, as shown in Figure 2.5.

At the National Association of Broadcasters conventions in 1998 and 1999, where TV station representatives go to make their equipment purchases for the upcoming year, most stations sent their chief engineer to begin putting together the hardware and software package that the station will need to make the digital transition. Over the next five years, virtually every U.S. TV station's operations will be largely digital.

Distribution: Digital Cinema

The topic of the digital distribution of television will be covered in a number of chapters in this book. But the digital distribution and exhibition of films is only a few years away as well. Although it has been talked about for twenty years, the technology to do this is actually here today. Indeed, George Lucas's *Star Wars: The Phantom Menace* was shown in a suburban Los Angeles theater digitally in the spring of 1999.

Earlier in the year at ShoWest '99, an annual convention of distributors and theater owners, attendees saw side-by-side demonstrations of a traditionally projected 35mm print and a digitally displayed image, put on by two companies: Cinecomm Digital Cinema, a joint venture of Hughes-JVC (projector) and QualComm (satellite delivery), and Texas Instruments. Even the most jaded exhibitors were astonished that the quality was as good—and sometimes better than—film images.[12]

"We'd screen the digital stuff every 6 months and tell them 'Not ready yet.' But now it *really is* as good as 35mm film. I've never said it before, but now we have to take it seriously," said Barrie Loeks, president of the Loeks-Star theater circuit.

One advantage that digital display offers is that instead of making multiple film prints of a movie, studios could simply make an original and transmit it to theaters via satellite. They would save the multimillions of dollars they must now spend on these prints.

Unfortunately, theater owners who would bear the burden of paying for the new technology see no reason why they should take it on. Film technology is reliable, well-understood, and lower-priced. What is needed is a business model that will get the expensive equipment into theaters.

Since it is too expensive for theater owners to bear the cost alone, some way of splitting the cost with distributors must be found, since the

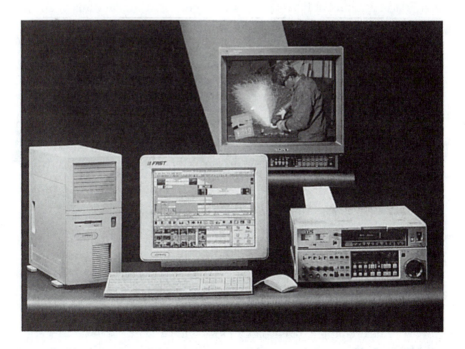

Figure 2.5 Fast Video editing systems can control analog and digital video tape machines and output digital masters. Source: Fast Video.

savings accrue to studios and distributors. Standards are required to prevent the expensive confusion that accompanied the addition of digital audio in theaters.

There are some advantages for theater owners as well, however. Today, if an exhibitor wants to hold over a print for a longer period because it is still drawing crowds, too bad, because that print needs to be sent to another theater that is waiting for it. In the digital world, this would be no problem, since each theater would receive its own print and store it on a server in the facility for as long as it was needed.

Another business opportunity is that once theaters were digitally outfitted, they would be able to exhibit other content besides films. There would be particular advantages to offering live events, which they cannot do now. They could show popular sports events, or fashion shows direct from Paris, Rome, or New York. They could rent their theaters in the daytime to organizations who want to hold virtual meetings and conventions. All these activities would bring in new revenue to exhibitors that in the end could exceed the profit they make showing films.

Both TI and Cinecomm Digital Cinema plan to be installed in several theaters by 2001, perhaps sooner. In 2000, TI will be running at least one test location. Once a digital distribution and exhibition infrastructure is in place, it is probably a matter of time before all-digital films begin to permeate the marketplace. High-resolution digital acquisition—shooting a motion picture—at a cost-effective price point is probably a number of years away.

Summary

In nearly every industry, digital technologies have brought about significant transformation. In media and entertainment businesses, the change is way beyond significant—think massive, think overhaul, and most of all, think fast! In a recent interview, former director of research for the National Association of Broadcasters Rick Ducey was asked what he thought broadcasters understood least about the Internet:

> The pace of change. People in the TV industry leave work on Friday, and when they return on Monday, things are still pretty much the same. In the Internet industry, people leave on Friday. And when they come back on Monday, they have no idea if they will still be in the same business, working for the same company, marketing their product to the same customers, or using the same technology.[13]

The next chapter, "Digitology 101," will present the basic concepts and terms that have brought about the digital revolution. Engineers can skip it, but business and creative people better not—or you will never get over that queasy feeling when somebody says, "You can't do that. You've only got 2 megabits per second here." It's not all that hard, and you can rest assured you'll be hearing this lingo for years to come.

Notes

1. B. J. Pine, J. H. Gilmore, and B. J. Pine, II, *The Experience Economy: Work Is Theatre and Every Business a Stage*, Cambridge, MA: Harvard Business School Press (1999).
2. C. Barish, "Superman's now super digital," *Videography* (October, 1993):30-32, 101-102.
3. J. Menn, "Pioneer Paints Portrait of Graphics' Future," *Los Angeles Times* (August 9, 1999):C5.
4. A. Crawford, Vice President of production development and corporate affairs, StationX, telephone interview, October 1998.
5. L. Kasanoff and A. Savitch, Threshold Entertainment, telephone interview, November 1998.
6. T. Mount, telephone interview, November 1998.
7. P. Mitchel, Novocom, telephone interview, November 1998.
8. J. Fancher, CTO, Pacific Ocean Post, personal interview, Santa Monica, October 1998.
9. S. Tolleson, telephone interview, October 1998.
10. F. Fletcher, telephone interview, November 1998.
11. R. Myerson, Senior Vice President and General Manager, WeB, telephone interview, October 1998.
12. B. Loeks, President, Loeks-Star Entertainment, telephone interview, February 1999.
13. R. Ducey, former Vice President, research, National Association of Broadcasters, telephone interview, April 1999.

Digitology 101

3

Chapter 1, "The Digital Destiny," presented the new world of digital video and looked at the how it differs from analog television. This chapter will now drill deeper into what makes information or technology digital and how an analog message becomes digital.

According to Webster's dictionary, the word digital comes from "digits," meaning fingers and toes, which were used as the first human calculator. "Digital" means counting, but it has also come to suggest other ideas, such as "discrete," "binary," "sampling," and "high-quality, clear sound." In modern usage, the word digital applies to the system of binary representation used in computer processing. The other word we have used to describe our current television system is "analog" (or analogue, as the British would write), and it is an alternative system for representing information. Analog is a shortened form of "analogous," which means similar to or comparable with.

The world itself is analog and the human perceptual system is also inherently analog. Thus, information about the environment is converted into digital data and must eventually be reconverted into analog representation so that people are able to sense it. Digital datastreams must be presented on an analog device—music must be reconverted to notes for human ears to enjoy, 0s and 1s must appear as alphabetic and numeric text, and images must be drawn in line, contrast, and color.

So why not just keep the information in the analog realm instead of all this movement between the two modes? The answer to this question is that translating information to digital for computer processing, then retranslating it back is currently the fastest, most efficient way to handle it. The future may hold advances in analog processing chips and computers, but for now, they are not as easy to design, or as inexpensive or efficient as their digital counterparts.

One way to understand the difference between analog and digital is to think of them as alternative ways of representing data. Consider the digital and analog watch faces in Figure 3.1. The analog watch face is

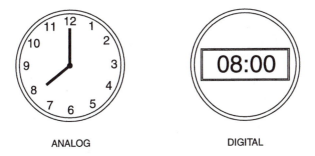

ANALOG   DIGITAL

Figure 3.1 Analog and digital: Two ways to measure and present the same underlying reality.

analogous to the apparent revolution of the sun. Allowing for a two-to-one compression (twelve hours of day, twelve hours of night), it is possible to calculate the position of the sun from the big hand on the dial. Given this code, people who wanted to know the time wouldn't need to be able to read Arabic numerals at all; they could estimate the time simply by the position of the big hand alone. By contrast, the digital watch face tells one time and one time only. If the reader isn't familiar with Arabic numerals, they are flat out of luck, since there are no point-for-point representational clues that an analogic system provides.

The analog watch gives the time as one point along a continuum of points, whereas the digital watch tells the time at one time only. This difference demonstrates an important distinction between analog and digital information: analog data is continuous; digital data is "discrete."

Discrete does not describe a person who doesn't gossip (spelled discreet); rather, it means separating something into distinct categories. Digitizing a continuous signal into discrete elements can be seen by putting a long, building blues note from Miles Davis' trumpet on a graph. The analog and digital measurements of the inimitable Davis are shown in Figure 3.2.

Sampling: The First Step to Convert Analog Signals to Digital Data

Since radio waves, sounds, paintings, text, and natural phenomena begin life in the continuous analog form, they require processing to turn them into digital data. The first step is to measure the analog material at regular intervals, say every second, and then to plot the resulting points. This process of measuring is called "sampling." The measurement could be taken every nanosecond, millisecond, quarter- or half-second, month, year; or every inch, yard, or mile; or every ounce, pound, or ton—the decision

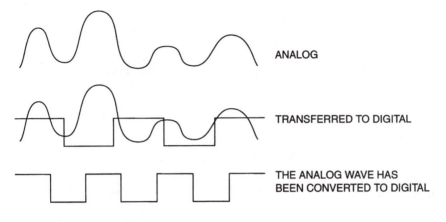

ANALOG

TRANSFERRED TO DIGITAL

THE ANALOG WAVE HAS
BEEN CONVERTED TO DIGITAL

Figure 3.2 Sampling an analog signal.

about the unit of measurement to be sampled depends on the meaning of the data and how they will be used.

Figure 3.2 depicts the same signal as measured and represented in the analog and digital domains. Neither representation alters the underlying reality: Both signals reflect the same Miles Davis blues note, radio wave, or Mariah Carey warble. However, the digital representation is one step further from the underlying reality because it consists of data taken from the analog representation: The discrete data points used to build a digital signal mark where the continuous analog signal is sampled.

The frequency with which the points occur is called the "sampling rate." The higher the sampling rate, the more the digital data will resemble the continuous analog data; the lower the sampling rate, the more information about the analog data will be lost. There are three important implications of a higher sampling rate when digitizing television pictures. The higher the sampling rate:

- The greater overall volume of information will result;
- The higher the quality of the picture (or sound) will be;
- The more resources must be expended for processing, error correction, and storage.

The sampling rate is absolutely critical to the resulting quality of digitized material. That is why music is "over-sampled" for Compact Disc—just to make sure the digital signal doesn't lose anything from the original analog sound. (Audiophiles may disagree, however. Many say they can hear the step difference between digital samples.)

A sampling rate of sufficient frequency is also crucial when digitizing video images. For this reason, the Consultative Committee for International

Radio (CCIR), an international standards organization, worked for more than two years to arrive at a sampling rate that would produce digitized pictures at predetermined levels of quality.[1] The CCIR found a rate that would translate video produced in both the U.S. NTSC (525/60) and European PAL analog television (625/50) standards into a nearly identical digital format. Essentially, after decades of incompatible equipment, the CCIR created a global digital standard that enables manufacturers to produce equipment for both HD and standard resolution DTV for the entire world market that varies only slightly.

To accomplish this aim, the CCIR adopted a family of rates, all based on sampling at multiples of the frequency of 3.375 MHz. This may seem like an awkward number, but it was really quite ingenious, because both U.S. and European TV material came out as a digital, broadcast-quality picture with 858 samples per line.[7]

Explaining the CCIR standard would require a technical discussion beyond the scope of this book. However, the following summary will explain an important term often encountered in the marketing brochures of digital video equipment: the 4:2:2 standard. Sampling standards for digital video are represented by a ratio; common standards are 4:2:2, 4:4:4, and 4:1:1.

Think of this ratio as a formula for sampling. The first number indicates how many samples should be taken for luminance, or brightness, sometimes called the Y component. The second and third numbers stand for the number of color samples, or chrominance, designated the as the U and V components. So the ratio 4:2:2 means that at each sampling interval, there will be four luminance samples (Y) taken, and two samples of each chrominance component (U and V) taken. A 4:4:4 standard would mean taking four luminance (Y) samples, and four U and four V color samples; 4:1:1 would mean drawing four luminance samples to 1 U and 1 V color sample, and so on.

It is obvious that a 4:4:4 sampling scheme will produce significantly more data than a 4:1:1, 4:1:0, or 3:1:0 procedure. Although the lowest sampling rates may compromise the color quality of the resulting picture, it is a method of reducing the size of the video datastream. For DV applications like audio/video streaming or videoconferencing, which are sent over low-capacity telephone lines, it is important to make the stream as small as possible.

Quantizing: The Second Step to Convert Analog to Digital Data

Sampling is the first step in converting continuous analog data to discrete digital data, whether it is time, temperature, music, or a television signal. The second step is "quantizing," or determining the number of measurement levels. Like sampling, decisions made about quantizing will determine the quality of the picture.

For example, suppose a television signal is sampled 858 times per line, using a 4:2:2 format. This sets the ratio of the luminance component to the chrominance components. But how many levels of chrominance or luminance are there? The sampling ratio leaves open the number of levels used to rate each sample. The luminance could have 100 intervals between black and white, or 50, or 4, depending on how coarsely or finely the data is quantized.

Sampling defines the points in time, along a horizontal axis. Quantizing determines how many vertical levels could define each sample. Sampling and quantization could be likened to the exams that students take throughout a semester. The number of tests is the sampling rate; the possible scores are the levels of quantization.

The highest possible quality of either a computer-generated or a camera-captured image is 8:8:8 RGB, standing for red, green, and blue, quantized at 24 bits per pixel (bpp). The blueness of the blue and the redness of the red are defined by quantization. Twenty-four bits, or 24 bpp, is the standard term used to state the level of quantization, specifying the amount of data provided about each picture element (pixel). It is a very high number of descriptors available to represent information about a pixel. Indeed, 24 bpp actually provides more data than are necessary, more information than the human visual system (HVS) can even perceive. Only the color palettes of the most high-end computer graphics systems offer 24 bpp still-image generation to allow pixel-level color correction.

Processors for professional television, which must manipulate moving pictures in real-time, quantize each sample at 10 bpp for the highest-quality applications, and at 8 bpp for less stringent applications. As the processing capacity of computers increases in the next few years, the depth of data may increase for video images such as footage from space, but it probably isn't necessary for most viewing situations.

Coding: The Last Step in Digitizing the Television Picture

Samples of the analog signal have been taken and then quantized at some level. Now it is time to "encode" the data into the language of computers, the standardized binary structure for representing data so that they can be processed, transported, transmitted, stored, retrieved, and ultimately reconverted into an analog image that the human visual system can perceive. It is possible to represent 10 bits per pixel using the ten Arabic numerals (1, 2, 3, and so on) with which we are so familiar, but they would not be processible by digital computers. The way DV becomes compatible with other digital data structures is by encoding it into binary code.

Binary means two, and binary code is based on two symbols. These symbols could be "y" and "n" (for yes and no), "a" and "b," or any other two arbitrary markings. By convention, binary code is composed of 0s and 1s. The important feature of 0 and 1 (or any other two-symbol set) is that one symbol represents "on" and the other "off," which translates quickly and easily into the presence or absence of electrical current. Computers use this binary code to represent all the numbers, alphabet letters, symbols, and images they process, and to signal the computer to perform millions of on-off computations per second. The rapid sequencing of electricity-on and electricity-off through processing chips gives digital computers their extraordinary power and usefulness.

If the only role of binary numbers, or binary digits (the full expression for the contraction "bits"), were to reside inside computers known only to engineers and programmers, there would be no need for most communications industry professionals to know about them. However, they have entered the public discourse through information theory. They are a constant topic of discussion; it is rare to enter a present-day discussion of video delivery without reference to them.

A Bit of Information Theory

A bit is a measure of information, just as an inch is a measure of space, a minute is a measure of time, and an ounce is a measure of weight. Measuring information like a commodity is a surprising and somewhat unsettling idea. It began in 1948, when Claude E. Shannon and William Weaver published a paper called, "A mathematical theory of information," which codified the knowledge about transmitting messages by telegraphy and telephony into a unified theory.[2] Shannon and Weaver started by identifying the parts of a communication system: source, message, channel, noise, and receiver. They also proposed an original definition of information as data that reduces uncertainty. The smallest measure of information, the basic unit of information theory, is the "bit"—a contraction for binary digit.

Any array of numbers or letters may be data, but it doesn't become information unless some uncertainty, somewhere, is reduced. For example, consider the following situation. Sixteen people named Kim Smith work in a huge, worldwide corporation. Two other coworkers are gossiping in the cafeteria when one of them mentions Kim. They quickly realize that there may be some confusion. They might discover which Kim they are talking about in the following way:

Is Kim blonde (= 0) or brunette (= 1)? 0 [1 bit]

Is Kim male (= 0) or female (= 1)? 1 [1 bit]

Is Kim over 30 (= 0) or under 30 (= 1)? 1 [1 bit]

Does Kim work in Accounting (= 0) or Sales (= 1) 1 [1 bit]

Bit by bit, it takes a total of four bits of information for the two conversationalists to establish that the subject of their confidences is the blonde Ms. Kim Smith, 28, who works in the Sales Department.

Information on the Move

A bit, the smallest unit of information, is represented by the smallest coding scheme, a binary code (usually by a 0 or a 1, although, as mentioned earlier, it could be any two symbols). Bits are grouped together and called a byte (pronounced bite), or a binary word. A byte is composed of eight bits.

When bits and bytes are transmitted through the air or transported over cable, they are referred to as a "bitstream" or a "datastream." The speed of the bitstream, its flow rate, is measured in bits per second, or bps. However, a single image of digital video contains millions, even billions of bits, depending on how frequently the original analog wave is sampled and how many levels at which the sample is quantized. To reduce the size of the numbers from long strings of unreadable zeros, there are standard prefixes to apply to the bits that describe radio wave cycles: kilo, mega, giga, and tera. Large numbers of bits become kilobits and kilobytes, to express thousands; megabits and megabytes to express millions; gigabits and gigabytes for billions; and terabits and terabytes for trillions of the little uncertainty reducers.

This nomenclature is used so often that there are common abbreviations for these terms. Capitalization matters; the letter for kilo is not capitalized but the letters standing for mega, giga, and tera are. The "b" in bit is not capitalized but the "B" for byte is. Thus, twenty thousand bits is "20 kb;" twenty thousand bytes is "20 kB." Twenty million bits is "20 Mb;" twenty million bytes is "20 MB," and so forth:

Information	Flow Rate	
20 kb	20 kbps (20,000 bits per second)	Kilobits
20 kB	20 kBps (20,000 bytes per second; 160,000 bits per second)	Kilobytes
20 Mb	20 Mbps (20,000,000 bits per second)	Megabits
20 MB	20 MBps (20,000,000 bytes per second; 160,000,000 bits per second)	Megabytes
20 Gb	20 Gbps (20,000,000,000 bits per second)	Gigabits
20 GB	20 GBps (20,000,000,000 bytes per second; 160,000,000,000 bits per second)	Gigabytes
20 Tb	20 Tbps (20,000,000,000,000 bits per second)	Terabits
20 TB	20 TBps (20,000,000,000,000 bytes per second; 160,000,000,000,000 bits per second)	Terabytes

Let us now turn to the direct relationship between the amount of information—the number of bits or the bit rate—and channel size or capacity. In 1924, Harry Nyquist, an early researcher of problems of telegraphy, first determined that this relationship is a ratio: channel capacity must be twice the bit rate.[3] Later, Shannon refined this proposition, making the ratio dependent on the type of encoding, the amount of accompanying noise, and the power used to transmit the signal—limitations that make up what is referred to as the Shannon Limit. Put another way, the Nyquist formulation and the Shannon Limit tell us that when information is transported or transmitted, the size of the channel determines how much information can move through it and how fast.

Remember that if the channel is an informational pipe, the bitstream is the "liquid" flowing through it. The huge amount of information in a sequence of video images must flow in real time and in the proper order for the viewer to receive a coherent picture; naturally, DV requires a large-bandwidth channel that can handle a very rapid bit rate. Such "big pipes," or conduits, are called broadband channels; and wired systems that are big enough to carry high-quality, full-motion video are called "broadband networks." Discussions of the future of television and digital video frequently refer to "pipes versus pictures" or "conduits versus content" to distinguish between the transmission or transport infrastructure and the programming it carries.

At present, particular conduits limit certain types of DTV content; in particular, narrowband telephone lines. If we start with Nyquist's 1924 formulation as a rough rule of thumb, that channel capacity must equal at least twice the bit rate, we can chart the improvements in encoding and the handling of the problem of noise. Today, Internet surfers routinely get information from the Net at 56 kbps, flowing over 30 kHz telephone lines. In other words, they have achieved a bit rate, or information flow, that is almost twice the size of the channel, an extremely important accomplishment when we consider the issues of the new television system.

Unfortunately, the improvement in the ratio of bit rate to channel size still isn't enough capacity for the enormous amount of information in a television picture because it is more difficult to achieve efficiency with video than with audio. The digital bit rate for a broadcast-quality NTSC picture is 128 megabits per second (Mb/s), which would require a channel with a bandwidth of about 64 MHz—more than ten times the 6 MHz per channel allotted by the FCC. It's even worse when considering the high-definition picture. It requires a bit rate of 1.5 gigabits per second, which would fill a 750 MHz-sized channel!

There is only one solution to this problem: reduce or compress the information in the television signal. To accomplish this compression, it is necessary to digitize the signal, because it is simply impossible to manipulate the analog signal to obtain the necessary reduction.

Digitizing Analog Video

How much information is in a single frame of digital video? Unfortunately, there is no straightforward answer to this simple question; this new digital world makes answers to such questions very elusive indeed. The reason is that it depends; it depends on how much information the original video contained, how often the original image is sampled, the depth to which it is quantized, and the size at which it is displayed. By specifying that the images should be very high broadcast quality and using today's vertical resolution, then the formula would be: 483 active lines, multiplied by 858 samples per line, resulting in about 400,000 pixels. When a lower quality image is acceptable, as is often the case, a lower sampling rate would result in fewer pixels.

In a search for the information content of very high-quality digital video, it would take 8 bits to represent the brightness level (luminance) of each pixel, and another 8 bits to represent the red, green, and blue colors (chrominance), resulting in about 6.5 megabits of information per frame. Since there are 30 frames in one second of television, the amount of information in that second totals approximately 200 megabits, for the highest quality broadcast images. In fact, professional studios work with even higher quality video at 270 Mbps.

For receivers to get all the quality of this image in real-time, the video must be transmitted or transported at 200 megabits per second. In fact, these numbers describe how pictures were sent back from Mars by the Mariner lander and from Jupiter and its moons by the Voyager spacecraft.[4] However, this is more information than digital video distribution systems actually require, whether delivered via satellite, terrestrial transmitter, cable, or telephone wire.

High-definition television poses an even greater challenge. The HDTV image has 1,050 active lines and 1,920 pixels per lines, for a total of about 2 million pixels. If each pixel requires 10 bits to represent it, every frame of HDTV involves about 20 million bits (Mb) of information, and one second of HDTV video contains about 600 Mb, over one-half a gigabit of information. This enormous mass of information presents a problem for signal transmission and transportation—or bringing HDTV to viewers—that are seemingly intractable. The solution is digital "compression."

Compressing Digital Video Images

The first step towards reducing the amount of information in video images is to digitize the analog signal to allow compression techniques to come into play. Compression does not mean making the images smaller; rather, it refers to encoding the signal more efficiently to result in the same image with fewer bits of information. It is the information, not the image, that is compressed!

Compression schemes and processes are divided into two types: "lossless" and "lossy." Lossless processes rearrange the data for ease or compactness in further processing; however, no data is actually dropped or lost. By contrast, lossy processes discard information that can never be retrieved. Sometimes the missing data can be reconstituted from the surrounding data, but most often it is lost forever. Whether or not the viewer can perceive the discarded data depends on which data is lost and how extensive the loss is.

The name for the equipment that actually carries out compression is a "codec," which stands for compressor-decompressor. Some codecs incorporate both hardware and software elements, while others rely only on software and use off-the-shelf, general-purpose equipment.

Types of Compression: Four Functions, Four Standards

There are four DTV compression algorithms, or ways of accomplishing the task, endorsed by standards-setting bodies. Standards are crucial because the equipment at the receiver end must be able to decode the material and convert it back to its original form.

All the officially sanctioned schemes are in common usage. They are frequently referred to in the literature, in lectures, and in conversations at the many seminars and conventions about new technical developments in the broadcast, cable, telephone, and computer industries. These four standards are H.261 and H.263 (or P*64), JPEG, MPEG-1, and MPEG-2, each designed to address a different media context.

P*64, H.261, and H.263: Compression Standard for Video over Telephone Networks

In 1990, the CCITT (Consultative Committee for International Telegraphy and Telephony) developed a compression standard for low bit-rate video-phone applications, that is, sending images over current telephone wires. The standard had two names, H.261 (pronounced "h dot two-six-one") and P*64 (pronounced "p by sixty-four"), which were joined by a third name, H.263 in 1994. H.263 pulled together new and even lower bit-rate standards with the existing P.261 standard.[5] It assumes a low-resolution, color image of a person's face or upper body, talking and gesturing against a minimally-changing or out-of-focus background, so its main application is videoconferencing. Some of the most popular videoconferencing software is based on P*64 algorithms (software processes).

One reason H.263 is so important is that it forms the basis for MPEG-4, the next generation of high-quality compression that will not be seen in products until 2000. Table 3.1 shows the various resolutions available for videoconferencing, starting with a standard modem speed of 28.8kbps and increasing in detail, depending on how much capacity the communication channel will allow. The bit rates of H.264 video range from 8 kbps to 1.5 Mbps.

Table 3.1 H.263 Resolutions, By Format

Compression Type	Size of Video Window
Sub-QCIF	128 x 96
QCIF	176 x 144
CIF	352 x 288
4CIF	704 x 576
16CIF	1408 x 1152

JPEG: A Standard for Still Images

The International Standards Organization established a standard for still images called JPEG (pronounced jay-peg), used primarily for applications using digitized images in a computer environment. JPEG stands for Joint Photographic Experts Group, the title of the technical committee that developed the standard. The compression scheme is based on a technique called the "discrete cosine transform," discussed later in this chapter.

JPEG applies to high-resolution pictures, including photographs, computer-enhanced photos, computer-generated images, and specialized medical images.[6] In 1993, several computer-based editing systems began using a version of JPEG compression called motion-JPEG or M-JPEG to edit moving images, because the method affords a high level of compression on each individual frame. Since every frame is whole and complete, motion-JPEG allows frame-accurate editing. (As we will see, other compression techniques for motion video involve providing only fragmentary information on some frames, so these schemes will not work for editing systems without additional processing.)

MPEG-1: A Standard for CD-ROM

The compression standard for moving images on CD-ROM is called MPEG-1 (pronounced em-peg one). MPEG stands for Moving Pictures Experts Group, a committee composed of technical representatives from film companies, video manufacturers, computer companies, and CD-ROM developers and makers.[7] Like JPEG, MPEG-1 compression uses the discrete cosine transform technique. MPEG-1 allows CD-ROMs to provide highly compressed moving images at a data rate of 1.5 kilobits per second (kbps).

Many companies abandoned the MPEG-1 standard in favor of proprietary compression algorithms, claiming that the process of setting the standard had become overly politicized.[8] Critics of the final MPEG-1 standard claim that representatives of the member companies jockeyed for a formula that would offer them a competitive advantage, rather than promoting the most technically superior process. As a result, detractors say that the compromised programming code is too clumsy and complicated for efficient compression.

The DIRECTV satellite-delivered, direct-broadcast-satellite (DBS) digital service, launched in the fall of 1994, first used MPEG-1 for the compression needed to bring consumers 150 channels. They would have preferred the higher quality promised by MPEG-2, but waiting would have delayed their 1994 launch by an entire year. Although the digital picture itself was relatively free of artifacts, some consumers complained about the quality of the decompressed video, saying it looked blocky and lacked detail. Later reports absolved the compression technique, claiming that the poor picture quality was really the result of a very narrowband satellite uplink. In any case, by 1996, all of DIRECTV's satellite-delivered programming was encoded in MPEG-2, and its images were the highest quality consumers could bring into their homes.

MPEG-2: A Standard for Broadcast-Quality Video

MPEG-2 is the standard for professional-quality digital video images. Versions of MPEG-2 are now in broad use for programming delivered via satellite, terrestrial digital broadcast, and DVDs, as shown in Figure 3.3. Like JPEG and MPEG-1, it is another discrete cosine transform procedure.[9] MPEG-2 is adequate for camera-generated images, but it has some serious drawbacks. It groups pixels into blocks, which can result in an unpleasant "blocky" look to the video after it is reconstituted.

The shortcomings of MPEG-2, combined with a confusing multiplicity of standards has led to the development of another standard, designed to overcome these problems. MPEG-4 combines discrete cosine transform with other algorithms to arrive at a more flexible, extensible compression technique.

MPEG-4: A Single Compression Standard in a Multifunction, Multiplatform Environment

The original work on MPEG-4 began with the intent of developing a standard for compressing low bit-rate video, with streamed audio/video over the Internet. The MPEG-4 Working Group changed its objectives as the

Figure 3.3 An MPEG-2 digital video processing and editing bay. Source: Optibase.

work progressed to embrace a much larger set of objectives. MPEG-4 draws compression formats for digital television, interactive computer graphics applications, and Internet protocols to provide variable compression for all widespread existing wired and wireless distribution networks and technologies: terrestrial broadcast, fixed wireless and mobile wireless, telephone, satellite, Internet protocol, telephone, DVD, and other optical discs.

The Final Draft of MPEG-4 Version 1 was presented at Atlantic City in October, 1998, and was accepted as an international standard in February, 1999, with the ISO (International Standards Organization).[10] The Working Group foresees applications in Internet multimedia, e-commerce, interactive games, videoconferencing and videotelephones, DVDs and CD-ROMs, multimedia E-mail, networked database services, remote video surveillance, wireless multimedia, and television broadcasting.

MPEG-4 provides the following tools for these various technologies and applications:

- Description of compressed, streamed 2-D and 3-D scenes;
- Identification and synchronization of streams;
- Version 2, released later in 1999, adds techniques for the management and protection of intellectual property.

MPEG-4 incorporates significantly different ways of describing scenes from the way it was done in MPEG-2. Rather than just compressing the entire scene, the new standard identifies "AV objects" and their relationships in both time and space. This method allows both producers and users to interact with specific objects in the space, rather than with the picture as a whole or with areas overlaid on the picture. Objects can be separated from the background, offering opportunities to compress a datastream by transmitting a background only once. It also permits the artifact-free integration of objects with different natures, such as video, graphics, text, and audio.

Proprietary Standards:
Standards with No Official Sanction

There are several proprietary compression algorithms that, although not sanctioned by official standards, are sufficiently well-known and promising to deserve mention. These algorithms are vector quantization, wavelet, and fractal compression.

Vector quantization (VQ) is a software-only compression technique, as opposed to P*64, JPEG, and MPEG, which all require both hardware and software. Schemes that use specialized hardware are faster but more expensive, while software-only compression is relatively inexpensive but significantly slower. Software-only techniques are typically severely asymmetrical, which means they take much longer to encode the video than they do to decode it, especially for high-quality images. As computer processors become faster, this asymmetry may lessen. Currently, it may take as little as ten minutes to encode one minute of video, or as much as an hour or two.

"Vector" is an engineering term for a straight line. A line drawn through the color spectrum of the television signal from black to white represents each color at a place along the line. Vector quantization allows very high compression by eliminating most of the color and luminance information from the datastream.

Unfortunately, the range of color is lost when it is replaced by such a smaller number of points along the vector. For example, although there may be 100 possible variations of a medium-to-dark blue, all will be shown as a single dark blue when VQ is used. As a result, although it is computationally simple and offers a high degree of compression, it doesn't deliver very satisfactory performance.

Wavelet compression is a sophisticated compression scheme based on the work of mathematician Isabel DeBauchies. By deconstructing and transforming the waves that carry video signals into wavelets (meaning "small waves"), this technique provides more efficient and appropriate procedures for compressing the information in waves of all types. It is the basis for several innovative codecs, including the low-cost, software-only Indeo by Intel and others. Objective and subjective tests indicate that wavelet-based compression may be superior to other compression schemes, including discrete cosine transform procedures.

Finally, fractal compression is another new technique based on the similarity of infinite detail of natural images, independent of the level of observation. For example, a coastline has a jagged outline, whether viewed from outer space, a high hill, on the beach, or staring at sand particles. Stephen Mandelbrot and others have produced images of extraordinary complexity and detail, using simple mathematical formulas for fractal images.[11] Iterated Systems, a company in Norcross, Georgia, does the most advanced work in fractal compression. The technique works best with complex natural forms; however, fractal-based codecs have improved as the processing power of computers has increased. (See Figure 3.4.)

Compression: A Complex, Multi-Step Process

Although there is some variation between various methods, Table 3.2 lists the common steps involved in compressing a television signal. The starting point is an analysis of the human visual system, HVS, which often guides the choice of which data is kept or discarded. Every scheme reduces the amount of data *within* each frame of video, a technique called intraframe or spatial compression. Intraframe compression relies on some combination of eight procedures: (1) perception characteristics, (2) filtering, (3) color-space conversion, (4) digitizing, (5) scaling, (6) transforms, (7) quantization, and (8) compaction encoding. Finally, the data in moving video is also reduced through eliminating redundant information from successive images, called interframe (*between* frame), or temporal compression.

Perception Characteristics and Compression

Exploiting properties of human perception has long guided efforts to compress the enormous information in video images. For example, people are much more sensitive to differences in brightness levels than they are to color information. In 1953, engineers on the National Television Standards Committee exploited this characteristic of perception when they converted the original black and white signal to the current color signal, by coding the signal with more luminance information than color information.[12] Today's 4:2:2 sampling standard works on the same principle: The ratio means that the luminance data is sampled at twice the rate as the color data. This sampling rate reduces the amount of data that would result from the most information-rich rate, 4:4:4.

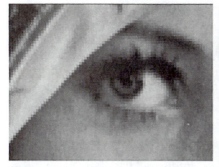

TIF

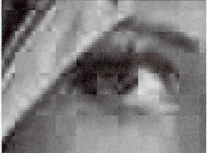

JPEG

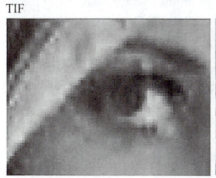

WAVELET

FRACTAL

TIF

JPEG

WAVELET

FRACTAL

Figure 3.4 The TIF Lena image as compressed by JPEG, wavelet, and fractal compression techniques. Image courtesy of *Playboy*.

Table 3.2 Preprocessing and Compressing the Television Signal

	Amount of Compression	Description
Preprocessing		
Filtering	None	Removes noise from signal
Color-space conversion	33%	Reduces 3 variables, red, blue, and green, to 2 variables
Digitizing	25%	Sampling, quantizing, and coding the analog signal
Decimating (Scaling)	75%	Subsample digitized signal and discard remaining information
Intraframe Compression		
Data transformation	None	Rearranges data for further compression
Quantizing	60%	Assign values to rearranged data
Encoding	31%	Rewrite data efficiently, using Huffman, variable-length, or arithmetic coding
Interframe Compression		
Predictive coding	See below	Codec looks at key frame, or I-frame, to predict next frame, or P-frame
Motion estimation and compensation	See below	I-frame is divided into 16 areas and amount of change in each area is estimated. Based on estimate, codec selects new I-frame or makes changes to P-frame.
Predictive Interpolation	Total, 3 steps 79%	MPEG creates a B-frame, which is a predicted frame, based on both past and future frames

Another important aspect of perception for compression is that research on vision pathways shows that there are two parallel but distinct channels to the brain from the retina of the eye.[13] Experiments to monitor the activity of single cells in the eye's retina indicate that the first channel, the sustained system, processes the details of a stationary scene. The second channel, the transitory system, detects motion and rapid changes in the ccene, and updates the information quickly and continually.

Based on an understanding of the HVS, researchers Robert and Karen Glenn developed a television transmission system that reduces the data in the television signal by taking advantage of these findings.[14] In the Glenn system, which is compatible with current broadcast technology, the detailed non-moving portions of a picture are sent less often than the fast-moving portions. A further refinement is that while motion must be updated rapidly, it doesn't have to be presented in as much detail as the non-moving parts of the picture.

Preprocessing the Signal before Compression

There are four steps that take place prior to intraframe (within frame) compression: (1) filtering, (2) color-space conversion, (3) digitizing and scaling, or (4) decimation. Filtering doesn't compress the information at all; it removes noise, information that is extraneous to the image and has been generated by the act of acquiring or processing it. Noise must be removed, because when the signal is compressed and decompressed several times, the noise will become an ever-larger part of the image until it severely degrades it.

Color-space conversion is a process that takes advantage of the greater perception of brightness levels than of color information. The three colors, red (R), green (G), and blue (B) are transformed into only two variables (U and V), reducing the amount of data by one-third.

At this point, the video is actually digitized. As discussed earlier, digitizing means to sample the analog image, then to quantize the samples, assigning numerical values to them, and finally to code the results into a datastream. Digitizing confers advantages for data processing, storage, and retrieval, and is necessary for further compression of the video.

As the name implies, decimation (or scaling) involves discarding as much as 75% to 80% of the information by subsampling the already sampled digitized data. Some codecs simply discard it; others retain data by storing average color and luminance information about neighboring pixels and lines. The codec then uses these averages to reconstruct the image during decompression.

Compressing the Processed Data

Now that the information in the video has been preprocessed, the codec can start the actual compression. It begins by transforming the data, quantizing it, and reencoding it into an even more compact, compressible format.

Transforms   Two types of transforms are widely used in video compression: the discrete cosine transform, the most common, and the wavelet transform. Discrete cosine transform, or DCT, is also called the Fourier Transform, discovered and developed by Michael Fourier in the nineteenth century. Engineers have used it since then to transform electrical pulses into frequencies.

The discrete cosine transform is complex. DCT first arranges the data about the video image into blocks of pixels, eight columns by eight rows. Comparing entire blocks of pixels, a single coefficient describes how often similar blocks appear. To say "how often does this happen?" is a matter of frequency or time. Thus, in DCT, pictures are translated from the spatial domain (space) into the temporal domain (time).

Although DCT is the most widely used transform, there are some disadvantages to it. It requires substantial computation to achieve compression, although its proponents have been proven correct that the advances in processing power following Moore's Law have delivered processors capable of manipulating DTV. When this power was weak, the method of breaking the pixel data into blocks created a similar appearance throughout the block, giving a quilt-like look to the overall picture, even at only moderate levels of compression. In the past few years, these artifacts have been lessened and are now not normally noticeable in professionally compressed MPEG-2 DTV.

The wavelet transform differs from the discrete cosine transform, the basis for the P*64, JPEG, and MPEG schemes, in two important ways. First, it preserves most of the spatial relationships between pixels in a wavelet that carries low-frequency picture information. At the same time, it places information about the image edges in repeated high-frequency wavelets. Second, because wavelets are computationally simpler, they contain information about the entire picture, not just sections of it. Intel's success with wavelet compression in the company's Indeo products has led to the inclusion of wavelet compression in the new MPEG-4 standard, which promises to offer a significant improvement over MPEG-2.

Quantization and Compaction Encoding   Recall that digitizing the video image involved sampling, quantizing and coding the analog signal. Now that the digital data has been transformed, it is re-quantized. In other words, the information is assigned new numerical values.

Quantization techniques are one way to differentiate between the various compression schemes. JPEG and MPEG compression require uniform quantization across blocks of data, while vector quantization techniques allow more variation in numerical assignment. For example, with DCT-data, the most frequently occurring elements of the picture are shown in the greatest detail. Recall that this transform converts the picture to blocks of 64 pixels (8 x 8), and a single value represents the entire block. Blocks that recur often are described with a greater number of quantization levels, while less-common blocks are quantized more crudely. This strategy conforms to the Glenn research. People process the stationary parts of the images (that appear in many successive frames) more thoroughly than the moving parts (that may appear in only one or a few frames).

After the data is requantized, it must be reencoded for efficiency. There are three techniques: run-length encoding, Huffman (or variable-length) encoding, and arithmetic coding. All three methods are mathematically sophisticated, allowing up to as much as a one-third reduction of the data.

Compression Steps for Moving Images

Up until now, we have considered how to compress the data within a single frame. Now we will look at ways to extract the redundant information from sequential frames. Suppose the viewer watches a video showing a plane landing. Fifty percent of the picture is a very slowly changing sky—the clouds are in a slightly different position as the camera follows the plane. The airport facilities change little. The plane itself is the only rapidly changing element in this picture.

Temporal means time. Temporal compression begins with video that has already been compressed as much as possible using the earlier mentioned techniques of spatial or intraframe compression. The process of interframe, or temporal compression between sequential frames, includes predictive coding, motion estimation and compensation, and picture interpolation. These steps now take advantage of the similarities between successive video frames. Since MPEG-2 is the dominant method used for professional entertainment applications, including the DVD, the following discussion will cover the MPEG-2 process.

However, note that MPEG-4 would use an entirely different method to compress this scene than called for by the MPEG-2 procedures outlined below. MPEG-4 would identify the plane as a moving object against a stationary panorama. This separation would make it easy for producers to extract the plane and to reuse it in many different environments. It would also enable higher rates of compression by limiting the number of times that the object and its background would need to be sent. In addition, MPEG-4 allows for intelligence in the receiving device, so the signal would include instructions to the processor in the receiver describing how to reconstitute the image locally.

Predictive Coding   The basis of temporal compression is the fact that, for video, the best predictor of any picture is the immediately preceding picture. One way to reduce information is to send only the information that has changed. (In our example of the plane landing, this would mean that only about 1/14th of the picture would need to be sent.)

The codec looks at each frame (already compressed using the intraframe techniques), called the I-frame, or key frame, and breaks it down into large blocks, called macroblocks. It then predicts that the next frame, the P-frame or predicted frame, will be the same. The codec compares the I-frame with the P-frame as macroblocks. If there is a small change, then only the "difference signal" is sent. If the frame is entirely different, then it cannot be predicted at all and the codec churns out a new I-frame, which it will use as the basis for its next prediction. If there is substantial change, but some elements are the same, then the codec will call upon motion estimation and compensation for further compression. (See Figure 3.5.)

Motion Estimation and Compensation   Motion estimation works from the assumption that many changes in the picture are quite predictable. Let's return to the example of the plane landing. The plane will move in a single direction, screen left or right, and the nose will travel across the screen, slowing at a relatively steady rate of deceleration.

Motion estimation and compensation identify those macroblocks that contain the changing pixels and predict the trajectory of motion as it travels. The predictions are translated into changes in macroblocks in successive frames. Motion estimation and compensation requires heavy information processing on the part of the codec, demanding continuous checks, trial and error runs, and comparisons with the actual and predicted frames.

Picture Interpolation   MPEG doesn't just predict P-frames from the previous picture. It also looks ahead to future frames and predicts an inter-

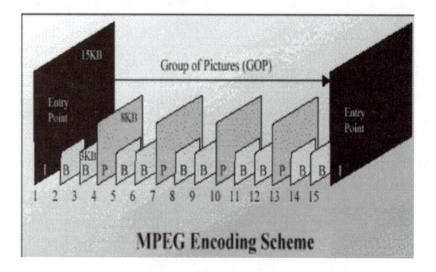

Figure 3.5 Predicting pictures in an MPEG group of pictures (GOP).

vening picture, called a B-frame, based on an average of the past and future images. The "B" in B-frame stands for bi-directional, so named for the dual sources—backwards and forwards—of its predictor frames. In order to accomplish this averaging, the codec must intake a few frames and hold them before it sends them out. This means that there is a necessary throughput delay, as much as 1/6 of a second, between the codec's input and output.

Compression for DVD   The MPEG standard that covers DVDs is called DVD-V. It adds a few special parameters that are specific to DVD. One example is that DVD-V restricts the coded size of the picture: The MPEG-2 specification that is used for broadcast allows any coded frame size between 16 and 720 pixels horizontally and 16 and 576 pixels vertically. DVD devices, however, require that frame sizes fit within a limited range. In the audio domain, MPEG-2 allows the sound to be coded at a sample rate of 32, 44.1 or 48 kHz. DVD-V sets the sampling rates for both Dolby AC-3 and MPEG-2 audio strictly at 48 kHz.

The Bandwidth Bandwagon

Although all the forms of digital video—DTV, DVD and CD-ROM, broadband, and streamed video—may eventually converge, today these distinct flavors exist. And to understand the differences between them, we need to look more closely at an idea introduced in Chapter 1, "The Digital Destiny": the carrying capacity of communication channels, often termed "bandwidth." This is such an important concept in the digital environment, especially with respect to any kind of networked communication, that it is essential to understand it. Textbox 3.1 explains bandwidth in some detail.

The illustration of the electromagnetic spectrum (EMS) in Figure 3.6 will aid in understanding bandwidth. Radio waves are electromagnetic in nature. They really are waves with electrical properties that travel through the air. The EMS arranges them by their frequencies from low to high. The properties of the different frequencies make them more or less useful for certain applications. Note the middle portions of the EMS which are used for over-the-air television broadcasting, and the higher range appropriate for terrestrial and satellite microwave transmission.

Take a look at the spectrum of a television signal, in this instance, Channel 2. (See Figure 3.8.) The carrier signal for any station designated as Channel 2 (in any market) is at 55.25 megahertz (MHz). Channel 2's AM video signal takes up 4.2 MHz of bandwidth on one side of its carrier signal (the upper sideband) and .75 MHz on the other side (the vestigial sideband), giving a total of 4.95 MHz. This 4.95 MHz occupies the frequency range from 54.50 MHz to 59.45 MHz.

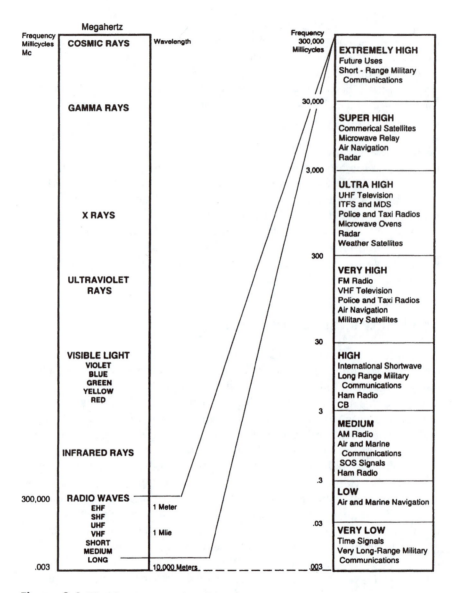

Figure 3.6 The electromagnetic spectrum.

Bandwidth refers to how much information a communication channel can transmit or transport. Sometimes bandwidth is called carrying-capacity. There are two aspects to consider: the channel and the information.

Think of a channel as a pipe that carries liquid, whether it is a wire or in the air, as shown in Figure 3.7. (People use the terms channel, pipe, conduit, line, and cable pretty much interchangeably.)

One way to describe a pipe is by its size: for example, a 1-inch pipe (one inch in diameter). An information pipe is measured in cycles, called hertz. This is a very small measure, so channel sizes usually come in thousands, millions, and billions of hertz. An over-the-air TV channel is a 6 megahertz (MHz) pipe, or a 6,000,000 hertz channel. Alternatively, another way to measure a pipe might be by the amount of liquid that could flow through it in some amount of time. So the description of the same pipe could be "a 50-gallons-per-hour pipe" or "a 3-megabits-per-second pipe."

Transmitting TV or radio involves modifying rapidly cycling radio waves, sometimes called sine waves, as they travel forward in the air. The waves are changed so that the variation in the cycles of the waves reflects the changing video, point for point. The waves in this part of the electromagnetic spectrum are very small, and they cycle very fast, which allows them to carry the millions of bits that make up DV images.

Another way of saying that the waves vibrate back and forth in a cyclic motion is to say that they oscillate. The oscillation occurs a particular

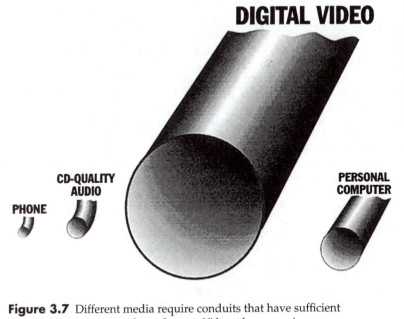

Figure 3.7 Different media require conduits that have sufficient bandwidth to transport them. Source: *Videomaker* magazine.

Textbox 3.1 Understanding bandwidth.

number of times per second; that rate of oscillation is called the wave's frequency. Conceivably, a wave might oscillate some number of times per minute or hour, or some other unit of time; but practically speaking, one second is the typical measure. If there are other signals in the same channel (air or wire), vibrating the same number of times (at the same frequencies), the result will be interference, preventing receivers from decoding the message.

The entire array of all frequencies is expressed in the electromagnetic spectrum (EMS), which ranges from heat to light, as shown earlier in Figure 3.6. Radio waves (including television and microwaves) are in the invisible portion of the EMS. They are measured in three ways: (1) by how many times they arc, or cycle in one second of time, or their frequency; (2) by how high and low they cycle, termed "amplitude"; and (3) by how long they are, called "wavelength."

The frequency of the cycles and the length of the wave are related: the faster the cycles, the shorter the wavelength; the slower the cycles, the longer the wavelength. If the EMS is displayed horizontally, as a person moves their finger from left to right, the number of cycles (frequency) increases and the wavelength decreases.

One cycle per second is called a "hertz" (Hz), named after Heinrich Hertz, who first measured the waves of light and energy of the electromagnetic spectrum (EMS). A wave that cycles ten times per second would have a frequency of 10 Hz and would be very long indeed.

In reality, radio waves that carry communication oscillate at much higher frequencies. The AM radio band starts at 540 kilohertz (kHz), or 540,000 hertz; both AM and FM radio are in the kilohertz portion of the EMS. Television signals are measured in megahertz (MHz), or millions of hertz (cycles) per second. Sine waves in the upper reaches of the radio portion of the electromagnetic spectrum—the 1 to 100 gigahertz (GHz) range—cycle a billion times per second; they are used for microwave transmission. They are used to carry out such diverse tasks as carrying the financial data of banks to cooking food in millions of households!

When radio waves are modified to transmit information (rather than to heat frozen dinners), they are called signals. The process of modifying them is called "modulation." At the center of a radio or television signal is the carrier wave, which is transmitted on a single frequency, a tiny fraction of the EMS. Traveling with the carrier wave are modulation signals that occupy additional frequencies on either side of the carrier, termed its "sidebands." The sidebands act on the carrier wave to vary its strength. The variation that modulation signals generate allows the carrier wave to be a precise representation of the information that is sent: the sound of music, a voice, or a television picture.

"Amplitude modulation" (AM) means that information in the sidebands is reflected by variations in the arcs and troughs (amplitude) of the carrier wave. "Frequency modulation" (FM) means that information is carried by variations in the number of cycles of the carrier wave, rather than the amplitude.

Textbox 3.1 *Continued*

> Video uses more bandwidth than audio because it takes more information to represent it. Television requires a bandwidth of 4.2 megahertz (4.2 million cycles) just for the NTSC picture; AM radio uses 10 kilohertz (10,000 cycles), and stereo FM radio takes up 150 kilohertz (150,000 cycles). The reason music sounds so much better on FM is because its greater bandwidth allows for two-channel stereophonic sound.

Textbox 3.1 *Continued*

Channel 2 also has a frequency-modulated (FM) audio signal 4.5 MHz above the visual carrier frequency at 59.75 MHz. The audio modulation adds sidebands to this signal that extend 25 kilohertz (kHz) on either side of the sound-carrier signal. This gives the sound signal a bandwidth of .050 MHz, extending from 59.70 MHz to 59.80 MHz. The total bandwidth is a fraction less than 6 MHz, because small slices of frequency on both sides of the main signal, called "guardbands," are left empty to avoid interference with other channels.

Broadcast television pipes invisible waves in the air. Wires and cables are also pipes, and they too have a carrying capacity. The bandwidth of the copper telephone wire typically used in homes varies greatly, depending on the electronics that are placed at either end and the way the information is arranged. Common copper will carry from about 30 kilobits (30 kbps, or 30,000 bits) per second to as much as 8 megabits (8,000,000 bits) per second.

Unless the customer pays much higher rates, the typical home telephone line will deliver a maximum of about 53 kbps. Compare this to the enormous pipes of cable TV systems, which carry signals from 550 to 750

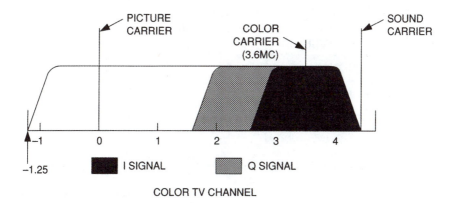

Figure 3.8 The bandwidth of a single television channel, here Channel 2.

MHz. These giant conduits are capable of carrying 3.375 trillion bits (terabits) per second—about 90 to 120+ channels of 6 MHz NTSC TV.

High-res, low-res, up-res, down-res, multi-res, and de-res . . . Resolution is the key determinant of image quality, a function of the amount of information captured in the original production and the bandwidth of the channel that carries it to the viewer. The more vertical lines there are displayed and the more horizontal elements there are present along each line, the more information will define the image, and the more detail it will contain—hence the term high-definition television, although it might just as well have been called high-resolution television. Here are the approximate bandwidths for the channels that carry the various forms of digital TV:

Table 3.3 Comparison of Approximate Throughput of Common DTV Channels

Channel	Throughput
Cable system	400 MHz to1 GHz
Fast Ethernet computer network	100 Mbps
Ethernet computer network	10 Mbps
Broadcast/cable TV channel	19.4 Mbps
Cable channel	27 Mbps
DVD	9.8 Mbps
Cable Modem	2.5–8 Mbps
xDSL	384 kbps–8 Mbps
T-1 telephone line	1.5 Mbps
1x CD-ROM	1.3–1.4 Mbps
Enhanced TV via TV signal: vertical blanking interval (VBI), horizontal overscan (HOS), and vestigial sideband (VSB)	20 kbps–600 kbps
Dual channel ISDN	128 kbps
Single channel ISDN	64 kbps
Standard phone line/modem	4 kHz, 28.8–56 kbps

Mixing and Matching: Modulation and Multiplexing

Modulation refers to how a signal is encoded for transmission with variations that reflect the information that is conveyed by it. At the reception end, the signal is demodulated so it can be played on devices such as TVs, PCs, screen phones, and other equipment. There are dozens of modulation algorithms, and each type of network employs schemes that match the specifics of its design and purpose. For example, QAM (quadrature amplification modulation) is a well-known public-domain modulation method, currently used in most cable systems. Telephone companies also use a QAM variation called CAP (carrierless amplitude phase) to implement high-speed data over their customers' telephone lines.

Multiplexing techniques allow message streams to be combined into larger streams for transport across networks. Put another way, the purpose of multiplexing is to load multiple message streams onto a single channel in the most efficient manner. It involves aggregating the signals that come from one source, although they may include different data types such as data, video, audio, and text, and weaving them all together into one thread to be carried over a conduit. The term "multiple access" is quite similar, except that it means bundling signals from different sources to go over the network.

Both multiplexing and multiple access are important in large networks. For example, in a cable system, television channels and digital services that come to the subscriber all originate from the headend; Internet data and the audio stream of a telephone conversation all come from the central office of the telephone company to the customer—so in both cases, the signals are multiplexed for transport downstream. The reverse is not true: when customers send signals from their homes, each signal originates from a different household. So multiple access is how they are bundled for transport upstream.

There are five ways to encode aggregated signals for transport:

1. Frequency-division multiplexing   FDM
   Frequency-division multiple access   FDMA
2. Time-division multiplexing   TDM
   Time-division multiple access   TDMA
3. Wavelength-division multiplexing   WDM
   Wavelength-division multiple access   WDMA
4. Dense wavelength-division multiplexing   DWDM
   (multiple wavelengths on a single fiber)
5. Code-division multiplexing   CDM
   Code-division multiple access   CDMA

Frequency-division, time-division, and wavelength-division techniques are well known and used in many systems. Most recently, another method has been developed called code-division multiple access (CDMA), a spread spectrum technology first invented by actress Hedy Lamarr. CDMA is now widely used in digital telephone systems and there is considerable discussion about using it for other applications.

The best way to understand multiplexing techniques is by two metaphors: the train and the footrace. Time-division multiplexing is like a train, with boxcars headed for different destinations behind the engine in some random order, as shown in Figure 3.9. After the engine, Signal A is going to the Smith's household, followed by Signal B headed for the Jones' house, followed by Signal C going to the Browns' home, and Signal D on its way to the Greens' TV. After these, we might find signals B, A, B, D, C, B, C, A, and so on. Time-division multiple access is the same train, except that the signals all converge from different locations, as well as move toward different destinations. TDM and TDMA are used for digital signals, especially with fiber optic cable.

Frequency-division multiplexing is like a footrace. The entrants run side by side, but stay within their assigned lanes, as in Figure 3.10. Frequency-division multiple access is similar, but all the runners come into their lane from different directions, already running. FDM and FDMA are used for analog signals.

In the late 1990s, Lucent Technologies introduced a series of products that incorporated a technology called "dense wave-division multiplexing" (DWDM). Wavelength division multiplexing and multiple access are similar to FDM and FDMA—the footrace metaphor works for WDM and WDMA as well. DWDM combines multiple streams of optical signals, allows them to be amplified as a group, and then transported over one fiber. Each signal can be carried at a different bit rate and in a different format, as its own wavelength, independent of other wavelengths moving down the fiber.

Lucent originally offered WDM over multiple wavelengths, as many as 16 channels of light, within a single fiber. At present, "muxes," multiplexers that combine streams, can send up to 100 light waves down one fiber, and Nortel is shipping test gear that pushes 160 light waves, each carrying 10 Gbps (10 billion bits per second). That capacity equals 1.6 Tbps (trillion bits per second), or enough bandwidth to let 28 million people surf the Web simultaneously at 56k—all carried by a single glass strand no thicker than a human hair. Another advantage of WDM and WDMA is that both digital and analog signals can be transmitted on a single fiber by using separate light-wave emitters if their frequencies are sufficiently separated. Network vendors such as AT&T/Network Systems, Northern Telecom, and NEC America began offering WDM for high-speed telephone systems in 1996.

This chapter has presented the basics of digital information and communication. The next chapter turns to an account of how the transformation from analog television to digital video occurred and is occurring now. It begins with HDTV—and HBO's cash flow problems.

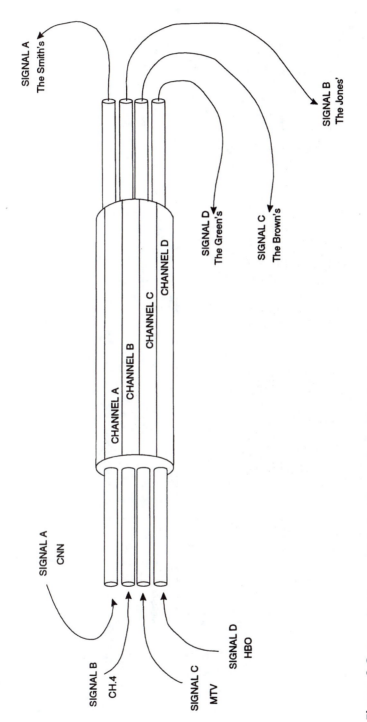

Figure 3.9 Time-division multiplexing. Source: James Bromley.

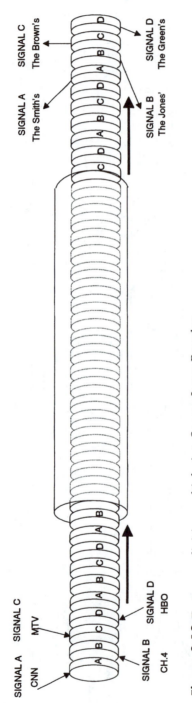

Figure 3.10 Frequency-division multiplexing. Source: James Bromley.

Notes

1. The CCIR is part of the International Telecommunications Union. The ITU Web site is at http://www.itu.int/.
2. C. E. Shannon and W. Weaver, *A Mathematical Theory of Information*, Urbana: University of Illinois Press (1949).
3. J. R. Pierce, *An Introduction to Information Theory: Symbols, Signals, and Noise*, 2d ed., New York: Dover Publications (1980).
4. Ibid.
5. Specifications for H.261 are available from the ITU for 20 Swiss francs, about $12.28, as of 6/15/00. See ITU publication at http://www.itu.int/itudoc/itu-t/rec/h/h261.html. For H.263, the site has a publicly downloadable document, "Table of Contents and Summary of Recommendation H.263 (02/98)" at: http://www.itu.int/itudoc/itu-t/rec/h/s_h263.html.
6. Information about JPEG standards may be found at http://www.jpeg.org/.
7. A good MPEG-1 site is http://drogo.cselt.it/mpeg/.
8. Comment is taken from an interview with a member of the MPEG-1 who requested anonymity.
9. Good MPEG-1, MPEG-2, and MPEG-4 sites are http://www.mpeg.org/MPEG/, http://www.mpeg.org/MPEG/, and http://www.whatis.com/mpegstan.htm.
10. A 1997 overview of the MPEG-4 standards is online at http://wwwam.hhi.de/mpeg-video/standards/mpeg-4.htm.
11. J. Briggs and F. D. Peat, *Turbulent Mirror*, New York: Harper & Row (1989).
12. E. B. Crutchfield, Ed., *Engineering Handbook*, 7th ed., Washington, DC: National Association of Broadcasters (1985).
13. E. Lane, "The next generation of TV," *Newsday* (April 5, 1988): Discovery section, 6.
14. Ibid.

4

High-Definition Television: Or How DTV Came Into Being Because HBO Needed Cash

Frequently it doesn't matter why a technology was invented, since the people who adopt it reinvent it anyway—using it in ways that are different from the original intentions of the people who originally created the technology. There are many instances of this occurrence. The radio was conceptualized as what we would now consider a telephone-like medium; the VCR was conceptualized as a recorder, but people mainly use it to rent and view movies at home at their own convenience.

However, the development of HDTV holds some particularly important lessons. Many of the issues this chapter covers that first arose in the late 1980s have proved to be perennial problems, still unresolved and problematic two decades later. Every technology comes into being within a social context, one that is almost always more political than might be imagined. HDTV is a particularly vivid illustration of the this tendency— certainly HDTV evoked strong cultural and nationalistic feelings in Japan and Europe as well as the United States that influenced the speed and direction of bringing it to fruition.

The quest to invent an HDTV system was an extraordinary worldwide effort of massive proportions. It was a nearly thirty-year-long journey that commenced in the late 1960s, ultimately involving more than twenty national governments, a host of commercial consortia, hundreds of business organizations, research laboratories, universities, and thousands of scientists, researchers, and skilled technicians. It began as an industrial goal to improve television technology. But in the mid-1980s, HDTV became inextricably intertwined with national and regional pride, economic autonomy, and the spirit of competitive pursuit. In countless news reports of that time, HDTV was evoked as proof of cultural domination or subordination, as confirmation of wise government policy, or, contrarily, as foolish anarchy.

During an era of extravagant corporate news stories, HDTV was one of the hot topics. The high-cost, high-tech, high-stakes global battle to establish an HDTV system was long and hard-fought, the field strewn with the detritus of wounded ideas and dead prototypes. More than once, the pace of technological change compelled companies and their research and development departments to execute an abrupt about-face or to leave the field altogether.

Today, the research still continues. It is quite possible, indeed likely, that there will be further casualties as new techniques and technologies become available. The roller-coaster development of high-definition television and other aspects of the new television system prompted Joseph Flaherty, the much-respected vice president of technology at CBS, to comment: "There are three ways to lose money. The most pleasurable is romance. The fastest is gambling. And the surest is technology."[1]

Even before implementation, HDTV cost billions of dollars just in research and development. Japan spent an estimated $1.3 billion.[2] The Europeans spent nearly that much, with estimates ranging from $994.5 million to $1.066 billion.[3] In the United States, the effort was private, in contrast to the combination of governmental and private funding provided to both the Japanese and European HDTV research teams. As a result, the U.S. program cost far less, approximately $200 million, expended by private industry with no federal funding.[4]

No matter whose checks were supporting it, HDTV was always a game for high-rollers. It cost $17 million to build the Advanced Television Test Center (ATTC) near Washington, D.C., a facility designed just for the testing of prototype HDTV systems, as shown in Figure 4.1. Although the test results were used by the FCC to set standards, the ATTC was paid for by private U.S. broadcasting, cable companies, and the high-tech organizations that actually developed the HDTV systems for testing.[5] Now that the implementation phase has arrived, billions are now being spent on changing over the existing television industry system to digital technology, with the cost of the transition estimated between $8 and $19 billion.[6]

HDTV Development, 1970–1986: The Era of Japanese Diligence and Western Complacence

The Japanese started work on HDTV in 1970 under the direction of Dr. Takashi Fujio of Nippon Hoso Kyokai (NHK), Japan's state broadcasting system. From the very beginning, the research aimed to create a better picture with greater detail and more vivid color. A major goal was to heighten the psychological effects of the medium to improve its use as a platform for advertising.[7]

Figure 4.1 April 5, 1995: Testing the latest version of the "Grand Alliance" Digital HDTV System at the ATTC. Source: ATTC. Photographer: David Poleski.

High-definition television came first as the focus of research because it would directly affect the viewing experience. From a technical perspective, just getting there was a daunting challenge. From an economic perspective, the successful realization of the goal would sell more toasters, toothpaste, and tacos—and more televisions. This last product was most important to the Japanese, who manufacture most of the world's TV sets.

The Japanese effort was based on industry-wide cooperation, coordination, and government encouragement. In the early 1970s two government offices, the Ministry of Posts and Telecommunications and the Japan Development Bank, began parceling out research to eleven Japanese companies. Sony and Ikegami developed cameras and the recorder; NEC, Mitsubishi, and others worked on projectors, tubes, and processing devices.[8] Through the Development Bank, the Japanese government began subsidizing the research. Since the psychological, social, and economic benefits they hoped would flow from HDTV rested on the delivery of a higher-resolution image, the concept guiding the effort was simply to increase the number of continuous scanning lines from 525 to 1,125. The improvement over NTSC was significant, but it was an extension of the old system, not a radical departure.

Nevertheless, the Japanese were quick to realize that any improvement at all would mean developing an entire array of technologies in order to produce, process, transmit, receive, and display the superior video and audio.[9] In 1981, Sony became the world's first supplier of HDTV equipment, which was demonstrated in Tokyo. That same year, Francis Ford Coppola used Sony's HDTV equipment in his film *One from the Heart*, utilizing the new technology for storyboarding scenes and as a database.[10]

In Europe and the United States, these developments went virtually unheralded. Among the three networks, only CBS' savvy and influential Senior Vice President of Technology, Joseph Flaherty, seemed aware of the implications of the Japanese advances. In 1981, to test the new technology, CBS applied for three channels on the new COMSAT satellite, and they also requested a waiver from the FCC (which was granted) for permission to operate a cable system of less than 90,000 subscribers.[11] CBS stipulated that they needed more than the usual space allotted for a channel on a satellite transponder, 27 MHz, rather than the usual 6 MHz given for TV signals.[12]

In the early 1980s, experts believed that transmitting HDTV would require a minimum of 27 MHz, possibly as much as 72 MHz—considerably more bandwidth, or channel space, than NTSC. (You can refer to Textbox 3.1 in the previous chapter for an explanation of bandwidth.) For that reason, the Japanese system was based on satellite transmission, which did not conflict with the centralized approach of Japanese broadcasting. But it did conflict with the television system in the United States where, unlike Japan and most of the other countries of the world, local broadcast is an historically important element of television. In fact, localism has been an established national policy since the TV industry's inception in the United States.[13]

In addition to offering network-supplied or purchased syndicated programming, local stations present local news, weather, and sports. They cover Main Street parades, school board elections, and council meetings, and the some of the anchors and reporters who appear in this kind of programming become popular, drawing audiences to their station's newscasts. To accommodate this locally originated programming, stations send their signals over the air from transmitters located within line-of-sight-of-receiver antennas, called terrestrial transmission. But in 1987, it was technically impossible to transmit the huge 27 MHz HDTV signal by terrestrial transmission.

For CBS, HDTV's reliance on satellite delivery made it an unattractive business prospect, because it could not be adopted within the framework of a local TV system. Throughout the early 1980s, CBS was the country's number one network, and its corporate image was deeply identified with the heritage of local broadcasting. It was out of the question for CBS to single-handedly transform the entire TV system and unclear that it would have been in their interest to do so—in spite of the lure of an improved picture.

Nevertheless, the Tiffany network, as CBS was then called, remained intrigued by the promise of a superior picture. In 1982, Dr. Takashi Fujio, then Deputy Director of Technical Research Laboratories at the Japan Broadcasting Corporation, paid a visit to CBS in New York and gave a demonstration of the Japanese HDTV system to CBS executives.[14] Despite their enthusiastic response, the sheer size of the HDTV signal and its inherent conflict with localism posed an insurmountable barrier to any imminent implementation in the United States.

It is less clear why the Europeans, with their already centralized television systems, failed to respond sooner to the Japanese advances in HDTV research, which proceeded at a furious pace. The Japanese called their line-doubling system "HiVision," and throughout the 1980s they continued to develop equipment that would displace the entire range of NTSC, PAL, and SECAM technology. Cameras, video recorders, switchers, routers, transmitters, and receivers rolled out of the laboratories of Japan's consumer electronic giants.

HiVision delivered a stunningly better picture that attracted crowds wherever it was demonstrated. By 1986, the Japanese were poised to extend their commercial domination of the television equipment industry, which was already significant, using entirely proprietary designs and technology. As a first step towards adoption of the HiVision production standard, NHK sent design plans and specifications to the Comite Consultatif International de Radio (CCIR), the technical evaluation apparatus for the International Telecommunications Union (ITU).[15] The ITU, part of the United Nations, is the most important global standards-setting organization, comprised of more than 150 member nations. Its mission is to propose, develop, revise, and administer worldwide technical standards and radio frequency spectrum allocations.

At first, it looked like the Japanese would have clear sailing. The United States and Canada both supported the NHK standard, called "1,125/60." (This nomenclature meant that the HiVision signal had 1,125 lines, progressively scanned 60 times per second.) The North American position was articulated by the Society of Motion Picture and Television Engineers standard, SMPTE 240M, a United States version of the 1,125/60 standard.

HDTV Development, 1986–1987: The Era of Confusion

If the Europeans had heretofore appeared complacent and unconcerned, the Japanese filing with the CCIR galvanized them into action. Influenced by vociferous French objections, Europe balked at the idea of allowing Japan to take over the next generation of television.[16] The Europeans believed that the Japanese technology posed a threat to their own consumer electronics industry, so rather than accept HiVision, they decided to subsidize the development of a domestic HDTV television system. In 1986, the European Community funded Eureka 95 with an initial $180 million.

A consortium of European universities, research institutions, and electronics firms, including Philips, Thomson, and Bosch, shouldered the mission of developing HD-MAC, standing for High-Definition-Multiplexed Analog Components. HD-MAC had originated in Britain at the Independent Broadcasting Authority, which had developed a television standard they called MAC. The proposed that the HD-MAC standard would call for 1,250 lines, transmitted at the European 50-frames-per-second rate and delivered via satellite.

The United States Giant Awakes

Like ripples in a pond, the European response opposing the acceptance of a Japanese HDTV standard provoked a reexamination of policy within the United States. Fifty-eight broadcast-related companies urged the Federal Communications Commission to consider the issues involved in establishing high-definition television in the United States. In July, 1987, the FCC opened an inquiry to explore the possible introduction of an advanced television service.[17]

The FCC had observed the storm of protest over HDTV that engulfed the regulatory authorities in Europe. Recognizing the potential for controversy and the need to proceed carefully, the FCC created an organization to study the difficult technical, economic, and social issues involved in HDTV adoption, a group they called the Advisory Committee on Advanced Television Service (ACATS). Composed of twenty-five volunteer members, one of the first steps on the agenda of ACATS was to establish a technical facility where high-definition television systems could be evaluated and tested.

Starting in September of 1987, supporting organizations, including all the broadcast networks, CableLabs, and others, spent $17 million to build the Advanced Television Testing Center in Alexandria, Virginia, near Washington, D.C.[18] In supporting these HDTV efforts, broadcasters were hedging their bets, making sure they would have its superior picture at the same time or before cable operators would. The competitive challenge that subscription-paid cable television had begun to present to free, over-the-air TV acted as a powerful motivator. Broadcasters feared that wired cable systems would be able to transport high-definition video more easily than TV stations could transmit it over the air. because of the enormous amount of information in the HDTV signal.[19]

Intense public discussion about the parameters of the new television system began almost immediately. The key issues during this early stage included compatibility, scalability, and interoperability. The argument over compatibility concerned whether or not sets would be capable of receiving both the new HDTV signal and the NTSC signal. This idea harkened back to 1954. When color TV was introduced, the engineers of the NTSC designed a signal that allowed people with color sets to receive a color picture, while those who had only black-and-white sets could still get a good picture. In the late 1980s, many observers hoped consumers would be able to use their existing sets to receive HDTV signals.

Scalability means that a system or a part will still fit if it is called upon to shrink or grow in capacity. It was one idea for obtaining backwards compatibility between HDTV and NTSC sets. In a scalable system, a television set would accept and display any number of lines; the more lines the set could receive, the better the picture it would display. An NTSC set would accept and display 525 lines, while an HDTV set would accept 1250 or 1125, or whatever the maximum number of lines set by the future HDTV standard.

The issue of interoperability addressed HDTV sets as one component in an overall system. For example, in a stereo sound system, all CD players connect to and function with all amplifiers and speakers so that consumers can assemble equipment from different product lines. Similarly, interoperability dictates that all TV sets will work with all audio speakers, set-top boxes, and VCRs, using standard connectors to link the components.

While U.S. interest in HDTV was high, in 1987 (and indeed for several years more) it seemed inevitable that a Japanese-designed satellite-based transmission system must surely triumph. Japan held a commanding two-decade lead that, to all appearances, was simply insurmountable. The only possible way to overtake the Japanese would be to develop a digital high-definition alternative to the analog HiVision. Knowledgeable people made public pronouncements to the effect that, unfortunately, a digital television design was at least ten years away.[20]

1988–1992: The Era of Competition

In spite of Japan's commanding lead and the apparently immature state of digital technology, the seemingly unequal battle was joined, and an enormous global corporate race for HDTV was on. In the United States, the David Sarnoff Research Center, Zenith Electronics, and BellLabs started work. In France, Thomson scientists began an intense effort, as did Dutch engineers at Philips.

In early 1988, the FCC laid out a plan for deciding how to adopt an advanced television system in the United States, but no one was sure the plan would work. In contrast to the cooperative Japanese effort, the FCC's procedure suited American business culture: The decision would be reached through a competitive process, pitting corporate giants against one another in order to reach the best possible system. The role of government was neither to coordinate nor subsidize; rather, it was to ensure a fair and level playing field for contenders.

By the end of 1988, the FCC had received twenty-three HDTV proposals, submitted by their developers, called proponents. It was a bewildering alphabet soup of competing systems, including IDTV (Improved Definition Television), EDTV (Enhanced Definition Television), and ACTV (Advanced Compatible Television). The different designs varied widely in how much they would improve the NTSC picture and how they would transmit the signal.[21]

Still, it seemed nothing could stop the Japanese juggernaut. While Europe and the United States belatedly tried to organize themselves to develop their own high-definition television, the Japanese did not hesitate to press their advantage. In February of 1988, they produced a live HDTV broadcast of the opening and closing ceremonies of the Olympics in Seoul, Korea, beaming pictures back to Japan via satellite. The public watched the broadcasts on 200 monitors displayed at 50 sites, including railroad stations and malls.

At the Shinshu University medical school, professors used HDTV pictures of brain surgery to teach operating procedures, an application made possible by the detail of the recorded HDTV image. At a museum, a "HiVision Gallery" allowed visitors to call up paintings and background information from individual viewing booths. The printing industry in Japan began designing equipment to allow them to make four-color separations and to print from HDTV images on tape. A restaurant started using HDTV pictures of its dishes in place of the plastic renditions usually on view in Japanese restaurants. Even a bookmaker used HDTV to show the horses racing that day.[22]

By October, 1988, NHK started to broadcast one hour a day in HDTV. The head of the Japanese Ministry of Posts and Telecommunications confidently predicted that there would be millions of HDTV sets in use by 1994 and 10 million by 2000.[23] As Japan looked at the slow progress in the West, mired in international politics, economic considerations, and technical issues, their optimism seemed more than justified.

Early in 1989, the Europeans entreated the United States to join them in blocking approval of the 1125/60 standard still awaiting approval from the CCIR.[24] Increasingly, European intransigence posed a barrier to acceptance of the NHK production standard. But within the United States, opinion was still divided. The broadcast networks disagreed among themselves as to which HDTV standards ought to prevail. Debate also raged within the National Telecommunications and Information Administration (NTIA), headed by a man who would soon go on to play an important role in the U.S. effort to develop HDTV, Alfred Sikes.[25]

In the spring of 1989, there were hearings in both the House of Representatives and the Senate on the standards-setting issues, a sign of the heightened interest in and concern over in HDTV.

The European influence proved decisive. In May, the Advanced Television Standards Committee (ATSC) reversed its support for the Japanese 1125/60 HDTV standard. Established back in 1983, the ATSC is a group composed of executives from the United States television industry. They come from broadcast networks, stations, cable companies, producers, electronics equipment manufacturers, and satellite companies. This influential group makes recommendations to the United States State Department on pending issues before international standards-setting organizations, such as the CCIR and the ITU.

There was little doubt that the ATSC's view would prevail. Shortly after their conclusion, the United States withdrew its support for NHK's standard—putting the issue of HDTV standards back to square one.[26] Moreover, the European position had hardened even more, and it became clear that they would never support Japanese-designed and controlled HDTV.

By mid-1989, the United States began to make substantial progress towards defining the country's parameters for HDTV. In July, the ATSC technical group approved production standards for HDTV:[27]

Aspect ratio: 16:9

Pixels per active line: 1920

Pixel arrangement: Orthogonal

Pixel aspect ratio: 1:1

Active lines per picture: 1080

Interlace: 2:1 (present), 1:1 (future)

Unresolved: frame rate and size of vertical blanking interval

Then, in October of 1989, all hell broke loose when the American Electronics Association (AEA) published their forecasts of HDTV's potential effects on the United States economy and technological prowess.[28] The report immediately ignited a firestorm of controversy. The smoldering conflicts over Japanese economic success and the fears it engendered broke out into open flames. The entire tone of the public discourse concerning television's next generation changed from an arcane, technical, insider style to one of immediacy, urgency, and action. Within a month, there were ten congressional hearings scheduled and nine bills concerning HDTV development on the docket.

The AEA is a trade association comprised solely of United States-based electronics firms. Their report considered the role that high-definition television would play in the overall future market for U.S. high-technology products and, by extension, in the U.S. economy. The AEA document reported one development in an ominous tone:

> *In mid-1988, several managers from a top U.S. semiconductor firm were able to preview the 40-chip set for HDTV receivers that has been developed by a Japanese firm. Validation that these chips already existed—a concern the U.S. executives had feared—was greatly overshadowed by the realization that their own domestic engineers probably did not even know how to engineer the chips. The level in sophistication in consumer electronics had taken another quantum leap forward.*[29]

The report predicted that the semiconductor-gobbling HDTV system would give the Japanese a competitive advantage that would allow them to produce integrated circuits more cheaply. This development was dangerous, concluded the report, because lower prices would enable Japan's companies to undercut American chips used in computers. The AEA formula was that if the United States held a 10% market share in HDTVs, then the country would keep 21% of semiconductor production and 35% of the personal computer market. By contrast, if the United States held a 50% market share in HDTVs, the nation's share of the semiconductor market would be 41% and its portion of the personal computer market would be 70%. The report concluded with a call for government to support the American electronics industry through subsidies, tax incentives, changes in patent procedures, and relaxation of antitrust laws.

Following the AEA bombshell, a rash of articles appeared in both the popular and serious press detailing Japanese superiority and prescribing what the United States should do to cure its trade ills. Robert Reich, then a professor at Harvard, wrote an article in *Scientific American*, where he translated the AEA figures into jobs.[30] The strong semiconductor scenario would deliver 331,000 jobs in the semiconductor industry itself, compared to only 83,000 in the weak scenario. In the computer industry, the strong scenario would result in 588,000 jobs, with the weak scenario offering only 147,000 jobs.

That same month, in October of 1989 at the annual convention of the Society of Motion Picture and Television Engineers (SMPTE), HDTV was the star of the show. The centerpiece was a live transmission from Tokyo to the convention's location in New York. Rank Cintel, LTD, a British manufacturer of equipment used to transfer film to videotape called a telecine, announced they were developing film-to-high-definition video-transfer equipment.

Not all references to the coming HDTV system were positive. At that same SMPTE Convention, broadcasters complained bitterly and loudly about the anticipated high costs of HDTV implementation for their stations, as outlined in the Ross Report. Bob Ross, then Director of Broadcast Operations and Engineering for Group W station, WJZ-TV, Baltimore, had prepared and presented his findings earlier at the National Association of Broadcasters' convention in the spring of 1989. The Ross Report showed that an HDTV station, built from scratch, would cost $38.5 million, $24.5 million more than an equivalent NTSC station.

While politicians were agonizing privately over whether there should be government support of the U.S. HDTV research effort and journalists stirred the pot publicly, European and American electronics companies continued active development on HDTV. The David Sarnoff Research Center had made sufficient progress on their Advanced Compatible TV (ACTV) to broadcast New York's St. Patrick's Day Parade over WNBC-TV, although only the researchers themselves had the HDTV reception equipment and set that were required to view the high-definition program. Zenith and AT&T were also both well along in their plans. By November, 1989, the FCC had twenty-one remaining proposals for some form of improved definition television, four terrestrial transmission concepts, and one satellite transmission plan.

In March of 1990, the Advisory Committee (ACATS) met to establish the fundamental determinations that would guide the development of HDTV. They reached three crucial decisions:[31]

1. They would give preference for high-definition designs, rather than enhanced or improved definition systems;
2. They would revisit technical questions again in 1992, in case radical new technologies had been discovered in the meantime;
3. They chose a simulcast rather than an augmented transmission plan, meaning that HDTV would have its own independent signals, rather than merely adding information to the existing NTSC signals. However, the committee directed that the simulcast signal not exceed 6 MHz bandwidth.

The Little Group That Could—and Did

These ACATS decisions were especially important to a small group of little-known engineers who worked for the out-of-the-mainstream VideoCipher division of General Instruments Corporation, in San Diego, California.[32] For the GI VideoCipher group, NHK's HDTV was a danger to the profitable business they had built around the existing satellite television transmission system because it demanded the replacement of GI's satellite hardware and software by Japanese-designed equipment.

As the global race to develop an alternative HDTV design that could compete with NHK's HDTV was reaching mach speed, a team of two

M.I.T.-trained engineers, Jerry Heller and Woo Paik, headed up General Instruments' VideoCipher division. They were laboring at a tangent to the HDTV efforts, working to perfect an innovation they called digital multi-channel NTSC that they thought would appeal to broadcasters. The NHK satellite-delivered HDTV menaced this new product as well as their lucrative VideoCipher business.

Jerry Heller and Woo Paik shared a background that enabled them to follow the HDTV race, even though they weren't actual participants. The two had worked together since 1978, part of an "invisible college" that originated at the Massachusetts Institute of Technology.[33] Invisible colleges, which permeate the upper ranks of American business and universities, are composed of people who form a relationship through their association with the same graduate school.

The M.I.T. invisible college that placed Jerry Heller and Woo Paik together at General Instruments had its roots in 1968, when Irwin Jacobs, an M.I.T. professor, supervised Jerry Heller's Ph.D. thesis there. Jacobs formed a company called Linkabit to do defense- related digital communication research and hired Heller. (Linkabit proved to be a fecund developmental incubator, ultimately resulting in more than nineteen derivative companies including Qualcomm, Jacobs' most recent achievement.)

In 1978, Jerry Heller had traveled back to M.I.T. and recruited Woo Paik to work for Linkabit. Heller recalls that he was shopping only for American citizens to work at a company whose business came mainly from defense contracts. At the end of the day, when he got up to leave, he found Korean-born Woo Paik waiting outside. Paik informed him, "I'm the one you're here to hire!" Despite the language barrier, Paik had a knack for communicating technically complex material in simple English. Since Heller's "employment test" was to ask candidates to explain their Ph.D theses to him in clear, plain language, he knew he had found his prospect in Paik.

Heller and Paik worked together for the next two years, then they both transferred to M/A-COM when that company bought out Linkabit in 1980. Robert Rast, now Vice President of HDTV Business Development at General Instruments, describes Heller as the "visionary" and Paik as "the one who makes it happen."

In 1985, Heller received a Request for Proposals from the Home Box Office (HBO) premium cable network, asking for ideas about how HBO could scramble their signal. The cable giant was at a crucial point in its growth, financing programming and buying expensive films. Cash flow was a huge problem, and the company was losing as much as $10 million a month in potential revenue to satellite dish owners who were bootlegging the unscrambled HBO signal as it was down-linked to cable head-ends for transport to subscribers' homes. HBO executives decided to try to sign up some of these freeloading viewers, and sent out a request for proposals to encode their signal to prevent unauthorized (and unpaid) viewing of their movies.

Heller and Paik responded to HBO's request because of their considerable backgrounds at both M.I.T. and Linkabit with digital signals and their knowledge of the advanced computers needed to process them. Although they didn't have direct experience with television, they brought to M/A-COM specialized knowledge about the rapid processing of massive amounts of data, similar to the quantity of information in a television picture. Based on the Heller-Paik proposal, HBO invited M/A-COM to meet with them before awarding the contract.

Heller sent Paik to the January, 1985 meeting with HBO. "I had to pretend I was some kind of expert in the scrambling area," recalled Paik. "Of course, I had some ideas about what could be done, but I'd never actually worked with television signals before, so my knowledge was theoretical. After we landed the contract, I went out and bought *Do It Yourself TV Repairs* by Robert Middleton, for $6.95 at a local drugstore, because I needed the actual numbers for color television processing," he said, pointing to a paperback on a bookshelf in his office.

M/A-COM won the HBO signal-scrambling contract from a crowded field of at least twenty other high-tech companies. The scheme Heller and Paik devised for HBO, called VideoCipher, became the basis of an extremely profitable business for M/A-COM. One of their competitors, General Instruments Corporation, watched these developments unfold and made a successful bid to buy that division, and only that division, from M/A-COM.

Heller and Paik now worked for GI in the VideoCipher Division, so-named after the product they had developed. Following the work they did for HBO, virtually every satellite broadcaster in the world encoded their satellite signals with VideoCipher, keeping the GI group busy, happy, and well-compensated. Thus, when the NHK HDTV system threatened to replace the entire satellite-delivery infrastructure with Japanese-designed and manufactured products, the negative implications caught the attention of General Instruments and, by extension, Heller and Paik.

The two M.I.T.-trained engineers had their sights on developing a product they called "multichannel NTSC." They reasoned that since sending a television signal over satellite transponders was very expensive, if they could figure out a way to make the NTSC signal more efficient, television networks could save money by piggybacking more channels on each transponder. Heller's and Paik's experience digitizing the HBO signals and their grounding in digital signals processing proved invaluable.

In order to reduce the size of the 6 MHz NTSC signal, they digitized it, invented ways to shrink the amount of information in it, sent it over the satellite, then, back on earth, de-digitized it, and reconstituted the complete 6 Mhz NTSC signal. By early 1990, Heller and Paik had simulated the solution and built a prototype. They invited potential customers, including several network executives, for a demonstration of their digital multichannel NTSC product. Since broadcasters followed the HDTV race closely, these knowledgeable network people informed the GI team that they believed the design elements used in M/C-NTSC would apply directly to the high-stakes HDTV sweepstakes currently underway.

May 31, 1990 was the deadline for proponents to submit HDTV plans for evaluation by ACATS and testing at the Advanced Television Testing Center. Inside General Instruments' VideoCipher Division, Woo Paik was pushing hard for an HDTV project, at least partly because he was certain he could digitize and compress a true high-definition television signal. Heller agreed that Paik's ideas were the only way an HDTV signal could be transmitted over the air in 6 MHz, thus meeting the preferences expressed by the ACATS in March.

Barely a month before the deadline, Heller gave Paik the go-ahead— but not, he warned Paik, at the expense of multichannel NTSC. Paik waited until the last possible moment, then cancelled his appointments. Leaving his colleagues at work on MC/NTSC, he went home for a week of isolated, uninterrupted thinking and writing. Despite the novelty of having their hard-working father home during the day, Paik's three children tiptoed around, staying far from the private study at the end of the house.

One week later, just days before the deadline, Paik had successfully designed the most advanced television system ever invented, a design that would put the United States in the forefront of HDTV, years ahead of the Japanese. He carried a yellow-lined pad into the General Instruments facility in north San Diego; written on it were the descriptions, formulas, block diagrams, and charts that comprised the basic design for digital, compressed HDTV that could be transmitted over-the-air using terrestrial transmitters.

Looking at Paik's work, Jerry Heller was so convinced that the design would work that he sat down and designed some of the electronic circuit boards himself. The week before the deadline, Heller and Larry Dunham, another GI executive, engaged in daily discussions about whether they should submit their system for HDTV testing precertification, as the FCC termed the first step. By the end of the week they hadn't resolved that question, so the two men continued talking over the weekend. On Sunday they called the president of General Instruments , Frank Hickey, in Chicago and requested the $130,000 they needed for submission. Hickey agreed and a check was cut on Monday.

The day of the deadline Heller and Dunham flew to Washington, D.C. and hand-delivered the check and application letter to the Advanced Television Testing Center (ATTC) in Alexandria, Virginia, minutes before the close of business. Joe Widoff, Deputy Executive Director of Finance of the Advanced Television Testing Center, remembers the shockwaves that

Figure 4.2 Dr. Woo Paik standing next to the digital HDTV he invented.

General Instruments' all-digital proposal sent through the advanced television development community: "In one day, the whole HDTV landscape changed."[34]

The Changed HDTV Landscape

The reaction from the media was immediate—Heller recalls being deluged by telephone calls from reporters. The response from the industry was less sanguine. In late June, Woo Paik traveled to Alexandria and presented more details about the design to ACATS. Skeptical of GI's claims that it had developed a workable digital television system, the representatives of the broadcast, cable, and manufacturing concerns questioned Paik closely. Only at the end of several hours were their doubts allayed. Although a core of skeptics continued to question the feasibility of GI's breakthrough for several months, the majority of those who looked closely at the system believed it would work.

Digitizing the TV signal answered all the issues raised at the beginning of HDTV development: compatibility with NTSC equipment; scalability as a means of accomplishing compatibility; and interoperability with other system components. The General Instruments digital coup added a new issue to the discussion of HDTV—extensibility. Extensibility means that the digital TV format could be extended to other technologies; specifically, to computer and telephone networks that processed digital data.

The GI digital proposal answered all these overarching questions at least in theory. Immediately, the whole ground of public discussion about HDTV adoption shifted from what the technology should do to how it could be realized in practice. It brought about a focus on specific standards for spectrum use, transmission and transport standards, frame rate, pixel size, scanning type, and cost.[35] Although still daunting, new figures showing that the cost of implementation would be $12 million more per station than NTSC (rather than the $24.5 million more that NTSC as predicted in the earlier Ross Report) helped allay broadcasters' fears.[36]

Proposals for HDTV were filed and studied, but they had not yet passed the scrutiny of actual operation. By August of 1990, the test bed was in place at the Advanced Television Test Center in Alexandria, Virginia, and testing was set for April, 1991. Unable to match GI's all-digital HDTV system, several proponents dropped out, leaving only six prototypes in the testing process. With strong encouragement from ACATS, most of the proponents who were left embarked on programs to develop their own all-digital systems.

For the Japanese, General Instruments' breakthrough was such a bitter pill to swallow that they did not respond quickly to the threat it posed. In any case, they may have been skeptical of the feasibility of the GI system in the real world, a belief still held by a few observers. A newer Japanese system, dubbed Narrow-MUSE, was a version of HiVision that could be transmitted terrestrially, developed to accommodate the United States broadcasting system. Having invested more than twenty years and a billion dollars in their products, Japan continued to push Narrow-MUSE. At its core, it is an analog system augmented with digital processing, a design referred to as hybrid analog/digital. In December, 1990, NHK shot the first opera ever taped in HDTV, Rossini's *Semiramide*, at Tokyo's Metropolitan Opera House.

Likewise, the Europeans still supported their own hybrid system, HD-MAC, although the EC countries experienced considerable difficulty getting the member nations to agree to it. Eventually, they reached accord on a two-step plan. First, they would launch D2-MAC, an improved-definition solution offering a better picture than the current system, in a 16:9 format, accompanied by four-channel digital sound. Then they would introduce HD-MAC, true high-definition television.[37]

Throughout the balance of 1991, research and development efforts on HDTV continued at a fever pitch throughout the world. In the United States, the proponents finally readied their operating prototypes for testing in April, 1992, at the Advanced Television Testing Center. The complex task, as well as the high costs involved, drove corporations to make alliances. In February, 1992, General Instruments teamed with M.I.T., and in March, Zenith Corporation joined hands with AT&T.[38]

Action at the Advanced Television Testing Center

The start of formal testing was by far the most important aspect of the HDTV story in 1992, despite several delays caused by the need to redesign the test site to accommodate the changeover to digital prototypes. The ground rules for testing decreed that each proponent bring in its own equipment to the Advanced Television Testing Center where technical experts would evaluate the picture across approximately thirty dimensions over a ten-week period. Proponents couldn't make any changes to their designs and breakdowns counted against their time. The General Instruments, Zenith, and AT&T, and the Thomson-Philips-NBC consortium all experienced at least one problem during the rigorous workout for both humans and machines.

Four of the six systems had become digital, with Japan's Narrow-MUSE and an enhanced-definition television design remaining the only hybrid entrants. In March, General Instruments successfully transmitted the first-ever over-the-air digital television signal at noon on March 23, 1992.[39] This transmission finally laid to rest any doubts about the viability of digital HDTV television. To most observers, the FCC's encouragement of a digital solution was vindicated. A few weeks later, General Instruments duplicated its transmission to the 20,000 broadcasters attending the annual National Association of Broadcasters in Las Vegas.[40]

By June, the enhanced-definition television prototype had dropped out and four of the remaining five had been tested. Only General Instruments' second entry, the single progressive-scan prototype developed jointly with M.I.T., awaited testing. But it was clear that digital design had prevailed.

The triumph of an all-digital television as a production standard inevitably led to consideration of the next step, transmission standards. Like the earlier standard, this would be a hard-fought, step-by-step battle. However, it marked a shift from a technical to a political sphere.

In early 1992, the FCC formally laid out its implementation plan for HDTV transmission, allowing a fifteen-year transition from NTSC to HDTV. From the time the commission set the transmission standard, current television broadcasters would have two years to apply for an HDTV license, entitling them to an additional 6 MHz channel. They would then have three years to build the HDTV transmitter. For the remaining years, broadcasters would simulcast NTSC programming on their current channels, along with available HDTV programming on their additional 6-MHz channel. At the end of the fifteen-year period, the broadcasters would have to turn their original NTSC 6-MHz channel back to the FCC.[41]

One long-range consideration that tilted the FCC towards the simulcast approach was that the released 6 MHz in the VHF portion of the spectrum would be freed up for reallocation. The principal beneficiary would be mobile communications. In addition, the change would make U.S. spectrum use compatible with international allocation.

Some called 1992 "the year of HDTV," and it did mark the beginning of the end of the standard-setting process for the new television system.[42] Three elements emerged that would recur in the future discussions and negotiations over HDTV. The first element was that cooperation could take place between the various proponents who had previously competed for FCC approval. The second trend was replacement of theory by hardware, allowing a more accurate assessment of implementation requirements and costs. The third circumstance resulted from this increased knowledge: a gradual disaffection and increased resistance of broadcasters to HDTV.

Behind the scenes, Richard Wiley, the head of the FCC's Advisory Committee on Advanced Television Service (ACATS), began encouraging the former competitors to form a "Grand Alliance." He argued that just as all the systems had become digital, their designs would come to resemble one another more and more as testing continued. This increased similarity would make it more and more difficult for ACATS to recommend any one system to the FCC. In addition, an alliance would allow the proponents to pool their resources, to share the risks, and to share the rewards. This last point was crucial. If one system were chosen, the remaining two proponents would lose all (General Instruments developed two of the four tested designs). Wiley, a telecommunications attorney, devoted his considerable diplomatic talents towards arranging an alliance that would benefit each of them.

1993–1996: The Era of Cooperation

As the standard-setting process for HDTV neared conclusion, Richard Wiley's corporate diplomacy among communications giants came to fruition in the Grand Alliance he had envisioned. However, even as the creative side of the equation came together, the implementation side grew increasingly uncertain. The testing of the HDTV systems was extremely expensive and continued evaluation of the four systems was going to take at least another year. Without an agreement, the result would be one winner and two losers. Worse, the losers would be likely to initiate costly, time-consuming lawsuits that challenged the process itself. With an accord, everyone would be a winner and would share in licensing proceeds.

With these incentives, preliminary negotiations to form a Grand Alliance began in 1992, and formally ensued in February of 1993. In May, just as another round of testing would have started, intense bargaining got underway in Washington, D.C. Wayne Luplow, vice president of Zenith, Robert Rast from General Instruments, and Peter Fannon, Executive Director of the Advanced Television Testing Center, all cancelled their participation on a panel at the Information Display Show in Seattle to rush back to Washington for the meetings.[43]

In order to forge the agreement, the participants postponed the hard decisions about progressive versus interlaced scanning, frame display rate, QAM versus VSB (vestigial sideband) transmission, and multiple interpretations of MPEG-2 compression. According to one report, Richard Wiley intervened personally to resolve some of the disagreements that threatened to prevent the alliance. Despite the difficulties, Wiley was successful, and the announcement of the formation of the Grand Alliance took place in the last days of May, 1993.[44]

The Grand Alliance Television Set

Once settled, the member organizations divided the tasks among themselves, sharing the projects in a style similar to that of the Japanese twenty years before. One area where standards were critical was the consumer television set. All parties agreed that the best solution would be a receiver that could display 1,125 progressively scanned lines at 60 frames per second. However, at the time, the technology to manufacture a set with these characteristics did not exist. Estimates suggested that it would be possible to produce them sometime between 1996 and 1999.[45]

Designing an HDTV TV set proved more complex than originally thought. The traditional TV set incorporated a cathode ray tube, the CRT, to display moving images. To get the maximum benefits from HDTV, viewers should have a large screen, measuring at least about thirty inches along the diagonal. Since large-screen televisions are the fastest-growing segment of the receiver business, the overall trend fits nicely with deriving the maximum benefit from the new HDTV technology.

But there was one problem. A thirty-inch CRT television set is power-hungry, hot, and monstrous, weighing 300 to 400 pounds. An even worse difficulty was that as the number of lines increased from 525 to 1,125, the screen became unacceptably dark because of low luminance.

Since adoption of HDTV depended on consumers being able to receive it, the Grand Alliance and the FCC put a priority on developing affordable, practical HDTV sets. Once again, U.S. companies were forced to play catch-up to a two-decade Japanese lead. Recognizing the importance of consumer displays the U.S. Defense Advanced Research Projects Agency (DARPA) had established a $30 million program of grants and loans for research and development of high-resolution displays and the underlying electronics to support them back in 1988.

Early on, the most likely candidate to replace the current television screen was active matrix liquid crystal display, or AMLCD, originally invented by U.S. researchers. Now simply called LCD, it is often used in large-size screens that depend on a design called "tiled AMLCD." Tiling refers to techniques that allow manufacturers to piece small sheets together into larger panels, in place of difficult-to-produce sheets of silicon. In 1994, Texas Instruments demonstrated a prototype HDTV display based on its micro-mirror technology at the ARPA High-Definition Systems Conference in Washington, D.C. The company claimed their system would deliver digital, high-definition pictures at a reasonable price within two years, but this goal proved unrealistic.

The Grand Alliance's vision of the consumer HDTV receiver was not universally accepted, and as time went on, it became an ever-greater problem, particularly as far as the computer industry was concerned. The digital companies wanted progressive scanning, square pixels, and a faster screen refresh rate. Their growing clout made their objections a serious barrier to setting standards, despite the agreement between the Grand Alliance members and consumer electronics companies.[46]

The Grand Alliance Television Signal

Another area where the Grand Alliance needed to get a consensus for standards was for the transmission signal that would be sent out by broadcasters and received by consumers. In February, 1994, the Grand Alliance announced the test results of two alternative ways of creating an HDTV signal for transmission, conducted at the Advanced Television Testing Center. The two proponents were Zenith Corporation, using a modulation method called VSB, vestigial sideband, and General Instruments' QAM, quadrature amplification modulation, which is used by the cable industry. The ATTC technical staff evaluated the signals for robustness, range, and potential for and vulnerability to interference.[47]

Zenith's VSB was the winner. In mid-September of 1994, after three months of field tests in Charlotte, North Carolina, the Advisory Committee reported that VSB "provided satisfactory reception where the NTSC service is presently available, and in many instances where NTSC reception was unaccepted. . . . . [It] performed well under real-world conditions of multipath and other propagation phenomenon."[48] Further tests continued to provide some support for VSB, but the question of the best modulation scheme for HDTV transmission remains a contentious issue. The perception that the approval of VSB was essentially political, an award to Zenith as a Grand Alliance member so the company could participate in the patent royalties, has proved difficult to shake.

Broadcasters' Resistance Grows

More than one observers has noted that HDTV is a good example of why it is important to be careful what you wish for. Broadcasters wanted the valuable additional 6 MHz of spectrum, and they accepted HDTV as a way to get it. But the rapidly changing competitive programming marketplace, the cost of the conversion, and uncertainty over how to actually use the bandwidth called into question whether or not viable business models could be developed.

Thus, during the same period that HDTV was coming closer to reality at the technical level, implementation seemed to grow ever further away. Broadcasters and cable companies alike continued to evince concern about the costs of implementing HDTV studio, transmission equipment, and transport infrastructure. By 1993, estimates of conversion had lessened to $1 million for each local station; however, this still represented a substantial investment in an era of declining audience share with no visible means of enticing advertisers to pay for it. And cable companies were not so sure they wanted to make the investment necessary to rebuild virtually every system in the United States.

On the broadcast side, many companies were also unhappy with the selection of the VSB modulation scheme recommended by the Grand Alliance and accepted by ACATS. They insisted that the FCC test a transmission method called COFDM, standing for Coded Orthogonal Frequency Division Multiplexing. COFDM is the basis for HDTV transmission in the rest of the world: Europe, Japan, and Australia.

Some cynics suggested that broadcasters' demands for the testing of COFDM was merely a ploy designed to delay the acceptance of HDTV standards, since it was not yet clear how stations could recoup the investment they would have to make for the transition to digital.[49] But broadcasters maintained that they had pragmatic reasons for wanting it— because it was well-known and inexpensive.[50]

COFDM has been used in France since the late 1980s for audio transmission, and its economics are well-understood. Another important benefit is that COFDM lends itself to transmission from small geographically-dispersed cells, instead of one giant, central transmitter. The dispersion means the signal can be propagated into valleys, canyons, and otherwise inaccessible locations.[51] The choice of COFDM by others, coupled with ongoing difficulties with DTV signal reception, reinforced the belief that VSB was a flawed decision. And, as we shall see further in this chapter, the VSB versus COFDM debate continued into the twenty-first century with no signs of abatement.

Spectrum flexibility is another area where the Grand Alliance and broadcasters were on a collision course. As broadcasters reviewed their assets, they realized that the additional 6 MHz of bandwidth they would receive for implementing HDTV might help fund the conversion to digital.[52] Recall that the FCC proposed to allocate existing broadcasters with 6 MHz in the UHF portion of the spectrum for simulcast HDTV transmission, bandwidth they would use in addition to their current VHF 6 MHz

NTSC assignment. This outright gift took an ironic turn as broadcasters started planning how to use this second 6 MHz of spectrum.

Even though broadcasters did not know precisely what they would do with the bandwidth, they realized that they could deliver a range of digital services, such as reproducing their regular NTSC channel four to six times to broadcast their programs at different times (multiplexing), or developing a suite of new channels of programming (multicasting). For example, Fox Broadcasting announced it was considering an all-news channel.[53] It was even possible that broadcasters might even offer pay services to viewers to bring in monthly subscription fees.

In order to effect these various money-making schemes and develop additional revenue streams, broadcasters organized their substantial lobbying apparatus. In 1994, Congress considered an omnibus effort to update the Communications Act of 1934. Although that bill ultimately failed to pass, broadcasters were successful in adding language giving them "spectrum flexibility" with the UHF 6 MHz intended for HDTV simulcast. Their proposal was that instead of using the UHF bandwidth solely for HDTV, which would require significant capital investment, they would have the freedom to use the extra 6 MHz for other purposes. These offerings, argued broadcasters, would help pay for HDTV—if they decided to implement it at all.[54]

Horrified at the mere suggestion that stations might not present HDTV programming, the companies developing HDTV appealed to ACATS, to the FCC, and ultimately to the Congress itself.[55] Everyone realized that the adoption of HDTV would depend on the availability of programming. Without something to watch, why would consumers buy sets?

ACATS acquiesced to a limited version of broadcasters' demands for spectrum flexibility, agreeing they should be allowed to use the additional UHF 6 MHz for other services when HDTV wasn't shown.[56] However, they also suggested to the FCC that if broadcasters didn't intend to use the additional bandwidth for HDTV at all, then the FCC should open up the additional 6 MHz allocation to all comers, not just existing broadcasters.

In mid-1996, a second effort to pass omnibus legislation to regulate the communications industry did pass.[57] The over-the-air lobbying team achieved two important objectives: they took over an additional 6 MHz of spectrum for free for fifteen years, and they enshrined spectrum flexibility in the new law so that the FCC would have to put it into effect. There were limits, however.

The intent of the arrangement was to preserve free, local, over-the-air television, and broadcasters would have to provide this service on at least one channel or face having to buy the right to use the spectrum at auction. Moreover, Congress reserved the right to establish new public services requirements. Within these parameters, broadcasters were set free to use the digital spectrum as they would, allocating the bandwidth to whatever mix of programming and services they believe would bring them the greatest return. The additional spectrum awarded to them in the Telecom Act of 1996 did not obligate them to air HDTV.

Japan and Europe Abandon Analog HDTV

Among communication experts, dissatisfaction in Europe with D2-MAC and HD-MAC had been brewing for two years, and, finally in January of 1994, the European community formally replaced the unsuccessful "MAC directive," 92/38/EC of May, 1992 with COM(93)-556. The new ruling adopted the 16:9 format with flexible transmission standards. France, the Netherlands, and Spain adopted analog D2-MAC as an interim standard, while Germany and Portugal chose the analog PAL-Plus system. Transmission of analog could be in D2-MAC or a system compatible with PAL or SECAM, and the transmission of analog HDTV had to meet HD-MAC standards.[58] There was broad agreement that Europe would aggressively pursue digital standards of their and would migrate to a digital system, including a high-definition version.

A private think tank of European technical experts, GERTN, began meeting to define the digital video standards. In 1995, they began reporting their findings to the EC Digital Video Broadcasting Group, a panel composed of broadcasters, manufacturers, satellite operators, and regulators. The DVBG made their recommendation to the European Community Telecommunications Council, paving the way for acceptance of the DVB (Digital Video Broadcasting) standard.[59]

The Japanese waited even longer than the Europeans to signal their surrender to digital technology. The abdication began in February, 1994, with a senior government official observing that HiVision was obsolete and that Japan would move towards a digital standard. The remark caused such an uproar in Tokyo—especially angering NHK, the public television system that developed HiVision—that the official was forced to recant. However, a few months later the NHK director of engineering, Shuichi Morikawa, conceded that the organization's research had shifted to more advanced digital systems. To make the point, NHK's laboratory opened up for a reporters tour to demonstrate a variety of digital research projects.[60]

Nevertheless, national pride dictated face-saving measures. Although Japan accepted the U.S. advanced system for export, internally the country continued to broadcast HiVision analog HDTV. In September, the Ministry of Posts and Telecommunications designated Hamamatsu City as the "HiVision City," the first of thirty-eight such municipalities. The Japanese built a citywide video HDTV network for use at concerts, international conferences, and academic meetings.[61] The MPT also announced they would license NHK and nine private stations to conduct trials of rectangular 16:9 enhanced-definition television (EDTV).[62]

In the United States, the Computer Industry Weighs In

By mid-1996, digital was the clear winner, and the standards-setting process was nearing a close. It was time to either mandate a transition or forget about it. As we have seen, although DTV was originally created by digitally-aware scientists and engineers, most of its actual development proceeded under the supervision of consumer electronics companies and television equipment manufacturers. HDTV, CD-ROM, and to some extent DVD were all products of the television industry at a time when the computer industry was still too young and weak to claim its place in the new technology landscape.

Schooled in the analog world of the consumer electronics industry, the developers of HD and DTV equipment believed that a detailed set of standards was appropriate. They also thought that the difference between analog TV and digital TV was evolutionary. In addition, they vastly underestimated the rapidity with which processing and networking would grow. As they designed the new generation of consumer equipment, they did not set out to exclude the computer industry—but they did.[63] As a result, serious consultation with advanced computer companies did not occur until the process of arriving at standards for the new TV system was nearly finished.

For their part, people in the computer industry were impatient and poor communicators of the digital revolution they could see so clearly. In industry conferences and trade shows held in the early 1990s, they blew off the TV industry's concern about image quality, which in the 1980s and much of the 1990s, was well out of reach of any but the most expensive computers. Computer technologists were content with "low-res" images because they knew the limitations were temporary; TV practitioners and film creators had spent lifetimes making beautiful, rich, pristine moving pictures—and the two sides remained divided by an abyss of frustrated objectives.

Throughout 1995 and 1996, the major computer companies became increasingly angry that they had not been involved in the development of HDTV standards.[64] They resented that the proposed standards had not resulted in greater compatibility with computer monitors, and they began doing some of their own lobbying. Then FCC-chairman Reed Hundt was sympathetic to the digital cause, and he paved the way for its proponents to engage in the process. Microsoft, Compaq, Intel, Apple, and a host of other prominent computer companies formed the Computer Industry Coalition on Advanced Television Services (CICATS) and took a seat at the standards-setting table.

Throughout the fall of 1996, representatives of CICATS, the broadcast, and consumer electronics industries worked to reach a compromise over the issues dividing them. Initially, the computer industry sought the elimination of interlaced scanning as an option for transmitting

DTV, a 2:1 aspect ratio to replace the 16:9 format, square pixels instead of round pixels, a change from a 60- to a 72-megahertz screen refresh rate, and greater error-correction in the transmitted digital bitstream.

The differences between television sets and computer monitors had been growing for some time. Computer displays were achieving more resolution in a smaller area, so that text was legible even as screens (and computers) became smaller, flatter, lighter, and more portable. Progressive scanning worked very well for displaying text and crisp still images. At the same time, the fastest growing segment of the TV-set market was in big-screen TVs, supporting home theaters and surround-sound audio systems. Interlaced scan used less bandwidth and performed better, displaying the fast-moving video images of action-adventure films and sports.

Towards the end of 1996, the computer, broadcasting, and consumer electronics industries reached a compromise of sorts that set the stage for the FCC to approve standards for advanced television in the United States. Fundamentally, it was an agreement to disagree—and to fight standards out in the marketplace. In a sense, this was a victory for the computer industry, because de facto marketplace standards-setting was the norm for them, while it represented a huge and unaccustomed risk to the other players that had developed under a fifty-year regime of strict formal standards.[65]

On December 24, 1996, the FCC voted to adopt a standard for the transmission of digital television in the Fourth Report and Standing Order, "Advanced Television Systems and Their Impact Upon the Existing Television Broadcast Service," MM Docket No. 87-268, shown in Table 4.1. Following the compromise agreement of the industries involved, the standard did not include requirements for scanning formats, aspect ratios, or lines of resolution. Rather, a "family of standards" was approved, that allowed for eighteen different combinations of these variables.[66]

Table 4.1 ATSC Transmission Standards

		Aspect Ratio					
Vertical Lines	*Horizontal Pixels*	*Wide Screen*	*Near-Square*		*Picture Rate*		
1080	1920	16:9		60I		30P	24P
720	1280	16:9			60P	30P	24P
480	704	16:9	4:3	60I	60P	30P	24P
480	704		4:3	60I	60P	30P	24P
480	640		4:3	60I	60P	30P	24P

A broadcaster can transmit in any of the eighteen accepted video scanning formats. Standards-compliant TV sets will receive all eighteen formats, but they may display only a subset of them, depending on the design specifications. No monitor is required to display any particular format, including HDTV. So, for example, a digital TV can receive a 1080i, or interlace HDTV feed but might display only a 720 progressive scan SDTV format. So in this case, the set must be capable of converting the scanning format and aspect ratio.

1997 to the Present: The Era of Implementation

The first widespread use of digital television actually began in 1994, starting with the launch of the Hughes Ku-band direct-broadcast satellite service, before DTV standards were approved. But it was the FCC's Christmas gift of an advanced-digital television system that set in motion the train of events that will take most of two decades to unfold. (And given Moore's Law, technological transformation may continue on into the indefinite future.) It will certainly take many years to actually implement a digital television system, whether it is considered on a national or global basis.

A little more than a year after the FCC promulgated standards for advanced television, the agency issued the Fifth Report and Order, "Rules for Digital Television Service," on April 3, 1997 to define the procedures for the transition to DTV.[67] It set forth FCC policy for the rollout of service and provided more detailed guidelines for the transition to digital. Spectrum flexibility merited special emphasis in the FCC summary, which noted that television stations could engage in

> *data transfer, subscription video, interactive materials, audio signals, and whatever other innovations broadcasters can promote and profit from. Giving broadcasters the flexibility in their use of their digital channel will allow them to put together the best mix of services and programming to stimulate consumer acceptance of digital technology and the purchase of digital receivers.*

The commission also clarified the position that such flexibility depended on broadcasters continuing to provide "a free digital video programming service that is at least comparable in resolution to today's service and aired during the same time periods as today's analog service."

The FCC order set a rollout schedule for DTV. The affiliates of the four major networks, ABC, CBS, FOX, and NBC in the top ten markets, were required to be on the air with a digital signal by May 1, 1999, reaching about 30% of TV households. Network affiliates in markets 11 to 30 had to be on the air by November 1, 1999, extending DTV to 53% of the United States. All other commercial stations must go digital by May, 2002, and PBS stations by May, 2003. By 2001, all commercial networks will broadcast in digital, and by 2002, all public broadcasting networks will broadcast in digital.

The original plan held that in 2006, the over-the-air VHF analog channels would go blank, and spectrum would be returned to the FCC for auction. However, in markets where fewer than 85% of the households have DTV sets, broadcasters in that market will be exempt from all-digital transmission. In addition, the FCC stated its willingness to extend deadlines where a licensed station has attempted to meet these deadlines but have been unable to for reasons that were unforeseeable or beyond their control.

Displays Get a New Look

In the 1990s, the research into display technologies resulted in considerable progress. A variety of different approaches were refined over the years, and several were developed for commercial production.[68] Table 4.2 shows the most promising products, including CRT (direct view and projection), liquid crystal display (LCD, direct view and projection), DLP (Texas Instruments' digital light processing, or micro-mirror technology), plasma display (PDP), image light amplification (Hughes-JVC), and reflective liquid crystal display (LCD). In 1998 and 1999, all the major television manufacturers announced HDTV or DTV product introductions or plans to do so within the next year or two.

By 2000, substantial progress had been made in reducing the price of sets. Konka, a Chinese manufacturer, announced in mid-1999 that it would introduce a product that included the decoder and the display for $3,000, although some felt the price was an accurate reflection of low quality in the production of the unit.[69] There were also many other models available, all costing at least $5,000 and some priced as high as $12,000. (See Figure 4.3.)

COFDM: The Signal That Wouldn't Die

While advances in TV-set technology—and manufacturing expertise as well—appeared to be getting closer to delivering viable products at affordable prices, disagreements about the shortcomings of the VSB transmission signal continued to roil the waters of DTV acceptance. As late as 1999, after more than forty stations were already on the air with DTV, Nat Ostroff of the Sinclair Broadcasting Group conducted a series of tests to try to settle the issue of whether COFDM would be enough of an improvement over VSB for the FCC to reconsider its adoption of the Grand Alliance specification.

HDTV maven Dale Cripps, a long-time analyst and manager of an influential Web site, wrote an open letter to the HD community, urging people to ponder this issue carefully, since the necessary testing of VSB could delay rollout of HDTV for as much as two years.[70] It is still not clear

Table 4.2 Comparing HDTV Display Technologies

	Parameters	Advantages	Disadvantages
CRT	Up to 42" screen Multi-sync Brightness 12–40 lumens Contrast 150:1	Small pixels Bright Good contrast Reasonable cost; $250–25,000 Long life; low maintenance	Heavy and big 42" maximum CRT burn-in
CRT projection	Up to 240" screen Native resolution: 640x480 to 1280x1024 Brightness 250–5000 lumens Contrast 150:1	Long life Ease of setup	Pixellated Low contrast Cost: $3,000–100,000
DLP (Digital Light Processing	Up to 240" screen Native resolution: 640x480 to 1280x1024 Brightness 600–3000 lumens Contrast 100:1	Bright Good resolution Fast response Long DMD life Ease of setup	Limited lamp life Cost: $10,000–30,000
Plasma Displays	40"–50" screen Native resolution: 640x480 Wide 160° viewing angle Brightness: 12–40 lumens Contrast: 50:1–150:1	Lightweight Thin Wide view angle Color accuracy Long life	Low efficiency Low luminance Limited contrast Limited resolution Contouring Cost: $10,000–15,000

Source: Based on presentation by Dave Taylor and Tom Brentnall at 1998 SMPTE seminar on Production and Display Technologies for HDTV, at University of Southern California, Los Angeles, May 16, 1998.[71]

Figure 4.3 The Sony Wega HD television set. Source: Sony.

how this issue will be resolved, although in mid-2000, the balance was tilt-ing towards COFDM, mainly because of the international market in equipment that has overwhelmingly adopted the DVB standard.

Technical efforts to save the ATSC VSB transmission standard con-tinue. In mid-1999, Motorola announced that it had developed a chip set that would eliminate one of VSB's problems, which was poor reception brought about by multipath interference in urban areas.[72] Multipath refers to conditions where radio signals bounce off the surfaces of buildings, cre-ating several signals to be extant. The set receives multiple signals and displays them both, resulting in one picture "ghosting" another.

ATSC Compared to DVB

The controversy in the United States over 8VSB and the European transmission standard COFDM points up the progress made in Europe
between 1994 and 1997. The Digital Video Broadcasting Project (DVB) is a
consortium of over 200 broadcasters, manufacturers, network operators,
and regulatory bodies from more than thirty countries worldwide.[73] The
group promulgated standards over a three-year period for cable, satellite,
and terrestrial transmission of digital and high-definition television.[74]

The two standards are not entirely dissimilar. Both systems use
MPEG-2 for compressing video, although the U.S. system includes eighteen different formats for digital television images, one of which (1080
progressive) is not supported by MPEG-2. The U.S. ATSC system incorporates Dolby AC-3 audio encoding system, and DVB uses the MPEG layer
II audio, sometimes called MUSICAM. This choice makes DVB 100%
MPEG-2 compliant, while the ATSC standards are not.

However, the modulation and transmission technology is the main
difference between the two standards in the VSB versus COFDM debate.
The VSB system is a single carrier technique, meaning that it originates
from one powerful transmitter. The COFDM system is a multiple carrier
approach, so the signal is sent out over a number of strategically-placed,
lower-power transmitters, giving it greater responsiveness to local terrain
and other transmission conditions. DVB also offers another advantage: It
is designed to provide a solution across the range of delivery media,
including cable, satellite, and terrestrial systems. ATSC does not have the
extensibility and technical coherence offered by DVB.

U.S. Broadcasters and Spectrum Flexibility

As allocated by the FCC under the digital transition plan, it was assumed
that broadcasters would simulcast analog TV on their VHF spectrum and
HDTV on the UHF frequency until 2006, when they would return the
VHF bandwidth to the government. But one of the achievements of the
broadcast lobby in the 1996 Telecommunications Act was that this spectrum return would only occur if 85% of the TV households in a given market had digital sets in their homes. Until that penetration level is achieved,
local stations will continue to hold 12 MHz of spectrum.

While Congress had intended for the UHF bandwidth to be used for
HDTV, broadcasters began to have other ideas. Spectrum flexibility really
became important as implementation came closer, and now it occupies a
pivotal place in the DTV rollout. The FCC's guidelines specify only that
broadcasters must provide a free digital video programming service that
is comparable in resolution and aired during the same time periods as
today's analog service.

One reason the transition to digital is so difficult is that it calls on station executives to conceptualize their function very differently than they did in the analog domain. In addition to being programmers, they must become bandwidth managers. Faced with actually going on the air with HDTV, stations began to take a closer look at their 6 MHz of digital bandwidth from a business and programming standpoint. This spectrum allocation will deliver approximately 19.39 Mbps of material. The HD signal accounts for 17.56 Mbps at 1080i/60, or 10.2 Mbps at 720p/30. NTSC-quality video requires 3.96 Mbps on average, and VHS-quality a bit less than that, between 2.5 and 4 mbps, depending on how much motion there is in the picture.

As the resolution of the main channel decreases, the number of channels that can be aired rises. Some content-rich broadcasters may well opt to present SDTV because of the number of ways it will let them develop additional revenue streams. Indeed, the over-the-air television companies have long envied cable operators their ability to generate monthly subscription fees in addition to advertising. Multiple SDTV channels offers them a way to do it with the new ability to add channels and market them as pay services. Figure 4.4 presents some alternative scenarios of how local stations might manage their bandwidth to match business strategies.

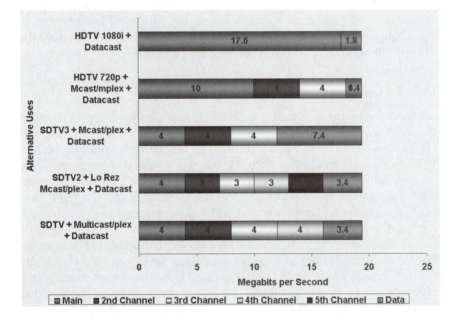

Figure 4.4 Alternative uses of broadcasters' 19.39 Mbps digital over-the-air bandwidth.

Take Fox. The network has developed a suite of cable channels—Fox Sports, Fox News, Fox Family, and Fox Movies—that would have a great deal of appeal to local audiences. They could air them all, as well as a high-quality digital signal, using the same bandwidth that previously afforded only one channel. This strategy is called multicast programming, meaning that different programming is presented on the available channels. Multicast programming will allow over-the-air broadcasters to reach many new, more targeted audiences, and help them compete more effectively against cable and satellite operators. Cable and satellite operators, for example, are presently the only outlets for ESPN and the Disney Channel.

Another strategy is called multiplexing. HBO has used multiplexing on cable and satellite platforms where there is enough channel capacity. Multiplexing means to transmit the same programming channel at different times, providing a kind of "near-video-on-demand." Just as satellite viewers can see an HBO film at staggered intervals (depending on how many channels are available for multiplex), so could a local station shift its programs by six-hour intervals, "stagger-casting" programming to make it more convenient for audience members to see their shows.

One kind of programming falls between multiplexing and multicasting, a sort of "multichannel-casting." It lets viewers to choose from among multiple versions of an athletic event that are being aired simultaneously, so they can select the particular camera angle they prefer at the moment. The production of football games frequently involves more than a dozen cameras, with the director selecting each shot—blimp, huddle, wide-angle, tight shot. Multichannel-casting makes it is possible for the consumer to make the selection, an especially useful feature for people who have a picture-in-picture option on their sets.

Finally, transmitting data, or datacasting, will probably be part of every strategy, but some stations will be much more aggressive about it. Right now, they are limited to analog datacasting, but when they become digital, they can do in-stream datacasting, where bits are mixed into the HD or SD bitstream. Or they can elect to transmit discrete channels, an activity called parallel datacasting. The next chapter will consider datacasting in more detail as PC (data) material on the TV screen (PConTV), or as TV or video material on the PC screen (TvonPC).[75]

To HD or Not to HD

Now let us consider the main channel, the high-definition picture that put the digital spectrum in broadcasters' hands. Each network and local station can choose from among the eighteen standards promulgated by the FCC, but as a practical matter, affiliates are likely to adopt the same one chosen by the network that supplies so much of the station's programming. CBS and NBC have decided to go with 1080i, ABC selected 720p, and FOX will not transmit HDTV at all, electing to air SDTV 480p, 480 x 704, in the 16:9 format.

Side-by-side comparisons reveal little difference between the 1080i and 720p HD formats; however, both are noticeably richer, sharper, and more colorful than any SDTV format. Some analysts believe that DTV will be successful, but in the 480p format. Gerry Kauffold of In-Stat research firm thinks that 1080i or 720p will not be able to gain traction as the transition gets underway, and that only high-end consumers will buy them for a long time to come.[76]

Kauffold argues that even when viewers can see the difference between the HD and SD formats, they won't think the price differential is worth the marginal improvement. However, they can see the higher quality of 480p over NTSC and are willing to buy sub-$1000 sets to get the better picture. Manufacturers can deliver 480p TV sets at lower cost than the HD formats because the SD standard has 480 lines, the same as NTSC monitors. This means that only the scanning mechanism of the design would need to be changed. (See Figures 4.5 and 4.6.)

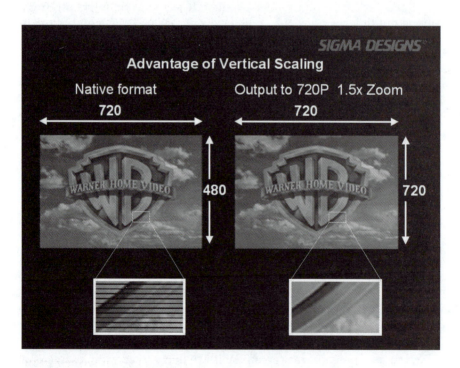

Figure 4.5 Warner Bros.' comparisons of different vertical sampling rates. Source: Warner Bros.

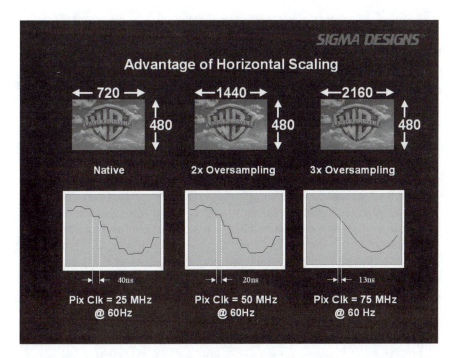

Figure 4.6 Warner Bros.' comparisons of different horizontal sampling rates. Source: Warner Bros.

It is unclear whether HDTV will ever be adopted. As FCC Commissioner Susan Ness remarked in January of 1998, "I am concerned that consumers will be confused. Even the most knowledgeable people in the industry can't agree on whether 1080 interlace or 720 progressive presents the best high-definition picture for the cost, or whether 480p represents the most cost-effective option."[77]

What's on? HDTV Programming

In more than twenty markets, crowds in museums and electronics stores watched the first live news event broadcast in HDTV—the October 29, 1998 launch of the space shuttle Discovery. In November, 1998, the ABC station in Detroit aired the film *Dalmations*, seen by hundreds of curious people who gathered in front of floor models at the Good Guys and other electronics stores. CBS aired the first football games in HD, showing four games in December, 1998 and January, 1999—although the only market where the signal went out in HD was in New York City. By mid-1999, CBS was on the air with HD programming for three hours a day in New York City, broadcasting from the top of the Empire State building and reaching a verified 80-mile reception radius. In November of 1998, PBS launched *PBS Digital Week*, four days of television and online events to raise the curtain on the next generation of broadcasting. The network showed a 90-minute special about the collaboration of renowned glass artist Dale Chihuly with master glassblowers in four countries in HDTV.

There are even some early HD commercials. Proctor and Gamble aired six high-definition commercials in an experimental HDTV broadcast in 1999 and had one more in the pipeline. One HD spot was for Mountain Spring Fragrance Tide with Bleach and was created by the company's ad agency, Saatchi and Saatchi. Creatively, the team exploited HDTV's wider aspect ratio and higher resolution to get the most from the commercial's setting in primeval Nature. The HDTV ads can be cropped and reformatted to fit analog TV sets. P&G reports that shooting in HD adds only about 5% more to the overall production cost, which typically run between $30,000 and $1 million.

The earliest HD programming focused on specials, including movies, sports, and events, because these programs capture an audience and are enhanced by the color, detail, and improved audio of advanced television.[78] Going forward, more regularly scheduled program genres will air in HDTV. HBO announced that it would add an HDTV tier, providing two channels, and DIRECTV would deliver them to subscribers.[79] NBC began transmitting *The Tonight Show with Jay Leno* in HDTV, and it will show *Titanic* and *Men in Black* in the 1080i format. In Los Angeles, KNBC-TV was the first station in the nation's second-largest market to offer a locally-produced HD show called *Travel Café*. The show was shot on Sony HDW700 digital cameras (HDCAM), part of an overall deal NBC made with Sony to implement new digital technology at the network. The popular program will be downconverted from HD for airing on the station's analog channel in its regular time slot.

From the beginning, everyone realized that compelling programming would be the key to the success of HDTV. Unless there are shows that people want to watch, the chances are not good that consumers will want to buy expensive HDTV sets. But while stations have the technological capability to transmit HDTV 24/7, it is likely to be a long time before they have sufficient incentive to do so, even though there is a significant amount of programming available.

The long development of HDTV and SDTV has allowed programming owners to digitize much of their material. Some of it is formatted for HDTV, although more of it has been transferred to high-quality digital SDTV. However, as of mid-2000, very little programming has been developed specifically for HDTV; rather existing drama, sitcoms, and specials have been remastered to a 16:9 digital format.

The QAM Story

So far, the emphasis on HDTV adoption has focused on over-the-air broadcast. They were the agents of influence who had pressed for its development since the 1980s, and they were the beneficiaries of free spectrum. The FCC concentrated on them as well because they are also the purveyors of free television.

But most Americans don't watch TV from their local stations. As of mid-1999, about 79% of U.S. TV households received their television through some form of subscription service—cable (about 65%), DBS (12%), or wireless cable (2%). Direct broadcast satellite operators have little difficulty making their systems HDTV-ready, but this is not true for cable companies. Cable systems are much more sensitive to the bandwidth required by HDTV. Bandwidth capacity is called "shelf space," the amount of room an operator has available for programming channels.

One burning question is, must cable systems carry both of the broadcasters' 6 MHz channels, VHF analog, and UHF digital? The issue is called "digital must-carry" and cable operators were outraged that the FCC might require two-channel carriage, especially in the beginning when there would be few, if any, TV sets capable of receiving HDTV.

This naturally led to whether or not cable-operators—or "cablers"— would have to carry broadcasters' multicast, multiplexed, or multichannel offerings.

Broadcasters may have lobbied successfully for spectrum flexibility, but they were unable to exert influence on the FCC to take their side in the dispute with cablers. In late 1998, the FCC dodged the bullet entirely, refusing to rule on digital must-carry. For one thing, cable systems do not use publicly-owned spectrum; they have paid handsomely to build their systems. If the FCC had ruled in favor of digital must-carry, it could have raised a constitutional question about whether the government was illegally taking private property, and there is considerable likelihood that the decision would have gone in favor of the cable companies.

Nevertheless, the cable industry realizes that DIRECTV and Dish Network, the industry leaders of their greatest competitor, DBS, will offer HDTV. These moves leave cable companies little choice but to match the competition, but it means massive upgrades of their systems. As a result, it could take them six to ten years before the majority of systems will be HDTV-capable.

Of course, cable companies also have the incentive to revamp their plants in order to offer high-speed Internet access. These considerations led the industry to accept HDTV as a key component of its future when it adopted the OpenCable architecture for its set-top boxes. The set-top standard will pass through all of the ATSC formats without impairing either the audio or video quality.[80]

Nevertheless, the reality is that HDTV poses technological problems for cable systems. It led CableLabs President Dr. Richard R. Green to call for cooperation in a technical keynote speech he made at the 1998 National Association of Broadcasters (NAB) convention. He asked that stations work out digital carriage business arrangements in individual markets, taking the capacity of local systems into account.

Green argued for cooperation among all the involved industries:

> We are mutually dependent. The cable industry needs computer software and hardware to put in the set-top box. The broadcast industry needs this same technology in providing advanced data and television service to the home. The computer industry needs connectivity. If the computer industry is to continue to grow at the double-digit rate, new applications—video, for example—need to be built into the PC. Who better to provide digital video and data services to computers than the broadcast and cable industries? All of us need the consumer electronics industry. The next generation of TV sets will certainly add to the entertainment value of services with clearer pictures and improved sound. It is likely that television sets will become display terminals for data supplied over the air and through the terrestrial networks.[81]

Winners—and Whiners

The advent of HDTV brings advantages and disadvantages to every segment of the TV industry, as shown in Table 4.3.

Technical Barriers to Implementation

Even at the beginning of the new millennium, there are some outstanding technical difficulties to bringing DTV into U.S. living rooms, including problems with the digital signal itself. There are challenges to developing and setting standards. And the high costs of replacing the installed base of analog equipment for production, transmission, transport, and reception remain a formidable obstacle to implementation.

The problems associated with picture quality, aside from those caused by compression, include pixellation, contouring, and aliasing.[82] Pixellation means that the viewer can see the pixels, revealing the tiny picture elements that make up the images. The problem can result from inadequate sampling, too few levels of quantization, or coarse encoding.

Contouring results when adjacent areas of the picture are quantized at the same level within each region, and the two are different from one another. In an analog picture, the range of values allow tiny differences within an area; with digital conversion, these small differences are quantized at the same value, giving the entire area a similar appearance. When two adjacent areas have two different, but internally identical areas, the line between them may mark an unsightly contour.

Table 4.3 The Two Faces of HDTV

Industry Segment	Advantages	Disadvantages
Broadcast Networks	Increase audience share	Cost of new equipment Higher cost for some Programming Increased revenue unlikely
Local Stations	Maintain audience share and parity with competition from cable, DBS Potential new revenue streams from added capacity	Cost of transition to digital Uncertain business models
Cable Networks	Greater picture quality Potential for new revenue	Higher cost of video-originated programming
Cable Operators	Parity with competitors, DBS, and local stations Potential for new revenue	Very high cost of system upgrade
Satellite Operators	Increase viewership and revenue through greater capacity to offer more HD channels	Takes away key competitive advantage of higher-quality picture
Consumer Electronics Manufacturers	Gives consumers incentive to buy new generation of TVs	Will cause short-term decline in home theater sales High cost of R&D and manufacturing
Computer Industry	PCs can become digital TV receivers Hastens convergence of PC and TV	Some formats difficult for PCs to decode Digital TVs can compete with PCs to receive digital material

Source: Based on compilation by Forrester Research, 1997.

Aliasing is similar to pixellation, in that the images acquire a "squared-off" look. In this case, however, it is not caused by coarse pixels. Rather, aliasing results from too-high a sampling rate, introducing unnecessary, spurious data.

Many of the problems with the digital signal will probably be the easiest to overcome, especially with increases in the processing power of chip sets for codecs. However, developing and setting standards is less amenable to solutions. They require a political process that entails intricate cooperation in the face of stiff competition and unrelenting technological advance.

Finally, implementing digital television in a cable system is a major challenge—and extraordinarily expensive. Engineers must address different formats for multiple layers: (1) analog-to-digital conversion, (2) compression, (3) encryption, (4) multiplex, (5) transport, (6) reception, and (7) hardware. Dr. Sadie Decker, Vice President, Tele-Communications Inc., demonstrates the software challenges, as shown in Table 4.4.

Table 4.4 Layers of Digital Formats in a Cable System

System component	Competing digital formats
Programming	MPEG-1, MPEG-2, Motion JPEG, DigiCipher, and other proprietary formats
Network communication	TCP/IP, Asynchronous Transfer Mode (ATM), SONET
File servers	C, C++, object-oriented formats
Set-top box	Oracle, Sybase, Informix, COTS, DOCSIS

Summary

High-definition television occupies a premier place in the emerging television system of the future—even if HDTV is never actually adopted. The search for a better television picture sped up the development of digital television by as much as ten years. Then the sheer enormity of the information in the HDTV picture forced researchers to refine signal compression techniques.

The advanced digital format will have a greater impact than the high-definition picture that inspired it. Digital video provides a format that allows switched, connected HDTV to be transported across cable, computer, microwave, and telephone networks. This universality promotes the convergence and interconnection of these different telecommunication architectures. The opportunity to create a new television system encouraged international standards-setting bodies to attempt the difficult goal of establishing a global standard.

The United States entered the race to develop a better television picture in 1987. It was an era when Americans felt an uncomfortable, unaccustomed sense of technological inadequacy and economic defeat. Dire predictions about the country's future abounded. In 1991, the Council on Competitiveness—a coalition of chief executives from business, labor, and higher education—warned that America's technological edge had eroded in one industry after another. On September 19, 1994, that same council reported that its membership believes that the United States has shown a dramatic improvement in several areas critical to developing an advanced communications network. They added that "new markets and applications are helping the United States to recoup its position in electronic components."[83]

As research solved the early issues of compatibility, interoperability, scalability, and extensibility, implementation issues came to the fore: spectrum use, conversion cost, frame rate, and display design. While technical answers exist to some aspects of these problems, their overall resolution lies in the political sphere and, finally, in the marketplace.

As of mid-2000, the National Association of Broadcasters maintains a Web page that lists 134 stations in 49 markets that are transmitting a digital TV signal. There is no mention of HDTV. Indeed, whether HDTV will ever become part of a mainstream TV system is still unclear. There are three camps of believers.

- HDTV will become mainstream over a period of ten years as prices for the sets decline and the amount of programming increases.
- HDTV will never become the standard. It will fail.
- HDTV will be very successful—but not on the TV. Rather, it will enter households via the computer.

Keep in mind that many computer monitors have 1280 x 1024 displays already, so they have the pixel density to display high-definition. This leads some computer advocates to say that the first viewers of HDTV could use computer monitors rather than television sets as the viewing device. The 1280 x 720 Grand Alliance format, for example, can fit on a 1280 x 1024 computer monitor with room left over for picture-in-picture (PIP) enhancements or added text. Hauppauge, a leading manufacturer of TV tuner cards that let PCs display TV signals, plans to come out with an HDTV-capable card sometime in 2000—for less than $200!

Many people have not considered that DTV could happen fastest and cheapest on PCs rather than TVs. A year before any consumer DTV sets came out, consumers could buy a DVD PC with a 34-inch VGA monitor and get beautiful progressive-scan movies for under $3000. The quality of a good DVD PC connected to a data-grade video projector beats $30,000 line-doubler systems.

These disparate viewpoints continue to wreak havoc with the implementation of HDTV, and the development of standards has probably caused more problems than it has solved—even though it was the search for them that accelerated the move to digital TV in the first place. The previous chapter made much of the flexibility of digital data. Nowhere has it proved more nettlesome and provocative than in the implementation of HDTV and SDTV. And as the millennium begins, the battle is barely joined.

In another way, the battle is already over. Digital processing has become so powerful that standards don't really matter. Plug any video into any processing board, use software to define the output, and it is converted in nanoseconds to any desired format. Start with near-square NTSC, convert it to PAL, SECAM, rectangular SDTV, MPEG-1, MPEG-2 for DVD, satellite, any one of several compressed such as Indeo or Cinepak, or a streaming format like Real (.rn), Microsoft (.asf), Apple Quicktime (.qt), Video for Windows (.avi), or Xing.

The triumph of processing power over standards will bring the Holy Grail of a unified worldwide television system within the grasp of its seekers. In 1985, when the Japanese applied to the CCIR for approval of the 1250/60 standard, there was great hope that the world would agree on one television system. As events unfolded, those dreams were dashed on the rocks of regional and national economic self-interest. Now, given the extraordinary advances in digital signal processing, there is renewed emphasis on simple conversion among systems, even if they do not utilize identical standards. A formula promulgated in directive CCIR-601 lays out the basic premises for such conversion and will continue as the CCIR moves through the process of setting HDTV standards.

Back in 1993, a bitter joke swept the National Cable Television Association: If a pollster asked viewers what they wanted from television, the two answers they would *not* hear would be "better picture quality" and "more channels to choose from." The point of joke is that although there is little demand pull for HDTV, there is a rapidly evolving supply push for it, because it is primarily the short supply of over-the-air bandwidth that is driving programming delivery companies to invest in compressed DTV. DBS programmers such as DIRECTV and EchoStar required the technology in order to offer more channels in its competition with cable, and now it is working in reverse to drive cable operators to digital as well.

Prognosticating the future is difficult and rarely accurate. However, DTV is well on its way to being implemented, and that process is likely to continue over the next decade. Even in the most favorable climate progress is slow. In 1994, 2% of television material was digital; in 1995, that number had risen to only 5%. Ultimately, it is the consumer who will determine whether HDTV or SDTV will prevail, or whether both formats will coexist—or if one or the other or both will stagnate as an interesting technological road not taken.

Notes

1. Recounted by Dr. Woo Paik, General Instruments Corp., in a personal interview, North San Diego County, CA, August 23, 1994.
2. J. Farrell and C. Shapiro, *Brookings Papers: Microelectronics*, Brookings Press (1992). Cited in Hong J., "High-definition television," in A. E. Grant and K. E. Wilkinson (eds.), *Communication Technology Update: 1993–1994*, 3rd ed., Austin: Technology Futures (1993):23. A good source for information about Japanese development of HDTV is R. Akhavan-Majid, "Public service broadcasting and the challenge of new technology: A case study of Japan's NHK," presented at International Communication Association, Miami, FL, May 21-29, 1992.
3. Reuters News Service Report, archived on NEXIS (June 17, 1994).
4. A. Lippmann, "HDTV sparks a digital revolution," *BYTE* (December 1990):297-305.
5. J. Widoff, Deputy Executive Director, Advanced Television Test Center, telephone interview, September 15, 1994.
6. D. I. Sheer, "Cost-effective DTV: Transmitter companies offer digital transition strategies," Digitaltelevision.com (December, 1998). Article at http://www.digitaltelevision.com/business1298bp.shtml.
7. T. Fujio, "High-Definition television systems," *Proceedings of the IEEE* (April 1, 1985):73:646-655.
8. A. Kupfer, "The U.S. wins one in high-tech TV," *Fortune* (April 8, 1991):60:123.
9. Ibid.
10. B. Winston, "HDTV in Hollywood," *Gannet Center Journal* 3:3 (1989):123-137.
11. E. L. Holsendorph, "CBS cable bid cleared by FCC," *New York Times* (August 5, 1981):D-1.
12. J. C. Lowndes, "14 seek direct broadcast rights," *Aviation Week and Space Technology* (August 10, 1981):60.
13. S. T. Eastman, *Broadcast/Cable Programming: Strategies and Practices*, 4th ed. Belmont, CA: Wadsworth (1993):29-30.
14. J. Flaherty, personal interview, New York, NY, April 1993.
15. R. I. Nickelson, "The evolution of HDTV in the work of the CCIR," *IEEE Transactions on Broadcasting*, (September 1989:35):250-258.
16. J. Farrell and C. Shapiro, op. cit.
17. R. E. Wiley, "High tech and the law," *American Lawyer* (July 2-6, 1994):6.
18. J. Widoff, personal interview, Alexandria, VA, June 1, 1992.
19. "HDTV: Broadcasters look before they leapfrog," *Broadcasting* 11:117 (September 11, 1989):24.
20. For example, I was in the audience during a keynote speech at a National Cable Television Association convention in the late 1980s when the speaker offered this opinion.
21. A. Lippman, "HDTV sparks a digital revolution," *BYTE* (December 1990):297-305.

22. "HDTV developments in Japan," *Financial Times* (archived on NEXIS, May 9, 1989).

23. "HDTV live broadcasts," *Japan Economic Newswire* (archived on NEXIS, February 23, 1988).

24. "HDTV cooperation asked," *Television Digest* 29 (May 22, 1989):9.

25. "HDTV production standard debated at NTIA," *Broadcasting* 116 (March 13, 1989):67. The role of HDTV in the U.S. economy is well-discussed in M Dupagne, "High-definition television: A policy framework to revive United States leadership in consumer electronics," *Information Society* 7 (1990):53-76.

26. J. Burgess, "United States withdraws support for studio HDTV standard: Japanese suffer setback in global effort," *The Washington Post* (May 6, 1989):D-12.

27. Engineering Report, Washington, DC: National Association of Broadcasters (September 4, 1989):1.

28. "Development of a United States-based ATV industry," Washington, DC: American Electronics Association (May 9, 1989).

29. Ibid.

30. R. Reich, "The quiet path to technological pre-eminence," *Scientific American* 261 (October 1989):41.

31. J. Stilson and P. Pagano, "May the best HDTV system win: An interview with R. Wiley," *Channels* 10 (August 13, 1990):54-55.

32. The story of the development of a digital HDTV design and its submission to the ATTC comes from individual interviews with R. Rast, W. Paik, and J. Heller, August 1994. See also: J. Brinkley, *Defining Vision*, New York: Harcourt Brace (1997).

33. D. Price et al., "Collaboration in an invisible college," *American Psychologist* 21 (1966):1011-1018.

34. J. Widoff, personal interview, op. cit.

35. "SMPTE: Seeking a universal, digital language," *Broadcasting* 121 (November 4, 1991):62.

36. "Refined HDTV cost estimates less daunting," *Broadcasting* 118 (April 9, 1990):40-41.

37. D. Tyrer, "The high-definition television programme in Europe," *European Trends*, 4 (1991):77-81.

38. "HDTV transmission tests set to begin next April," *Broadcasting* 119 (November 19, 1990):52-53.

39. M. Lewyn, L. Thierren, and P. Coy, "Sweating out the HDTV contest," *Business Week* 6:33 (February 22, 1993):92-93.

40. P. Lambert, "First ever HDTV transmission," *Broadcasting* 122 (March 2, 1992):8. For a good contemporaneous account of the Las Vegas transmission, see C. Patton, "Digital HDTV: on-the-air!" ATM 1 *Advanced Television Markets* 5 (April 1992):1-2. The first simulcast occurred on WRC-TV in Washington, D.C., detailed by F. Jacobi, "High-definition television: At the starting gate or still an expensive dream?" *Television Quarterly* Winter 16 (1993):5-16.

41. H. Fantel, "HDTV faces its future," *New York Times* (February 2, 1992):H17.

42. Ibid.

43. B. Santo and J. Yoshida, "Grand alliance near?" *Electronic Engineering Times* (May 24, 1993):1, 8.

44. B. Robinson, "HDTV grand alliance faces tough road," *Electronic Engineering Times* (May 31, 1993):1, 8.
45. R. Rast, Vice President, HDTV Business Development, General Instruments, personal interview, North San Diego County, CA, August 23, 1994.
46. J. Van Tassel, "The computer industry re-boots the U.S. regulatory system," *New Telecom Quarterly* 1 (1997):21-40.
47. P. Lambert, "ACATS orders issue of broadcast multichannels," *Multichannel News* 15 (February 28, 1994):3.
48. System Subcommittee Working Party 2, Document SS/WP2-1354, Washington, DC: Advisory Committee on Advanced Television Service (September 1994).
49. M. Levine, "Critical time nears for setting HDTV standard," *Multichannel News* (June 12, 1995):12A.
50. M. DeSonne, personal interview, Los Angeles, CA, February 1995.
51. M. Sablatash, "Transmission of all-digital advanced television: State of the art and future directions," *IEEE Transactions on Broadcasting* 40 (June 1994):2.
52. R. Rast, Statement of Robert M. Rast, Vice President, HDTV Business Development, General Instruments Corporation, Communications Division. Washington, DC: FDCH Congressional Testimony (Archived on NEXIS, March 17, 1994). For the perspective of TV stations, see D. Halonen, "FCC: Who pays for advanced TV?" *Electronic Media* (March 1995):1, 75.
53. E. Rosenthal, "FBC studies multiplexing strategies," *Electronic Media* (February 28, 1994):26.
54. J. Hontz, "Infohighway bill passes Senate panel," *Electronic Media* (August 15, 1994):1.
55. D. Wharton, "HDTV org threatened by flexibility," *Daily Variety* (August 8, 1994):32.
56. P. Lambert, "Abel says multichannel options could pay for HDTV," *Broadcasting* 122 (October 26, 1992):44. See also R. Wiley, "Entertainment for tonight and tomorrow too," *The Recorder* (July 26, 1994):6-10.
57. J. Van Tassel, "Act Two: The Curtain Rises on the 1996 Telecommunications Reform Act," *The Hollywood Reporter*, Special Issue on the New Regulatory Environment (April 30, 1996):S1-3, 4.
58. Reuters News Service, "Commission background note on digital TV," (archived on NEXIS, December 6, 1993). For country-by-country transmission scheme selection, see, Commission of the European Communities, "Widescreen television lifts off," RAPID press release IP:94-21 (archived on NEXIS, January 14, 1994).
59. "Member states ready to impose EU norm on digital TV," *European Insight* (June 10, 1994):615.
60. "Will shift to digital HDTV, Japan firm says," *Los Angeles Times* (June 9, 1994): C3.
61. "Hamamatsu City designated HiVision City by Ministry of Posts," *COMLine Daily News Telecommunication* (archived on NEXIS, September 6, 1994).
62. "NHK and 9 private stations for trials of EDTV," *COMLine Daily News Telecommunication* (archived on NEXIS, September 1, 1994).
63. J. Van Tassel, *New Telecom Quarterly*, op. cit.

64. J. Brinkley, *Defining Vision*, New York: Harcourt Brace (1997):359.

65. Ibid.:393.

66. U.S. Federal Communications Commission, "In the Matter of Advanced Television Systems and Their Impact Upon the Existing Television Broadcast Service," FR, MM Docket No. 87-268. Adopted: December 24, 1996; released: December 27, 1996; archived at: http://www.fcc.gov/Bureaus/Mass_Media/Orders/1996/fcc96493.txt.

67. U.S. Federal Communications Commission, Report No. MM 97-8 Mass Media Action April 3, 1997, "Commission Adopts Rules For Digital Television Service," (MM Docket No. 87-268). Archived at http://www.fcc.gov/Bureaus/Mass_Media/News_Releases/1997/nrmm7005.html.

68. F. Dawson, "The state of the display: Flat-panel screens coming soon to a PDA or computer near you," *Digital Media* 3 (February 1994):11. For a good discussion of plasma flat panel display technology, see S. W. Depp and W. E. Howard, "Flat-panel displays," *Scientific American* (March 1993):90-97.

69. M. Schubin, telephone interview, November 1999.

70. D. Cripps, "An open letter on Sinclair's challenge to VSB," archived at: http://web-star.com/hdtv/cofdmvs8vsb.html.

71. D. Taylor and T. Brentnall, "Display Technologies for Advanced Television Overview," presentation at 1998 Society of Motion Picture and Television Engineers Seminar on Production and Display Technologies for HDTV, held at the University of Southern California, Los Angeles on May 16, 1998.

72. "Motorola corporate press release," September 6, 1999. Archived at: http://www.apspg.com/press/090199/8vsb.html.

73. J. Yoshida, "DVB digital TV format gains on ATSC Standard," *Electronic Engineering Times* (June 5, 2000).

74. The organization's Web site is at http://www.dvb.org/.

75. PricewaterhouseCoopers, *Digital Television '99: Navigating the Transition in the United States* (1999). Available from: PricewaterhouseCoopers Entertainment and Media Marketing Group (212)596-3737.

76. G. Kaufhold, telephone interview, November 1999.

77. Remarks of Commissioner Susan Ness, "DTV in the Desert," Symposium Consumer Electronics Show in Las Vegas, Nevada, on January 8, 1998. Archived at http://www.fcc.gov/Speeches/Ness/spsn808.html.

78. J. Nickell, "Tune In, Turn On . . . To What?" *Wired News* (September 10, 1998). Archived at http://www.wired.com/news/news/culture/story/14781.html.

79. D. Cripps, "DIRECTV goes HD," *HDTV News* (January 22, 1998).

80. Specifications for OpenCable are available at http://www.opencable.com/public_docs.html.

81. Technical keynote speech by Dr. Richard Green, President, CableLabs, delivered at the National Association of Broadcasters' convention, Las Vegas, NV, April 1998.

82. A. C. Luther, *Digital Video in the PC Environment*, New York: McGraw-Hill (1991).

83. "U.S. Staging Comeback in Technology," *Los Angeles Times* (September 19, 1994):C3.

5

Digital Shape-Shifting: The Many Forms of DTV

The immediate ancestry of digital video can be traced to advances in television, audio, and computer technologies. In the 1970s, the "frame grabber" first came into popular use through sports programming, where it was used for freeze frames. In the beginning, it took a huge computer to grab and save one frame, but throughout the 1980s, Moore's law increased processing so that ever smaller computers could do more and more. By the end of that decade, one computer box could save several hundred frames, enough to create sophisticated postproduction effects in real time—in multimillion dollar studios.

A second technological predecessor was the audio CD. The first 16-bit digital recording in the United States was done by Dr. Thomas G. Stockham, who used a hand-built Soundstream digital tape recorder to capture a performance of the Santa Fe Opera in 1976. Two years later, Sony and Philips introduced a type of digital encoding that enabled commercial recording and playback of digital audio. The two companies finalized the standard in 1980 and separately launched consumer products in 1982.[1]

The CD was coated with layers of aluminum and lacquer, holding enough information to play for 74 minutes. It rotated at variable speeds of 200 rpm near the edge to 500 rpm near the center. The size was 120mm in diameter, and used 16-bit encoding with a sampling frequency of 44,100 per second. The maximum playing time was 74 minutes, long enough to hold Beethoven's 9th Symphony.

In 1985, Sony and Philips allied again to produce a standard for a compact disc that would store and play computer data, the compact disc read-only memory (CD-ROM), using essentially the same technology as the audio CD. Philips went on to develop the CD-I, or compact disc interactive, that ran on a player connected to a TV. An on-screen menu allowed user navigation, and the video and audio were seen and heard on the TV set.

The final source of digital video was computer technology, which contributed the last, major step to realizing the full promise of digital moving images. This was the 1991 breakthrough by Woo Paik and the General Instruments team. As described in Chapter 4, "High-Definition Television: Or How DTV Came into Being Because HBO Needed Cash," they brought what was at the time the most advanced processing and compression technologies off the battlefield and into the civilian sphere of commercial television.

The rest of this chapter will describe the forms of digital television that have been developed in the last fifteen years. Some, like Internet TV, have only come into general use in the last five years.

Digital Television Systems

It is really impossible to neatly separate the various types of digital video by any consistent methodology. The list below represents an attempt to do so, but the fluidity of digital innovation quickly makes obsolete most efforts to categorize. Here is a quick review of emerging TV's alphabet soup, followed by Table 5.1, which provides a comparison of the parameters that define them.

- HDTV—high-definition television
- SDTV—standard-definition television
- eTV—enhanced television
- PConTV—Internet content on the television screen
- TVonPC—television content on the computer monitor
- DVD—digital versatile disc or digital video disc
- CD-ROM—compact disc, read-only memory
- BBTV—broadband television, transported over cable and telephone networks
- ITV—Internet television, streamed over computer and telephone networks
- TTV—telephone television, such as video telephony and video-conferencing over telephone networks

Today, these are nearly distinct categories of digital television, but in the ever-shifting digital world, that situation won't last long. Indeed, there are already signs of overlap; for example, eTV is indistinguishable from PConTV in some implementations, serving to provide instructions for the PConTV programming. Moreover, BBTV programs may well become part of over-the-air television signals from local broadcasters as they begin to develop a digital tier of services. Then the lines between datacasting/PConTV, BBTV, and eTV will become moot as well.

Table 5.1 Comparison of Forms of Digital Television

Form of DTV	Bandwidth	Resolution	Compression	Receiver
HDTV—High-Definition TV	19 Mbps	1080 x 1920i, 720 x 1280p	MPEG-2 (Audio: Dolby AC3)	Digital TV set or PC
SDTV—Standard Definition TV	3–5 Mbps	704 x 480 640 x 480	MPEG-2	Digital TV set or PC
CD-ROM	1.4 Mbps	360 x 242	MPEG-1	PC
BBTV—Broadband TV	Cable: 1.5–3 Mbps xDSL: 300 kbps–8 Mbps	Cable: 360 x 242–640 x 480 xDSL: Up to 640 x 480	MPEG-1, MPEG-2, H.323 wavelet, fractal, other proprietary schemes	Cable: PC or TV XDSL: PC
PConTV—Data on the TV	21 kbps–500 kbps	Low resolution: up to 140 x 120	Uncompressed, or sender-defined	TV
TVonPC—Television signals on the PC	Up to 19 Mbps	Resolution agnostic: User-defined	Compression agnostic: Sender-defined	PC
eTV—Enhanced TV	21 kbps–500 kbps	Low resolution, up to 140 x 120	ATVEF standards–Still image compression: .gif and .jpg	TV
ITV—Internet TV	28.8 kbps–1.5 Mbps	SubQCIF: 128 x 96 QCIF: 176 x 144 CIF: 352 x 288 4CIF: 704 x 576 16CIF: 1408 x 1152	H.261, H.263, H.264 Proprietary	PC
TTV—Telephone TV	28.8 kbps–8 Mbps		H.261, H.263 Proprietary Uncompressed	PC or specialized telephone device

Streams within the Television Signal: HDTV, SDTV, Digital VCR, eTV, PConTV, and TVonPC

All these types of digital television involve some use of the traditional television signal (whether it is from a broadcast station, cable headend, or satellite) to carry video content or other information. They share the same underlying technological infrastructure, and the emerging television system will be shaped by the choices that companies in the TV industry make to offer some mix of these services to their audiences. A communication system that incorporates technology in the message flow requires three components that all share the same communication protocols: (1) production facilities and equipment to create and encode material; (2) transmission facilities and equipment to send it; and (3) reception devices to receive and decode it.

At the turn of the millennium, the industry is in the midst of migrating from digital to analog equipment in transmission, and to a lesser extent, in production. By mid-2000, 134 local stations in 49 U.S. markets had made the transition. Over the next decade, most programming networks, stations, and cable systems will have converted their systems to carry digital content. Many will embrace digital production as well, although analysts predict that analog cameras, video tape recorders, editing, and switching equipment will be around for a long time, particularly in small organizations and in small markets.

The slowest part of the system to mature is the acceptance of reception devices on the consumer side. Consumers have to pay for them. The initial high cost of HDTV and SDTV sets place them outside the reach of the majority of viewers, in spite of the fact that consumers were supposed to be the beneficiaries of both HD and SD television.

HDTV

High-definition television offers many benefits by bringing several movie-like properties to the small screen. The most obvious visible characteristic is the changed aspect ratio from today's almost square 4:3 screen to a horizontally rectangular 16:9 shape, approximating the look of films as seen in theaters. Research shows that the wider screen takes advantage of the fact that the eyes see more along the horizontal plane than the vertical plane, resulting in a picture that seems more life-like. When a viewer watches NTSC 4:3 television, the eyes move from side to side, right and left, only about 10 degrees. By contrast, the wider 16:9 aspect ratio causes the eyes to move from side to side about 30%, a scan similar to everyday vision.[2] (See Figure 5.1.)

Undoubtedly the most important improvement HDTV offers is superior picture quality, especially on large-screen television sets. Due to

STANDARD

HDTV

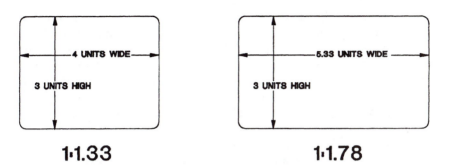

1·1.33

1·1.78

Figure 5.1 Aspect ratios of NTSC and HDTV television.

its greater resolution, the HDTV picture is much more detailed than the current TV picture, and its color is richer, deeper, and more complex. Since large TV sets are the fastest-growing segment of the television set market, there is considerable demand for the higher-quality picture that HDTV can deliver.[3]

One way to appreciate the difference higher resolution will bring is to experiment with the distance you sit from your TV set. In order to see the picture without lines, the sitting distance must be approximately seven times the screen height. If the screen is eighteen inches high, you need to be ten and a half feet away. HDTV allows the viewer to sit much closer to the screen, at a distance of only three times the screen height. In the case of an eighteen-inch screen, one could sit four and a half feet away and see a line-free picture.

In addition to higher quality images, HDTV brings with it a whole new sound. If seeing is believing, hearing is all-involving. Many people don't realize that most film budgets nearly as much money towards the sound track as they do for the picture. Seamless audio is often as important as a believable picture for the audience to become psychologically involved, or to suspend disbelief, as the movie industry refers to this phenomenon.

Every HDTV standard will include improved audio. The United States has adopted 5.1 channel Dolby AC-3 audio technology to provide compact-disc-quality, digital, stereophonic surround-sound, which expands the enjoyment of television entertainment dramatically. This innovative new technology offers better sound separation and potential for wider range dynamics, encoding up to six audio channels, in less than 50% of the space required for a single channel on a CD.[4]

The entire television industry infrastructure must change in order to create, carry, and receive HDTV. Standard-definition analog video will not look good on an HDTV system. Even the production components themselves are not adequate. For example, news sets are not usually finished with any degree of sophistication. Nor is TV make-up nearly as refined as make-up for film. The greater resolution of HDTV will call for more finely designed and created sets, props, make-up, and costumes. And the CD-quality audio will expose poorly produced music and special effects as well. This means that for a television facility to produce material that looks good on an HDTV set, it will have to use HDTV cameras and editing equipment in the production process.

The transmission of the TV signal requires a tower and electronic equipment to send it out, and stations will have to buy all new equipment. HDTV's greater information also takes more electric power. And digital signals have different properties, such as a sharp drop-off between "good" and "bad" reception at viewer homes, and "good" and "none," as well.

The cost of production and transmission equipment is onerous enough, even when relatively large organizations are doing the spending. But customers have to pay for the reception device, the TV set, or display unit. Manufacturers are producing two types of appliances. The first is an HDTV set. Reception ready, it must only be hooked up to an antenna or cable system. The second option is a component system, where the display is only a display, and the tuner and other electronics come in a companion set-top box. Most HDTV displays and set-top boxes cost between $7,000 and $12,000. Both types of products require buyers to purchase a special antenna designed to pick up digital signals.

The consumer antenna has been a matter of some concern as broadcasters become more familiar with the reception characteristics of digital television. Manufacturers have provided retail outlets with printed material to help them assist consumers choose the appropriate antenna for the location of their home. Nevertheless, there is considerable consumer confusion about what they have to buy and how to make it work.

SDTV

Standard-definition television refers to a digital television picture that has approximately the same resolution as today's analog NTSC picture, transmitted and displayed in either a 16:9 or 4:3 aspect ratio. The same digital transmission and transport systems underlie both HDTV and SDTV. Once the transition to a digital system is made, the difference between the two formats comes down to the amount of information in the bitstream, the bandwidth. Thus, HDTV infrastructure will also work for SDTV, but not in the reverse; that is, the greater information of HDTV will accommodate lower-resolution signals, but lower-capacity infrastructure will not carry HDTV.

Analog production equipment, such as cameras, editing machines, switchers, and sets, is much more compatible with SDTV than it is with HDTV. Some people argue that SDTV is a more realistic goal than HDTV; they believe that HDTV's marginally better picture is not worth its enormous cost. Indeed, the Europeans have taken this approach and started by perfecting an SDTV standard called DVB, with the idea of turning to HDTV sometime in the future.

SDTV is the format used by the current satellite services, DIRECTV and Echostar's Dish Network, and they appear to have added HDTV channels to their programming lineup with little difficulty. Consumers report that the SD images are clearer than analog over-the-air TV, and that the quality of the audio is much higher, appealing to viewers who own large-screen TV sets and home theater equipment. Even without the higher resolution of HDTV, SD digital images eliminate some of the problems of NTSC; there is no shimmer, jitter, dot-crawl, or false color display with digital TV.

eTV and PConTV

These two formats rely on similar technologies, although PConTV typically uses more bandwidth capacity. eTV is a stream of data that is sent along with a TV picture that refers to and complements the programming it accompanies. Sometimes eTV is called "parallel programming" or "parallel datacasting." PConTV is Internet content viewed on a television screen. It may not be related to the main video channel at all—it's up to the viewer or the service provider. Some allow the full range of Web surfing; others restrict the sites a subscriber can visit.

The technology design for the two types of digital TV is similar. Data is embedded in unused portions of the current analog TV signal. Capacity can be as much as 600 kbps, enough to carry a moderate quality TV picture. The unused parts of the signal are the vertical blanking interval (VBI, the black bar at the end of every TV picture that viewers can see when there is vertical roll of the picture); the vestigial sideband (VSB), a part of the wave that does not need to be used for viewers to get a good picture; and the horizontal overscan (HOS), room left at the end of each line to be sure that all parts of the video image are included in the TV's "safe" picture zone.

The VBI contains 21 lines; the first 9 lines are used to tune the field, leaving lines 10 through 21 to transmit information. Line 21 is used by closed-captioning services for the hearing impaired, leaving lines 10 to 20 for sending information at about 150 kbps. Putting together every possible unused part of the TV signal, various schemes have been able to reach a combined bandwidth of approximately 600 kbps.

Both eTV and PConTV must be created in some kind of production facility, usually a room full of computers. It has to be formatted in software and provided to a TV distributor—a local TV station, a satellite broadcaster, or a cable headend—for inclusion in the combined signal sent to the audience. Today, there are many companies that provide the creative input to design the information and digital postproduction facilities to perform the encoding.

On the consumer side, there must be some kind of decoding capability, which can be in a separate set-top box, packaged in a generic set-top box or personal video recorder, included in the TV set itself, and (someday) in a PC. Essentially, TV set-top boxes (STBs) are nothing more than slow computers with a small amount of memory. Like a computer, the STB has a software operating system and programs to perform various functions, sometimes to decode the incoming material, and to combine it with TV programming. Finally, the viewer needs a way to navigate and to communicate with the set-top box—a remote control device, keyboard, or joystick.

eTV

Enhanced TV, or eTV, is a professionally-produced product. Different proprietary technical schemes allow for a specific mix of instructions and creative elements—text, graphics, animation, audio, and video—displayed on the TV screen. They might provide additional content, such as graphics and text that give more information about whatever is going on in the TV program. Sometimes the eTV content is a transparent overlay and the video shows through; in other designs, it is opaque and covers the picture. Often, the TV picture is placed in its own window, with text and additional graphics appearing in their own frames. Recurring enhancements are icons, banners, labels, menus, information about the program, data the user can print, open text fields where users insert information, or forms to fill out to order a product.

The content for eTV centers on the provision of information and e-commerce activity, perhaps because both require only limited interactive capability. Some systems also allow Internet access so that viewers can chat with others as they watch a TV show. eTV platforms differ in what they let customers do.

One service that has made significant progress towards acceptance is Wink Communications. The company's technology embeds an interactive graphic interface into the over-the-air television signal. A set-top box decodes the data and overlays it on the TV picture. Using the Wink remote control devices, viewers respond to various interactive opportunities, typically offers of additional information or purchasing transactions. A modem dials up an online server that processes the viewer's request and sends information back.

Some enhancements aren't content at all, but merely "triggers" that call up digital material from the Internet. In this instance, the eTV portion will have only a few creative elements. It will consist mainly of an Internet address where the rest of the material is located and instructions about how it is to be placed on the screen relative to the program video and audio. The rest of the of the program will come from the Net. And now we have crossed the line into a different kind of digital television, PConTV, such as WebTV, AOL TV, and NetBox, which will be covered in the next section.

One of the earliest eTV trials occurred in late July, 1997, when Microsoft tested a co-broadcast of interactive data with television programming. Microsoft provided its own production staff at a cost of about $250,000 per episode for scheduled TV shows on FOX's *Moesha* and USA Networks' *Pacific Blue*. Microsoft is also packaging content for enhanced TV viewing, including Big Ticket, Ministry of Film, New Digital System, Rysher Entertainment, and the Sci-Fi Channel; by 1999, there were close to 100 eTV programs in development. NBC, for example, is working closely with Intel to create new content to go along with some of its shows.

The first nationwide eTV broadcast took place on November 10, 1998. A collaboration between PBS and Intel, the additional digital programming was aired by PBS that complemented Ken Burns's documentary about the life and work of architect Frank Lloyd Wright. Intel contributed the technologies that enabled the network to broadcast the video-plus-data (V+D) content and the receiver PCs to decode and display the digital signals.[5] (See Figure 5.2.)

Six PBS stations sent the programming to Intel, as well as PBS employees, engineers, and station managers, who provided technical feedback about the enhanced broadcast. Participants had DTV-enabled PCs in their homes, or viewed the program at the local PBS station. The companion data was transmitted in the television signal as the documentary aired. However, the audience could use their PCs to explore and interact with video, audio, and a rich array of additional information about Frank Lloyd Wright only after the broadcast, because Burns wanted people to see the program only as he intended it to be seen.

The enhanced content included interviews with Wright by CBS correspondent Mike Wallace, recorded in the 1950s; virtual tours of three historic Wright creations—Fallingwater, the Guggenheim Museum, and Unity Temple; and other material that could not be included in the broadcast documentary because of time constraints. For example, Eric Lloyd Wright, Frank's grandson, and others provided stories, descriptions of Wright's innovative architecture, and analyses of his style.

"This event signals the birth of a new medium," said John Hollar, executive vice president, PBS Learning Ventures. "Imagine being able to transmit a superb documentary by Ken Burns and the equivalent of a companion DVD brimming with educational content, all within the same signal. This is PBS's vision of enhanced digital television."

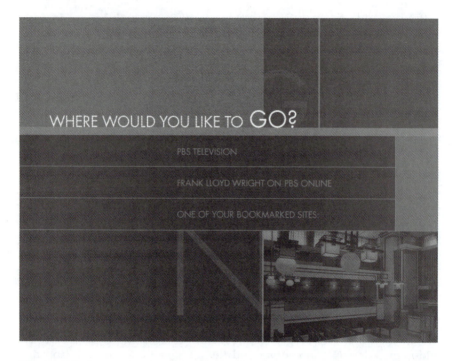

Figure 5.2 Screen shot from accompanying eTV material. Source: R/GA.

In early 1999, PBS and Intel Corporation teamed together for another eTV project, a children's program called *Zoboomafoo*. It is a series for preschoolers about wildlife that lets kids interact with on-screen animations on their PCs. There are games—Animal Alphabet Scrapbook, Animal Matching, Who's in the Hole, and Closet Adventure—and an audio/video album, where the youngsters collect enhanced audio, video, or graphical media in the form of souvenirs, which they can interact with after the broadcast.

Starting in April 2000, PBS and Intel Corporation broadcast eTV content 24 hours a day, using local stations' digital bandwidth. It gives viewers additional video, graphics, and educational applications in addition to the main program. Special emphasis is placed on primetime and children's programming.

The instructions that describe how the elements are put together are written in HTML (hypertext markup language), the language used to create pages for the World Wide Web. The standards for the enhancements are under development by an industry group, the Advanced Television Enhancement Forum, or ATVEF.[6] In late 1999, Intel, which had provided its Intercast technology to PBS for the Frank Lloyd Wright datacast and other projects, discontinued it in favor of the ATVEF standards. This move was not a rejection of Intercast, rather a recognition that the standards for eTV, TVonPC, and other forms of datacasting had developed sufficiently that the company no longer needed to provide protocols for content providers to use.

PConTV

About 99% of TV households have a television set. As of 2000, about 50% of them have a computer, and between 35% and 40% are online.[7] No wonder Internet companies want to put their content on the television set! When the Web is linked with television TV on a single monitor, many possibilities exist for new forms of entertainment and information that emphasize the use of video.

A distinguishing characteristic of PConTV is that it always has a return path. A return path is a communication channel between the TV set and the source of information, an Internet Service Provider or some other data aggregator. Most often, that channel is an ordinary telephone line, but increasingly it will be a cable modem, high-speed telephone line, or even a wireless transmitter. However the signal goes back upstream, it gives the subscriber a way to request Web pages, drill down for further information, move from one Web site to another, engage in transactions, send or receive E-mail, participate in chats, or some other desired activity.

Given the converged nature of PConTV, it is difficult to say if audience members are TV viewers or computer users. Two new terms that are being bandied about are "viewser" and "telewebber," which convey the converged nature of the platform. Whatever they are called, navigation for these user-viewers is typically provided by on-screen buttons, icons, or links. Viewers use some kind of handheld remote control device to click on the part of the screen they want to activate.

One of the most impressive features of the Internet is its support for communities of interest. So one of the first uses of PConTV is to combine chat sessions with TV reception. As viewers watch a program, they click a window on the TV and join a chat session with others watching the same show, or with production people who can provide insight into how the show was produced. This activity requires two-way capability and presents some technical challenges, but it is very attractive to TV programmers. It allows them to create loyal fans around shows, and gives them viewer-provided information for further marketing and PR opportunities.

PConTV brings something to advertisers. Consumers can drill deeper down into information about products they see. As viewers watch TV, they can go to the advertiser's Web site and learn more about the products and pricing. Or they might see an advertisement they would like to keep and refer to later. In this case, the viewers could bookmark the commercial, similarly to how they bookmark Web pages on a browser. Later, data could be transmitted from the browser to the content provider to activate the commercial and possibly add some other features, such as initiating a purchase order. The whole universe of order processing is a subject of great interest to advertisers and programmers.

Then there's the virtual channel, a way of transmitting some kinds of programming that takes up only a fraction of the bandwidth required for a real TV channel. A broadcaster could transmit computer commands that instruct a PC or a digital set-top box to create pictures. A cartoon channel would be ideal for such a service, allowing a programmer to transmit the equivalent of PostScript commands to the PC, or even a TV if it had a set-top box with the appropriate software. The receiver would accept these computer commands and render the images and sounds as directed. This means that the images and sounds themselves are not transmitted, nor is the data that composes them. Rather, it is the instructions about how to create them that are sent.

A market leader in the PConTV space is WebTV, which is owned by Microsoft. It brings Internet content to the television via telephone wires or satellite, routing it through a dialup modem in the set-top box. For satellite delivery, WebTV comes into the living room through the EchoStar dish, along with digital TV programming. For the return path, the system sends the customer's messages from the home over telephone lines to the WebTV server site.

Essentially, WebTV is an ISP (Internet Service Provider) that markets its set-top box and charges a monthly subscription fee for service. It is marketed with EchoStar's digital TV service, providing high-speed Internet downloads. Viewers also receive their digital TV programming guides through WebTV. The special set-top box gives customers digital recording capabilities, competing with TiVo and Replay Networks Inc., services that allow digital recording and replay of TV programs.

WebTV has expended substantial resources to make text designed for computer screens readable on TV sets. They have designed an attractive, well-designed interface. As a result, receiving Internet Web sites works well for people who do not have a PC. Computer users, however, are likely to become frustrated. There may be restrictions on the number and type of Web sites they can surf. Unless the material arrives via satellite, the downloads can be slow, and there may be no easy way to open up the set-top box and replace obsolete components, as can be done with PCs.

America Online, the Internet service, also has offerings in the PConTV space. As part of its AOL Anywhere initiative, the company has partnerships with Philips Electronics and Liberate Technologies to develop a range of reception devices. It is also developed AOL TV and allied with the DIRECTV service to compete directly with the WebTV/EchoStar partnership. Essentially, subscribers will be able to get AOL services on their television sets. In June, 2000, AOL hooked up with TiVo to provide AOL TV on the personal video recorder platform.[8]

The AOL TV platform was created by a host of companies that are loosely allied to compete with Microsoft in the computer software arena. Liberate, Netscape, and Oracle contributed the core technologies with the goal of making the Internet as commonplace as electricity. The technology brings Internet data to the television screen from any source, including both wired and wireless distribution technologies.

Sun is another company that has the technology to enter the PConTV market if it decides to do so. It acquired a company called Diba that produced an end-to-end system including servers, system software, and consumer hardware and software. In addition, Diba designed single-purpose "information appliances" that centered around a specific functionality, such as an Internet-TV appliance, an E-mail appliance, a remote-banking appliance, and so forth. Diba developed about forty-two such appliances that would support many different input, processing, output, and storage options, depending on the application. The Internet Set-Top Starter included a modem and Ethernet connection, infrared remote control, infrared keyboard, power supply, phone cable, video cable, and left/right audio cable. The processor automatically reformatted Web content for maximum readability on the television set. Today, this work is carried out by Sun's Consumer Electronics Solutions. Not much has been heard of the products, but Sun's Java software and network-centric philosophy could lead the company to develop information appliances along these lines.

PlanetWeb is a software player in this space with a platform-independent browser that takes less than 1MB (megabyte) of space. The software runs on Internet appliances with only a small amount of memory, including TV sets, screen telephones, and future devices. The company plans to market it to hardware original equipment manufacturers (OEMs) whose Internet appliances fulfill different design specifications.

NetGem's NetBox is the European entry. Headquartered in France, the system is similar to other Internet appliances designed to display Web content on television sets. The technology allows full access to Internet sites and E-mail services. It is equipped with a navigation joystick, a virtual keyboard, and an optional infrared keyboard is available for text-intensive functions. The box has an X86 processor, a 33.6 modem, a smart card reader, and software that takes less than 1MB of flash RAM.

TVonPC

Now we turn from technologies that put digital video on the TV set. TVonPC means bringing in a television signal—but displaying it on PC monitors rather than the TV screen. The basic technology is a signal processor, typically a TV tuner card that is installed into the PC. An antenna plugs into the device or card to bring in the broadcast, or some kind of cable carries signals from the cable company or satellite dish. Importing the TV signal into the PC using a card is typically an action taken by individual consumers because they want to watch TV on their PC monitor.

There are many fine TV tuner cards available: WinTV, Hauppage, and ATI All-in-Wonder are just a few. Planetville is a service built around this technology. It launched its business by giving away the tuner card and MPEG decoding software. When consumers fire up the card, the software directs them to the company's Web site, where they can get information about channels and program listings. The business model depends on attracting traffic through the product giveaway, then aggregating the consumers and marketing them to other sites and advertisers.

The other TVonPC technology is datacasting, where TV broadcasters use part of their spectrum to deliver data to PCs. It has special importance because it is likely to become an important part of the business models used by local television stations to pay for the upgrade to DTV. Like eTV and PConTV, early configurations of this kind of service transmit information in unused portions of the analog TV signal, the vertical blanking interval (VBI), the vestigial sideband (VSB), and horizontal overscan (HOS).

To give an example of how TVonPC works, a pioneer in this arena is GeoCast. The company provides equipment free to broadcasters and allows them to beam data in their local coverage area. TV stations are signing on with GeoCast because they want to be the dominant news brand in their market, and if they do not reach PC users, they fear their now-prominent brand will be lost to online companies like CNNInteractive. Belo, a large TV station group, partnered with GeoCast for local datacasting. As Flory Bramnick, Belo's VP of strategic alliances and business development, commented:

> *We were looking for ways to monetize our spectrum and to present content to our viewers' PCs. Anytime anybody wants our content in any form on any device, it is incumbent on us to provide it. GeoCast gives us a way to do that, and we were very impressed with their knowledge of broadcasting and the quality of images consumers receive.[9]*

Another company in this space is iBlast, backed by the Tribune Company, Gannett, Cox, Post-Newsweek Stations, the E.W. Scripps Company, Meredith Corporation, Media General, Lee Enterprises, *The New York Times* Company, McGraw Hill, Smith Broadcasting, and Northwest Broadcasting. The startup will charge TV stations for the delivery of bits and bytes on a per megabit, per second, per month basis.

According to Michael Lambert, iBlast founder and CEO,

> *We have put together a new national network very much the same way that the over-the-air broadcast networks assembled theirs. They get TV programming from*

three sources—the network, the material they produce, and syndication. They'll get content for their digital spectrum the same way—from the iBlast network, the material they produce, and from third party suppliers.[10]

One barrier to the success of TVonPC and datacasting services is the nature of the computer itself. Many people doubt that audiences will be satisfied watching entertainment on its small screen. So one of the first steps in making TVonPC a rewarding experience is to transform the home computer from a "lean forward" screen, like the computer monitor, to a "lean back" screen, like the television set. Intel is working with several different manufacturers to design what it calls "interactive PC theaters," designed to make computers entertainment-friendly. A PC-for-fun has an extra-large monitor, high-quality speakers, Web access, and wireless peripherals that work from anyplace in the same room, or even around the house. Several computer manufacturers have developed and even produced TV-set like computers to meet the perceived need for an "entertainment PC."

Microsoft has drafted specifications for what it calls the PC 98 system. It would feature a fast processor, built-in TV capability, a port for digital satellite reception, a DVD-ROM player, a high-speed connection to a video camera, a FireWire port, eTV reception, a wireless keyboard, and a large display. A Universal Serial Bus connects speakers, keyboards, and printers. Compaq now builds a $5,000 PC Theater PCTV that draws a small group of enthusiast buyers. However, the company plans to build a new lower-priced version that would receive digital television signals and interactive services.

Gateway 2000 manufactures its Destination series system. The company intends to develop less-expensive versions of the PCTVs that now cost between $2,800 to $5,000. Destination currently connects to a 31-inch Mitsubishi monitor and lets you hook up your own amplifier and speakers. The least expensive model has a 120 MHz processor, wireless keyboard, 16MB RAM, a 1.2GB hard drive, and a 6X CD-ROM drive. NetTV's Worldvision 2900 has a 29-, 33-, or 37-inch Thomson monitor that supports up to 1024-by-768 resolutions and accepts both analog and digital signals. It also allows the capture of snapshots from the TV display. The model also has built-in stereo speakers. The least expensive Worldvision has a 100 MHz Pentium processor, infrared keyboard, 8MB RAM, 1GB hard drive, and a four-speed CD-ROM for $2,995.

Until the PC-for-pleasure gets here, information services will probably dominate the TVonPC space. One example is WavePhore. It is the first national service of its kind, transmitting data in unused portions of the TV signal to send information at about 500 kbps. In July, 1997, WavePhore contracted with ADS Channel Surfer to bring PC add-in cards to users' computers. This technology allows users to switch between TV, cable, and Internet video, while simultaneously receiving WavePhore data.

WavePhore initially distributed digital data and information over satellite and FM radio subcarriers. In 1997, the company made a deal with PBS to transmit over their 250 member stations, reaching 99% of U.S. TV households. Subscribers get news and information from Dow Jones, Reuters, the Associated Press, Knight-Ridder/Tribune Business News, Thomson First Call, Comtex, and Federal News Service. The service goes to more than 45,000 business users across North America, providing individually profiled information to users of LAN E-mail and groupware, including Lotus Notes, Microsoft Exchange, and other widely used desktop productivity/communications applications.

Other services include Hughes DIRECPC service. A partnership between DIRECTV and Microsoft started in July, 1997, delivering data to the PC via satellite to desktop PCs at a rate of about 400 Kbps. When the service was introduced in 1996, users needed both a DIRECTV and DIRECPC dish; however, in July, 1997, the DIRECDuo dish was launched, a 21-inch hybrid dish that receives both types of signals. For the return link, the user needs a modem and telephone return line to an Internet Service Provider.

SpeedUS.com datacasts using a different kind of over-the-air signal. Originally formed to operate as a "cellular" television service, SpeedUS.com decided to enter the Internet service provider business in mid-1996. The system was beta-tested in late 1996 and deployed in 1997. LMDS, local multipoint distribution service, operates in the very high 28 GHz portion of the spectrum. The signal is robust over short distances, so this service works especially well in areas where population is dense. The user has a six-inch antenna and a high-speed modem. The return link is through the telephone lines over a regular modem to an Internet Service Provider. The cost of the service is about $49.00 per month, plus telephone and ISP charges.

Disk-Based DV Devices: DVD and CD-ROM

The first group of digital technologies were all related to the television signal in some way. This section now examines disk-based DVD and CD-ROM. They are often used as standalone devices attached to a PC, TV, or game console. But in a networked world, they won't stand alone for long because the digital material can be stored and streamed over networks.

DVD: Digital Versatile Disc or Digital Video Disc

Work on a high-capacity CD-ROM began in 1972 when Matsushita Electric developed a technology called phase-change recording that enabled a new kind of optical storage. However, at that time, the data was analog, embedded into the disc in micro-pits that were read by the laser. In 1989,

Matsushita introduced a data disc, and the company kept introducing variants of their patents throughout the 1980s. At the same time, Sony and Philips were working along a parallel track, trying to extend CD technology.[11]

Demand for DVD from studios became clear in 1994, when the Hollywood Digital Video Disc Advisory Group (HDVDAG) was formed, which represented the entertainment industry's stake in using a more robust distribution medium than tape. Tapes not only wear out after a few hundred plays, they are also bulky and expensive to produce. But Hollywood insisted there be one standard, and one standard only. After the debacle a decade earlier with VHS and Sony Beta, they knew only too well the pitfalls of dueling devices, and they had resolved that they would not license their content unless manufacturers acceded to their demand.

Hollywood holds the trump cards in this negotiation because most successful entertainment platforms for visual material—from movie theaters to television, cable, and satellite—have depended on the industry's films for sales appeal. Game platforms are the first devices that break this long-standing dominance. The HDVDAG had other requirements as well, including a single standard that would be used by all drive, media, and content providers. Their requirements:

- Capacity to store feature length film on one side of disc
- Video quality superior to existing VHS and laser disc
- Support for high-quality digital audio
- Multiple language soundtracks
- Multiple aspect ratios
- Copy protection
- Multiple versions of content with parental lockout

In December of 1994, the Philips/Sony product came out, introducing the MultimediaCD (MMCD) format, an extension of the existing CD technology. A whole host of committees formed to set forth the requirements of the various industries. For example, the computer camp wanted a single, common file format and systems that would operate across both PCs and home entertainment systems, would be backwards compatible with CD standards, and would include writable and rewritable technologies as well.

After much jousting over standards, compromises were reached that resulted in the creation of the final DVD standards group, the DVD Forum, with the leaders of both factions, Sony and Toshiba, as members. The organization established the baseline protocols for interactivity, copy protection, DVD rewritable standards, and DVD-ROM. And, like the CD before it, the standards for long-playing data types like video and audio were different from those required by computer data. By 1996, most of the issues were resolved and producers began developing content for the

DVD, priced at about $25 for a disc. In November of 1997, Pioneer launched the first DVD-Recordable drives, and consumer players from Toshiba, Panasonic, and Hitachi came on the market at about $799.[12]

DVD is the next generation of optical disc storage technology, faster and more complex than its CD predecessor. It also has much more capacity, able to store enormous amounts of any type of digital data—numerical, text, graphics, video, and audio. Many analysts believe it will eventually replace audio CD, videotape, laserdisc, CD-ROM, and video game cartridges.

DVD players may be installed as a drive on a PC or used as a standalone player. There are even portable DVD players that will become smaller and less expensive over time. Replication of discs is inexpensive. The discs themselves are compact, easy to handle, store, and ship, and fairly durable. (See Figure 5.3.) Of course there is no wear from playing. They are heat-resistant, and they are not susceptible to magnetic fields. Like CD-ROMs, however, they will become unplayable if scratched.

DVD capacity and features depend on how the disc is produced, and what the developers choose to include. At minimum, consumers will get more than two hours of high-quality digital video, and more than eight if the disc is dual-layer and double-sided. The playback machines will tailor the video to play on standard (4:3) or wide-screen (16:9) TVs. The sound comes from up to eight tracks of digital audio, which can be used for multiple languages or other commentary. Each track can present as many as eight channels.[13]

DVD players allow for rapid, invisible branching of the video, a feature usually referred to as "seamless." Branching allows for multiple story lines, endings, and versions, including Motion Picture Association of America (MPAA) ratings-based changes that parents can select for their

Figure 5.3 DVD discs look like their CD-ROM predecessors. Source: Panasonic.

childrens' viewing. In addition, users can select up to nine different points of view, or camera angles, from which to watch the material.

Random access offers VCR functionality (pause, rewind, fast forward) without the wait. Data embedded on the disc allows search by title, chapter, track, timecode, and sometimes special search terms that relate specifically to the content on the disc. Data about productions usually includes multilanguage credits, including title name, album name, song name, cast, crew, and production notes. Although everything on the disc must be specifically programmed, the features commonly available are multilanguage support for menus and searchable keywords; VCR functionality; parental control and lock; programmable playback of selected sections; random and repeat play; digital audio out, including Dolby Digital and pulse code modulation (PCM) stereo; and backwards compatibility with audio CDs and CD-ROMs.

The early predictions of DVD sales were often wildly optimistic. For example, in 1997, IDC said there would be 10 million DVD-ROM drives sold in that year, 70 million sold in 2000 (overtaking the penetration of CD-ROMs), and 118 million sold in 2001. The report estimated that over 13% of all software would be available on DVD-ROM in 1998. Later prognostications were more reasonable. In 1998, InfoTech forecast that there would be 99 million DVD-ROM drives worldwide in 2005. They did not expect there to be more than 500 DVD-ROM titles available by the end of 1998, but the market would explode to about 80,000 titles by 2005.

What actually happened was that in 1997, manufacturers shipped 347,000 DVD-Video players to U.S. consumers, and in 1998, there were about 900 DVD-Video titles available. In 1998, 906,000 DVD-Video players shipped, and by mid-1999, there were more than 2 million players in U.S. homes, making the DVD one of the most successful launches of entertainment technologies. As of 1999, there were 400 DVD-Video titles in Europe (135 movie and music titles) and 3,000 DVD-Video U.S. titles (2000 movie and music titles). In all, consumers have purchased about 7.2 million DVD-Video discs.[14]

Since 1997, DVD technology has achieved widespread support from every major electronics company, all computer hardware companies, and most of the movie and music studios. This level of acceptance is unprecedented and bodes well for DVD's likelihood of success. One reason studios have chosen to buy into DVD is because of its capability to produce high-quality video and audio, better than any other consumer appliance on the market for either data type. DVD is vastly superior to videotape, and typically better than laserdisc. However, DVD quality depends on many production factors, and since such work is a costly, labor-intensive, and complex business, there is considerable variation between the quality of DVDs.

This is one instance where one technology may spell the extinction of another, rather than merely adding new equipment choices. It appears that the DVD could replace CD-ROM over time. But it is already wiping out an earlier analog technology, the laserdisc. Mostly used in professional applications for editing and storage, the laserdisc did achieve some minor consumer use. Now that is ending. As of August, 1999, new DVD releases outnumbered new laserdisc releases by a ratio of ten or more. The major players in the laserdisc space, Criterion and Pioneer, stopped releasing new laserdisc titles, and existing laserdiscs were selling at 70% or higher discounts.

Even though DVD-ROM drive sales are beginning to make inroads on the market share of CD-ROM drives, the makers of CD-ROM drives continue to design and produce ever-faster equipment. In mid-1999, two vendors, Kenwood Technologies (U.S.A.) Inc. and Asus Computer International, released models that broke the 50X speed barrier.

CD-ROM

A CD is a small plastic disc that stores digital data. The information is encoded on the 4 3/4-inch (12 cm) disc as a series of microscopic pits on an otherwise polished surface, and the disc is covered with a transparent coating so that it can be read by a laser beam. Since nothing touches it, the CD is not worn or damaged by playing it. When the audio CD was introduced in 1982, consumers were quick to grasp the other advantages it offered over the phonograph record and recording tape—compact size, more dynamic range, low distortion, better sound. People snapped them up, and by 1991 CD sales exceeded those of audiocassettes.

CDs also store and play digital video. The drive merely reads 0s and 1s—from a laser's point of view, video is simply another form of computer data. The standard for such data is called CD-ROM, standing for Compact Disc-Read Only Memory. A CD-ROM is both a disc and a drive that will read any data type; the disc is played back on the CD-ROM drive as part of a computer.

Philips and Sony brought the CD-ROM into being when they announced the Yellow Book standard in 1983. The Yellow Book built upon an earlier standard called the Red Book, which had defined the CD digital audio standard. The new format was necessary because of the differences between audio data and other data types.

Audio data is linear and can extend over considerable lengths of time, making long data strings that are only occasionally divided into songs, movements, and other breaks. By contrast, computer data is composed of much smaller files that must be searchable in tiny bits—key words. There's also a difference between how the different types of material stand up to missing data. Audio and video can tolerate small amounts of missing bits of data, but computer data can not. Think about it: Missing numbers from a column of checks or dates may well render an entire document meaningless. But a disc with video is more like a computer CD than an audio CD, because this format is rarely used for long video sequences. It is more common for very short segments to appear, interspersed with text and graphic elements.

The requirements of different material led to the development of several different CD standards. These include CD-ROM XA (eXtended Architecture), that added features and capabilities to the Yellow Book CD-ROM, enabling Photo CD, CD-i players, and CD-ROM/Recordable and Rewritable multisession recording. This is just the beginning.

CD+G is a special standard for compact disc plus graphics, making advantage of unused space on audio discs to add computer graphics that are put together and displayed while the music is played. Like CD+G, CD Text uses blank portions of the disc to store additional textual information. El Torito is a bootable CD specification, and Enhanced CD is an umbrella term that incorporates some alternative techniques to store multimedia data on audio CDs. Mixed Mode CDs make room for data tracks between one or more audio tracks. Hybrid discs are readable in native form on more than one operating system, particularly those that can be read by both Macs and PCs. HFS discs use Apple's Hierarchical File Structure.

Except for simple data storage, the production of CDs for entertainment purposes is a complex, time-consuming, and difficult process that requires significant expertise. It is very much a group activity that can take more than a year to complete, even where there is a generous budget. Until very recently, the distribution of discs required a marketing organization, but in the last two years, the Internet has emerged as a vehicle for getting material to consumers. However, anyone who has used the Internet realizes that downloading even the content from a 650MB CD-ROM would be quite a daunting prospect and would take many hours over a dialup modem.

The real limitation of CD-ROM drives is not their speed, an uninspiring 1.5 Mbps. Rather, it is their limited data capacity. No matter how fast the equipment runs, the discs still hold only 650MB of information. In the data domain, that's a huge repository. But in the world of video, it could be less than a second of video—depending on the amount and technique of compression.

The CD-ROM is nearly universal, a component drive of millions of PCs around the world. However, DVD-ROM not only establishes a new market for optical discs—the full-length movie—it also eats away at CD-ROM sales in two of its most important markets, video games and music. Gamers want both high-quality graphics and speed. The new CD-ROMs can deliver on the speed, but they don't have the "real estate," the available storage capacity to match DVD-ROMs.

Most digital video on CD-ROM is not so much used for linear programs as it is for games and educational material, presented in highly interactive formats. The CD-ROM market is enormous, as Tables 5.2 and 5.3 show.

**Table 5.2 Total Video Games:
Dollars (Millions) by Individual Category**

CD-ROM Products	Market Revenues (millions $)		
	1995	1996	1997
TV Video Hardware	$513	$909	$1,214
8-Bit TV Video Cartridges	17	5	1
16-Bit TV Video Cartridges	848	525	270
16-BIT TV VIDEO CDs	20	8	2
32/64-Bit TV Video Cartridges	17	180	784
32/64-BIT TV VIDEO CDs	121	437	813
8-Bit TV Video Accessories	6	2	1
16-Bit TV Video Accessories	111	73	44
32/64-Bit TV Video Accessories	39	148	417
Portable Video Hw	148	102	109
Portable Video Cartridges	159	113	125
Portable Video Accessories	35	20	15
Total	$2,034	$2,522	$3,795

Source: NPD Group/NPD TRSTS Video Game, 1998; data collected at twenty retail chains, representing 75% of U.S. sales.[15]

Audio CDs account for about 65% of the $38.7 billion spent for pre-recorded music in 1999, according to the International Federation of the Phonographic Industry.[16] DVD technology is also a threat to this market. Not only can more music be stored on the same disc, but videos, interviews, alternative arrangements, notes, scores, and other information can be produced along with the music, creating a much richer environment than would ever be possible on CD or CD-ROM.

BBTV: Broadband Digital Television

BBTV is designed for the digital cable tier of programming and the midband DSL telephone company services. It ranges from sub-VHS (less than 1.5 Mbps) to over-the-air broadcast TV quality (3-5 Mbps). It makes more sense

Table 5.3 Total Video Games:
Dollar (Millions) Growth by Category

Product Category	% Change 1995/1996	% Change 1996/1997
32/64-Bit TV Video SW	349	159
TV Video Hardware	77	34
All Accessories	27	96
Portable Video SW	−29	11
Portable Video HW	−31	7
8/16-Bit TV Video SW	−39	−49
Total	24	51

Source: NPD Group/NPD TRSTS Video Game, 1998; data collected at twenty retail chains representing 75% U.S. sales.

to call it midband programming, and a few companies do refer to it this way, but for now, the term "broadband" seems to be the popular designation.

The reason producers are beginning to create broadband content is because about half of all U.S. TV households will have the ability to receive it by 2003 or 2004. People may not sign up for a broadband connective as quickly as one might think. According to the Television Bureau of Advertising, there are about 100 million U.S. households that have a television, or 98.2% of all households. As of February 2, 1999, Jupiter Communications made the following estimates of how many people will be online, and even by 2002, it may only comprise a little more than 61%, and only about 11% would have more bandwidth than that provided by a POTS (Plain Old Television System) connection, as shown in Table 5.4.

Most broadband programming will be interactive in some way, but it doesn't have to be. Much of the material being produced is reality-based in some way, such as news, sports, shopping, and advertising. At the 1999 Broadband Developers' Conference, held at Sony Studios in Los Angeles, a number of speakers confirmed that they found it much easier to design information services rather than entertainment programs.

Going against the trend, however, Showtime Network commissioned Big Band Media to produce a broadband episode of *Star Gate* that included additional video streams to show interested viewers parallel story lines. In the story, a small band of time travelers are captured while others escape. The main story line is about the efforts of those who fled to

**Table 5.4 U.S. Online Households
by Access Technology (in Millions)**

Access Technology	1997	1998	1999	2000	2001	2002
ISDN	0.2	0.3	0.5	0.8	0.5	0.3
Satellite/Wireless	—	0.1	0.2	0.4	0.7	1.2
DSL	—	0.2	0.5	1.4	2.1	3.4
Cable Modem	0.1	0.5	1.2	2.9	4.9	6.8
Dial-up	27.4	35.1	41.5	43.7	46.8	49.6
Total	27.4	36.1	43.9	49.2	55.0	61.3

Source: Jupiter Communications.

free their jailed captain. In the broadband version, viewers can visit the captain in his cell whenever they want to see what he is doing while the others try to get him out.

Even when broadband DV is entirely interactive, the viewer can choose to be passive and simply watch the video as it is presented. In most of the applications under development in late 1999, the "main" video stops while the interaction takes place and resumes when the viewer comes back to it. For the creators of programming, this ability to make sure that a story or presentation unfolds the way they designed it and that everyone sees some common material is very important.

Looking at video over the Internet will give viewers a sense of what broadband content is like, if they can make the imaginative leap to pristine, colorful, vibrant, smooth, full-motion, full-screen video and CD-quality sound. Like Internet video, the footage is shown in a window, surrounded by data, usually with at least intermittent opportunities for interaction. Broadband digital television may be streamed like Internet TV, or it may simply be encoded as an MPEG bitstream. The systems it plays over, digital cable tier and telco DSL services, do not have unlimited capacity, so bandwidth is still a concern. Producers need to combine the appeal of TV with many Internet tools to compress and format AV streams to match the platforms over which they reach consumers.

BBTV is very difficult to create because it is a new medium. It combines the professional requirements of television with interactive elements that are not yet well-understood. One reason is that only a few consumers in system trials have had the opportunity to receive this kind of content. Program creation requires an entire team, if not the interaction of two or more aesthetically-oriented teams. The audio and video portions of the content are assembled by people with good traditional production skills, and the interactions and the programming required to realize them are usually put together by a technology-savvy group.

For BBTV to reach an audience, a high-speed connection must be in place. The specific platform—telephone line, cable network, or wireless reception—matters less than the carrying capacity of the connection into the home. The customer needs a high-speed modem to receive and decode material. While the system setup calls for BBTV to be displayed on the PC, there are several products that allow consumers to port it to their TV as well.

Digital Television over Telephone Lines: ITV and TTV

Telephone networks are the most pervasive of all communications infrastructure, reaching into virtually every corner of the world. One reason for the rapid global penetration of the Internet is the result of its having the ubiquitous telephone system as its transport. The two types of digital television over telephone lines, Internet TV (ITV) and telephone TV (TTV), began from completely different directions. But what they have in common is that massive amounts of information have to be squeezed to fit narrow copper telephone wires.

It's not surprising that compression protocols developed for narrowband videoconferencing were quickly pressed into service for Internet TV. Other factors have brought the two telephone TV types together as well. The Web browser put a standard communications interface on PC desktops, sweeping aside many of the proprietary videoconferencing programs. And both TV types now occur mostly at the PC, as expensive wired videoconferencing facilities were replaced by desktop connectivity to company-wide local area networks.

The price tag for videoconferencing has declined from several million to a few hundred dollars. And Internet television, which began as mere slide shows, improved as modems got faster, compression algorithms got better, and digital cameras got cheaper. The result is that while there is still a distinction between the two forms of digital television, the gap is closing quickly.

Internet Television: ITV

Many Internet activists believe that eventually all television will be delivered over the Internet. And they may be right. It won't happen in the next five years, but it could happen in the next decade. At minimum, more and more digital video will be available over the Internet, probably including films as well as TV. As early as September, 1999, copies of a popular movie, *The Blair Witch Project*, were floating on the Net.

ITV is just in its infancy. In 1993, the World Wide Web brought graphics to the Internet. Within a year, there were slide shows.

Netcasting began with "webcams" delivering a series of still photos to users' computers. The first known webcam was the famed coffee pot at Cambridge University in the UK. A video camera was focused on a coffee pot used by generations of grad students at the university, famed for its commitment to technological innovation. It can still be seen on the Web.[17] By 1995, low-resolution, low-speed video appeared in the RealAudio player. At the same time, dialup modem technology was gradually improving, moving from 14.4 kbps to 28.8 kbps and, finally, to 56 kbps.

Today, every kind of video imaginable is available on the Internet. Sports, talk shows, interviews, news events, even drama and comedy shows have proliferated. Probably there are thousands of choices each day. In 1996, the disastrous Mt. Everest climb unfolded online. NASA has shown many liftoffs. The birth of a baby, a young woman's daily life reported from her apartment by a continuously running camera—all these video streams have drawn an Internet audience.

There is also substantial business use of ITV, particularly when timeliness is an important consideration. Many companies are putting live video of stockholders' meetings, product announcements, news conferences, and other events. One popular text application, chat, is going audio. Up until 1999, people in chat rooms typed in their comments and the written comments of others flowed across their screens. In September of 1999, a large portal site, Excite, created a set of audio chat rooms, based on existing standards for Internet telephony. The free service let users join a chat or create their own. The browser automatically downloads any needed software. It works with any Windows 95 or 98 machine that has a microphone and speakers installed; people speak by pushing the Control key. The quality, even at 28.8 dialup modem speed is adequate. Of course, it is clear that it will not be long before graphics and video capabilities follow.

Back in 1994 and 1995, when people first wanted to send out audio and video over the Internet, not only did telephone lines seem too narrow to send AV material, but also the speed of 1200 baud and 14.4 kbps modems was too slow to bring the data into the computer. Until 1996, the only way to get AV was to get it with the "download and play" method. This technique means the entire audio or video file is transported across the Internet to the client computer and stored on the hard drive.

When the whole file is transferred, the user plays it from the hard drive. This method allows the delivery of reasonably high-quality audio and video, even over a low bandwidth network. Unfortunately, at 1996 speeds, the receiver had to wait for a long time (sometimes hours) for the download to occur, plus it required what at that time was a huge amount of disk storage space. (In 1996, a 450-MB hard drive was state-of-the-art in the consumer market.) Moreover, the entire process raised serious copyright issues, since an actual digital copy existed on users' hard drives.

These problems provided incentives for online wizards to develop new techniques to deliver AV material. It was clear that the video either had be digital originally, or analog video had to be digitized in order to take advantage of extreme compression. The extreme compression techniques came from work in fractal and wavelet compression, as well as the H.320 family of standards, which advanced to become H.323 and H.324.

The technique for sending DV across the Internet or other computer network is called "streaming," as shown in Figure 5.4. Streamed video is inherently interactive. As it arrives on PCs, users can stop, pause, rewind, or fast-forward it, just as they would videocassette recorders (VCRs) on their TV sets. In addition, streamed video is played in a window on the PC screen. Frames on either side of the window, and often above and below it as well, contain text and links to other material relating to the video. The user can also go to another Web site altogether, reading something else while the audio/video stream continues to play.

Streamed media is played out from a server. On the user side, enough material is buffered to allow smooth playing until more arrives.

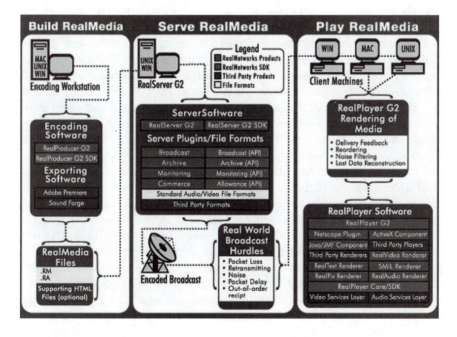

Figure 5.4 The process of streaming video over the Internet.

This lets video and synchronized audio begin playing on the user's machine before the whole file is received. Due to variability in Net congestion, buffering may or may not work, and there can be delays while more material is buffered. Many owners of intellectual property like this kind of streaming because it plays the content almost immediately, and a complete copy may never reside on the receiver's hard drive at any one time. Finally, the streamed video and audio are integrated with text, graphics, and photos, and it all plays inside the Web browser.

Smooth-playing video requires that the streamed data arrives continuously, in the proper sequence without interruption. The term used to describe this kind of time-sensitive data is isochronous. "Iso" means equal and "chronous" means related to time. So isochronous means that the time increments must be equal; that is, smoothly delivered.

It is no easy task to deliver continuous data over a computer network. This type of network is packet-switched, which means that a message is broken down into packets and sent every which way. They take different paths, arrive at different times and in a different sequences than the order in which the original material was arranged. Sometimes packets are lost. Or there are mistakes in them. On shared networks, the packets may lose out to the transport of other, more high-priority packets. The way engineers solved this problem was to create a "playout" buffer. The technique involves measuring the amount of average packet delay and then storing just enough material in a buffer to insure that by the time the buffered video has all played, new packets will have arrived that can play.

In order to display streamed video, the user needs a "player." This is software that accepts the incoming stream and decodes it. It also includes an interface that lets the user customize the way the stream will be displayed—volume, size of video window, and stop, start, pause, fast forward, and rewind controls, as well as a slider bar to access different parts of the stream. The three most popular players are the RealPlayer from Real Networks, Windows Media Player from Microsoft, and QuickTime from Apple Computers.

Creating streamed video is fairly simple, well within the capability and financial reach of the average video hobbyist. The process starts with the video, which must either be recorded with a digital camera or digitized. The video is compressed and encoded in the same format as the software players that the desired receivers have on their PCs. The two most popular formats and players are RealPlayer and the Windows Media Player, although many others exist as well. Once encoded in these formats, the material can go to users around the world. (See Figure 5.5.)

Internet video is one reason consumers want to get residential broadband services. This factor is leading to the dominance of the Web browser across multiple DTV markets such as eTV, PConTV, and BBTV, as well as ITV, positioning it as the broadband graphical user interface (GUI)

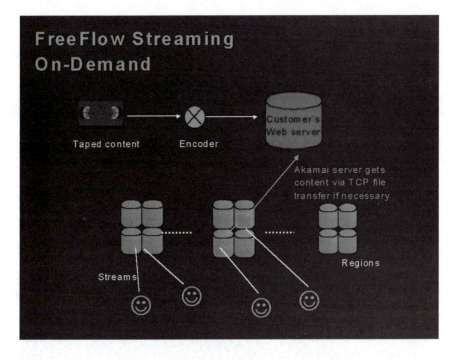

Figure 5.5 Streaming video around the world. Source: Akamai.

of the future. Web browsers are already entrenched on the desktop as universal GUIs for current services such as E-mail, file transfers, transaction processing, and Web surfing. And they are easily extended to function as broadband GUIs for accessing electronic program guides (EPGs), video on demand (VOD) content orders, and control over VOD material, such as rewind, pause, and fast forward.

ITV has some problems. The biggest one is the inefficient way video reaches users over the Internet. For the most part, every stream of video that a user requests goes out as an individual stream from the server to the user: one request, one user, one stream. If 100,000 users want to see the video, 100,000 streams must be sent out. This method is called "unicast," or "video on demand" (VOD), as shown in Figure 5.6.

It makes sense for little-requested material to be unicast, but it is a wildly inefficient use of bandwidth to unicast popular material. For example, when Victoria's Secret webcast a fashion show, it crashed the server at the outset. When the promotional trailer for the movie *Star Wars* was released on the Internet, many university networks were brought to a standstill and the entire Internet was slowed.

The alternative to unicasting is multicasting, a technique that falls between broadcasting and unicasting. Its way of distributing video brings the streams to servers that are closer to end-users, usually the Internet Service Provider (ISP). The ISP notifies people when the material is "live" and they "tune in" to the stream. I will cover this topic more extensively in Chapter 9, "Wired Bitpipes," but for now, it suffices to say that multicast reduces bandwidth requirements through the network because the minimum number of packets are replicated to service a multiplicity of receivers. Mulitcasting also reduces connection requirements, which alleviates overhead imposed on the network and the server.

Telephone Television (TTV)

The dream of sending images over telephone wires dates back several decades. In 1927, the first television pictures were transmitted over telephone lines, and the first commercial TV transmission between Philadelphia and New York was also carried that way. In 1956, Bell Laboratories announced that they had succeeded with an experiment to transmit pictures and sound over a regular telephone line, which became the product called the AT&T Picturephone. Ultimately, the Picturephone was the centerpiece of AT&T's Picturephone Meeting Service, that linked people in two or more locations, providing a mix of audio, video, fax, and data.

Businesses in particular considered the potential of being able to conduct meetings without paying the expenses associated with travel. Today, most videoconferencing is still business-related, with the most common uses being departmental meetings, interviews of prospective

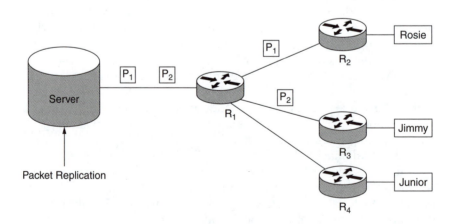

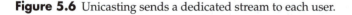

Figure 5.6 Unicasting sends a dedicated stream to each user.

employees, corporate training, sales and marketing meetings, product development meetings, and depositions. In the beginning, videoconferencing costs were prohibitive and the image quality was poor.

In 1984, it cost $1,500 per hour to operate a point-to-point videoconference over a specially set-up leased T-1 digital telephone circuit. Today, a company can lease a dedicated T-1 line for $1,500 a month—and even less through an competitive local exchange carrier (CLEC). The price of equipment has followed the same precipitous downward curve as leasing costs.

Throughout the 1980s and 1990s, many videoconferencing systems were developed to satisfy this market. Initially, companies spent millions on special meeting rooms with built-in cameras, then paid for expensive telephone lines to link the facility with another similar room. Over the decade of the 1990s, equipment costs dropped from millions, to hundreds of thousands, to thousands of dollars, and facilities shrank from whole rooms, to "huddleware," to nothing more than the computer and small camera on the desktop. There are even videoconferencing kits for consumers for under $200.

Many parts of this market are now being absorbed by the Internet. A good example of the way digital innovation has moved the videoconferencing market is provided by VTEL, which originally marketed proprietary videoconferencing systems. In 1991, VTEL introduced the first open PC-based videoconferencing system enabling application sharing over videoconferencing. In 1992, the company introduced a H.320 system that standardized communication modes throughout the industry, and in 1996, it brought out a Windows-based PC videoconferencing platform. In 1999, the company launched its Galaxy systems that use an HTML-based GUI running on Windows, allowing it to operate over the Internet and other networks. (See Figure 5.7.)

Summary

The proliferation of distribution systems has led to an explosion of formats of digital television, although several of them are likely to merge together in the next few years. Such convergence is to be expected. When a platform emerges, its promoters do what they need to do to get content for it in order to differentiate it from their competitors. But as the technology becomes established, its similarities to other platforms appear, and the content can cross from one to the other with little re-working. Ultimately, standards evolve that make it possible to create content once and send it across multiple platforms.

So far, we have looked at the digital future and how most of our communication, information, and entertainment needs will be delivered to us by digital technology. Armed with a basic understanding of digital basics

Figure 5.7 VTEL's Galaxy, used for application sharing and videoconferencing. Source: VTEL.

and the varieties of DTV, let us now turn to an exploration of interactivity, one of the significant benefits that digital communication brings us.

Notes

1. S. Schoenherr, "Recording Technology History Online," at http:// history.acusd.edu/gen/recording/digital.html (revised February 24, 2000).

2. F. Takashi, "High-definition television systems," *Proceedings of the IEEE* 74:4 (April 1985):646-655.
3. "Video product sales picture perfect for the month of July." Press release from the Consumer Electronics Manufacturers Association (August 11, 1999). Available at http://www.ce.org/.
4. "DMX Music Service first DBS user of AC-3 audio." *PRNewswire* (January 17, 1994).
5. "Intel and PBS collaborate on digital broadcast programming," press release from PBS (April 6, 1998). Available at http://www.pbs.org/insidepbs/news/intelpbs.html.
6. T. Swedlow, *Enhanced Television: A Historical and Critical Perspective*, commissioned by the AFI-Intel Television Television Workshop (July 1999). Copy available through Intel Corporation at http://www.intel.com.

   Tracking Internet usage is like describing the location of a speeding bullet. See http://www.nua.ie/surveys/ for up-to-date information about Internet usage and demographic data.
7. Reuters, "AOL, TiVo hook up on interactive TV service," *Los Angeles Times* (June 15, 2000):C3.
8. Flory Bramnick, Vice President, Strategic Alliances and Business Development, Belo. Television Group, telephone interview, June 2000.
9. Michael Lambert, founder and CEO, iBlast, telephone interview, June 2000.
10. Panasonic maintains a history of DVD at http://www.panasonic.com/industrial_oem/computer/storage/dvd-ram/about/history.htm.
11. J. Wilcox, "Who will win the DVD standards fight?" *CNET News* (August 10, 1999).
12. For technical data and links to other buckets of information about DVD, see http://www.videodiscovery.com/vdyweb/dvd/dvdfaq.html#6.4.
13. J. Brinkley, "Disk vs. Disk: The fight for the ears of America," *New York Times* on the Web (August 8, 1999). Available at http://www.nytimes.com/library/tech/99/08/biztech/articles/08disk.html.
14. For a good resource for links to sites with information about CD-ROM technology, see http://www.tardis.ed.ac.uk/~psyche/cdrom/.
15. Information about the NPD Group may be found at http://www.npd.com/.
16. These and other statistics about the music industry are posted at http://www.ifpi.org.
17. Check out the Trojan Room coffee pot at Cambridge University at http://www.cl.cam.ac.uk/coffee/coffee.html.

6

Two-Facing TV:
The Interactive Promise

> ... it is the potential for interaction that is one of the things that distinguishes the computer from the cinematic mode, and that transforms the small electronic [screen] from a novelty to a powerfully gripping force. Interaction is the physical concretization of a desire to escape the flatness and merge into the created system. It is the sense in which the "spectator" is more than a participant, but becomes both participant in and creator of the simulation. In brief, it is the sense of unlimited power which the dis/embodied simulation produces, and the different ways in which socialization has led those always-embodied participants confronted with the sign of unlimited power to respond.[1]
>
> —Allucquere Rosanne Stone

At the Digital Coast 1999 conference, held at the Directors' Guild in Los Angeles, a panelist demonstrated a broadband program that his company had designed. When he reached a certain point, he stopped the presentation and told the audience that viewers would be offered an "interactive e-commerce opportunity." The audience of Hollywood creative people groaned. For most of the companies that fund interactive content for any of the digital television platforms, e-commerce is the point. Executives get excited, their faces become animated, and their voices fervent: "Imagine watching *Ally McBeal* and being able to click on what she's wearing, see it in different colors, and order it!"

Even for a medium where interaction is native, like the Internet, creating interactions is difficult. Even transactions, a truly fundamental interaction type—give me this for that—are often so poorly designed that the majority of online shoppers abandon their efforts to make a purchase. Some sites have long, cumbersome forms to fill out. Others fail to make their prices easy to find, so shoppers must go into the purchase process to get basic information, and back out again when they get it. Many sites fail to offer policy pages that will overcome consumer concerns about product returns, privacy, and transactional security.

Interactivity has proved to be one of the most elusive elements of effective design of digital media interfaces, content, and services. One reason for the difficulties is that interaction is so instinctive, it is invisible to most people, most of the time. Face-to-face interaction is so inherent and natural that individuals who fail to engage in it—hermits, the mentally ill, and substance abusers—are viewed with suspicion, even alarm. Interactivity also implies a bewildering array of potential experiences, technologies, contexts, and users. For example, consider the following examples:

- A child in San Francisco races in real time against three other kids who are at home in Virginia, Finland, and Singapore—it's *Sega Rally2* time, on a Sega Dreamcast player that is hooked up to the television set for display and to the Internet for connectivity.
- On her computer screen, an art director in New York can see a director in Los Angeles, her own image, and a mutually-shared "whiteboard" where the TV commercial they are working on is playing. They can draw and make notes over the video and see one another's comments—and they can talk to each other on the telephone at the same time.
- A student puts an encyclopedia on CD-ROM into the drive on his computer and plays the listing under Romania, listening to music and watching a short video clip of the capitol city, Bucharest.
- A financial analyst has CNBC playing in a small window of her computer screen. A fan of the Los Angeles Raiders pushes a button to watch the tight end in a close-up shot, in place of the wide-angle shot aired by the network.
- A Smothers Brothers fan watches shows that were censored by CBS in the 1960s on the Internet, seeing the show on demand whenever he wants.
- While watching a boxing match on Showtime, a viewer goes to the Showtime Web site and scores the fight along with the judges. A few minutes later, the judges' scores and viewers' scores are compared over the air.
- A boating enthusiast bids on an outboard motor on eBay.com.

Remarkably little effort has been expended in exploring what makes interactions appealing, fun, exciting, and satisfying experiences. Yet people will only repeat interactions that bring them pleasure, the foundation for long-term success in any communication situation.

Many current creative efforts do not go beyond the work done in the early 1990s for the interactive TV test trials and CD-ROMs. Sometimes, when there is little progress in a given field of endeavor, it helps to go back and look at basic assumptions. This chapter will explore what is known about interactivity, with the goal of raising levels of understanding and provoking greater experimentation to create better, more challenging, and more involving interaction sequences.

Dimensions of Interactivity

It is difficult to bring much conceptual clarity to the subject of interactivity. The term is used by companies that design and manufacture equipment, by content providers, and by creative people. And interactivity is a complex process that has prompted research in more than a dozen fields, including artificial intelligence, business, communication, computer science, cybernetics, education, electrical engineering, ergonomics, human factors, industrial engineering, information science, library science, linguistics, medicine, and psychology.[2]

The word itself provides a starting place. "Inter" is a prefix that means between or among. "Interact" means to act mutually or to engage in reciprocal acts. Reciprocal does not mean identical. You bow, I curtsey; she argues vociferously, he assents quickly. In the same way, although interaction involves reciprocal actions, they may vary greatly depending on the situation, the interactors, and the technology of communication.

Note that there must be at least two entities for interaction to take place. In addition, the interaction takes place within some context: historical, social, political, economic, architectural, and so forth. The context is one of the main determinants of who has relationships with whom, and of the rules under which the relationships are initiated, developed, conducted, and terminated.

There are many other ways to consider the characteristics of the interactions: Is response rapid or slow? Is the overall length of each response long, short, or varied? Does one party have control, is control shared, or is it passed back and forth from one interactor to the other? Do responses relate meaningfully to initiations? Is the mode of interaction verbal, textual, visual, or acoustic? What is the purpose of the interaction and is it related to the content? What is the result of the interaction?

Compare broadcast television (a minimally interactive medium) with the telephone (a highly interactive medium). When a viewer watches a program, there really is no interaction except for the ability to change channels or turn off the TV set. Neither participant has much control over the interaction. The programmer has the opportunity to air a show in a vivid visual format, and hopes for the best. Viewers are restricted to text. They can write to the network and will receive a form letter in return. To the extent that this can be called interaction at all, it is slow. While the viewer's initial message may be highly personal, the broadcaster's response will be an impersonal, corporate letter. The viewer may be responsive to the network, but the network is more responsive to advertisers and viewers in the aggregate than to the individual viewer. On the part of the network, the purpose of the interaction is to aggregate an audience. For the viewer, it is to relax and be entertained or informed.

By contrast, the telephone allows rapid interactions of variable lengths. Control is accessible to both parties and they are able to be highly responsive to one another. The format is audio and the purpose is connection and exchange.

The above examples both involve communication exchanges where technology links the communicators, a form called "mediated communication." When a technology is interposed between two interactors, it often makes interaction possible—but it also imposes constraints. For example, two-way radios only allow one person to talk at once, allowing no reassuring "um-hm, um-hms" in the back channel; the telephone as it is typically used eliminates the visual element from communication.

The influence of a particular technology in mediated communication has led some scholars to define interactivity as a characteristic of a medium. For example, one researcher offers this definition: "a measure of a media's potential ability to let the user exert an influence on the content and/or form of the mediated communication."[3] Subdimensions that fall under this explanation are:

- Transmissional interactivity—The potential offered by a one-way platform for viewers to choose from a continuous stream of material. Interactivity is very restricted and includes such actions as changing channels, or reading teletext or datacast material.
- Consultational interactivity—The potential offered by a two-way system for viewers to choose and receive material from an existing selection: content on-demand, CD-ROM material, FTP, WWW, information services, etc.
- Conversational interactivity—The potential for a two-way system to allow people to create and distribute their own content, such as chat, E-mail, news groups, voice telephone, etc.
- Registrational interactivity—The potential of a system both to accept input from a user and to respond to that input. This category includes: home shopping and banking; intelligent agents, guides, and interfaces; and surveillance systems.

One important way a particular medium or communications technology influences interactivity is its channel capacity. A given channel may offer participants a greater or lesser ability to communicate within a specific interaction, allowing them to send or receive certain types of information. When they have the same channel capacity for expression, it is called "symmetrical." Symmetrical capacity (or bandwidth) gives both parties an equal ability to send and receive information. The opposite condition is "asymmetrical bandwidth," where one person has a channel that permits more information than the channel used by the other person.

Another characteristic of a medium is whether the interactors are networked together or are using a standalone or stranded device. The telephone is a networked reception device; television sets and radios are stranded devices. The computer can be either networked or stranded, and users move back and forth between the two conditions with a few keystrokes. A stranded technology is "offline," a networked one is "online."

You might think that it isn't possible to have interactivity without being networked, but it is. Interaction can take place via the pony express, the mail, semaphore, and smoke signals. It may take a long time for the interaction to take place, and it may be unsatisfactory in some respects, but it is still interaction. And that's just human-to-human communication; there is also human-computer interaction (HCI), where the user interacts with software and data through some kind of interface. In a sense, this communication really takes place between humans, too, in that software programs are artifacts of someone's endeavor; however, HCI is the usual term for this type of interaction.

Feedback: The Heart of Interactivity

Feedback is a term that came out of early studies of whole systems, a field called cybernetics. Feedback is a signal that goes back to the original communicator, informing the initiator about the effect of the message. It is a relatively new addition to the venerated communication model published in 1949 by Shannon and Weaver.[4] Originally formulated to help design more efficient telephone networks, the model was linear and one-way, which made sense for that application. Over time, it became clear that Shannon and Weaver's work could be extended to all forms of communication, and their famous representation was dubbed the SMCR model, an acronym for the elements in the process:

Noise
|
SOURCE —>MESSAGE —>CHANNEL —>RECEIVER

In this formulation, a message from one source passes through a channel and reaches a receiver. There may be some noise on the line, interfering with the transmission—that's it, end of story. It's a simple and lucid model. The model worked well for analyzing the adequacy of telephone technology, because technicians working out of their local central offices concerned themselves with only the channel and its potential to pick up noise, not the sources and receivers. There was no need to consider feedback.

However, when researchers began studying real messages, it became clear that the SMCR model needed modification to include feedback. In a conversation, both parties to the interaction alternate as sources and receivers. The stream of communication is composed of sequential messages; the latest message is a response to the previous message, and earlier messages affect later messages. The messages are not independent of one another because some of them serve as feedback to previous ones. Of course, not all messages are direct feedback—people do change the topic to introduce entirely new avenues of conversation.

Here is how researchers might categorize a set of statements between two people to incorporate feedback:

Message 1: Hi!	Source to 2
Message 2: Hi!	Receiver to 1/feedback to 1, source to 3
Message 3: How are you?	Receiver to 2/feedback to 2, source to 4
Message 4: Fine. You?	Receiver to 3/feedback to 3, source to 5
Message 5: Good. How's your family?	Receiver to 4/feedback to 4, source to 6

Feedback is the key to understanding the above exchange, and it unlocks the mysteries of interactivity. It is central to any framework for sequential interaction. In some ways, feedback is somewhat mysterious. It is a message, but it's a message with a difference: It is a change agent that automagically turns the source into a receiver and the receiver into a source. Feedback differs in how much of it there is, how often and how quickly it occurs, and what format (text, numeric, aural, or graphic) it takes. It can be positive (deviation amplifying), negative (deviation correcting), neutral, or nonexistent (which is another kind of feedback, too). Feedback can be verbal, vocal, and nonverbal.

Current Interactive Communication Technologies

The next section offers some descriptions of technologies defined by the kind of feedback they enable, in terms of bandwidth symmetry and connectivity, as shown in Table 6.1.

Table 6.1 Media Bandwidth and Connectivity Characteristics

Connectivity	Symmetrical (Bandwidth up = Bandwidth down)	Asymmetrical (Bandwidth up ≠ Bandwidth down)
Networked	Communicative Media Telephone, fax , chat, LANs and some WANs, some high-speed cable and DSL implementations, 21st century game consoles.	On-Demand Media Some high-speed cable and some DSL implementations. Wink, Intertainer, WebTV via satellite or asymmetrical wired system.
Stranded (Standalone)	Information Media PC hard drive, CD-ROM, DVD, offline E-mail.	Entertainment Media Broadcast, cable, and satellite TV, DVD, game consoles.

Symmetrical, Networked Interactive Channels: The Communication Media

The telephone, fax, E-mail, bulletin boards, Internet chats, and online multi-user domains, some local and wide area networks, and some DSL implementations fit the description of symmetrical media. All these uses involve some kind of direct interpersonal communication. It's not surprising; symmetrical, networked channels make people "equal in the eyes of the technology," so to speak. Each party is equally constrained or empowered. Inequalities may result from the individuals involved or the situation in which they find themselves, but not from the technology itself.

One symmetrical activity—face-to-face interaction—is the most common model for all communication, the very standard against which communication and interaction are judged. During World War II, Alan Turing helped develop prototype computers used to decode German radio traffic. In 1947 he wrote an influential paper, "Computing Machinery and Intelligence," in which he introduced the "Turing test."[5] He argued that the ultimate test for a computer would be for a human being, interacting with a computer, to be unable to tell that the machine wasn't another human being. The "Turing test," the ability to mimic face-to-face communication, is still the accepted measure of the intelligence of a machine.

The most important symmetrical interactive technology is the telephone, and, of course, the same technology underlies the Internet. The telephone provides for person-to-person interactions; the Internet makes

Figure 6.1 Net phones are just one new way people will access the interactive Internet. Source: Comtek.

possible person-to-machine and machine-to-machine interaction as well. (See Figure 6.1.)

Broadband networks extend this capability by enabling people to see and hear one another in real-time—recreating many of the characteristics of face-to-face communication.

The Silicon Graphics (SG) Studio network is an advanced implementation of a broadband symmetrical, connected network that demonstrates how people will use such platforms in the future. Designed for professional creative people in the entertainment industry, including producers, directors, and editors, they allow people to collaborate on visual material from multiple locations. They are able to watch footage or graphics at the same time, drawing on the material on-screen and discussing it on the telephone simultaneously. If their computer has a camera attached to it, they can also see themselves and the person on the other end in small "thumbnail" windows along the side of their monitor screens. In the middle of their screens, they see the "virtual projector," where they view the footage they want to work on. They then "cut" a few frames at a time and "paste" them on a shared "whiteboard." Both parties can draw on the whiteboard and see what the other has drawn. In addition to the visual interaction, the link allows the interactors to carry on a voice telephone conversation.

Symmetrical, Stranded Interactive Channels: The Information Media

The most common use of this kind of channel is PC computing, including such activities as information retrieval, processing, and formatting. Accessing statistics from a database stored on a PC's disk drive or a CD-ROM are other examples of this type of interaction. In a sense, even some E-mail behavior fits here, since people sometimes download their messages to their local machine, and read and compose responses some time after the original E-mail was received by their ISP.

The cost of networking bandwidth drives ISPs is to move information as rapidly as possible to the local machine. As long as the cost of bandwidth is higher than the cost of local processing, there will be a tendency to move information to the edge. However, in the next decade, there may well be an abundance of bandwidth, altering the ratio of costs. In that case, many stranded operations will move to the network because it will be cheaper to do it that way. In the 1990s, when bandwidth was expensive relative to processing, many libraries subscribed to services that sent them new CD-ROMs with updated information on a regular basis. Students accessed the material on the CD-ROM through a local campus network, saving the cost of connecting with an off-campus location like Nexis-Lexis or other repository. In the future, such searches will probably move online.

Asymmetrical, Networked Platforms: The On-Demand Media

Asymmetrical networks are common, and anyone who has ever ordered pay-per-view over a cable system has experienced using this type of platform. Cable video on-demand and high-speed Internet access services, DSL implementations, Internet access over satellite, and multiplayer video games are other implementations. When a channel is asymmetrical, it always means there is more downstream capacity than upstream capacity. If there's more capacity going upstream than down, there is a slow server or net congestion, sometimes called "net l-a-a-a-a-a-g." This is because, from a network architecture perspective, network backbones are enormous, like tree trunks, and as the branches fan out further and further towards the final destination, they become ever smaller.

There are degrees of asymmetry. Upstream capability may range from a customer's ability to send a few bits to a fairly robust bitstream of

1.5 megabits. Systems that support high-speed Internet access are likely to have more upstream capacity, while those that carry only video-on-demand or e-commerce need only a very narrow, low capacity return path. Shopping, for example, requires that the consumer be able to see color photos, and video is likely to be even more persuasive. But information from the buyer involves little more than a way to indicate the chosen product and its desired features—size, color, quantity, and so on—and the buyer's information, such as name, credit card number, delivery address, and phone number.

The current architecture of most cable systems allows for some small amount of interaction, enough to order pay-per-view movies. Of course, the systems that still don't have any upstream capacity at all ask viewers to use the telephone. They dial in to a computer that sends a pulse to their set-top box, opening the pay-per-view channel in the customer's home at the right time. The ability of the computer to make changes in the consumer's set-top box is called "addressability," which means the operator can send a message to the STB, providing or denying access to any of the channels on the cable system.

Telephone companies face a different network design problem. They can support interactivity, but their networks were built to handle symmetrical narrowband messaging. To increase capacity in either or both directions, they must install costly electronic equipment on the telephone line and at their central office where the line terminates. From the central office, they usually offload data traffic (as opposed to voice traffic) onto a network specifically designed to handle it. All this retrofitting costs money, but telephone companies are spending it because two-way cable systems will be able to handle telephone traffic as well as broadband programming. In order to compete, the telcos must provide for broadband interactivity as well.

Intertainer is an example of a company that is working in this space. The company's user interface is shown in Figure 6.2. It offers a video-on-demand service to operators of asymmetrical networked platforms like the cable digital tier and telephone xDSL networks. The high downstream bandwidth will transmit the films; the smaller upstream bandwidth is used by customers to order them.

Intertainer requires dedicated broadband capacity, transmission and interface software, an advanced set-top box, and storage on the consumer premises. The service will price recent-released films at $3.95, and hopes to grab part of the lucrative $8 billion video rental market brought in by retail stores like Blockbuster. In addition to films, Intertainer offers music concerts and original NBC programming, such as news, *Saturday Night Live*, and *The Tonight Show with Jay Leno*. Intertainer has made deals with every major Hollywood studio, except Paramount (owned by Viacom, which also owns Blockbuster). These content owners will get 50% of the revenue; Intertainer will then split the remaining half with the platform provider, the cable or telephone company.[6]

This company has chosen to put its material on this type of platform because the management does not believe that the Internet will be able to handle video, even with high-speed access. The Internet sends material in bursts rather than a steady stream, which Intertainer execs believe adversely affects the Web's ability to transport high-quality video images. Nevertheless, the company has been surprised to discover that about half of its early subscribers watch its programming on their PCs rather than TVs.

Asymmetrical, Stranded Interactive Channels: The Entertainment Media

The premier models of this kind of channel are the familiar broadcast media, radio and television, and other entertainment devices, DVD players, and game consoles. It is probably arguable that broadcasting is not interactive at all. About all viewers and listeners can do is to mail their responses to programmers, which involves a three-to-five-day return loop.

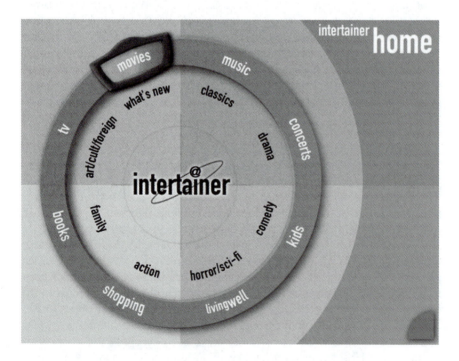

Figure 6.2 Intertainer video-on-demand consumer interface.

DVDs and game consoles are also stranded machines that deliver high-quality animations to the TV screen, allowing users to input only small datastreams. However, game players won't be standalone for long. The new Sega Dreamcast and the Sony PlayStation II will both incorporate modems that connect them to the Internet so their owners can hook up to multiplayer games. They will also have significant multi-gigabyte storage capacity that will increase over time until they are essentially computers dedicated to game play. Some analysts believe that Sega and Sony will eventually make a play to replace existing products in both the TV set-top box and PC markets with their high-powered machines.

Other stranded devices are getting connected as well. Take CD-ROMs. Many of them are standalone devices—the players we carry around with us to play music wherever we go. And they are also drives in the desktop computer, playing music and data on local machines that may or may not be networked. In the fall of 1999, the Comcast cable system began trials of the Arepa service on some of their networks, which allows users to download information or entertainment from CD-ROMs that sit at the cable headend.

Stranded devices wouldn't necessarily need to be networked to incorporate more interactive features than they currently do. They could keep a record of user moves and characteristics and use them to determine the direction of the story or game. But little use has been made of the ability of computers to "learn" about their human users—our "personal computers" aren't very personal at all.

In the future, computer programs will incorporate better and more complex dimensions of interactivity. One stumbling block is the lack of understanding of how to design better interactions. We will consider this problem later in the chapter.

Ground Zero for Interactivity: The PC and the Internet

We have now considered some of the important interactive technologies and their characteristics. By far the most important of them are the PC and the Internet. The telephone system is undeniably powerful as well. However, its effects have already been absorbed into every facet of modern life, while the final tale of the new technologies is still unfolding.

Until the last few years, most people thought of the PC as a computational device. But since the emergence of the Web, it is clear that PCs are also a powerful communication platform. The public acceptance of the Internet has moved computing into a new era: computers as broadcast, point-to-multipoint devices. The stages in the evolution of the personal computer and the characteristics of those stages are shown in Table 6.2.

The rise of the PC and the Internet have put a spotlight on interactivity. The next section will explore the variety of technology-mediated interactions and the programming that provides these experiences.

Table 6.2 The Evolution of Computing

	STAGE 1 300BC *Abacus-Babbage-PC*	STAGE 2 1961 *DARPA network*	STAGE 3 1994 *WWW*	STAGE 4 2015? *N-Dimensional Networking*
Form of usage	Thinking: Standalone computation	Interpersonal communication	Combines interpersonal and mass communication	Integrated communication
"Killer apps": Popular implementations	Data processing and storage	Text E-mail	Web surfing, webcasting: push and rich A/V media	Sun's concept of Java; Motorola's LON networks
Sender/ receiver format	User/programmer	Point-to-point: one-to-one, one-to-few	Point-to-point or multipoint: one-to one, or one-to-many	Combines all: adds multipoint-to-multipoint or many-to-many
Interface	OS and application software	Browser	Interim: plug-in, media player Final: TBA	TBA
Hardware and infrastructure	A computational device	Computer, modem, telephone line	Computer, modem, high-speed network	Ubiquitous information appliances and richly connected high-speed networks
Image	Extension of individual brain and nervous system	Extension of a personal network: wheel and spokes	Extension of a social network: clustered network	Unknown: A geodesic sphere?

The Human Factor in Interactivity

Understanding interactivity leads to an exploration of some of the most basic of all human behaviors. At the same time, it is not at all simple. Systems theory offers a special perspective on the interaction that involves mediated communication, because it bridges both biological and technological systems.

Earlier I stressed the importance of feedback in interaction and communication. In systems theory, feedback is one of the principal aspects of any living system. Researchers of human-computer interaction have defined two types of systems that are based on feedback: cybernetic and homeostatic systems.[7]

A good example of a technology that incorporates a cybernetic feedback loop is a thermostat. The goal is to maintain a room's atmosphere at some specified temperature, say 72 degrees. When the temperature drops below 72, the furnace goes on; when the temperature moves above 72, the furnace shuts off. Homeostatic feedback loops define a balance between two inputs. The balance may fluctuate within some range, and is determined by the interaction itself. Feedback guides the system towards this balance.

A relationship between two people provides a good example of how homeostatic feedback loops work. Both members provide input. The relationship fluctuates between intimacy and distance, openness and exclusion, and other emotional poles determined by the participants. Goals play a big part in systems theory, and the purpose of any given communication is also a major determinant of the form and meaning of messages. Research in interpersonal communication finds that individuals interact for two overarching reasons: affiliation and dominance.[8] Another way of stating these constructs is love and power. People affiliate because it is the only way they can raise themselves above the level of hermits, preoccupied only with feeding, clothing, and housing themselves. They desire power because it gives them feelings of control and security.

Affiliation refers to closeness and intimacy in a relationship. When individuals seek to affiliate, they establish commonalities, similarities, and links between one another. Although differences in social status and personal characteristics exist, there is a movement towards equality of participation and commitment on the part of those in the relationship. Dominance refers to the distance and differences between people in relationship. The dominant person seeks to control the other, based on inequalities in power and status. In everyday life, we expect parents to dominate their young children, supervisors to control their staff, and stronger personalities to influence weaker ones.

What Are Interactions?

So far, we've learned a bit about interactive media, some of the dimensions of interactivity, and a little about what it might mean in human communication. But before moving ahead, we need to step back just for a moment and address some other considerations of what interactions are and what they do. Essentially, interaction is the way people create relationships, which are formed and sustained through repeated interactions over time. Some researchers have gone so far as to say that relationships are no more and no less than the total set of interactions between people.

Interactions occur in some format: linguistic, nonverbal, face-to-face, and mediated. For the most important interactions, people tend to meet for face-to-face conversation, but they also employ many other forms, such as letter-writing, E-mail, signing, gestures, and even Morse code. There is a structural similarity between relationship, interaction, and conversation. All involve at least two distinct entities engaged in mutually coordinated activities over time. In the theoretical terms of communication, we have a source, a receiver, a message, and feedback. Of course, there may also be channels, or media, and the noise that accompanies such mediation.

What kinds of interactions are there? A small group of bright graduate students or a convention of screenwriters could come up with hundreds of variations of possible interactions, but the number of types is much smaller. When the two kinds of feedback are put together with the two reasons people have for interacting, a typology of interaction types results, as shown in Table 6.3.

Transactions

Transactions are the most simple and predictable of all interactions. They're exchanges where people buy, sell, and trade. In today's convergent media world, they are already quite common, on the Internet, on cable with pay-per-view services, and even on broadcast TV shopping channels. The appeal of a transaction is simple: The participants get something they didn't have before.

Table 6.3 Typology of Interactions

	Types of Feedback	
Human Motives to Interact	*Cybernetic*	*Homeostatic*
Affiliation	1. Transaction	3. Correspondence
Dominance	2. Simulation	4. Comparison

Transactions fill the basic human need to have and to hold. People eat, work, earn, and learn through contact and exchange with others. They are affiliative—people come together to arrive at an agreement. The structure of an interaction is cybernetic because the two parties negotiate their way to a mutually rewarding point then stop. (The overall structure of a market is homeostatic in that prices tend to hover around a particular point in time, governed by forces of supply and demand. But individual interactions can vary considerably.)

The initial stages of a transaction require the most information and the most bandwidth. Transactions can be divided into three phases: initiation, negotiation, and exchange. From a message point of view, each stage requires progressively less descriptive information than the last. In a mediated transaction, pictorial information crucial. Successful marketing requires that the buyer be able to examine the merchandise, and sellers have an incentive to provide the best-looking images they can. In TV shopping programs, hosts enthuse at length about the wonderful qualities of the products on display.

The negotiation and exchange don't require anything like the bandwidth that the initial stage does. (See Figure 6.3 showing Wink's narrowband service.) Pictorial information is not nearly so important in the negotiation phase. The language at this second stage is direct and relatively terse. The final phase, coordination of the actual exchange, entails discourse that is virtually devoid of fancy rhetoric. On e-commerce sites, the last stage is particularly bleak, usually consisting of a series of sterile form fields.

Consider a transaction on eBay.com. A user puts up a product for sale, usually supplying color photos. Prospective buyers looking for this kind of product examine the pictures and, if they are interested in the item, bid on it. The introduction is a color photo; the bid is nothing more than an amount of money. The exchange is a money order. In the case of physical goods, there is another stage—fulfillment—that needs a "bandwidth" only the Post Office or a delivery service can provide.

Considering the communication requirements for a transaction, it is probably not an accident that interactive shopping has begun over broadband broadcast and cable television with the telephone providing the interactive feedback medium. Nor is it surprising that e-commerce did not take off until the advent of the World Wide Web brought graphics to the Net.

The problems of transaction interactions are the classic difficulties inherent in commerce and trade: consumer protection, misrepresentation, fraud, and transactional security. Managers of such services must establish policies to address these concerns before launching services or severe legal and public relations problems and other difficulties can result. Indeed, these very issues are the most formidable barriers to e-commerce.

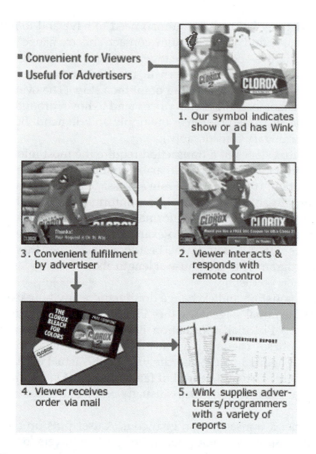

- Convenient for Viewers
- Useful for Advertisers

1. Our symbol indicates show or ad has Wink

3. Convenient fulfillment by advertiser

2. Viewer interacts & responds with remote control

4. Viewer receives order via mail

5. Wink supplies advertisers/programmers with a variety of reports

Figure 6.3 Buying on Wink's system using narrowband remote control communication upstream.

Simulation

Simulations include such activities as games, computer-aided instruction (CAI) and design (CAD), and environments for exploration. They are cybernetic in that there is some desired state, that has been defined beforehand. Feedback guides participants to create, find, or maintain fidelity to that standard. They are dominance oriented because there are external criteria for success and failure, good and bad, and almost always some kind of rating mechanism for measuring outcomes.

Simulation interactions fulfill the human need for challenge, mastery, and control. Their formats range from simple question-answer dialogues to complex choices in detailed fantasy environments. As a result, the language of simulation is so varied as to defy description. However, just like real life, control and dominance may be given to the users as a result of their roles, such as computer engineer or company executive, or the power may be earned through strategy and expert knowledge, as structured in some games.

Images make or break simulation interactions. When they are well-executed in a well-designed environment, simulation experiences can be extraordinary. In commercial products, the higher quality the images, the more likely a product is to succeed. However, they are not essential. Early text-based multi-user domain (MUD) games were very involving, and evoked detailed images in the minds of the players using only words.

Simulations re-create some reality (even an imagined one). In a broadband environment, the means to achieve fidelity is extraordinarily powerful. For example, players of a game in which they are virtual fighter pilots will prefer one that conveys a rich, detailed environment and realistic action and speed. Similarly, computer-aided design is more valuable when the designer can incorporate as many features as possible of the constructed object.

Interactions are always better when it is possible to provide rapid feedback as it is needed. But it is especially important in simulations.[9] In game playing, for example, the more sensitive and rapid the feedback, the more involving and intense the experience of the game will be for the player. Game players have a goal, so they always want to know how they're doing with respect to their goal, through feedback. You make a move . . . good news, you're winning. You make another move . . . bad news: you're losing.

Games are among the most popular Internet sites—Gamesville.com, NTN.com, Shockwave.com, WebRPG.com, and the portal sites, Yahoo!, Go.com, and Entertaindom.com all offer many games online. They are also important as elements to programmers of interactive lineups on broadband system. The Sega Channel has a channel for broadband networks, which downloads Sega games to players over the Internet, and is developing multi-user networked games for the new DreamCast console.

Computer-aided instruction is another example of the action in this quadrant. You progress through a module. At the end, you are asked questions. You answer and get a message: "Congratulations! You're right, you can advance to the next module," or "Oops, you're wrong—repeat this module, please." In computer-aided design, the user's work must conform to physical, architectural, or ergonomic standards. The sooner the system lets users know they are violating those standards, the more useful the program will be.[10]

Learning is a big business on the Internet. Offline, students pay $50 billion in tuition every year. Now dozens of colleges and universities are offering virtual classes, and there are a variety of sites that feature learning for work and for hobbies. Learn.com, Learn2.com, 7thstreet.com, and

Lightspan.com are all services that hope to be part of a market that International Data Corporation estimates will be worth $6 billion by 2002.[11]

An interesting feature of simulations is that users, players, and viewers believe they have control—the interactions are designed to create that belief. But in fact, control resides with the system, because fidelity requires that the interaction follow within the parameters of the mimicked reality as represented by the system. Violating the parameters will reveal who (or what) is in charge very quickly. However, designers should recognize that users are a canny, cunning lot who will invest many hours in figuring out how to defeat and humiliate tyrannical systems.

The problems of simulation are those of excess: addiction and violence. Addiction is a social and personal problem that has significance beyond the management of interactive services. For interactive managers, addiction is a special problem because it puts users and providers in an adversarial relationship when users spend more time on the system than their budget allows. Violence in interactive media has already become an issue with respect to video games and CD-ROM titles. An ongoing stream of news reports about excessive violence keeps this issue in the public mind. Indeed, the potential for addictive involvement, combined with realistic violence, is cause for some concern. Questions about these effects, particularly on children, will continue to be raised about interactive media, perhaps even more vociferously than they have been about television.

Correspondence

Correspondence has two meanings here. First, it means to send messages to someone over time, as in an "active correspondence" with a pen pal. Second, it means to move towards increased similarity, as in "his quietness corresponds to her talkativeness." Here, both meanings apply.

Correspondence interactions are affiliative in that they encourage and facilitate social contact. They are homeostatic because there is no preset outcome. Users talk, engage in therapy, teaching, collaboration, confrontation, and conflict. There is no exterior judgment of success, and the participants themselves make their own evaluations at the conclusion of their interaction. The interactors decide when a conflict is resolved, a lesson learned, a therapy session concluded. If users want to stay up all night, there is no score at the end to show who "won" the discussion. The goals of correspondence interactions are determined by both of the participants, emergent from the interaction itself, and adapted to their personal needs. This flexible outcome has no fixed goal, as contrasted with transactions and simulations, which have goals like winning a game, retrieving information from a database, buying a compact disc, or passing a course.

The basic human need that correspondence serves is that of personal identity. People learn who they are through contact with others. The "handles" adopted by users are an interesting index of this function. On one bulletin board, for example, one user calls himself "A_Cool_Guy," and countless others who go by "Terminator," "Liberator," and "Sexy_Lady" populate chat rooms.

Like correspondence outcomes, correspondence language is emergent and adaptive, based on the user's knowledge and mental image of the communicative partner. Users evaluate others' motives and abilities and construct a language register to accommodate them. The role of pictorial information in correspondence is to establish and enhance intimacy. However, there are questions about whether the images serve that purpose. In an early broadband experiment in Japan called Hi-Ovis, people didn't like being on-camera, as it required them to arrange their appearance for a performance.[12] The lack of consumer acceptance of AT&T's Picturephone in the 1960s also creates questions about what forms of visual communication can be successful.

However, it is clear there are some applications where images are welcome. In business dealings, being able to see the other person you are negotiating with is important enough for companies to spend thousands of dollars on travel for a single meeting. Further, while people may not want video phones for ordering the Thanksgiving day turkey from the butcher, they are eager to see their loved ones who may live far away.

In other applications, such as routine transactions, play, or exploratory communication, it is possible for subscribers to have their own graphic representations, such as avatars. Some networking programs allow a subscriber to create a personal avatar offline in a drawing program, then import it into a message. Once the avatar is sent to the other participants, only changes to the avatar are sent: smiling, frowning, crying, jumping up and down, bowing, and so on.

One of the most interesting applications of correspondence interaction comes from Starbright, a nonprofit group that delivers high-tech entertainment to sick children and their caregivers. (See Figure 6.4.) Starbright began by providing bedside carts with videogames. However, one of the biggest problems sick kids face is isolation.[13] Headed by producer Steven Spielberg and retired General Norman Schwartzkopf, the group expanded to develop a worldwide online network, linking children, their parents, and their medical caregivers.

The Starbright Network first launched in November, 1995, linking together five U.S. hospitals. The two-way broadband network now reaches more than 100 hospitals, providing software that allows the children to design their own personal avatars to represent them in communication with children located in any of the network locations. Together, the kids roam (or at least their personal avatars do) around sophisticated rendered environments, communicating with one another through voice and text.

Figure 6.4 Steven Spielberg shares some fun on the Starbright Network.

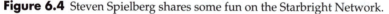

Interpersonal problems arise for services that provide correspondence interactions, such as harassment, electronic stalking, and domination by "netbozos." Netbozos, the derisive term that "netizens" use to describe offensive users, launch personal attacks, hog conversations, digress from preestablished conversational threads, and send many, lengthy messages in annoying communicational blitzes. Uncontrolled, netbozos will drive sensitive, thoughtful, and efficiency-oriented users from correspondence services. For example, netbozos caused many Santa Monica public officials to stop logging into the nation's first governmentally-sponsored network, the Public Electronic Network (PEN).[14] Stalking has been another difficulty. Online services have developed policies to protect their (mostly female) subscribers from others (mostly male), including message-tracking that prevents anonymous messaging and identity and address protection.

Comparison

A metaphor for this quadrant might be the Wicked Queen questioning her mirror, "Mirror, mirror on the wall, who is fairest of us all?" People search the social landscape to understand who they are and how they fit in.

While correspondence is a search for personal identity, comparison is quest for social identity. People learn how they fit (or don't) through comparison interactions that allow them to register and compare their ideas, opinions, answers, votes, predictions, decisions, and traits with those of others. Subscribers choose populations, subgroups, and other individuals with whom to compare themselves.

There are no examples of broadband interactive services that are comprised purely of comparison interactions, although many services offer some opportunities for comparison within the framework of other types of interactions. The Cincinnati public opinion Qube system and electronic town halls with telephone call-ins are examples of programming that feature comparison interactions. On the Internet, there are some fascinating examples. Some sites let people take hundreds of tests online—on personality, values, sexuality, leadership, and so forth. Two sites that offer many such tests are www.queendom.com and www.psychtests.com. Flexibility in comparison is important. In some comparisons, the value rises with the number of respondents. If a subscriber wants to know how their online use compares with that of other Internet users, it tells them more if the system logs all users, not just a few. However, the subscriber may also wish to make more local comparisons, for example, with members of their department or with just one other member.

Unlike the other types of interactions, the language of comparison can be numeric as well as verbal and graphic. Some people have difficulty understanding numerically-represented information. Statistical results will be more interpretable if they are verbally explained or represented by such devices as bar and pie charts to make them more comprehensible, so sufficient bandwidth to allow graphics is important to the comparison activity. In addition, there is a trend in many business organizations to illustrate research results with video. Presenters will introduce a category of consumers, then show a short video of actual consumers talking about their preferences. Although bandwidth for high-resolution graphics and full-motion video may not be as crucial for comparison interactions as it is for simulations and transactions, pictorial information has its uses here too.

The problems with comparison are those of manipulation. Data distortion, fraud and misrepresentation, and a false view of public opinion due to biased, self-selected samples can all occur. For example, the temptation to influence public opinion in a high-speed communication environment, especially when commercial stakes are high, may prove irresistible. System managers should prepare for attempts to manipulate comparisons by means fair and foul. Already, manufacturers have been hurt badly by highly contagious, Net-generated public opinion. One familiar example is the controversy over the flaw in Intel's Pentium chip, which a company official acknowledged was greatly inflamed by communication over the Net. Table 6.4 summarizes the four interaction types.

Table 6.4 Comparison Chart of the Four Interaction Types

Model Type	Activity/ Programs	Need	Communication Style	Uses for High-Res Video	Problems
Transaction	Buy, sell, and trade	To possess	Moves from florid and descriptive to terse	Sales appeal	Trade problems Fraud and misrepresentation Consumer protection
Correspondence	Communicate with others: Talk, teach, co-orient, collaborate	Social contact Personal identity	Attribute of participants	Intimacy	Interpersonal problems: Netbozos Harassment
Simulation	Games Computer-aided instruction and design Information retrieval	To win, master, control, and achieve	Too varied to classify	Verisimilitude	Problems of excess: Addiction Violence
Comparison	To compare with others: Opinions, traits, votes, predictions	Social identity	Information must be processed, aggregated, and presented: Text explanations, charts graphs, tables	Credibility	Problems of manipulation: Public opinion, voting fraud, data distortion

Combined Interaction Types

On the Internet, there may be some sites that specialize in a single interaction type, but it is far more likely that they are combined in some way. It turns out that mixing and matching may turn out to be good business practice. AT&T operated an early prototype interactive network using 140 employees and their families. The project studied what kinds of interactions would appeal to consumers.[15]

Research showed that the type of interaction was less important than the overall implementation. Successful interactions included a combination of four features: entertainment, information, transaction, and communication. In other words, it's fun, you learn something, you get something, and you relate it to someone else.

The AT&T sample isn't representative of consumers in general and must be interpreted with care. However, the findings do suggest that putting together an interactive system calls for multiple interaction types rather than a single style. At the same time, an important point to consider is that when interaction types are combined, both the positive appeals and the complicating problems of each of the combined types will appear in new forms.

A fascinating example of a popular Web site that combines interactions is at www.thepalace.com. The site offers users free software to build their own "mansion" on the site. There are tens of thousands of such domains designed by users. Corporations, entertainment companies, and celebrities have all put material on Thepalace.com. Comparison plays a greater role on Thepalace.com than it does on most sites as creators show off their rooms.

From the beginning, the company anticipated that chat (a correspondence activity) would be the "killer app," the activity most appealing to users that would draw them to the site. Executives have been surprised as they have watched people's behavior evolve into something they would call "collaboration," rather than pure chat: It's not so much communication as it is group activities. This change is especially marked among younger visitors. One example is where a group of young girls will "make a baby" upon request, drawing a picture that can be used as an avatar by the person who receives it.[16]

Revisiting Cybernetic Feedback

Cybernetic feedback, shown in the first column in Table 6.3, includes the two goal-driven interactions, transactions and simulations. These two types give participants a sense of success and failure, but in different ways. With a transaction, the parties reach a deal together—or they don't reach one at all. If they do, it means that they have found an equilibrium point where an exchange is mutually beneficial. But in simulations, results are most often posed on a continuum—how many points attained out of 10,000, for example. Also, users often reach their goals independently, although not necessarily so.

Given the goal-orientation of this column, it is not surprising that these interactions are often framed as tasks, even when the job is to rescue a maiden from a virtual dungeon in a simulated castle. The human reward from cybernetic interaction is a sense of accomplishment from a goal achieved, a job well-done, or a game intensely played and, perhaps, won.

The Homeostatic Dimension

Correspondence and comparison are characterized by a user's search for identity. These interactions are characterized by emergent properties, which means that the participants probably have internal goals, but they are not structured in advance by the force of interactions themselves. Rather, users identify, alter, and realize their goals through emergent outcomes, and the optimal system adapts to the needs and desires of the users. Since the participants continuously negotiate both what the communication is about and its purpose, only in retrospect can interactors (or observers) identify and label what actually happened. If the activity is correspondence, users will involve themselves in teaching, therapy, collaboration, and plain old conversation, but this does not mean that they do not argue, confront, conflict, and engage in hostile discursive forms such as flames and rants. In comparison interactions, users seek information that allows them to compare their own ideas, opinions, and other attributes and characteristics with those of others.

The Affiliative Orientation

Different researchers will argue that affiliation results from drives, desires, needs, or expectancies rather than goals, but there is general agreement that sociality is fundamental to communication. People join together to fulfill their material, psychological, social, aesthetic, ludic, sexual, and novelty needs and desires. People who affiliate together usually already have something in common—culture, social class, education, and outlook Often, they share interests, bonds, network ties, and occupation as well.

Historically, most exchanges have taken place on a one-to-one basis—one person to one person, one group to one group, or one organization to one organization. When a group exchange occurred, one individual acted as a spokesperson for the group. Networked communication differs from this classic model because the technology allows both one-to-many and many-to-many conversations, as seen in chat sessions and games.

The main feature of affiliative communication is the search for agreement, coordination, and co-orientation. "We agree to disagree," is a familiar example of a conflict that cannot be resolved at the substantive level, yet affiliation continues at the relational level. Exchanges occur only when mutual benefits exist, and they do not continue for very long in their absence.

There are few conversational markers for affiliative behavior. Some examples are the mutual deference of greeting and leave-taking behaviors ("Hi, how are you? Hope you're well"; "Yes, I'm fine, I hope you're also well"), reciprocal compliments ("I like your new haircut"; "Thanks. Nice jacket you're wearing"), and expressions of liking and caring. Conversational exchanges between equals exhibit considerable reciprocal discourse about the communication process itself, shared interests, personal topics, and direct and indirect signals of affection.

The Control Orientation

Like affiliation, there is disagreement about whether control is a need, drive, goal, or expectation. However, there is agreement that it is basic to understanding why people form relationships and communicate with each other, as well as with computers. In the affiliative dimension, the activity is exchange; in the control dimension, the distinctive activity is the creation of social distance.

In many instances, evaluation is the mechanism for invoking distance between actors. Evaluation is a judgment with success/fail or rank-ordering criteria. In a two-player game, one player wins and the other loses (or the user loses and the system wins, or vice versa). Or the scores of several players may be ranked, as in a multiplayer game. In an opinion poll, respondents' views fall along a normal curve and users discover whether they are in the majority or the minority.

Notes on Passing the Turing Test

Recall that the Turing test is a measure of how well a machine holds a conversation: if a human being cannot tell if he or she is talking to a computer, the computer passes. Generally speaking, interactive services from content and network providers and computer software are not even close to getting a "C."

Considering what it might take to implement interactivity better, some general lessons have been demonstrated. Take the experience of the early Prodigy online service in the late 1980s. At that time, Prodigy conceptualized users as information-starved masses, yearning for data from a centralized source. The company designed a complex, elegant system to deliver such data.

Unfortunately, the growing online community was just discovering the delights of group discussion on Usenet, chat lines on local BBSs, and the convenience and close contact of E-mail.[17] Users rejected the centralized Prodigy model in no uncertain terms, demanding the lateral, user-to-user communication the company had ignored. Prodigy lost nearly a billion dollars and most of its advanced users to more flexibly-designed competitors, especially America Online. AOL itself was designed before the explosion of the Internet and the World Wide Web, and their architecture continues to have problems interconnecting with Internet applications and activities.

Some lessons can still be learned from these early experiences.

- People would rather communicate with others than be communicated to by others;
- Interactive services should consider a bundle of offerings rather than a single one;
- Locking system design into a particular mode of service is dangerous in this era of rapid technological advance and heightened consumer awareness;
- Embedded design flaws are exceedingly costly.

One way to find out how to create better interactive programs is to ask people who use them. For example, in one study, the people in the research received health information from a computer.[18] The subjects' responses presaged what interactive content providers are struggling to offer online customers now. They said that they valued the computers ability to gather information about them and to monitor their learning and responses to the material. People wanted the structure of the interaction to be similar to a face-to-face interaction. But the most important aspect of the PC was its ability to "provide the information wanted, when it is wanted, in the order wanted, and at a desirable speed."

The notion of structuring interaction like face-to-face communication holds a great deal of promise for today's designers of interfaces, content, and applications. There is a vast reservoir of untapped research about interpersonal communication, relationships, and conversational interaction that has been compiled over five decades that should be brought to bear on these issues. The power of today's computers makes it possible to build a profile of users that would account for a wide range of user preferences and behaviors—and to update them over time. Cognitive and affective styles, actual versus stated preferences, weaknesses in understanding, and skills to be cultivated are all areas to identify, track, and incorporate into interactions with users.

Addressing the way consumers communicate (not just their credit rating, psychology, and educational level), as well as understanding their deepest needs and desires, offers service providers immense benefits. For instance, studies of interpersonal communication indicate that interactors would find the characteristics of a highly involved communication style—immediate language, greater certainty, and more relational pronouns—very attractive.[19] Interactive environments rarely (if ever) incorporate what is known about the temporal aspects of relationship and communication. Time is crucial to the processes of forming, developing, and maintaining interactions.[20] Even though the relationship may be between a computer program and a human being, people are accustomed to a relational progression that affects how they expect interactions to change.

Initial interactions differ greatly from those that take place after a relationship has been established. First impressions give rise to expectations that influence the way one person approaches the other. If that impression creates insecurity, the relationship may never move to deeper levels. The communication in initial interactions tends to be about surface facts, rather than intimate thoughts, beliefs, or dreams. There is some discussion about the communication process itself and about each person's goals and intentions in the situation.

As the relationship progresses, participants reveal more intimate information. They discuss negative thoughts about themselves and others. For example, they may admit to shortcomings, disappointments, and critical appraisals of others. Interaction requires coordination. The later additions to the information stock about the other allows the interactors to anticipate and shape their actions to fit their partner's knowledge, wants, attitudes, beliefs, goals, and perspectives.

Communication continues to change as relationships become established and intimate. A private language evolves that allows minimal responses, curtailed verbal expressions, and a rich gestural and paralinguistic repertoire. A wide gap develops between conversations that take place in public and in private. Tensions arise along the predictable fault lines of change and stability, connectedness and autonomy, affection and control. Some people may dominate private discussion, while others may disappear from it altogether in a mutual decision to repress potentially explosive disagreement.

People have distinct individual preferences when it comes to conversation. They like to be able to interrupt—but they dislike being interrupted. They like short pauses, and they want the amount of talk coordinated with the amount they talk. Finally, people prefer predictable patterns in both the amount and style of talk, particularly in the punctuation of speech and silence.[21] This last point is easily demonstrated by some international calls that are controlled by TASI (time-assigned speech interpolation) technology. TASI computers eliminate silences, making it very difficult for speakers to interpret nonverbal turn-taking cues and to coordinate their conversations.

System designers should take advantage of the enormous publicly-available research base in relationship-building and conversational analysis to create interactions that are natural, involving, and pleasurable—because people like participating in interactions that are easy, fun, and engaging.

Special care must be taken with the popular transaction and correspondence interactions. Designers need to make exchanges between users as flexible as possible, enabling participants to shape the system to their own uses and ends. Here, too, they will find a large research base to guide them in understanding how people coordinate, coorient, and collaborate.[22]

The design of interactive programming and services is still rather primitive. Interface designer guru Bill Buxton notes that the facilities in the men's room at O'Hare Airport can sense when customers are using them, and automatically run water when they've finished. As Buxton puts it, "Why should the urinals at O'Hare be more interactive than your PC?"[23] The short answer to Buxton's question is: The PC (and online interactions) should be infinitely more interactive—and personal—than they now are.

Stephen Acker makes a strong case for two design features: flexibility and incompleteness.[24] Services need to be flexible because we humans are a creative, clever, opinionated, and stubborn lot. We want it our way and we will do whatever we have to in order to get it. In the end, we will abandon a system that fails to address our needs. The second principle is that of incompleteness. This practice lets users participate in the design of their own system. We need only look at our own works to realize that we love and we cling to nothing so much as our own creations.

One great advantage that interactive systems have over one-way systems is the ability for users to participate, to create, to invent, and to reinvent. History is rife with the creative ability of end-users to alter technology to their own purposes, including such familiar communication devices as the telephone and the radio. Providers can turn user creativity to their own advantage and put users to work for them. Yogi Berra coined the phrase, "faced with an insurmountable opportunity." This sentiment accurately describes our initial attempts to harness the extraordinary computational power of today's systems.

Multidimensional connectivity between people and each other and users and computers are innovations that will sweep us into the twenty-first century and change us and our world. We must design for the future as well as the present. As Michael Schrage wrote: "The web of technology tacitly or overtly complicates our relationships with others. If technology isn't busy framing our interactions, it's busy mediating them. To a very real extent, the quality of technology determines the quality of our interactions."[25]

Summary

Interactivity is a fascinating topic that reaches into human beings' deepest desires for personal connection and communication. It also meets many of our most pragmatic needs, from shopping to orienting information that we need to survive. At the beginning of the new millennium, the majority of U.S. households receive many channels of TV but they are able to engage only in the limited interactivity provided by the telephone. This severely asymmetrical condition prompted Mitch Kapor, developer of the popular computer program, Lotus 1-2-3, to describe the American home's communication system as one that had "a multilane highway going in and a cowpath coming out."

Over the next decade, broadband connectivity will be available to almost everyone, making mediated visual telepresence a reality. Connections between people have consequences, both wonderful and terrible, that occur over technological platforms, just as they do in face-to-face relationships. It will bring in its wake difficult issues of privacy, misbehavior, consumer protection, copyright and intellectual property rights, and system security. Already, all these problems have arisen on the Internet, and they will continue to grow and evolve as broadband connectivity becomes more common.

Notes

1. A. R. Stone, "Will the real body please stand up?: Boundary stories about virtual cultures," in M. Benedikt, ed., *Cyberspace: First Steps*, Cambridge, MA: The MIT Press (1991):81–118.
2. C. L. Borgman, "Theoretical approaches to the study of human interaction with computers," unpublished paper, Institute for Communication Research, Stanford University (1982).
3. J. F. Jensen, "The concept of interactivity," *Interactive Television*, Aalborg, Denmark: Aalborg University Press (1999):25–66.
4. C. E. Shannon and W. Weaver, *A Mathematical Theory of Information*, Urbana: University of Illinois Press (1949).
5. A, Turing, "Computing machinery and intelligence," *Mind LIX* (1950):11–35. A good contemporary discussion is found in M. Shapiro, "Control and interactivity in perceptions of computerized information systems." Paper presented at the International Communication Association, Montreal, May 1987.
6. Jonathan Taplin. Telephone interview, August 1999.
7. S. Acker, "Designing communication systems for human systems: Values and assumptions of 'socially open architecture." *Communication Yearbook 12*, Newbury Park, CA: Sage Publications (1989):498–532. The original source of the categories was N. Wiener, *Cybernetics: Control and Communication in the Animal and the Machine*, New York: Wiley (1948).
8. M. Argyle, *Social Interaction*, Chicago: Aldine (1969).

9. G. Arlen and P. Krasilovsky, "Got a minute? And the need for speed?" *Convergence* 10 (June 1995).

10. C. L. Blaschke, "CAI effectiveness and advancing technologies: An update," paper presented at International Communication Association (1987).

11. For access to this proprietary research, contact International Data Corporation, Five Speen Street, Framingham, MA 01701, (508) 872–8200, http://www.idc.com.

12. L. Becker, "A decade of research on interactive cable," in W. Dutton, J. Blumer, and K. Kraemer, eds., *Wired cities: Shaping the future of communications*, Boston: G. K. Hall (1987):75–101.

13. S. Spielberg, telephone interview, August 1996.

14. J. Van Tassel, "Yakkety-yak, do talk back. Santa Monica's PEN System," *WiReD* (January 1994).

15. J. K. Keller, "AT&T secret multimedia trials offer clues to capturing interactive audiences," *Los Angeles Times* (October 6, 1993):C1.

16. L. Samuels, telephone interview, September 1999.

17. M. L. James, C. E. Wotring, and E. J. Forrest, "An exploratory study of the perceived benefits of electronic bulletin board use and their impact on other communication activities," *Journal of Broadcasting and Electronic Media* 39:1(1995):30–50.

18. D. Gagnon, "Toward an open architecture and user-centered approach to media design," *Communication Yearbook 12*, Newbury Park: Sage (1988):547–555.

19. R. M. Warner, "Speaker, partner and observer evaluations of affect during social interaction as a function of interaction tempo," *Journal of Language and Social Psychology* 11:4 (1992):253–266.

20. I. Altman and D. A. Taylor, *Social penetration*, New York: Holt (1973).

21. J. Capella, "The Structure of Speech-Silence Sequences," *Human Communication Research* (1982).

22. J. K. Burgoon and J. L. Hale, "The fundamental topoi of relational communication," *Communication Monographs* 51 (1984):193–214.

23. J. Mountford, "Essential interface design," *Interactivity* (June 1995):60–64.

24. S. Acker, op. Cit.

25. M. Schrage, *Shared Minds*, New York: Random House (1988).

7

From Network TV to Networked TV: The Convergence of Television and Telecommunication

Whether by one telephonic tendril or millions, they are all connected to one another. Collectively, they form what their inhabitants call the Net. It extends across that immense region of electron states, microwaves, magnetic fields, light pulses, and thought which sci-fi writer William Gibson named Cyberspace. Cyberspace, in its present condition, has a lot in common with the nineteenth-century West. It is vast, unmapped, culturally and legally ambiguous, verbally terse (unless you happen to be a court stenographer), hard to get around in, and up for grabs. Large institutions already claim to own the place, but most of the actual natives are solitary and independent, sometimes to the point of sociopathy. It is, of course, a perfect breeding ground for both outlaws and new ideas about liberty.

—John Perry Barlow[1]

Flash forward to the year 2015. The ubiquitous broadband network is a reality: Schools, hospitals, most businesses, and the majority of U.S. households are connected to a network capable of carrying all manner of information: encrypted data, voice, graphics, and video. (See Figure 7.1.) Other technologically advanced nations, including the United Kingdom, Holland, Germany, Denmark, the Scandinavian countries, France, Japan, Singapore, and Australia are similarly wired.

Laptops are a product of the past because in most public places in these countries there are kiosks where people can log on to the network, retrieve information from their own home computers, and send information to it. If there is information they need to have with them, it is stored in a small personal organizer that incorporates one of the new dense storage technologies that fit in the pocket-sized information "e-ppliance." Inexpensive videocams are everywhere. People shop at local stores and e-

Residential Broadband Access, 1998-2007

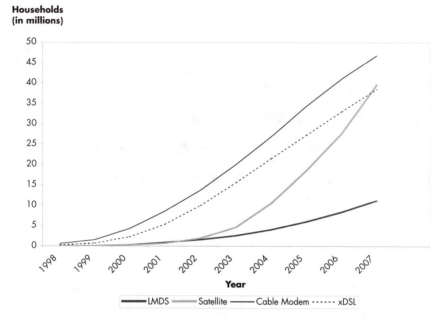

Figure 7.1 By 2007 the majority of U.S. households will have access to a broadband network.

storefronts from their living rooms. They hold two-way video conversations with friends, relatives, and business associates, and use video databases to download and save material. (See Figure 7.2.)

Information flows easily both ways on this broadband network that now links many of the world's businesses and about 10% of its people. In well-connected countries, that percentage approaches 75% of the population. For many people, it's a connected world. But why? What do they get out of it?

The network networkers want to meet people. They use chats to meet others and E-mail to maintain a large circle of electronic acquaintances, colleagues, and even friends. A large network like the Internet also offers an astonishing range of information. Beyond the material that can be found posted to sites, there are discussion lists for every possible aspect of existence. A person can go online with a very specific problem and receive practical suggestions, instruction, pointers to further information, and emotional support. The Net enables a vibrant peer-to-peer connection, powerfully distinguishing it from the more typical top-down

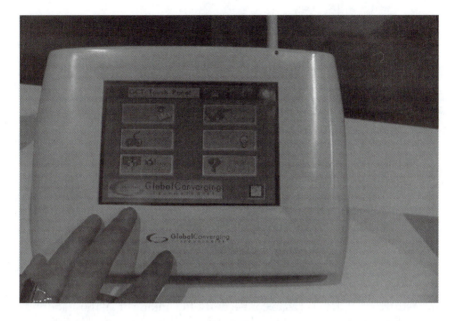

Figure 7.2 Wireless e-ppliances will be located throughout people's homes for always-on access.

communication of the parent-child, superior-subordinate, media expert-audience interactions that occupy so much of people's time.

Finally, there is the sheer pleasure of escapism. Sharing a virtual space with other people feels like a real place in The Real World (TRW). It's eerie; sometimes you might realize you've been online for five hours only when your neck starts to ache.

This chapter looks at the forces of convergence that are putting our lives online. Within the next few years, your car will be a mobile IP address on wheels; you will E-mail your stove, and watch your kids on their school playground during recess via the webcam mounted on the side of the building. The emerging online world will come about through the convergence of video, telecommunication, and network technologies, as shown in Figure 7.2. Two decades ago, convergence was a theory—today, many believe it is becoming a reality. In a few short years, these previously separate communication fields have begun to merge, blur, and come together. Observers have concluded that this convergence is occurring at all levels: physical (systems and equipment), functional (processes), organizational, managerial, and individual. William Stallings provides an overview of the convergence of computers and communication:[2]

- There is no fundamental difference between data processing (computers) and data communications (transmission and switching equipment).
- There are no fundamental differences among data, voice, and video communications.
- The lines between single-processor computer, multiprocessor computer, local network, metropolitan network, and long-haul network have blurred.

One effect of these trends has been a growing overlap of the computer and communications industries, from component fabrication to system integration. Another result is the development of integrated systems that transmit and process all types of data and information. Both the technology and the technical standards organizations are driving toward a single public system that integrates all communications and makes virtually all data and information sources around the world easily and uniformly accessible. (See Figure 7.3.)

This chapter begins with a description of the components of network infrastructure. It then turns to connectivity, the linking of all those individuals who wish to participate in the communication system as individuals and as members of groups, organizations, communities, and societies. It will also explore how network elements and protocols affect the way individuals use and are affected by its use.

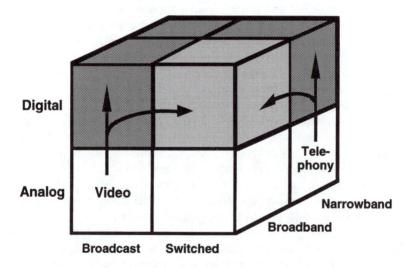

Figure 7.3 Industry convergence to switched digital format. Source: Carl Podlesney, Scientific Atlanta.

The Communication Infrastructure: Hardware, Software, Firmware, and Wetware

"Infrastructure" means the underlying physical, technical, and human organization necessary to carry out some activity. Two methods for delivering television are over-the-air and over wires. Both systems are complex, composed of four types of parts: "hardware," "software," "firmware," and "wetware." Hardware refers to the actual physical components of an information processing system: the "stuff"—boxes, connectors, and cables—composed of metals, plastics, silicon, and other materials.

Software programs control how the hardware operates. However, the usage of the term "software" has changed over time so that today it refers to anything that the user sees on the screen. Thus, software includes Microsoft's Windows program, the Myst CD-ROM-based game, and a movie, such as *Natural Born Killers*. Firmware has characteristics of both hardware and software. It is software that is permanently loaded into memory so that it is like part of the hardware; however, firmware can be reprogrammed so that it is flexible like software. "Wetware" means people, human beings. It is a literary term, coined from cyberpunk futurist fictional genre, and involves a play on a Cold War KGB euphemism for murder, or "wet work." Naturally, using "-ware" as a suffix has led to a proliferation of such words. For example, "bioware" describes technology that gets built into wetware, such as a pacemaker. "Vaporware" is software that doesn't exist, but a company releases a blitz of press releases to trumpet their future achievements, a good example of "hypeware." A disgruntled employee might refer to her unpleasant supervisor as a "sicko piece of nastyware" and split pea soup might be called "glopware."

Although it is possible to say that each of these types of parts—hardware, software, firmware, and wetware—will play an important role in the evolving high-speed, image-capable, two-way communication infrastructure, no one is sure how any one element will finally become integrated into the whole. The character of the whole system is even less clear. In fact, there is a cottage industry of prognostication that feeds off the aura of uncertainty. However, as the Internet gets faster, becomes more visually adept and ubiquitous, it is possible to observe features that have heretofore existed only in theory.

Connectivity>Complexity

Familiarity breeds contempt, familiarity breeds . . . if nothing else, complication. Connectivity permits people to develop intimacy, closeness, and familiarity by linking them together. When people are close, they often form relationships. Indeed, for most of us, few aspects of our lives are more important than the interpersonal relationships we share with others. By making it possible for people to develop relationships, connectivity leads to social complexity and multidimensional interdependency. This mouthful of syllables simply means that in socially complex networks, people depend on each other in many ways to fill a variety of wants and needs.

Mathematician John Casti introduces a discussion of complex systems with Mark Twain's story of Chang and Eng.[3] Chang was a heavy drinker and his connected brother, Eng, was a teetotaler.[4] Notes Casti:

> *What makes them special even today is the fact that they were two humans linked together in a very unusual fashion, a connection that led to interesting consequences. This connective structure led to a system considerably more complicated than that representing the typical man or woman. So in trying to understand the complicated system "Chang and Eng," it's essential to take this connectivity into account; Chang and Eng can't be understood by thinking of them as two disconnected individuals.*

Connectivity leads to complexity by linking the members of a set to one another along one or more dimensions. The number of contacts that members of the set make is only one dimension along which members could be linked. They could be multidimensionally connected by type of relationship, type of communication content, depth of emotional bond, amount of liking, proximity, and many other variables, such as how often they lift a brew, shop, play video games, rob banks, make love, and so on.

When people first connect, their relationship exists only as potential. However, over time, if individuals continue to communicate, they become linked in ways that depend on their characters and the nature of the experiences they share. Generally, the greater the number of unique (or, as Casti puts it, nonequivalent) dimensions that link people together, the greater the complexity of the relationship. When individuals are members of a group of people, the extent to which they are connected—the greater the number of dimensions that link them—the more complex their social system is.

Consequences of Complexity

Ever think about getting rich by becoming a day trader? Who hasn't? Read this section carefully to learn why this new form of transaction has brought a new level of instability to the stock market.

Complex systems differ from simple systems by the presence of frequent feedback. Compare a simple small pond with the huge, complex ocean. Throw a stone into the pond and ripples from its impact make small concentric waves for a few minutes. Then the motion stops. A physicist could probably describe this motion with just a few equations. By contrast, the ocean is in such continuous motion that, even though the concentric waves are present, it is difficult to see them. The motion of the ocean exists, at least partly, because there is constant interaction between the different wave motions. No number of equations could explain oceanic movement because the system is too complex. In other words, the way the ocean moves is unpredictable.

Complex social systems are something like complex biological systems; their complexity doesn't so much result from their size as from the sheer density of connection. Even in a very small social system, when people are multidimensionally connected there are opportunities for constant feedback, which is itself fed back into the system. (See Figure 7.4.) (In systems that have a goal, feedback is positive or negative information about the outcome of an action that allows the system to correct its course.)

Classic systems theory only accounted for feedback, but there's another special kind of time-related loop—"feedforward." Feedforward is anticipatory information about the likely outcome of a potential action. For example, suppose Steven is getting ready to graduate in June and thinks he might ask Fred to help him get a job. In December, he sends Fred a Christmas card. When he runs into Fred at the hardware store, he chats with him about the fascinating class he's taking in parameter estimation in discrete time series models. He checks out Fred's personal home page on the Web and reads in his bio that he was involved in track, so he starts reading about local track events and sends Fred an E-mail, "What about those awesome Framingham High runners?" In short, the anticipatory feedforward communication of a world class suck-up is thoroughly engaged.

Complex social systems also anticipate the results of actions and consider the results of past actions, perhaps many times over. This "feeding back" of feedback is called "iteration." Densely connected social systems are characterized by iterative feedback, so they grow, change, develop, differentiate, and evolve, depending on the type and amount of iterative feedback. Although this process can be broadly understood, it is too complicated to allow detailed prediction of the effects of iterated feedback.

Casti points out other consequences of complexity: hierarchy, increased number of possible outcomes, influence from indirectly connected set members, and unpredictable behavior. In describing hierarchy, he uses the term "cover set" to mean the highest level and "subset" to refer to the lower levels. Thus, in the hierarchy of the family, the cover set is "family," the parents are one subset, and the siblings are another subset. Much of the unpredictability and richness of complex systems results from the vertical interactions between cover sets and subsets, and from the horizontal relationships between subsets, writes Casti. Everyone who has grown up in a close-knit family or been part of a highly coordinated team can attest to the validity of his observation.

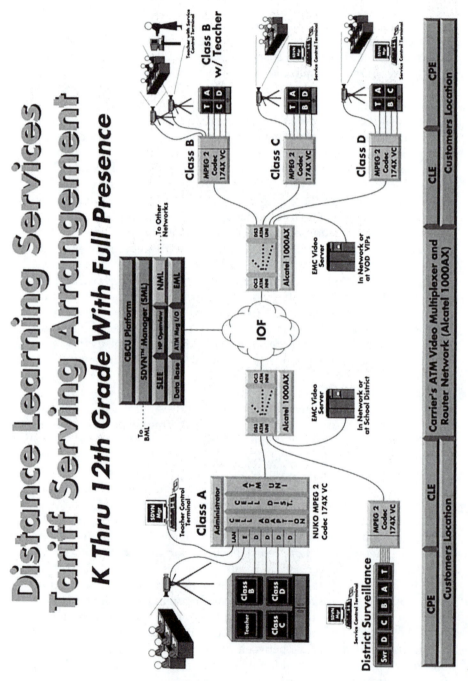

Figure 7.4 Schools are complex social systems that will become even more complex when classrooms are linked to each other as well as the outside world, as in this Alcatel project. Source: Alcatel.

Another potential source of unpredictability is the indirect influence of people not actually in the set, but connected to those who are. A good example of this? An influential financial news service on the Internet publishes a critical article of a company's management—and you can bet lots of day traders will be keying in their sell orders as fast as they can. This influence increases as the number of relationships rises and leads to an expansion of possible interactional options and outcomes for both individual members and the system as a whole. A further consequence of dense connection is that unexpected relationships may emerge from the rich fabric of associations, leading to counterintuitive and surprising outcomes.

Jonathan Gill, a former professor at M.I.T. who served as an advisor on high technology in the Clinton White House, says that another reason connectivity creates additional outcome possibilities is due to an increased "idea space."[5] The greater the number of people who can contribute their thoughts about a topic, the greater the range of ideas for potential action, which enriches the entire social sphere. Likewise, the number of alternatives for individuals is also increased because they are influenced by others to whom they are only indirectly related as well as by those with whom they are directly related.

A mathematical technique called q-analysis demonstrates that a member of a set can be affected by another member who may be separated by several intervening members. The abstract findings of q-analysis turn out to be surprisingly true: Just about anyone in the world is closer than you might think. One famous experiment tested the "Small World phenomenon."[6] The results demonstrated that an individual in one city can find someone they don't know in another city quite easily. Subjects in Omaha and Boston were given a letter with the name of a person who they did not know written on the envelope and told to record the steps they took to locate the mailing address of that person. On the average, it only took 5.2 steps to locate the missing address. For many, it took even less effort: 48% of the searchers were able to locate the unknown person with three calls. Imagine redoing this experiment over the Internet. How many clicks away are you from someone you don't know in Finland? In Alabama? In the urban center 300 miles away? Or twenty blocks from your house?

In addition to the influences of indirect links, there are other differences between complex and simple systems, such as limited divisibility, distributed decision-making, and multiple communication loops. Simple systems can be decomposed into their parts without doing much damage to the overall system because the linkages between them are so minimal. However, taking a complex system apart may well destroy the system. For example, it isn't possible to eliminate the nervous subsystem of a mammal without killing the animal, because the nerves are richly connected and crucial to the operation of the overall organism. Similarly, if just part of a company's Intranet goes down, it can bring the entire organization to a standstill.

Simple systems typically have centralized decision-making, while complex systems operate with many such centers. The pattern of distributed decisions makes for stability in complex systems because any one choice will affect only a small arena. Political scientists have observed the instability in dictatorships and monarchies because of the weaknesses of centralized decision-making.

However, by far the most important difference between simple and complex systems is the fact that simple systems have few components that interact with one another. Complex systems have a large number of components with a high level of interaction, including many feedback and feedforward loops. The presence of so many communication paths may contribute to overall system collapse when internal stresses are present.

Except for distributed decision-making, all the above properties add to the likelihood that complex systems may become unbalanced. According to scientist Ilya Prigogine, when a system grows far from equilibrium, the relationship between cause and effect may cease to be linear, so that small causes may engender large effects and the system becomes unpredictable.[7] This is the problem that worries regulators who are concerned about the effects of day trading.

An example of social unpredictability caused by the presence of dense connection is the Pentium chip debacle experienced by the semiconductor manufacturer Intel. As mentioned in the previous chapter, the company marketed the chip knowing there was a flaw in the math processing, but presumed it would almost never occur, or that if it did, there would be no serious consequences. When a newspaper printed a report of a professor's discovery of the error, discussion of the problematic Pentium raged in every nook and cranny of the Internet.

Initially, Intel said the flaw was unimportant and the company would do nothing. However, Intel's position collapsed in the face of rapid contagion of anger among computer users. Afterwards, observers noted that the extraordinary speed and reach of communication over the Internet forced the company to change its policy. In other words, a richly connected system of computer users increased the instability of Intel's operating environment.[8]

All this means that the connectivity enabled by advanced networks has critical implications for everyone in our society. We are currently connected by the moderately narrowband telephone system, which has influenced virtually every aspect of social life. The emergence of an omnipresent Internet, even though it is still narrowband communication, has changed many aspects of life already. People can meet and contact many more individuals than ever before, exchange information (especially fact-based data), act with greater efficiency due to increased possibilities for timely coordination, and maintain vibrant, ongoing connections between previously isolated individuals. When this capability becomes broadband, telepresence will bring an amazing escalation in the kinds of distance relationships that will exist between people all over the world.

Broadband Connection— Bandwidth>Proximity, Proximity>Intimacy

Consultant Steve Rose coined the phrase "bandwidth equals proximity," and the implications of this aphorism are startling, giving some indication why ubiquitous broadband networks are likely to be so powerful. The more bandwidth a channel has, the more the people using it will find that the communication experience is like being physically close to their partner. Rose's observation conceptually captures a decade-long effort by a British research team at the Communication Studies Group (CSG) to understand how using technology to communicate changes people's messages. The results of their research suggest predictions about how we will use advanced broadband networks and how these systems will affect our patterns of communication.

The CSG research, often called the "Short, Williams, and Christie studies" began with the most mundane of objectives. The researchers were tasked to investigate how Britain could reduce the cost of governmental communication.[9] They asked office workers to rate the various communication technologies they used in the course of carrying out their everyday duties, and to evaluate their responses to several interaction modalities— face-to-face, video, audio, two-way, real-time teletype (now morphed into computer conferencing), and written documents. They classified media as sociable/unsociable, sensitive/insensitive, personal/impersonal, and cold/warm. Finally, office workers assessed each medium for its perceived appropriateness and effectiveness for such tasks as giving and receiving information, idea creation, problem-solving, bargaining, persuading, conflict resolution, and so on.

The studies found that people preferred face-to-face communication, followed by video, audio, and written communication modalities. The higher a medium's rating, the more it was judged to be sociable, sensitive, personal, and warm. The researchers dubbed this cluster of characteristics as a medium's ability to convey people's "social presence," including their verbal and nonverbal communication. These findings suggest that people will probably like communicating over broadband networks.

Social Presence>Social Cues

The CSG researchers quickly recognized that they were onto something much more fundamental and important than just how the British government could save money (important as that was). They were on the track of a key variable that explains why people like using some communication technologies more than others. They called the variable "social presence." But what exactly is social presence and how do participants perceive it? A CSG colleague, Derek Rutter, defined it more precisely as "the number of social cues that can be communicated across a given channel."[10]

In social science parlance, social cues include observable verbal, paralinguistic, and nonverbal behaviors. Translating this shoptalk into plain English, verbal behaviors are words, organized into phrases, sentences, and paragraphs. Paralinguistic behaviors are communicative sounds that accompany words, such as "um-hmm," "hunh," "yeah," "oooooh," throat-clearing, laughs, chuckles, giggles, gulps, sighs, and so forth. Nonverbal behaviors include facial expressions and body gestures.

The following dialogue is an example of the various components of a communication:

Sara: "Bring me that book, please."	Verbal
Sara directs her eyes to Gary, nods head, points toward book on mantel, smiles	Nonverbal
Gary: "Um-hmm."	Paralinguistic
Gary nods in understanding, moves to get book	Nonverbal
Gary: "No problem."	Verbal

Rutter concluded that the more total cues a communication technology or channel provides, the more people will like it. However, there turned out to be another piece to the puzzle. It's not just the total number of cues, it's also the specific kind of cues transmitted. The answer to this final, clever knot took most of a decade to understand and requires a reading of anthropologist Gregory Bateson.

Bateson was married to another famous anthropologist, Margaret Mead, and they were preeminent in the academic world. He was a fascinating thinker in his own right and formulated a novel way of approaching how people communicate to one another beyond the apparent content of the messages they exchange.

Bateson wrote that one level of a message is what people say—the "report" level: The weather is pleasant, I'm feeling tired, and so forth. The report level is characterized by cognitive or factual information.[11] The second level is called the "command" level, which provides information about the relationship between sender and receiver, and how the receiver should interpret the sender's message. This relational data is sometimes called socio-emotional information.

It turns out that there is a connection between the form of expression of a message and the information conveyed. For example, most factual data (the report level) is communicated verbally. By contrast, at the command level, people don't usually use words at all to express socio-emotional information. Instead, they employ paralinguistic cues, the sounds that accompany words, and nonverbal cues, facial expressions and bodily gestures, to convey socio-emotional messages.

This finding confirmed earlier research. In 1971, Meherabian found that people base their evaluation of a communicator's intent on tone of voice and facial expression, rather than words.[12] Later work by Burgoon confirmed that verbal cues are more important for factual, abstract, and persuasive messages, while nonverbal cues predominate in relational and emotional messages. The results also indicated that adults place more reliance on nonverbal than verbal cues to determine social meaning, and that this reliance is greatest when there is a conflict between the two modes. Finally, people differ in the reliance they place on the two kinds of cues—people believe nonverbal cues over words. So if someone tells you they feel happy but their face looks glum, we will believe their facial expression, not their verbal claim.[13]

Short, Williams, and Christie were the first to emphasize this relationship between types of cues and communication channel bandwidth. They concluded that people value channels that allow them to receive social cues. The greater the number of received cues, the more accurately the receiver can decode the sender's intended message—if the message is socio-emotional in nature. To put this conclusion another way: Face-to-face communication (and broadband channels) provide both visual and auditory cues that give people the socio-emotional information about how the speaker regards them and how they intend the message to be taken. The ability to receive paralinguistic and nonverbal cues increases the confidence people feel about the accuracy of their understanding, and makes them more secure about constructing an appropriate response to the message.

These findings can be summarized by the following dichotomies:

Report Level	Command Level
Cognitive	Emotional
Factual	Evaluative
Verbal	Nonverbal and paralinguistic
Abstract	Grounded

Not only does the number of cues a channel permits (its bandwidth) help receivers interpret messages aimed at them, it also profoundly affects the communication patterns of message senders. Further study by the CSG group revealed that as channel capacity decreased, people became more aggressive in their language, even hostile. They theorized that communicators increase the level of these behaviors to compensate for the barriers to social presence imposed by narrowband channels. In other words, interactors became more strident when trying to express their messages across channels that restrict the number of cues they can send.

The highly rated face-to-face and video modalities that allow communication partners to receive a complete repertoire of social cues (such as television and films) are sometimes referred to as "rich" media. The telephone stands midway between visual media and textual media—it conveys verbal and paralinguistic cues, but filters out facial expressions and bodily gestures. Written documents severely restrict the cues available to participants, permitting only verbal and graphic information. Text-based computer communication permits the fewest cues, not even transmitting the identifying clues that can be gleaned from a letterhead, quality of the stationery, typeface, and format. In chat rooms, for example, people express the way they want others to think of them through imaginative self-referential user names.

The work of the CSG is further supported by research that looks at how people communicate over computer networks. Kiesler, Siegel, and McGuire observed the tendency for online computer users to "flame," a writing style that employs intemperate, usually angry, language when addressing others. They believe that the "emoticons" (the smiley [:-)], "frowny" [:-( ], and "winky" faces [;-)]) they employed to indicate how they feel are a way that Net-communicators overcome the limitations imposed by the narrow bandwidth of computer interaction.[14] (Read these emoticons sideways to see their "faces.")

The implications of all this research for understanding how people will use broadband networks are clear. While no technological communication replaces face-to-face interaction—after all, they still don't convey olfactory or tactile information—broadband networks will approach in-person experience more than any other channels we have adopted so far because they will permit the simultaneous exchange of socio-emotional information. Broadband infrastructure will enable people to communicate sensitive, delicate communications that are emotionally charged. They will be able to assess the truthfulness and sincerity of their communication partners, and that information will help them feel comfortable that they are responding appropriately. They will be taken in by gifted actors. They will decide not to hire that shifty-eyed accountant. They will fall in love.

Face Time in Cyberspace

In a networked world, is there an audience? More to the point, when is a participant an audience member? People in this environment move between being producers, participants, and audience members with a fluidity unhampered by technology.[15] But there are mental costs associated with these changes in perspective. As audience members, we are accustomed to relative privacy, anonymity, and the control of our appearance; we often believe these attributes extend to our messages. But once disseminated, messages become public property that can be further transmitted, altered, and manipulated so that they do not reflect the intention of the originator. Almost every organization has an apocryphal anecdote about E-mail messages that escaped from a private sphere into a public venue to embarrass the originators.

In the narrowband environment of the current Internet, it is easy to confuse whether we are in a private or public space in chat rooms and on discussion lists. There is a certain anonymity because people know us only by our user-IDs or avatars. But in the future broadband environment, this anonymity will be challenged in many such activities, especially chat. More creative participants may be able to hide behind animated avatars that disguise their identities, but many people will just turn on their webcams and face the world.

Public Offerings

As communication moves from private to more public venues, the nature of messages and what is significant about them changes. While conveying emotional information is not as important in organizational settings, there is tremendous emphasis on the communication of status. Remember that control and dominance are also included in the "socio" part of socio-emotional information.

Up until now, technology has only begun to deliver on its promise to make telecommuting a part of the workplace. It is likely that broadband connectivity will increase the willingness of supervisors to let people to work at home. The visual capability of advanced networks allows them to carry status data and may make superiors more comfortable having off-site workers. They will be able to see their subordinates at work in their home offices, and to assess the degree of effort involved.[16] Extended, complex negotiations will be possible because participants will believe they have a good feel for the situation and for how much they can trust the people on the other side. Indeed, one witty observer noted that the needed functionality for videoconferencing is that they must be good enough "to tell if the smile is sincere."

Virtual Communities

The association of many people in groups is one of the most rewarding of all human activities, enabling achievements that can be accomplished only by the coordinated efforts of large numbers of people. The communicative capabilities of computers is just another outlet for this tendency, one that became apparent in the earliest uses of networked computers, back in 1969. That year is when the ARPANET began. Funded by the United States Department of Defense, it was the first long-haul computer network connecting machines at the University of California campuses at Los Angeles and Santa Barbara and the University of Utah.

Virtual communities sprang up in the 1970s. However, there were no Internet Service Providers at that time. Users dialed up local bulletin boards and messages were stored and forwarded across the Internet to members on other bulletin boards, to universities, or to one of the few companies that were on the Internet. By the 1980s, online interest groups were commonplace, although at that time the only form of communication was text. Writers, researchers, and observers of these first phases of the development of online communities spent much time considering whether they were like geographically-based communities and neighborhoods. More heat than light was shed on the subject, but Table 7.1 compiles some of the ways they made their comparisons. The column labeled "broadband virtual communities" is speculative, unless otherwise annotated.

When users interact in a visual environment, even though they may be communicating with text, the graphics seem to make the site "stickier," meaning that people stay there for longer periods of time. This involving effect of visual stimulation was first observed in the late 1980s George Lucas online project, Habitat, which gave each inhabitant a two-dimensional social space that could be customized, and a personal avatar. The community of users set up a church where marriages were performed and divorces granted. Habitat had a sheriff, a newspaper, lawyers, and thieves. Clearly, this project was a predecessor to the contemporary Web site, The Palace (www.thepalace.com), and the successor technology will be virtual reality immersive environments.

Virtual communities may be thought of as individuals who voluntarily choose to associate with one another in a shared environment—but only if new definitions of "individual," "associate with," and "environment" are invoked. The individuals are dematerialized, self-created, verbal personalities that represent only a small portion of themselves, however vital and meaningful that part may be. The association with others is purely electronic and only rarely includes shared joy and pain. This is not to say that sharing reports of joy and pain is insignificant, but it is not your father's idea of association. Similarly, the shared electronic environment (composed entirely of recycled electrons) is a mental as well as a physical space, but one where participants find it difficult to gauge one another's understandings of the environment or other individuals.[17]

Current online communities tend to be linked by shared interests, and communication is dominated by the exchange of cognitive information or

Table 7.1 Characteristics of Online Communities

Dimension	Geographic Communities	Narrowband Virtual Communities	Broadband Virtual Communities
Visual Discriminatory Cues (Race, Status, Etc.)	Yes	No	Yes
Anonymity	No	Yes	No
Commonalities	Concerns	Interests	Interests
Context of Communication	Necessity	Choice	Choice
Community Based On . . .	Proximity	Affinity	Affinity
Origination	Spontaneous emergence	Created	Created
Membership	Finite, relatively fixed Easy to join Nearly impossible to leave	Infinite, relatively transient Economic and educational barriers to entry Easy to leave	Unknown
Attributes of Members	Diversity of age, gender, educational, and economic status	Diversity of race, national origin, lifestyle, cultural background	Unknown
Community Hours	Approximately 6:00 A.M. – 2:00 A.M. Mostly synchronous, face to face or phone	24/7 Synchronous or asynchronous, multimedia	24/7 Synchronous or asynchronous, multimedia

Table 7.1 Characteristics of Online Communities (Continued)

Dimension	Geographic Communities	Narrowband Virtual Communities	Broadband Virtual Communities
Self-Representation	Constrained by social norms, interactions over time, dense connection of interpersonal networks	Constrained by imagination, willingness to experiment, and inclination Deconstruction of social boundaries	Constrained by visual impression, artistic ability, or animation software
Communicative Style	Rich texture of verbal, paralinguistic and nonverbal behavior	Verbal, often terse Uninhibited (flames, rants, and raves) Irreverent, ironic	Unknown
Social Sanctions for Misbehavior	Ostracism Trial, fine, imprisonment	Banishment Trial, fine, imprisonment	Banishment Trial, fine, imprisonment
What's Missing?	We are hostage to molecular existence which makes radical shape-shifting and time travel more difficult than in the pixel world.	Dressing up, exchanging business cards, etc. " . . . body language, sex, death, violence, vegetation, animals, architecture, music, smells, sunlight, that old harvest moon, the old, the poor, the blind, prana . . ." —John Perry Barlow	Touch, smells, death, physical violence, exchanging business cards, etc.

just egotistical bit-blasting. Communities in touch over richer media will also have the capacity for collaborative interaction and emotional bonding, which are likely to make them much more durable and resilient than today's computer interest groups.

A study of Japanese interest groups examined the phenomenon of lurking. Many people who are in a chat room or on a discussion list could participate but do not. In the United States, we call being present but not participating "lurking." The process of identifying and introducing oneself is sometimes called "de-lurking."

In Japan, passive and active participants are called ROMs (read-only members) and RAMs (radically active members). The research found that the ROMs had a number of reasons why they didn't participate:[18]

1. They were reluctant to communicate with people they didn't know.
2. They were hesitant to enter an already-formed group.
3. They felt they lacked expertise and feared evaluation by others.
4. They were uncertain about how much to self-disclose.
5. They worried about making themselves understood.
6. They were afraid of public criticism.

There is a dark side to "cyberian" community, so some of the expressed concerns are not unrealistic.[19] A variety of experiences in Net history have demonstrated that, as in TRW (the real world), some amount of surveillance and social control is necessary for online communities to survive. The CommuniTree in San Francisco in 1983 was destroyed by ungoverned free expression. In 1990, Santa Monica city officials withdrew from the useful online feedback of some constituents because of the abusive language of others. And in 1999, eBay removed items such as marijuana, pharmaceuticals, pornographic material, and other inappropriate items from the site. In spite of the problems, the ongoing search for community is easy to understand. Alarm over environmental pollution in TRW, concerns about crime and violence, and the commercialization of most public spaces make understandable the ongoing search for satisfying communities.

The Wild, Wild Web: They'd Rather Route Than Switch

We have now considered the effects of broadband connectivity: complexity, intimacy, and socio-emotional interaction at the personal, organizational, and social levels. However, broadband networks of the future are switched and routed as well as connected. The next section considers how they are complementary to connection. This is a highly speculative discussion of centralized versus distributed networks, which is a topic of much debate among people interested in social effects of networks.

Switching>Centralized Control; Routing>Distributed Control

While connectivity is affiliative, providing the means for people to form relationships and to get close to each other, switching and routing control messaging, allowing senders to bypass and exclude others. Without these functions, connectivity would result in chaos. All messages would go to everyone in a cacophony of communication. In other words, switching and routing are a complement to connectivity, providing the necessary filtering of incoming messages and targeting of outgoing messages—discretionary control that makes connectivity functional and useful.

By allowing users to choose who (or what) they contact, the switching and routing of high bandwidth material over a network allows people to design their own media menu. They become their own programmers, tailoring the information they receive to their special wants and needs. Whether they like to watch mysteries, action-adventure, cartoons, comedy, music videos, access political or financial information—the new advanced networks will allow them to access material exactly to their taste.

Switching and routing exercise control in different ways. A switch sits in the middle of a network and connects one circuit with another, as occurs in the telephone system. When a person places a telephone call to a specific number, the call is sent to the device dialed and only that device. Switches are expensive, so they are few and far between, and as centralized as traffic will allow. Switches are Old Technology, in the view of supporters of IP and edge intelligence. Control emanates from the center of the network.

Routers are less expensive than switches, so there is an incentive to locate them throughout the network. When a message is transported across the network, it passes through many routers as part of a stream of material composed of many messages. As the message gets near to the intended receiver, it is still embedded in a stream that contains hundreds, thousands, even millions of communications. To reach a specific user at the edge of the network, the person's ISP mail server (or local machine, if it is a node on the network) "reads" the entire stream, and picks out messages addressed to it and forwarding the rest. Routers empower the edge of the network, laying responsibility for who gets which messages on the last processor sitting on the far boundary of the net. Control, and therefore power, are moved to the edge, not the center, of the network.

Switching creates a dedicated channel. In broadcast or cable, a viewer's choice of a channel is a way of activating a switch on the receiver side (in their TV set), while the senders just keep on pushing out the signal. Routing dechannelizes the media landscape. The Internet doesn't mean 500 channels; it means a gazillion channels.[20] People access video material from sources of their choice, not just their (channelized) over-the-air TV or cable system. Libraries, museums, record companies, individual artists, program producers, musical groups, and performers of all kinds— they all offer their video and audio products directly to the consumer bypassing the intermediaries of programming services and packagers.

Just as routed networks change the way individuals respond to the entertainment media environment, it also alters organizational behavior and communication in significant ways. Although connectivity creates complexity, which in turn promotes hierarchy, routing acts in the opposite way—it flattens hierarchies. It doesn't eliminate them, but there is evidence that it does reduce the number of layers.

The flattening of hierarchy was first noticed in the 1970s, when organizational researchers predicted that routed computer networks would reduce the need for middle management. They observed that top-level managers increasingly accessed shared databases to get up-to-the-minute information that had previously been provided to them by data-processing middle-level supervisors.[21] In the 1990s, the white collar layoffs of corporate downsizing were in part a response to the more horizontal organizational structure that switched computer networks made possible. The structure of commercial transactions has been similarly affected, believe some observers. In this domain, the elimination of middle layers is called disintermediation. The transaction is reintermediated as the "middle man" function is picked up by networked information handlers, or cybermediaries.

Connectivity is about to revolutionize the political world as well, in two ways. Similar to the commercial transaction just outlined, people can get information quickly and easily. In the 2000 election cycle, C-SPAN is sending cameras to virtually all candidates' campaign appearances. The cable industry's public service network plans to log and archive the material and to make it searchable over the Internet. Anyone who can access C-SPAN's Web site will be able to look for candidates' positions across locations, audiences, points in time, and topics. The days of politicians being able to say one thing in one setting and something else in another instance are almost over.

On the other hand, the power of switching should not be underestimated. It provides the ability to reach some people while excluding others. For example, consider the capacity for coordination it enables for previously disorganized and disenfranchised groups. Campaigns are coordinated. Government agencies, companies, institutions, and well-funded lobbying groups—they are all coordinated, capable of mounting directed efforts to achieve their goals. Voters, audiences, taxpayers, and nonunion workers—they have been isolated and alienated. The advent of massive switched connectivity will change the way all these organizations relate to their constituency groups because of people's new ability to coordinate their actions as well. The public will be empowered as never before to acquire patterned knowledge and to engage in patterned action. Switching provides the ability to reach individuals with different messages, so that each can engage in specialized, patterned action.

Patterned knowledge is different from random knowledge. A computer course on emergency medicine provides some insight into how this difference works. The user sees a video of a patient on a gurney rushed into the emergency room. The video stops and a screen of alternative tests pops up. The user selects the appropriate tests, and a pop-up of proposed treatments appears. The user chooses the appropriate treatments. The computer tracks the patient's progress in real time, offering the user full opportunity to change the treatment protocol.

At one demonstration, the first person in line was a graduate student in communications—his patient died in less than five minutes. The second participant was a dermatologist who had little emergency room training. She stabilized her patient in less than five minutes. The difference between these two individuals is that the physician understood the basics principles involved enough to enable her to take patterned action based on her knowledge. "I knew I had to get as many fluids as quickly as possible into the patient. I didn't worry about the tests until later," said the doctor in the exit interview.[22]

Patterned knowledge leads to patterned behavior—and both of these are enabled by ubiquitous connectivity. Grass roots organization received a big boost from the telephone, and it is a well-organized phenomenon seen during every election cycle. However, coordinated political action among computer network users is a new phenomenon and is likely to be just as powerful—we have only seen the outlines of how it may evolve.

One documented example occurred with the nation's first government-supported computer network, the Santa Monica Public Electronic Network (PEN). The system allowed residents to communicate with city departments and each other. One of the first acts of PEN-users was to use the system to successfully oppose the privatization of an ocean beach facility. Broadband, two-way switched networks will offer all these features, plus rich interaction.[23]

Table 7.2 summarizes the effects of connectivity and routing on communication at different levels of social aggregation. Person-to-person refers to communication between individuals, while person-to-station interaction occurs when an individual contacts an institutional network node or retrieves information from an archive or database. Organizational communication occurs between linked individuals who share some form

Table 7.2 Conceptualizing Packet-Switched Connectivity

Type of Connection	Person-to-Person	Person-to-Station	Organizational Communication	Broadcast Mode
Narrowband Connection	Cognition-based relationship	Efficient data upload and download	Data exchange	Communities: attraction based on interest
Broadband Connection	Emotion-bonded relationships	Environmental immersion	Trusted, dispersed work groups	Communities: attraction based on emotion
Ubiquitous Switching	Discretionary contact	Channelization	Enforces and reinforces hierarchy	Social coordination
Ubiquitous Routing	General contact: call to arms	Dechannelization	Flattened social hierarchies	Social action and influence

of membership tie and, usually, a shared database that can be accessed by anyone in the group. Members can send messages to the group as a whole, to any combination of members, or to just one member of the group.

Person to-person is one-to-one messaging. Person-to-station occurs when an individual visits the Web site of an organization or group. Broadcast mode is invoked when people communicate to large numbers of people with a single keystroke, reaching from a few people to a few million, both within and without a particular group.

Some of the effects noted in Table 7.2 have been demonstrated by research. Others are based on anecdotal reports. The limited bandwidth and number of online users mean that we have only seen the beginning of how ubiquitous connection will affect social life—and we are probably in for quite a few surprises as it unfolds over the next two decades.

The preceding chapters have described the features of next-generation networks: digital, broadband, interactive, and ubiquitously connected delivery systems. The actual construction of this infrastructure is underway in many regions of the world. The next chapter will took at the global proliferation of such communication networks.

Notes

1. J. P. Barlow, "Crime and punishment: Desperadoes of the datasphere," published on the Internet at http://www.eff.org/pub/Legal/Cases/SJG/crime_and_puzzlement1.
2. W. Stallings, *Data and Computer Communications*, 4th ed., New York: Macmillan Publishing (1992):803.
3. J. Casti, *Complexification*, New York: HarperCollins (1994).
4. M. Twain, *Pudd'n Head Wilson and Those Extraordinary Twins*, New York: Harper (1922).
5. J. Gill, telephone interview, April 1994.
6. J. Travers and S. Milgram, "An experimental study of the 'Small World Problem,'" *Sociometry* 32 (1969):425-443.
7. G. Nicolis and I. Prigogine, *Self-Organization in Non-Equilibrium Systems*, New York: Wiley (1977).
8. L. McLaughlin, "Pentium flaw: A wake-up call?" *PC World* 13:3 (March 1995):50-51.
9. J. Short, E. Williams, and B. Christie, *The Social Psychology of Telecommunications*, New York: Wiley (1976).
10. D. Rutter, *Looking and Seeing: The role of Visual Communication in Social Interaction,* New York: Wiley (1984).
11. G. Bateson, *Steps to an Ecology of Mind,* New York: Ballantine Books (1972).
12. A. Meherabian, *Silent Messages,* Belmont, CA: Wadsworth (1971).
13. J. K. Burgoon, *Nonverbal Signals: Handbook of Interpersonal Communication,* eds. M. L. Knapp and G. R. Miller, Beverly Hills: Sage (1985):344-390.

14. S. Kiesler, J. Siegel, and T. W. McGuire, "Social psychological aspects of computer-mediated communication," *American Psychologist* 39:10 (1984):1123-1134.

15. D. M. Davis, "Illusions and ambiguities in the telemedia environment: An exploration of the transformation of social roles," *Journal of Broadcasting and Electronic Media* 39 (1995):517-554.

16. M. Hayes, "Working online, or wasting time?" *Information Week* 525 (May 1, 1995):38.

17. J. P. Barlow, "Is there a there in cyberspace?" *Utne Reader* 68 (March-April 1995):54.

18. K. Aoki, "Virtual Communities in Japan," paper presented at Pacific Telecommunications Council, University of Hawaii, Honolulu, HI (1994).

19. H. Rheingold, *The Virtual Community: Homesteading on the Electronic Frontier*, Reading, MA: Addison-Wesley (1993).

20. J. Quittner, "500 TV channels? Make it 500 million," *Los Angeles Times* (June 29, 1995):D2, 12.

21. N. M. Carter and J. B. Cullen, *Computerization of newspaper organizations: The impact of technology on organizational structuring*, Lanham, MD: University Press of America (1983).

22. Personal interview with anonymous subject in research study, 1987.

23. J. Van Tassel, "Yakkety-Yak," op. cit.

8

Fiber Fever: The Growth of Global Networks

Globalization is an oft-heard buzzword. It's a clear trend, with daily stories about multinational mergers and acquisitions and corporate worldwide operations in the business pages of the newspaper. In the twentieth century, globalization accelerated enormously, as measured in "the increase in the numbers of international agencies and institutions, the increasing global forms of communication, the acceptance of unified global time, the development of global competitions and prizes, the development of standard notions of citizenship, and the rights and conception of humankind."[1]

But no matter how small the world may appear to be getting, it's still quite a large place—immense, in fact: There are approximately 5.9 billion people on earth, as shown in Table 8.1.

Table 8.1 World Population Growth (in millions)

Region	1950	1998	2050
World	2,521	5,901	8,909
Africa	221	749	1,766
Asia	1,402	3,585	5,268
Europe	547	729	628
Latin America and the Caribbean	167	504	809
Northern America	172	305	392
Oceania	13	30	46

Source: United Nations Population Division, *World Population Prospects: The 1998 Revision.*

Wired and wireless telephones, television, cable, satellites, and the Internet are now commonplace not only in countries with complex, state-of-the-art communication infrastructures, but also those that, until now, have been relatively isolated. Even when one considers that about 500 million of the world's households (33%) have telephones and 925 million of them (61%) have television sets, the figures are impressive.[2]

This chapter will look at the buildout of infrastructures throughout the world that support international communications, particularly those that will provide broadband access. It will also examine the worldwide distribution of television and other video programming, and the barriers to a global communication systems that will deliver moving images. Communication infrastructure includes: wired and wireless transport systems that support telephone and computer communication, submarine cables, satellite networks, terrestrial VHF and UHF broadcasting, and terrestrial microwave systems like MMDS (multichannel/multipoint distribution service) and LMDS (local multichannel distribution service).

The rate of growth of these infrastructures is dizzying. According to FCC chairman William E. Kennard, aggregate private investment in global networks between 1994 and 1999 was about $600 billion. Networks are crammed with new traffic and applications. International traffic, as measured in telephone minutes, more than doubled between 1990 and 1996 to reach 70 billion minutes. In 1990, fiber optic cable provided 11,000 64-kbps transatlantic circuits; satellites handled 100,000 of them. Just five years later in 1995, the capacity of fiber combined with satellites reached 200,000 64-kbps connections. In 2000, those numbers have increased to 350,000 transatlantic fiber circuits and 250,000 satellite circuits.[3]

While some of the communication is voice traffic, which has been increasing more than 13% per year, international data transport has expanded at ten times the rate of voice usage. Bandwidth-intensive applications such as corporate Intranets and videoconferencing—plus the sheer volume of the Internet—are responsible for the growth. The traffic on telephone networks continues to gain almost everywhere, and the rise of international data traffic is especially strong. The world's busiest international telecommunications route is the transatlantic corridor. It is experiencing a doubling of Internet traffic every year, which surpassed voice traffic volume in September of 1997. By 2000, Internet messages are expected to account for 75% of all transatlantic communication.[4]

Satellite capacity has grown so rapidly that there were more satellites launched in the 1990s than in the past five decades combined.[5] In the next few years, the pace will continue, with about 35 to 38 launches scheduled per year. DBS subscription television service is an expanding revenue source for the satellite industry, accounting for $8.2 of the satellite industry's 1998 income of about $26 billion, and there will be 94.7 million worldwide digital satellite subscribers in 2005.[6]

In the air as over the wire, Internet traffic played a significant role in demand for communication services, providing international Internet backbone capacity. Indeed, according to a panel study commissioned by U.S. NASA and the National Science Foundation, nearly 60% of all overseas communications is routed through satellites. The study estimated that more than 200 countries used satellites for domestic, regional, and global connectivity, defense communications, direct broadcast services, navigation, data collection, and mobile communications.[7] In addition to the geo-synchronous satellites that are now in service, there are several satellite fleets planned that will provide additional bandwidth, and because they will circle the globe at lower altitudes, they will offer Internet services as well. These constellations of "birds" include Globalstar, Teledesic, Skybridge, ICO, Aries, and Elipso, among others.

Another way to get a sense of the explosion of intercontinental communication is to look at the number of undersea cables carrying communications between continents. A decade ago, there were fewer than a dozen submarine cables; today, there are more than 600 of them, accounting for 18% of the total worldwide installation of fiber optic cable. In 1996 there were approximately 46 million fiber kilometers (fiber km). Projections from *U.S. Industry and Trade Outlook 1998* pegged fiber kilometers at 70 million.[8] But this projection turned out to be substantially lower than the astounding growth that actually took place.

A 1999 study by Pioneer Consulting estimates that 289,000 km of submarine fiber optic cable and 127,000 km of terrestrial fiber optic cable will be deployed by 2004 to serve the international telecommunications market, costing billions of dollars, as shown in Figure 8.1.[9] These figures exclude domestic fiber deployment, which is growing at more than 20% per year. All told, Pioneer predicts that a total of 5.1 million submarine fiber kilometers will come into service between 1999 and 2004. The investment to bring this much bandwidth online is staggering: about $22.1 billion between 1998 and 2001. Pioneer summarizes the growth of submarine cable in regions of the world with the following estimates:

- Compound annual growth of transpacific demand will be 116%.
- Compound annual growth of North America–South America demand will be 88%.
- Compound annual growth of North America–Australia demand will be 97%.
- Compound annual growth of Western Europe–South Africa demand will be 108%.
- Compound annual growth rate of Europe–Asia demand will be 59%.

There are several reasons why there is so much activity, but the development of new technologies is an important one. In 1956, the cost of

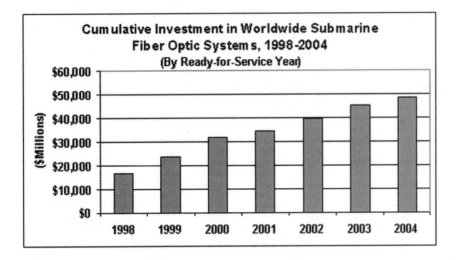

Figure 8.1 Investment in submarine fiber optic cable. Source: Pioneer Consulting.

a transpacific voice path (the infrastructure to handle a dedicated connection between two people across the Pacific from one another) was $378,000. The cost per voice path of new fiber optic telecom cables today is about $1,500.[10]

Terrestrial cable capacity is growing too, with investment reaching $9.53 billion. International fiber optic networks, like the WorldCom/MCI pan-European system, are growing at annual rates of 85% to 95% per year in every part of the world. Much of this astonishing growth is driven by the explosive expansion of the Internet. Pioneer Consulting tracks global Internet usage and reported growth between 1996 and 1998; their estimates of the number of Internet users by region through 2003 is shown in Table 8.2.

The Bandwidth Blowout: Case Studies

In 1998, George Gilder put on a conference in Squaw Valley, California, which he named "The Bandwidth Blowout." More than anyone else, Gilder has insisted that capacity is exploding. The following case studies give some idea of how that explosion came about.

Global Information Infrastructure

The idea of a nationwide, interconnected, broadband, two-way network, capable of carrying voice, data, and video, was captured in a phrase

Table 8.2 Growth of the Internet, 1996–2003

	North America	Latin American/ Caribbean	Asia-Pacific	Europe	Africa	Total
2003	181.07	21.66	137.24	129.17	4.79	473.9
2002	164.61	14.94	109.79	103.34	3.69	396.4
2001	149.65	10.30	87.83	76.55	2.84	327.2
2000	133.61	7.10	58.55	56.70	2.18	258.2
1999	111.35	4.90	33.46	42.00	1.75	193.4
1998	0.69	3.68	19.12	30.00	1.45	134.9
1997	48.90	1.47	9.81	17.00	1.27	78.4
1996	33.30	0.42	5.30	9.70	1.10	49.8

Source: Pioneer Consulting.

coined in the late 1980s—the Information Superhighway or the National Information Infrastructure (NII). This concept was extended to a worldwide context, and became a project to create a global information network called the Global Information Infrastructure, or GII.

The GII began in February of 1995 at the Ministerial Conference in Brussels, Belgium. The G7 nations (United States, Canada, Japan, Germany, Britain, France, and Italy) pledged to work together to develop a GII.[12] The group urged the European countries to dismantle communication trade barriers, ensure competition, agree on technical standards, and develop intellectual property rules, and approved eleven pilot projects to start the effort despite opposition from some nations, especially France.[13] The United States' adherence to a private enterprise economic model was voiced by the National Telecommunications and Information Administration (NTIA) of the U.S. Department of Commerce.

NTIA proposed five essential principles of telecommunications that the agency believed should apply to the GII: A flexible regulatory framework, competition, open access, private investment, and universal service. A discussion of these points revealed a belief that a successful rollout of the GII must accommodate economic realities. It was agreed that there must be paying customers with sufficient buying capacity, and that suppliers of services must be able to deliver them at a cost that can be afforded by those customers.[14]

Ambitious plans were laid. An optimistic rollout schedule called for initial tests in June, 1995, initial links by December, 1995, additional links by June, 1996, and deployment and final testing through mid-1997. But the GII never materialized.

The original objective to interconnect the major high-speed (34 to 155 megabits per second) facilities of all the G7 countries by 1997 was overtaken by aggressive moves by private sector companies, hell-bent on building high-speed networks. (See Figure 8.2.) Under the rapid onslaught of these well-funded, fast-moving concerns, the GII lost its steam, and now has become little more than a forum for international corporations to coordinate their efforts to establish systems for electronic commerce. There is some doubt that the "consortium cables," where governments, agencies, and sometimes private enterprises come together to create some large project, are viable at all. As private networks proliferate, there seems to be little room for the slow, relatively inefficient bureaucratic approach that such efforts entail. In late 1999, the U.S. Federal Communications Commission was hearing arguments from both private operators and consortium cables about how each should be regulated.

Fiber-optic Link Around the Globe (FLAG)

FLAG bills itself as the broadband superhighway, offering nearly three-fourths of the world's population access to international communications services.[15] FLAG is the longest undersea fiber optic cable in the world, 27,300 km (17,000 miles). It was conceived in 1990, construction began in 1996, and it opened for business on November 22, 1997 with two fiber pairs, offering 10 Gigabit per second capacity—the equivalent of up to 600,000 simultaneous conversations. Three companies received credit for building the behemoth, Tyco Submarine Systems, a U.S. company that was formerly AT&T Submarine Systems, KDD Submarine Cable Systems of Japan, and Cable and Wireless Marine from the United Kingdom. The cost was $1.5 billion.

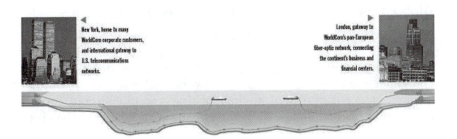

Figure 8.2 Projects like WorldCom's Gemini undersea cable quickly outstripped the ability of governments to build a GII. Source: WorldCom.

FLAG was the first venture of its kind to be financed entirely by investors rather than by the "old boys' club" of international carriers. At the time it started commercial service, there was much doubt if the company could remain in business, since about one-third of the capacity needed to be taken in order to make debt payments. Despite an economic slump in Asia and wavering demand from Malaysia, India, and the Middle East, it appears that FLAG's construction of this monster network was a good business decision, riding on the crest of worldwide demand for bandwidth.

By 1999, FLAG's route from the United Kingdom to Japan had 16 landing points in 13 countries: the U.K., Spain, Italy, Egypt (2), United Arab Emirates, Saudi Arabia, Jordan, India, Malaysia, Thailand, China (2), South Korea, and Japan (2). (FLAG was the first submarine cable to obtain landing rights in China.) Each landing point is served by a "landing party," often the sole or leading international telephone company in the landing country, and is responsible for operating and maintaining the onshore equipment.

In 1999, the company also announced that it would add another link to its network, the FLAG Atlantic-1 (FA-1), to be completed in 2001. The first transoceanic 1.28 Terabits per second cable system, FA-1 will extend FLAG to 40,000 km, offering direct access to the two largest telecommunications and Internet markets, the United States and Japan. This link was designed as a new generation network composed of four-fiber pairs. It takes advantage of technical advancements, particularly a new technology called dense wavelength-division multiplexing (DWDM) to increase the capacity in the busy Europe-United States corridor. In 1998, even existing submarine networks were able to quadruple the speed of information transported on individual fibers by using wave-division multiplexing (WDM), upping their capacity from 2.5 Gbps to 10 Gbps. By the end of the same year, individual fiber ratings had increased to 40 Gbps with the new DWDM specification.

Global Crossing—Up, Down, Around, and Everywhere

Global Crossing (GC) was the brainchild of a Beverly Hills, California investment firm, Pacific Capital Group, led by Gary Winnick.[16] In 1996, he put together a deal with Tyco (then AT&T Submarine Systems) to build a $750-million transatlantic undersea network. GC went a step further towards independence from multinational telecommunications companies than FLAG had done—GC had no financial backing from the major carriers.

In the beginning, though, there were doubts that GC would be able to fill its capacity and make money. There was a sense that the carriers who shared no financial ties to GC would resist its aggressive attitude and style, and punish its outsider status. These concerns proved unfounded; GC was even more successful than FLAG. Traffic on the Atlantic corridor exploded, and the profits weren't offset by lower performing network segments. Atlantic Crossing 1 (AC-1) began in May 1998, and on December 31 of that year, GC reported $51.6 million net income applicable to common stockholders.

At this point, the entire scope of GC's vision blossomed. The company embarked on a mission to build an advanced global IP-based, data-centric network, an end-to-end fiber optic platform for data, voice, video, and Internet communication. The blueprint calls for connecting the world's 100 largest metropolitan areas, spanning five continents across corridors that account for 80% of the world's international traffic. To carry out its plans, GC bought the Cable & Wireless submarine cable construction division, creating Global Marine Systems Limited. GMSL operates the largest fleet of cable laying and maintenance vessels in the world and currently services more than one-third of the world's undersea cable kilometers.

GC also raised buckets of money, so with its newly-acquired construction capability and investor cash, the company announced it was putting it all to work to build six undersea systems, a terrestrial system, and a hybrid submarine/land cable. The undersea systems are: (1) Atlantic Crossing (AC-1); (2) Pacific Crossing (PC-1); (3) Mid-Atlantic Crossing (MAC); (4) Pan-American Crossing (PAC); (5) South American Crossing (SAC); and (6) Atlantic Crossing 2 (AC-2). The terrestrial system is the Pan-European Crossing (PEC), and the East Asia Crossing (EAC) covering east Asia will be a hybrid terrestrial/undersea cable.

- Global Crossing's Atlantic Crossing (AC-1) links the United States, the United Kingdom, the Netherlands, and Germany. It began commercial service in May, 1998, initially offering 40 Gbps capacity. Strong demand in the Atlantic corridor led GC to increase capacity by another 40 Gbps in July 1999, and again in August to contract for a second upgrade of an additional 60 Gbps for March 2000.
- Pacific Crossing (PC-1) is a joint venture between GC and Japanese trading company Marubeni Corporation to build a four-fiber, 80 Gbps, 21,000 kilometer cable connecting the United States to Japan.
- Mid-Atlantic Crossing (MAC) is a two-fiber pair self-healing ring connecting New York (at the AC-1 Brookhaven cable station), the Caribbean, and Florida. Initial service began in December 1999.
- Pan-American Crossing (PAC) brings traffic from California, Mexico, and Panama to the Caribbean MAC cable after February 2000. The Panama/St. Croix junction interconnects Asia, the Americas, and Europe.

- Pan-European Crossing (PEC) will be a $850 million, 10,000 kilometer, 30 fiber-pair network connecting 24 major commercial centers in Europe, including London, Amsterdam, and Frankfurt, with the United States, Asia, and Latin America.
- South American Crossing (SAC) is a four-fiber-pair system between the U.S., Virgin Islands, Brazil, Argentina, Chile, Peru, Colombia, and Panama. The network will be deployed in several phases, and completed in 2001. It will use WDM to achieve 40 Gbps capacity.
- East Asia Crossing (EAC) is a partnership between GC, Tokyo-based Softbank, and Microsoft. The 17,700-kilometer terrestrial and undersea broadband network will cost $1.3 billion. It will link Japan, China, Singapore, Hong Kong, Korea, Taiwan, the Philippines, and Malaysia to GC's global network, including North America, Europe and South America. Phase One will see the completion of 10,200 kilometers by year-end 2000, linking with Pacific Crossing 1 at a landing station in Japan, and connecting with Taiwan, Singapore, Hong Kong, Malaysia, and the Philippines. The remaining 7,500 kilometers will be completed in Phase Two, coming online in June 2001. It will tie together Japan to two stations in China, one in South Korea, and one in Taiwan.

EAC will be a four fiber pair, 80-Gbps cable, upgradeable to 1.2 Tbps with DWDM. Future plans call for building out both the networks to connect Thailand, Indonesia, and key Asian cities.

And Global Crossing just keeps on growing. It has entered into an agreement to take over Frontier Corporation, resulting in a company that owns and operates an end-to-end global fiber-optic network that connects 88,000 route miles, 1.25 million fiber miles, and offers ultra-high bandwidth to 170 major cities in 24 countries.

The increased demand for international bandwidth is occurring because of a number of factors. It's not just that the Internet is growing. It's also because the ways people use communication networks are fundamentally changing. For example, FLAG Telecom believes that 40% of the people in the United States will be online by 2001, and that 30% of all U.S. business will have an online presence by then. Activities that are now local, like banking, shopping, and entertainment, will be offered to users around the world. Finally, broadband access will also result in higher-bandwidth applications, such as video telephony, and audio video streaming.

So while capacity is increasing very quickly, doubling every year to eighteen months, demand seems to be growing along with it. Nevertheless, forecasting future bandwidth is hazardous duty. Estimates fluctuate from one analyst to another, and the stakes become extraordinarily high when multibillion dollar systems with terabits of capacity are involved.

There's another important issue. Nobody wants to be a wallflower at the big dance. For a long time, a good portion of the world has been left out of the communications loop. But in the next decade, every continent will be served, and in the next quarter century, there will be ubiquitous wireless communication to and from almost any place on earth.

A World of Difference

It is common knowledge that a huge gap separates the rich and the poor, both within and between countries. The enormous inequalities extend to communication infrastructures as well. Communication infrastructures come sooner to some areas than to others—and not at all to a few. The nations with the most advanced communication infrastructures are Singapore, France, Germany, and Japan, followed by the United Kingdom and the United States. According to a report compiled in the mid-1990s, the comparison of these media-rich societies with the rest of the world is stark.[17] Europe and North America constitute about 15% of the global population, yet they:

- own 65% of all the TVs and radios
- read 50% of the newspapers
- publish 67% of the world's books

The industrialized nations have six times the number of radios per capita as the underdeveloped nations, and nine times as many TVs. The industrialized nations publish about 500 book titles per million persons each year, whereas underdeveloped nations publish 55 titles per million persons. A comparison of the penetration of telephones is even worse. In the developed nations, there are 246 telephones per thousand people, 1.1 in China, 0.7 in India, and 0.7 in Nigeria.

Factoid: There Are More Internet Hosts in New York City Than in All of Africa[18]

Contrast this communications wealth with Africa, the continent with the least developed information infrastructures. There are 739 million people in Africa and approximately 14 million phone lines—80% of which are in six countries. Africa has 12% of the world's population, but only 2% of its telephone lines. In 1990, Africa had 37 TVs and 172 radios per 1,000, and published 1% of the world's newspapers.

The Internet is the worst case scenario of all. In central and sub-Saharan Africa, there is less than one Internet host per 100,000 people, and only about one million Internet users on the entire continent.[19] Actually, about 97% of the world's Internet users come from the 15% of the population that lives in high income countries—more than 85% of the world's people have yet to go online. As of January 1997, Asia had 6.3% of the Internet hosts in existence, Latin America and the Caribbean, 1%, and Africa, less than 1%.[20] More than 75% of Internet content resides in North America, although this disparity in information resources will change over the next decade, as shown in Table 8.3.

Currently, more than 90% of all Internet traffic originates, terminates, or passes through the United States, because that is where 94 of the most visited 100 Web sites and many of the major Internet backbone providers are located. The imbalance of information in the United States causes some problems for network management. In a typical Web-surfing session, a small bandwidth request is sent to a content server, followed by a high-bandwidth response. Since so much data is located on U.S. servers, very little traffic flows into the country, yet a substantial amount flows outward. When ISPs or private network operators seek to purchase international capacity to provide global customers access to the United States, the asymmetric data flow results in a dearth of data flowing back and a great deal of unused capacity.

Getting over the Continental Divides

It's a classic chicken-and-egg problem. The huge global networks move traffic between high traffic locations. There isn't sufficient bandwidth demand in low traffic locations to justify the expense of putting in fiber optic networks, and without them, access to them is most expensive for those least able to afford it. The information rich get richer.

The communications infrastructure of a country is largely determined by how much its economy can afford to invest in it. The usual measure for this capacity is called per capita income. The World Bank classifies economies by taking the country's gross national product (GNP), the total of all the goods and services produced there, and dividing the GNP by the number of citizens. The categories based on 1998 GNP data are: low income, $760 or less; lower middle income, $761-$3,030; upper middle income, $3,031- $9,360; and high income, $9,361 or more.

The private development of cable systems requires a population with a relatively high disposable income. In countries where per capita income is low, cable systems are too expensive, costing an average of $400 to $1500 per household to build. However, wireless cable (MMDS), which costs approximately $300 per household, may be possible. Satellite systems also cost about $300, but DBS programming is expensive because of payments for transponder time. And with DBS, reception costs are shifted to receivers, who must buy their own downlink dish. As a result, satellite-delivered TV is largely restricted to the wealthier citizens, and the poorer

Table 8.3 Worldwide Internet Host Computers

Region	1/94	1/95	1/96	1/97	1/98	1/99	CAGR[a] 1994–1998
Africa	11,639	29,042	53,175	109,929	144,379	182,901	73.5%
Americas	1,467,062	3,080,753	5,913,686	9,918,654	19,394,554	28,522,639	81.0%
Asia	87,009	193,44	473,678	1,149,020	2,032,630	3,167,004	105.2%
Europe	614,721	1,195,352	2,569,217	4,238,025	7,213,688	10,570,691	76.6%
Oceania	99,024	201,319	382,863	632,757	915,355	1,055,767	60.5%
WORLD	2,283,978	4,711,226	9,417,531	16,135,992	29,802,017	43,498,699	80.3%

[a]CAGR = compound annual growth rate

ones receive TV via terrestrial broadcast channels or MMDS, rather than via cable or satellite.

As for Internet access, poor countries can barely get a dial tone, much less a Web tone. But there is room for some optimism for the future. Countries can get together to provide a minimum usage guarantee and, as the cost of constructing networks comes down, these arrangements could become attractive to operators and service providers. Another strategy that a developing economy can adopt is to build a fiber optic loop all around the country. These are called "festoon" systems, and if the area allows it, they can be built without using submarine repeaters, which add significantly to the cost of a network.

Then there are the satellites. They can provide ubiquitous, efficient, and effective coverage at relatively low infrastructure costs, because the cost is amortized across high-traffic areas as well. The global satellite fleets mentioned earlier will go far to bring communications services to unserved and underserved markets around the world, even if fiber optic cables are never built in these developing regions.

Finally, the Internet may prove to be an impressive educational resource. One important project is The Alliance for Global Learning, composed of three nonprofit groups: Schools Online, I*Earn, and the World Links for Development (a program of the World Bank Institute). They are working together to provide Internet access to schools in underdeveloped areas around the world by 2010. The Alliance will wire remote locations of Africa, in Uganda, South Africa, Mozambique, and Zimbabwe. In the Americas, it plans to work in Brazil, Peru, Paraguay, and Mexico. In the Middle East, the group plans to establish programs in Israel, Palestine, Egypt, Lebanon, and Jordan.[21]

Even Africa has its prosperous nations, Egypt and South Africa. Even the poorest country has its well-off citizens. And even the wealthiest continents are characterized by great differences in wealth. The next section looks at the communications infrastructure and media access that exist around the globe.

Around the World in 80 Paragraphs (or so)

The United States

The United States has a sophisticated communications infrastructure that nevertheless presents substantial difficulties in both the present and the future. Actually, it has four infrastructure systems, and that's something of a problem, because they differ widely by geography, standards, business culture, lobbying capabilities, and techniques for profitable foot-dragging. The four infrastructures are (1) terrestrial local broadcasting, (2) cable television, (3) telephone, and (4) satellite.

Start with the two major wired systems: cable television networks and the public switched telephone network (PSTN). About 10,700 local cable networks pass more than 95% of the U.S. population. About 67% to 68% of U.S. TV households subscribe to them.[22] Constructed over the past thirty years, most systems are now in their third developmental generation, offering fifty or more channels and pay-per-view capability.

The cable system is ubiquitous, but it is somewhat fragmented into local headends due to the system of local franchising that originally established cable television in the United States. Despite considerable consolidation within the industry and some system swapping between MSOs (multisystem operators), local systems remain difficult to interconnect into a seamless national structure, especially those outside major markets. However, in the last few years "interconnects" have been built to facilitate simultaneous advertising coverage across systems.

Finding a way to link their systems together became important to cablers when they found it difficult to get the share of advertising budgets that their levels of viewership would have suggested they should. Advertisers wanted them to guarantee same-time, same-network coverage against a given target audience, and the ability to measure advertising effectiveness. Now, the cable headends of each major metropolitan area are wired to an interconnect facility that drops commercials into cable network feeds. However, the intersystem network is not sufficient to handle Internet traffic, so in most cases, cable systems still have to build, buy, or lease bandwidth to move subscribers' Internet data in and out of the cable system.

Cable pipes are big—but they are also one-way. TV programming flows downstream from the headend to customer homes, and the networks were built with little or no return-path capability. Currently, many of the largest operators are upgrading their systems to carry digital television programming and, at the same time, to retrofit them for two-way traffic. It's expensive and technically difficult, but it will enable cable companies to offer high-speed Internet access and telephone service to their customers.

The Public Switched Telephone Network (PSTN) in the United States lags well behind similar networks in Europe, Japan, and Singapore, which have predominantly digital service to private residences. The fifteen-year long deployment of Integrated Services Digital Network over copper wire in the United States took so long that the technology was surpassed by the more advanced DSL (digital subscriber line) equipment before there was even much of an installed base of ISDN customers. (Scoffing at the tortoise-like speed of implementation, cynics say ISDN stands for "It Still Does Nothing.")

Telephone companies have invested in their networks for 150 years, and the infrastructure has changed more in the last decade than the previous 50 years. The backbones are all fiber optic cable, and high-level switches are electronic. Some local exchange carriers have upgraded their central offices (where neighborhood calls aggregate) so that they can off-load Internet traffic onto a separate data network, while handling voice calls over the PSTN (remember, the public switched telephone network).

Turning to the wireless technologies, only Canada and the United States established a locally-based system of television delivery. About 1,594 over-the-air local TV stations broadcast delivery-free service to 98% of the U.S. population. Stations and programming networks market the mass audience to advertisers, a profitable business for nearly fifty years. In the past decade, the availability of multichannel subscription services delivered first by cable and then by satellite have undercut the ability of terrestrial broadcasters to aggregate audiences. Now, facing an expensive, government-mandated transition to digital transmission, stations are unsure how they will pay for it, since advertisers are disinclined to pay higher rates.

Satellite services are the new kid in the air. TV stations transmit to their local market, but the rapidly growing, subscription-supported, satel-lite-delivered programming is national in scope. (See Table 8.4.) It is beamed to about 1.8 million owners of large C-band satellite dishes (TVRO, transmitted at 4 GHz and 6 GHz), and more than 12 million users of the smaller Ku-Band dishes (DBS service, transmitted at 17 GHz and 31 GHz).

The first DBS service in the United States was offered by Hughes DIRECTV, which combined its product with the USSB programming pro-vider. A year later, AlphaStar also entered the market, and in 1996, two more services came online, the cable-company-financed service, Prime-Star, and EchoStar's DISH service. Just two years later, PrimeStar and AlphaStar were both defunct. Only EchoStar and DIRECTV were still standing, and DIRECTV inherited (or purchased) most of the subscribers, as shown in Table 8.5. In 1999, Hughes bought its long-time programming partner, USSB as well. The revenue from packaging programming is much greater than it is from signal transport. DIRECTV estimates it will bring in between $12 to $12.50 per month per subscriber, or $800 to $900 million annually. In addition, the acquisition will save $160 to $180 mil-lion from the resulting operations cost efficiencies.

Table 8.4 Subscription Revenues for Satellite Television Service ($ billions)

1993	1994/ % growth	1995/ % growth	1996/ % growth	1997/ % growth	1998/ % (estimate) growth
$12.1	$13.3/9.9%	$15.2/14.3%	$17.0/11.8%	$18.4/8.2%	$20.3/10.3%

Compound annual growth rate: 11.0%.
Source: DIRECTV.

Table 8.5 Direct-to-Home U.S. Market Share, 1996 and 1999

	Company	*# Subscribers*	*Market Share %*
1996	DIRECTV	2,800,000	88.0
	EchoStar DISH	350,000	11.0
	AlphaStar	31,000	> 1.0
1999 (June)	DIRECTV (Includes former Primestar sold to DIRECTV, 1999)	7,462,000	74.1
	EchoStar DISH	2,616,000	25.9

Source: DIRECTV.

Canada

Canada has one of the most modern communications infrastructures in the world. The nation's telephone infrastructure is more than 95% digitized and has more than 15 million lines, reaching more than 98% of its population. International traffic is served by five coaxial submarine cables. And a domestic satellite system that includes about 300 earth stations receives signals from 1 Pacific Ocean and 6 Atlantic Ocean satellites.

A fiber optic long haul network that will stretch across the country with landings in the United States as well is under construction by Metro-Net, now part of AT&T Canada.[23] The first segment connected Vancouver and Seattle, Washington, followed by deployment to Toronto. MetroNet operates fiber networks in major business centers across Canada, and has acquired the fiber to connect its metropolitan networks. Some Hollywood studios are now using this infrastructure to get overnight delivery of film dailies to view in California. Just two years ago, film was flown to supervising executives on the early morning "fish flight" that also carried the fresh catch for Los Angeles restaurants.

Several telephone companies are planning to provide broadband service via DSL. They are running tests before rolling out service more generally:[24]

- Manitoba Tel Sys (MTS) Winnipeg, Manitoba Down: 1.5 Mbps; Up: 64 kbps
  Tech Trial: Nov 1996—Ongoing
  Internet Access, Mutimedia, Interactive Video, VoD
  Will provide ADSL service to most Winnipeg. In March 1998, partnered with 4 ISPs and 12 computer stores, which have demo ADSL lines and act as one-stop shops.

- New Brunswick Telephone Co. St. John, Fredericton & Moncton
  Tech Trial: December 1996—Ongoing
  Down: 1.5 Mbps; Up: 64kbps
  Internet Access, TV over IP
- Quebectel (CLEC) Rimouski, Quebec Down: 640Kbps; Up: 272Kbps
  Technical and Market Trials: September 1997–September 1998
  September 1998 in Rimouski. In 1999 service will be expanded to
  Sept-Iles, St. Georges, Baie-Comeau, Hauterive, Port Cartier,
  Gaspe, and Matane

The Canadian television system more closely resembles the U.S. system than that of any other country in the world. Canada has a strong system of about eighty local television stations and, since most of its 10 million TV households are located within 50 miles of the U.S. border, most receive American television programming as well. Canada is one of the most cabled nations in the world. In 1997, cable systems passed more than 90% of homes, reaching nearly 8 million subscribers, as shown in Table 8.6.[25]

Canadian cable companies receive programming from the United States via wire as well as satellite. Vyvx, a private television carrier, was the first to bring broadcast-quality, fiber-optic television distribution to both the United States and Canada. In addition, British Telecommunications provides direct transoceanic connectivity between U.K. broadcasters and Canadian distributors. For the most part, Canadian cable companies' plans to upgrade their systems to carry digital signals or provide high-speed Internet access are well below expectations. Cablers are required to file regular reports with the Canadian television regulatory agency, the CRTC. They indicate that most cable systems outside of Quebec have limited, if any, digital capacity.[26]

As in the United States, DBS is the most credible competition to cable in the near future. Canada's leading DBS provider is Bell ExpressVu. The parent company, BCE, has invested more than $1 billion to put up a new high-power DBS satellite that will deliver digital television programming in the near future. In mid-1999, satellite and terrestrial MDS services served a combined total of approximately 500,000 subscribers. However, the number of subscribers does not include unauthorized U.S. DBS reception dishes

Table 8.6 Growth of Cable Penetration in Canada (in millions)

	1993	1994	1995	1996	1997
Subscribers	7,657	7,833	7,791	7,867	7,957
Homes passed	9,765	9,935	10,019	10,248	10,437

Source: Canada Statistics.

owned by Canadian residents. Estimates of U.S. DBS presence, including both DIRECTV and EchoStar, place the number of Canadian subscribers at about 300,000. There are about 400,000 C-band dish owners as well.

Canada is very wired into the Internet. As early as 1997, an AC Nielsen survey indicated that about 31% of adults use the Internet, and about 12% use it daily.[27] The government actively seeks to promote Internet usage and sponsors a program to provide universal access to the Internet in public and school libraries.

Mexico, and Latin America

South of the United States, communication infrastructures are far less sophisticated. Mexico has a teledensity of only 10 lines per 100 inhabitants, and progress towards building out the system is inhibited by regulatory disputes, lawsuits, and anticompetitive practices. Some analysts think that Mexicans are more likely to use wireless services than wired lines.[28] Throughout Latin America, the percentage of the population who have telephones ranges from single digits for Bolivia and Brazil to 18.3% for Costa Rica and 21.1% for Uruguay.[29]

High-capacity telephone networks are going up in other countries in the region. Cisco Systems gave a $10 million credit to help finance First-Com Corporations infrastructure buildout in Chile and Peru. FirstCom will deploy ATM networks in those companies where it already operates a 120 km fiber loop in Santiago and a competitive long distance service in Peru.[30]

Broadcast television in the region is highly controlled by governments, and throughout the region there has been only nominal deregulation. In Mexico, TV only extended to the entire country in the mid-1980s, and it is still dominated by a single family-owned company, Televisa, closely allied with the government. For the most part, the per capita incomes of Mexico, Central, and Latin America, ranging from $1,022 in Ecuador to $5,632 in Argentina, are too low to support the construction of cable systems.

In Mexico City, lower-cost wireless cable was initially popular, with several systems providing service, although growth stalled in the late 1990s. However, there are neighborhoods within larger cities where wealthier people live, and they often do have cable. In Mexico, cable service is available in some parts of Mexico City, Guadalajara, and Acapulco. At the other end of the spectrum, in Argentina, more than 50% of the population watches multichannel cable or satellite TV, and about 30,000 subscribers have high-speed Internet access via cable modems. A few systems in Mexico and Brazil launched Internet service in 1999.

Two international DBS services are battling for the eyeballs of Latin customers, DIRECTV's Galaxy service and NewsCorp.'s Sky Latin America.[31] In late 1998, they were running neck and neck with about 500,000 subscribers each. These numbers are lower than either company would like. Galaxy served twenty markets at that time and says it will break even when it reaches about 2 million subscribers. Sky beams only to four markets (Mexico, Brazil, Columbia, and Chile), and leads in two of the most important ones, Mexico and Brazil. In Mexico, Sky has about 200,000 subscribers, while Galaxy has about 100,000 customers. The service is strengthened by the presence of strong partners in each of those countries: Televisa in Mexico and Globo in Brazil. NewsCorp is aiming to have 5 million subs by 2006. The region also has some satellites that send signals only into one country. Argentina has Television Directa la Hogar (TDH) with more than 12,000 subscribers. Brazil's Tecsat Video high-power DBS service reports it has 40,000 subscribers.

There is some movement towards deregulation in the communications sectors. The growth of satellite programming has led to additional licenses for cable TV systems, and existing operators have upgraded their systems to enable cable modem and cable telephony services. If Latin American countries continue to liberalize their regulatory restrictions on communication industries, people in these markets will be able to choose between competitive providers.

The lackluster telephone infrastructure of Mexico and Latin America does not bode well for Internet connectivity, and the market is miniscule at present. There are only about 2 million PCs and between 3 million and 6 million people using the Internet—3.3% to 6.6% of Latin America's 90 million population.[32] Internet usage is growing 300% or more per year on the continent, with most of the activity coming from Brazil, Mexico, Colombia, Chile, and Argentina. A Saatchi & Saatchi study found that nine out of ten of these Web users come from the upper-middle and high social classes, who can well afford to participate in e-commerce. In Brazil, an estimated 40% of all online users have made purchases.[33]

The most prominent dot-com company that provides content and service for Spanish and Portuguese language Net users is StarMedia.com. At the end of 1998, the company received 471 page views, the number of Web pages viewed by visitors to the site. In one year, that figure grew 675% to 3.6 billion page views at the end of 1999.[34] In the same period, E-mail accounts grew from 293,000 to 2.5 million. According to StarMedia.com's 1999 Annual Report, the future of this and other Latin American ventures depend on continued improvements in regional communication policies and infrastructures: "In the past, Internet service providers in Latin America charged an average of more than $80 per month for basic Internet access. Today, monthly access fees have decreased to as low as $9 in Peru and under $20 in Brazil, Chile, Colombia, Puerto Rico, and Venezuela. A significant number of free access providers have entered the region as well."

Europe

The European telecommunications and media markets are huge, amounting to nearly a third of global activity. In 1994, the European Union working group, presided over by Commissioner Bangemann, drafted a report that urged the EU to recognize that the globalization of communication was inevitable and urged Europeans to move quickly to avoid being overwhelmed by American and Japanese competitors.

With a speed and aggressive attitude that belies notions of "ye olde worlde" charm, Europe has moved towards privatization so that, in some ways and in some nations, communications industries are less regulated than in the United States. This is remarkable because historically telephone services were provided by government telecommunications departments, referred to as PTTs, or public telegraph and telephone agencies since their establishment in the nineteenth century.

The countries of northern and western Europe have an enviable telecommunications infrastructure that is almost entirely digital, as shown in Figure 8.3. In Germany, the majority of telephone subscribers have ISDN.

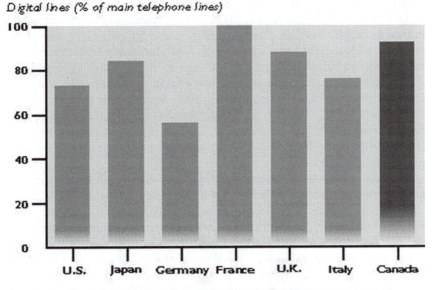

Digital lines (% of main telephone lines)

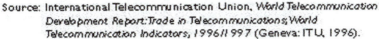
Source: International Telecommunication Union, *World Telecommunication Development Report: Trade in Telecommunications; World Telecommunication Indicators, 1996/1997* (Geneva: ITU, 1996).

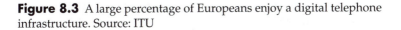
Figure 8.3 A large percentage of Europeans enjoy a digital telephone infrastructure. Source: ITU

The fiber networks that crisscross the continent are too numerous to mention, but new transborder European networks that have recently launched or will soon do so include:

Project	Main Investor(s)	Major nodes
Hermes Europe	Railtel Global Telesystems Group	Antwerp, Brussels, Amsterdam, London
Ulysses	Worldcom	Amsterdam, Brussels, Frankfurt, Paris, London
Circe	Viatel	Not released
Concerto	Flute	Belgium, Netherlands, United Kingdom
Euro	Telenor, IXC	Not released

As one travels south and east, the European picture changes and the communications infrastructure declines. Table 8.7 demonstrates the enormous variation between countries.

Terrestrial television channels have a limited reach but generally extend to neighboring countries where the same language is spoken. Germany's national public service channel, ARD, for example, is transmitted terrestrially to Belgium, Denmark, and Luxembourg. Danish channel, DR TV, is likewise received in Sweden. The advent of satellite television has extended the coverage of channels originally intended for local broadcast. For example, BBC channels are seen on cable by about 70% of Belgian viewers and 90% of Dutch viewers. RTL, a major private German channel, is transmitted by cable and satellite into Austria, Denmark, Luxembourg, and the Netherlands.

These channels are not really pan-European, in the sense that they are exported products that were intended to appeal only to the domestic market, such as the United Kingdom's BBC1 channel. The language barrier has long been a barrier to pan-European programming; however, in the past few years, a number of programs have begun to appear. They are generally broadcast in English, like Eurosport and MTV Europe, both created for the European market as a whole rather than for any particular country. There is also new programming designed to reach linguistic communities within foreign countries, such as TV5 Europe, which broadcasts to the French expatriate community living in other countries.[35]

All European countries have some system of terrestrial broadcasting, but there is tremendous variation in the extent of their cable infrastructures. At the high end are Switzerland and the Netherlands with 90% or more homes passed. The United Kingdom, Germany, and Scandinavian are in the

Table 8.7 Penetration of Telephone, Television, Cable, and DBS in Europe, by Country (in millions)

(All in millions)	Total Households or Population	Telephone Lines	Total TV Households	Cable: Homes Passed	Cable: Homes Subscribed	DTH: Homes Subscribed
Austria	3.1	3.6	3.1	1.75	1.28	1.27
Belgium	4	4.5	3.85	4.3	3.9	.35
Bulgaria	8.6 (pop.)	2.5	3.3	—	.50 (approx.)	13
CIS (Russia)	160.0	27.0	75.3	—	23.0	31.0
Croatia	4.8 (pop.)	1.3	1.1	—	.12	45
Czech Republic	10.3 (pop.)	2.5	3.98	2.5	.83	.91
Denmark	2.5	3.5	2.2	1.65	1.5	.9
Finland	2.3	2.9	2.1	1.3	.92	.17
France	22.9	32.4	21.6	6.65	1.55	1.4
Germany	37.45	40.1	32.5	24.3	19.2	9.8
Greece	3.65	5.0	3.3	0	0	.85
Hungary	10.3 (pop.)	3.5	3.9	2.45	1.95	1.3
Ireland	1.1	1.11	1.0	.95	.58	.11
Italy	24.0	22.0	23.6	—	.07	.64

Table 8.7 Penetration of Telephone, Television, Cable, and DBS in Europe, by Country (in millions) (Continued)

(All in millions)	Total Households or Population	Telephone Lines	Total TV Households	Cable: Homes Passed	Cable: Homes Subscribed	DTH: Homes Subscribed
Luxembourg	.18	.21	.18	.13	.17[a]	.04
Netherlands	6.8	7.8	6.6	6.1	6.2	.375
Norway	2.1	2.36	1.77	.87	.78	.37
Poland	13.1	6.6	12.6	3.0	2.65	2.2
Portugal	4.3	3.2	3.2	48	.31	.28
Romania	24.1 (pop.)	8.1	3.5	2.3	1.9	.85
Slovakia	2.25(pop.)	1.3	2.05	.68	.475	.715
Slovenia	2.5 (pop.)	.6	.65	.45	.27	.2
Spain	39.3 (pop.)	16.9	11.8	1.2	.45	1.69
Sweden	8.8	6.1	4.2	2.2	1.9	.94
Switzerland	3.1	4.25	2.97	2.5	2.3	.26
Turkey	15.3	10.0	11.9	1.8	.425	.425
United Kingdom	20.4	26.21	—	12.1	4.2	4.0

[a]Includes subscriptions to wireless cable
Source: Compiled from data provided by European Union Information Society.[36]

middle with systems passing upwards of 60% of homes. Then there are Italy, Greece, and some Eastern European countries with little or no cable. Like terrestrial broadcasting, cable companies operate in their own countries.

One exception to the rule is in the United Kingdom, which developed its extensive cable infrastructure through partnerships with U.S. companies. One large cable operator is Telewest, owned mainly by Tele-Communications Inc. (now owned by AT&T) and United States West (which merged with Qwest). Cable and Wireless Communications is partly owned by Nynex and BCI. In France, there are two major cable network operators, Lyonnaise Communications and Compagnie Générale des Eaux, and they are both private French companies. In the Netherlands, cable systems were held by local authorities until they were sold to private, mostly Dutch, companies. In Germany, Spain, Denmark, and Sweden, the largest cable provider is still the national telecommunications company.

Pan-European television has become the purview of the satellite services. The two major companies that own and operate satellites are Eutelsat, which carries 550 video and audio channels, and SES (Astra), which transmits 880 channels. Both of these companies can send signals all across Europe. Companies that package and market transborder programming are called satellite television providers, such as Canal Plus/ Nethold, BskyB, and Kirch. Canal Plus (CP) is the giant in this arena, serving 8.5 million subscribers. Since it acquired the European operations of Nethold, CP holds a commanding position in France, Belgium, the Netherlands, Scandinavia, Italy, and Spain.[37]

Multichannel television service has been remarkably successful in Europe. Within five years after it was introduced, some public broadcast channels lost 47% of their audience. It generated a substantial pay-per-view (PPV) market of $135 million in 1998, a 350% increase over 1997. PPV revenues will grow to $940 million by 2004.

Europe is also developing some very high-tech film and television production centers. The United Kingdom is rapidly becoming the Hollywood of the region, and neighboring countries regularly send their footage to London for some of the most sophisticated postproduction work in the world. Director Steven Spielberg frequently works with facilities there. The West End area of the city has a media network, SohoNet, which is used to squirt material to various postproduction houses for collaborative work.[38] SohoNet also has a high-speed connection to Los Angeles to serve its U.S. clients.

Film buffs will be familiar with the famous Babelsberg studio just outside of Berlin, Germany. Today, the old lot is called Medienstadt (Media City). It houses a state-of-the-art facility with sound stages, digital postproduction, multimedia production, and even housing. Of course it is wired by a high-speed fiber optic cable network that ties in directly to Germany's sophisticated telephone system.[39]

The use of the Internet in Europe is soaring, and Finland has the highest rate of usage in the world, about 50%. (See Figure 8.4.) In Sweden, 30% of the adult population of 8 million people are online. Overall, the number of users is growing at an annual rate of 37.3% and will reach 82 million in 2002, with e-commerce sales revenue projected to reach $55.4 million by that year.[40]

France has had its own network called Minitel for fifteen years. About 3 million businesses and 6 million people use it to make more than $1 billion in purchases per year in 1998. However, reluctantly, the French have decided to link Minitel with the Internet, and France Telecom signed a deal with IBM in 1998 to help with the migration. Now the Web has caught up with Minitel and has almost 6 million users. Internet e-commerce, which Jupiter Communications estimates about $120 million in 1999, however, lags far behind spending over Minitel.[41]

The European Commission continues to encourage telephone companies to establish conditions for increased Internet participation. And Net companies are a vibrant area of entrepreneurial growth. In the United Kingdom, Cambridge has emerged as the region's Silicon Valley. The area, known as Silicon Fen, hosts more than 1,200 high-tech companies.[42]

Asia

Asia Pacific is third after North America and Europe in terms of its telecommunications development. The region will need to invest as much as $25 billion a year for several years in order to build everything from basic telecom networks to sophisticated services. At present, telephone line penetration ranges from a teledensity in Japan that is the equal to any other developed country, to a figure of around one per 100 inhabitants in countries like China. In between are countries such as Taiwan. Within

	U.S.	U.K.	France	Germany	Europe (Net)
Heard of it (Net)	79%	93%	45%	79%	72%
Know a lot about it	5%	8%	3%	3%	5%
Know a little about it	24%	35%	17%	28%	27%
Heard of it but know nothing about it	51%	50%	25%	48%	41%
Never heard of it	20%	6%	55%	21%	27%
(Number of respondents)	(802)	(800)	(800)	(802)	(2402)

Figure 8.4 The United States and European countries have a similar awareness of the Net, compared to many other areas of the world. Source: PricewaterhouseCoopers.

China, development plans call for a projected teledensity of 67.48 (per 100) in Guangdong province by 2000. A few cities, like Singapore and Hong Kong, are totally wired for digital. China's investment is massive, as shown in Table 8.8.[43]

China's window on the entrepreneurial world, Hong Kong, is impressively wired. Speaking at a telecommunications forum, the Hong Kong secretary for information technology and broadcasting cited statistics to profile Hong Kong's telecommunications infrastructure and services: fixed telecommunications networks are 100% digital; 36% of households and businesses have connection to broadband networks, and this figure doubled to about 70% in 1999. About 3.6 million exchange lines serve the population, or 108 exchange lines per 100 households. Mobile telephone penetration is 37%, one of the highest in the world. Homes with PCs now constitute about 34%, and video-on-demand services were launched commercially in March of 1998 by the city's Wharf Cable system.[44]

The region is well-served by submarine cables, especially Japan. In the past five years, there have also been several landings in China, and more are planned. Overall in Asia, total telephone lines have increased by 11% since 1990, and Asian countries are spending as much as they can afford. There is wide recognition that improving telecommunications will underlie economic gains. Malaysia continues to invest in its Multi-Media Super Corridor project, and China will spend $62 billion by 2000 on fiber optic cable to double the number of phone lines in the country. The developing nations of Vietnam, Burma (Myanmar), and Cambodia are still limited by financial resources in establishing sophisticated telecommunications infrastructures, although Vietnam has a $300 million investment program to expand its telecom service.[45]

The Pacific Rim is currently the world's fastest-growing television market. The economic growth of the area, combined with the potential

Table 8.8 China's Telecom Growth, 1999

Key Items	Growth, Jan.–June, 1999	Accumulated Capacity
Capacity of central switching system	8.16 million lines	143 million lines
Capacity of GSM mobile switching system	10.47 million lines	39.88 million lines
Telephone installed base	16.47 million sets	148 million sets

Source: China Research, July 5, 1999.

audience size, have made these countries a source of revenue for hardware manufacturers and programming suppliers.

There are four ways Pacific Rim countries receive television: local broadcast transmitters, cable, wireless satellite (microwave), and direct-to-home satellite services. As mentioned earlier, the level of economic development, roughly approximated by per capita income, determines which type of delivery prevails. Asia, Hong Kong, Japan, Singapore, and Taiwan are high-income areas. Malaysia is upper middle income. Thailand and the Philippines are lower middle income, and the rest of the Asian countries fall in the low-income category.

Modern fiber optic cable systems exist in Hong Kong, Singapore, Taiwan, and China. To a lesser degree, they also exist in Japan, South Korea, and China. China is an exception to the general rule of per capita income determining delivery mode, because the government is paying to build modern cable facilities.

Japan could certainly afford to deploy a sophisticated cable infrastructure, but the number of headends is still limited. The public turns to satellites to consume multichannel fare. Foreign services provided by DIRECTV and News Corp. have not been accepted at the levels the companies expected. DIRECTV's internal research indicated that the paucity of Japanese programming was the reason. Satellite delivery of television is well-developed throughout the region. (Besides Japan, News Corp. is beaming programming into Malaysia, Indonesia, and many other Asian countries.)

In China, access to foreign satellite channels is prohibited except in star-rated hotels and foreign compounds. Only domestic channels like CCTV and some provincial satellite broadcasters are allowed to deliver their services through satellite to remote areas. However, DBS is coming to China in a big way. APT Satellite Holdings Limited received a government grant to provide satellite television uplink, downlink, and broadcasting services.[46] APSTAR-IIID high-transmission-power DBS satellite launched in late 1999 and broadcasts about 100 programs to the People's Republic of China (PRC). China is adding satellite service to carry Internet traffic as well. In December of 1998 Intelsat announced it has a five-year contract to provide China Telecom with an 8 Mbps link to the Internet backbone.

Asia has about 4 million Internet users.[47] *Fortune* magazine ranked wired-ness in Asia with Japan at the top, followed by Singapore, Hong Kong, South Korea, Taiwan, Malaysia, Thailand, Indonesia, Philippines, and China. In Japan, more than 1.5 million people are online.

A growing percentage of Singapore's 3 million population, and is far along on its goal to network all businesses and homes into an "intelligent city." Nearly 11% of Hong Kong's citizens are online as well. In Thailand, the Asian economic crisis slowed development there; nor has the country deregulated services, so growth is still inhibited by high charges that make it impossible for average citizens to afford Internet access.

Malaysia is using the Internet to bootstrap its information economy.[48] University Tun Abdul Razak (Unitar) is a cyber-university that is part of an ambitious national agenda for transforming Malaysia into a technology superpower. It uses the Net to teach a new generation of engineers and technically sophisticated workers for the country's $40 billion Multimedia Super Corridor project, a nine-mile by thirty-mile broadband networked zone dedicated to cutting-edge digital technology.

Oceania: Australia and New Zealand

Australia has a modern telephone infrastructure from its two major providers, Telstra and Optus, with about 9.41 million subscribers; the number of lines per 100 is 46.5. Aussies are enthusiastic international communicators, and a new pan-Pacific submarine cable, the Southern Cross, will soon provide increased capacity.

For terrestrial television, the government funds multiple channels, including national, regional network, and community stations. There is some requirement for Australian programming, and even commercial television networks increased their Australian content in 1998 in line with the Australia Broadcast Authority's requirement for a minimum 55% Australian transmission quota.[49]

Pay satellite TV started in Australia in 1995, with Galaxy being the first company in the market. In 1996, Galaxy was joined by two heavyweight telecom companies backed satellite operations, Optus Vision and Foxtel. With 1997 per capita income standing at $21,202, it is not surprising that Australia has a good cable infrastructure. About 18% or 3.4 million Australians subscribe to some kind of multichannel television service. The same companies that provide satellite service also own the largest cable systems. Providers include: Optus (cable and satellite), Foxtel (cable and satellite), Austar (cable, MDS, satellite), Neighborhood Cable (cable), Tarbs (MDS, satellite), and TBG Internet (satellite).[50]

Australians are high users of the Internet and computers. About 18% of Australians have accessed the Internet and three-quarters of these recently.[51] According to the OECD, Australians are ranked fourth for home and business usage of the computer, behind the United States, Hong Kong, and Canada. Australia has twice the OECD average number of computers per 1,000 population connected to the Internet, and has the third largest number of servers per 1 million inhabitants in the world.

Nearby New Zealand is also well-served for international communication. Two of the four Intelsat series satellites, VII and VIII, stationed in the Pacific Ocean region, provide Australia and New Zealand with primary- and major-path satellite support for virtually all international telephone and private services traffic to and from these countries. Submarine cables landing in New Zealand are:

- PacRimEast digital submarine fiber optic cable, NZ—Hawaii
- Tasman 2 submarine fiber optic cable, NZ—Australia
- Tasman 1 submarine coaxial cable fitted with digital modems, NZ—Australia
- ANZCAN submarine coaxial cable fitted with digital modems, NZ—Australia—Fiji
- Telecom New Zealand is also a major sponsor of the planned Southern Cross fiber optic cable network

TVNZ, TV3, and News Corp.'s Sky service currently offer a total of ten terrestrial-based channels.[52] Prime commenced broadcasting one nearly nationwide channel in August, 1998. Cable is not a major force in New Zealand, as shown in Table 8.9. Saturn does provide a cable service in parts of Wellington, and New Zealand Telecom installed cable in areas of Wellington and Auckland but ceased expanding its network. The telecom is now investigating the services it might provide using ADSL (asynchronous digital subscriber line) technology on its copper pair telephone lines. Sky broadcasts 30 digital channels over the Optus B2 satellite.

Table 8.9 Television Service in New Zealand

New Zealand Television Industry				
Operator	*Delivery*	*Main revenue*	*Homes Passed*	*Ownership*
TVNZ	VHF/UHF	Advertising	100%	65% Government
TV One				
TV2				
MTV				
TV3	VHF	Advertising	96%	CanWest
TV4			29%	
Sky	Satellite	Subscription	100%	INL/ News Corp
	UHF	Subscription	73%	
New Regional	UHF	Advertising	89%	Prime
First Media	Cable	Subscription	65,000	Telecom
Saturn	Cable	Subscription	40,000	UIH

Source: AC Nielson, channel share of 18- to 49-year-old viewers, 6:00 P.M. to 10:30 P.M., January to March, 1998.

Currently there are an estimated 560,000 people using the Internet in New Zealand. This accounts for 15% of the total population. New Zealand is placed among the top twenty countries in the percentage of Internet population, the total population, and many of the common measures of development.

India and the Middle East

As a whole, India has a very low teledensity, about 2 telephones per 1,000 people in the most wired areas, and less than 1 (.49) in the least connected ones. With a population of close to 1 billion people, the estimated 19.1 million phone lines represent only a superficial market penetration. For awhile, it looked as though deregulation would bring investment from foreign companies that would begin to improve the country's infrastructure. Unfortunately, this has not happened. The telephone system itself is mainly copper, although fiber has replaced switch interconnects in the major cities, and efforts to install a nationwide trunk fiber optic cable are underway.[53]

The policies of India's Department of Telecommunications caused a flight of the outside companies that might have made a difference, including exorbitant license fees, high interconnection charges, prohibition again private operators handling intercity or value-added traffic, and saddling the private operators with economically nonviable deployment of telecom facilities in villages. Four years after telecom reform, only two private networks were launched, and there were no private sector takers for more than a dozen telecom market availabilities.[54]

Indian audiences currently receive television from a government-owned and operated service, Doordarshan. In 1995, Indian television underwent major change. Doordarshan TV was granted autonomy, and private broadcasters were allowed to both send and receive programming via satellite. India also passed new regulations over its chaotic cable industry. At present, 25 million Indian homes have cable TV, so there are more cable television customers than telephone subscribers in India. The industry is comprised of more than 600,000 cable operators in India, most of which have a small customer base of 50 to 100 households, and photographs of cable lines running up and over fences, down alleys, and into basement windows show just how basic "basic cable" can be.

Naturally, the dearth of phone lines constitutes the greatest immediate hurdle to the growth of the Internet. As noted, there is still considerable dependency on the government to provide service, and the Department of Telecommunications retains control of domestic transmission capacity and responsibility for intercity Net connections. Until such time as free market ISPs are in a position to set up their own international gateways, they remain dependent on the former state-owned provider VSNL. In early

1999, the communications agency licensed 75 domestic ISPs to operate, dramatically lowering access costs, although it did not address the high price of telephone service.[55] Another barrier to Internet access is that the total number of PCs in India is estimated at just 3 million nationwide.

In early 1998, there were about 50,000 Internet connections in India, and an estimated 300,000 Hotmail.com accounts. The number of domain hosts grew from 7,175 in January, 1998, to 13,253 in January, 1999, according to Network Wizards. Network Solutions reports that India is the fourth most active source of domain name registrations from Asia, and the tenth most active in the world.[56]

Many of India's PCs are in universities, and the country, like Malaysia, is using the Net to teach students technology skills. They can earn a bachelor's degree in information technology online, operated by the Indira Gandhi National Open University (IGNOU), a distance learning institution that has wired classrooms in telecenters in twelve cities across the country. The Quantum Institute also offers U.S. degrees online to students in India, as well as a master's degree in computer science from the University of Illinois at Urbana-Champagne. It is a twelve-month program that costs about $8,000.

The regions least-served by international telecommunication and television services are the Middle East and Africa. In the Middle East, this situation exists largely because governments enforce religious values that prohibit their citizens from participating in international communications or viewing programming that is seen by people in the rest of the world.[57] However, there is an enormous and ever-increasing demand for local communications. Estimates are that the Middle Eastern countries will spend around $30 billion on telecommunications by the year 2000 and will invest as much as 4% of their gross domestic product in communications infrastructure over the next fifteen years. Satellites and mobile systems are launching at an unprecedented rate across the region.

Bahrain is the major communications center in the Gulf and has invested heavily in telecommunications since 1960. The country will spend another $300 million per year for the next five years to further expand and upgrade its network. It is increasing the number of wireless lines from 35,000 to 80,000, which will reach more than 10% of the nation's citizens. In addition, Bahrain is allocating $250 million over the next five years for broadband services such as global messaging, videoconferencing, cable TV, video-on-demand, and premium audio-text services.

The development of satellite communication services such as Iridium, Globalstar, Thuraya, and ICO Global (Inmarsat) will provide people in the Middle East with access to the global world of communication. The FLAG (Fibre-optic Link Around the Globe) now connects the Fibre-optic Gulf Cable (FOG), linking Bahrain, Kuwait, the United Arab Emirates, and Qatar, and will now support the huge investments that are required to improve telecommunications in the region. All this will soon give the Middle East fiber optic connectivity of the highest quality.

The convergence of telecommunications and computer communication technologies is opening up huge opportunities throughout the Middle East. Internet penetration in the Middle East is probably among the lowest in all regions of the world. This is likely to change when the region has access to the bandwidth required to transport huge volumes of Internet data. This in turn will fuel the demand for Internet access providers. Telephony and multimedia is another growth area in the Middle East.

The region's countries have governmentally operated terrestrial broadcast stations; for the most part, these countries do not have cable systems. One exception is Israel. It has 1.5 million TV households, and almost all of them, 1.35 million, are passed by cable. About 910,000 subscribe, and another 10,000 pay for satellite multichannel service.

As with television, countries in the Middle East are inhibiting Internet usage through strict censorship, taxes, and outright bans on some foreign Web sites, says a report from the Human Rights Watch.[58] The most restrictive policies are exercised by Iran and Libya, which have no Internet connection at all. However, these actions are not universal across the Middle East. Jordan's government has stressed the benefits of the Internet and has imposed few restrictions. For example, users do not have to register to visit or create a Web site, although even here they are so heavily taxed that an account costs about $70 per month.

The Human Rights Watch report said that Morocco and Egypt were even less regulated, but that Saudi Arabia mounts the region's most ambitious plan to block the flow of what it believes is inappropriate data. In May 1998, Saudi Arabia prohibited Internet Service Providers and users from using the Net for pornography and gambling or any activities violating the social, cultural, political, media, economic, and religious values of the Kingdom of Saudi Arabia.

Africa

Africa's connectivity to the rest of the world is provided mainly by geostationary satellite links, and telecommunications traffic between countries is often sent through space to Europe and then forwarded to the terminal number. This "double hop" costs millions of dollars a year.[59] Now, however, fiber optic links are beginning to complement and, in some cases, replace these services. The SAT-2 submarine cable is already in place, tying South Africa to Europe, the United States, and South America.

The Africa One cable, instigated by the International Telecommunications Union in a bid to develop a solution for interconnecting African nations, was mainly constructed by AT&T. Owned and operated by African telecom companies, Africa One launched in 1999 and connects 41 landing points in Africa using a 39,000 km fiber optic network. It links the continent directly with Italy, Saudi Arabia, Portugal, and Spain, and interconnects with other global undersea cable systems. Africa One was originally projected to cost $2 billion, but the specification of the network has been adjusted so that the ring and the coastal links will now cost $1.3 billion, releasing more capital to interconnect landlocked countries.

Another new cable for southern and western Africa is the SAT-3/WASC/SAFE cable. It will significantly expand bandwidth capacities to Africa when it enters commercial service in 2001. Plans call for it to have a 40-Gbps capacity, upgradeable to 80 and 120 Gbps along a route connecting Portugal and Malaysia. Landing points include Senegal, Cote d'Ivoire, Ghana, Benin, Nigeria, Cameroon, Gabon, Angola, South Africa, Mauritius, Reunion, and Malaysia. In addition, the East African Cooperation (EAC) has established a $60 million dollar project for a high-speed digital fiber optic backbone linking Kampala, Nairobi, and Dar es Salaam.

Africa has such a legacy of poor infrastructure, especially in rural areas, that three-quarters of Africans have never made a phone call. In 1994, over 4 million people were on official waiting lists for phone lines. High demand is one reason why Africa's telecom utilities, nearly all of which remain government monopolies, are among the most profitable in the world. The average revenue per line in Africa is $1,225 compared to the world average of $735. There tends to be a large urban-rural gap in telephone service as well, because more than 50% of phones are in capital cities, while as much as 80% of the population lives in rural areas.[60]

Like other continents, there is variability between African countries in infrastructure. Some African countries have made telecommunications a priority and are building fiber optic intercity backbones with digital switching and new cellular and mobile technologies. Sophisticated national networks include those in Botswana and Rwanda where 100% of the main lines are digital. At the other end of the scale are countries like Madagascar and Uganda with unreliable, fragmentary systems.

Terrestrial broadcast television stations are operated by governments throughout the region. However, most African countries also get some kind of multichannel subscription service. In southern Africa, Multichoice offers TV over MMDS (multichannel multipoint distribution service) to 900,000 customers and satellite television to 400,000 subscribers.

PanAmSat, the world's first private satellite service provider, now also has two satellites serving Africa, one of which is being used by the South African Broadcasting Corporation to beam TV to rural villages in southern Africa. Long-time plans by the 41-member RASCOM (Regional African Satellite Communications Organization and Comsat World Systems) also came to fruition when Intelsat 804 launched in December, 1997, and now carries TV signals into Tanzania, Mozambique, and Uganda.

Next on the agenda is Intelsat 805, which will orbit above Zaire as the first dedicated land-mass African satellite.[61]

RASCOM has more on its agenda. By the end of 1999, the agency was testing ground systems using signals beamed down from existing satellites. If the stations work well, two more satellites should be launched in 2001 at a cost of about $400 million. An additional $800 million will be spent over the next seven years to implement 500,000 rural ground stations. The facilities will incorporate solar-powered telephone booths where people can place and receive satellite calls. Alcatel is a partner with RASCOM in this effort.

One use of satellites is already helping. Healthnet, for example, is an organization that employs a low earth orbit (LEO) satellite, simple ground stations, and radio and telephone-based computer networks to bring more effective and timely healthcare to rural areas of the developing world.[62] With ground stations in fourteen African countries, the system is being used by physicians to schedule visits, to communicate with experts in larger cities or overseas, and to gather and share data. However, operations are limited because each "pass" of the Healthnet Satellite, HealthSat-2, lasts only 13 minutes at present.

South Africa is the eighteenth largest user of Intelsat services in the world and by far the biggest user of satellites in Africa. The country's transponder use is growing rapidly, and is very expensive for both business and government. South Africa contributed towards the 804 satellite and will also do so towards 805; its national provider, Telkom, plans to integrate its high-capacity VSAT (very small aperture terminal) ground station network with the satellites and its upcoming South Africa to Far East (SAFE) cable.

The poor telecom infrastructure naturally prevents people from accessing the Internet, and the small number of PCs in Africa is also a problem. Because the Internet is growing and the telecommunications infrastructure in Africa is changing, statistics change quickly. In 1998, estimates of the number of African Internet users ranged from about 800,000 to 1,000,000.[63] Most of them were (and are) in South Africa, approximately 700,000 at the time, leaving about a 100,000 among the remaining 700 million people on the continent. This would mean that Africa had about one Internet user for every 5,000 people. The world average is about 1 user for every 400 people; in North America and Europe, it's about 1 for every 46 people, depending on the country.

Mike Jensen who has led efforts to bring the Internet to Africa and tracks usage for the Internet Society, reported in 1997 that "more than three-quarters of the capital cities in African countries have developed some form of Internet access." Thus, at that time 43 of 52 countries were connected. By comparison, in 1989, only Egypt and South Africa were connected. By 1998, most capitals had more than one Internet service provider, and 11 countries had active markets, including the Ivory Coast, Egypt, Ghana, Kenya, Morocco, Nigeria, Senegal, Tanzania, Uganda, Zimbabwe, and South Africa.

High connection cost is another reason for the low number of users. Jensen gathered prices from around the continent and found that the average cost per year for 5 hours per month connection time is $722, ranging from lows of $136 in Botswana and $226 in South Africa, to highs of $880 in Algeria, $1105 in Uganda, and $1740 in Angola. Average monthly cost in Africa is $60. A joint ITU/UNESCO project developed in 1997 sought to improve Internet connectivity in Africa, but nothing seems to have come of it.[64]

As in Asia and India, the Internet is being used for education. The World Bank created and now manages the African Virtual University, where students from some of the poorest countries in the world can get an education. AVU hopes eventually to reach 500,000 students during an academic year, with the goal of educating a technologically skilled workforce that can participate in a digital economy.[65]

The Global Television Market in Programming

We have seen that there are great differences in communications infrastructure and usage. Yet it seems clear that in the next decade, almost every place on earth will have at least a minimal level of connectivity, far more than has existed in the past. Much of the content is communication between people, messages that come out of the interactions between them. But entertainment fare is typically created by the giant entertainment industry. This chapter now turns to the global programming created by that industry: who watches it and the market mechanisms that deliver it to them.

Eight of the largest ten media and entertainment companies in the world are U.S. entities; the exceptions are Sony and the German publishing giant, Bertelsmann. Feature films, television programs, and recorded music are premier American products, and entertainment has been a strong U.S. export since the 1920s, when American-made movies achieved worldwide popularity. Studios and distributors built on this success with U.S. television programs, whose licensing in other countries has given the nation's television industry a strong global presence since the 1960s.

Ambitions for a worldwide television market emerged during the Kennedy administration's New Frontier policy when "U.S. policymakers first envisioned a global television system linked by satellite technology. Their utopian discourse suggested that, in the face of Third World 'unrest' and growing Soviet competition, television would play an important role in promoting an 'imagined community' of citizens throughout the Free World."[66] Then and now, entertainment is one of the most largest and most important U.S. industries.

Today, the forces propelling the globalization of entertainment are mostly economic, rather than political or ideological. One such force is the change in the financing of television programs in the United States. Prior to 1978, TV networks owned outright the programs they aired. In that year, Congress passed the "financial interest and syndication rule" that prohibited networks from such ownership.

At first, the TV networks financed programs in their entirety. However, as the network audience declined under the onslaught of cable viewing and VCR penetration, networks gradually reduced the amount of funding they would provide to a production company. Studios and production companies, receiving only about two-thirds of the money needed to produce an episode, turned to distributors and syndicators to make up the difference, a process known as "deficit financing."

By 1999, international sales of programming accounted for about 50% of content producers' revenue, according to Bruce Johansen, president of the National Association of Television Programming Executives.[67] As a result, in recent years, producers have turned to international coproduction arrangements, increasing the number of programs designed to appeal to a multinational audience. Another factor encouraging a global TV program market is the privatization of television systems around the world, leading to the creation of many additional television channels. To fill the increased demand, there are now six international markets for the purchase of television programs: NATPE (National Association of Television Programming Executives), Monte Carlo, MIP-TV (Marche International des programmes de Television), MIPCOM (Marche International des Filmes et des Programmes Pour la TV, la Video, le Cable et le Satellite), May Screenings, and MIP Asia.

NATPE is held in the United States in January of each year. Primarily a market for domestic syndication where original and off-network reruns are marketed to over-the-air network affiliate stations, independent stations, and cable networks, NATPE has become a popular market for Latin American and Canadian customers. For European buyers, the early-February Monte Carlo market is probably the most important venue, where producers and syndicators introduce midseason shows to the European market. MIP-TV, a March market, used to be where Europeans took their first look at U.S. television pilots for the fall. Lately this market has declined in importance as U.S. networks have moved the announcements for their fall schedules to June. However, foreign television sales generate as much profit as a film's domestic box office, so increasingly Hollywood film producers who come to Europe for the Cannes Film Festival in May check out the MIP-TV market in April on their way to Cannes.[68] The new timing has made the New York May Screenings show more important, and more than 400 buyers from around the world descend on the Big Apple to look at new programs from producers.

MIPCOM and MIP-TV were established by U.S. program producers and distributors as European sales venues for their products. MIPCOM is a mid-October market, while MIP-TV, held in April, is geared to U.S. network-style programs. MIP-TV is especially useful because the U.S. fall television season has shaken out the successful programs, allowing buyers to predict their popularity, consistency, and quality. MIP Asia is a new market, held for the first time in Hong Kong in late November of 1994. It caters to the new television stations and satellite delivery services in the Far East.

The growth of digital satellite services and other multichannel pay services that include a pay-per-view tier of programming is becoming another important revenue stream for studios and production companies that distribute feature films. And motion pictures are important in premium channels and terrestrial broadcast well. The largest deal of this type exists between Paramount Television Group Television Par Satellite (TPS) and TCM Droit Audiovisuels (TCM). Paramount will license its current motion picture product for TPS and TCM's pay TV, free television, basic cable, and pay-per-view services. The two companies will also have access to Paramount's extensive library of motion picture and television product for their terrestrial, cable and satellite channels.[69]

Trends in the International TV Program Market

By far the most obvious trend is growth. The last decade has seen an enormous expansion in the number of channels worldwide. This larger market has resulted in greater demand for programming, of which the United States has been the primary beneficiary. Overall, analysts estimate that Hollywood will generate more than $3 billion in annual sales to TV channels outside the United States in 1998, a figure that grew rapidly throughout the 1990s. As far as home video rentals and sales go, just as in the domestic market, international sales and rental of popular movies is a lucrative market. For example, in Japan, Paramount Home Video sold more than 900,000 units of Paramount home videos in 1996 alone.[70]

Channel Marketing In past decades, the market for programming has been individual films and programs, or packages of them as they aged and became less valuable. However, the market has moved towards the packaging, buying, selling, and distribution of entire channels. The lineup of channels offered over digital satellite services are sometimes called a "digital bouquet." MTV, CNN, and Discovery stand out as especially successful, and many more are scheduled to follow. The marketing of channels has substantially reduced the value of individual works.

CNN transmits five different services: CNN, CNNfn (a financially-oriented channel), Headline News, CNN International, and CNN en Español. Because of different time zones, sending out these five channels actually involves the distribution of 27 "screens" or transmissions.[71] CNN has also spawned considerable competition. Spanish-language multi-channel subscription homes in Mexico and Latin America receive Telenoticias, an all-news channel prepared by Reuters TV. Discovery Networks has eight services. By far the biggest, most watched channel in the world is MTV. This Viacom product—MTV Networks Asia, MTV Australia, MTV Brasil, MTV Networks Europe, MTV Networks Latin America, and MTV Russia—is available in 298 million households in 82 countries.

There are many other examples of channels for export. ESPN Latin America airs in 41 countries. It is the most widely distributed pan-regional cable network in Latin America over satellite, cable, and MMDS, reaching 11.3 million households in Mexico, Central America, the Caribbean, and South America. The Country Music Channel is beamed to Latin America and Thailand. The Discovery Channel is seen by Latin American viewers, and HBO is a favorite of the Czech audience.

TV shopping is an extraordinarily popular exportable channel format. From Canada and the Caribbean, to China and Japan, the United Kingdom and Israel, there are hundreds of millions of viewers buying merchandise they see on television. The business success of this kind of marketing depends on a preexisting credit card economy and a delivery infrastructure for rapid, accurate fulfillment, however, which limits the countries where these ventures can be profitable.

Not all attempts to market channels have been successful. NBC tried to take its successful CNBC to the Far East as ANBC, a 24-hour Asian business channel that was projected to go to about 24 million households in 15 countries.[72] However, the venture became bogged down in cultural, economic, and institutional difficulties, all of which led to too-high operating costs and organizational upheaval. Two years later, the joint venture was merged with Dow Jones, re-branded, and moved to Singapore.

Program Format Export   Many program types simply do not travel well, including comedy and game shows. Producers are becoming adept at marketing franchises to shows that have proved popular elsewhere. They sell the format—the way the show is put together—to a production entity in a foreign territory, and help that company fashion a local version of the program. For example, *Wheel of Fortune* in Brazil replaces Vanna White with a Portuguese-speaking Brazilian. The host, graphics, music, products, contestants, and other aspects of the production all come from local sources.

Internet Program Creation   At the moment, most servers and Web site development is in the United States, although some production is occurring in the United Kingdom and European centers, as well as Australia and Asia. The Internet is the premier training ground for learning how to create digital and interactive programming for both the Web and broadband systems. Although content creation is not always profitable, it is an essential starting point for companies to acquire the necessary expertise. There is some urgency in non-United States production centers to get online, so that they will not cede the next generation of content to traditional American centers of cultural manufacturing.

Barriers to the Globalization of Broadband Communication

In spite of the enormous growth of world networks, there are factors that could prevent, or at least slow down, a global broadband network. These factors include: infrastructure disparity; censorship and other restrictions by regional and national governing bodies; and violations of intellectual property rights. The disparity in the quality of communications infrastructures not only limits the economic development of these areas, it also slows the globalization of broadband networks and other rich communication modalities.[73] The solution to this problem is difficult, given the high cost of constructing sophisticated communication infrastructures, but it is particularly difficult to construct new pathways into consumer households—sometimes called the "last mile" problem.

The satellite distribution of television programming to elites anywhere in the world already occurs. However, the vision of a truly global communication system, especially one that is broadband and interactive, will not come about solely through wired networks. At an approximate cost of $400 to $1,500 per household connection, the world would need to invest more than a trillion billion dollars. Therefore, hopes for ubiquitous broadband networks rest with some form of wireless infrastructure and their lower implementation costs.

Nor does infrastructure pose the only barrier. The need for international communication to support economic, political, and cultural activities motivates nation-states toward allowing their citizens and institutions to engage in transborder data flows. But social and political considerations often lead them to restrict such interaction. We have already presented some of the limits placed on communication, television viewing, and Internet access by specific governments.

Typically, the American zeitgeist encourages open communication policies. However, in fairness it must be said that too much openness can have serious negative consequences, resulting in political instability; the devaluation of local culture, religion, and customs; the consolidation of the yearnings of ethnic and linguistic groups that transcend national boundaries; and the importation of images that encourage violence and rebellion. These concerns often seem unimportant to those who hold the value of free speech and other associated communicative freedoms as their highest ideal, but they are not trivial—it is no accident that a successful coup is always accompanied by the capture of the governmental palace, army headquarters, and the national television and radio facilities.

Some countries worry about a potential loss of cultural identity. Canada, France, and Australia are examples of nations that have recurrent concerns about a flood of international information, particularly from the United States. One strategy they have adopted is to place quotas on the amount of foreign programming television distributors may show. Led by the French, in 1989 the European Union's fifteen member states agreed to restrict foreign programming to 50% of non-news and sports programming, which had to be of European origin where practicable.

In February, 1994, the French successfully blocked movie and television programming from being included in the free trade provisions of the General Agreement on Trade and Tariffs (GATT). In February, 1995, the French were able to influence the EU Executive Commission to close the loophole created by the words "where practicable." The group placed less stringent restrictions on thematic channels, such as cartoon and all-movie channels, which must have 25% of their programming coming from European production companies. Video-on-demand and other new technology services were exempted altogether from quotas.[74]

The French position is controversial, even within the European Union. For example, Germany and the United Kingdom are far less concerned. The London *Economist* magazine noted that 85% of Europe's film directors are over 50 years old, and the editorial page of the *Munich Süddeutsche Zietung* observed: "Prime time on European television still belongs to Hollywood and the wee hours will belong to a lonely, doubtless very 'cultural' European movie. As it circulates through the sprockets, it will discharge only one function: to meet the quota."[75]

Similarly, the Australians established content quotas in 1998. The government monitored channels to insure that they were met. However, according to Bruce Johansen, president of the National Association of Television Programming Executives, quotas are going by the wayside. In a multichannel world and one where the Internet courses through the International bloodstream, quotas are difficult to maintain, and efforts to enforce them are largely ignored by operators and consumers alike.[76] Even aided by computerized filtering, the sheer logistical difficulty of monitoring an infinite number of television channels and Internet usage when software programs can insure anonymous logins and address masking is discouraging to the most zealous censor.

Another barrier to the distribution of valuable information and entertainment across global networks is the lack of an effective global legal system. Enforcing intellectual property rights against theft has surfaced as a severe problem of world trade, as exemplified by the conflict between the United States and China, first over CD-ROMs and now over video CDs and computer software programs. The World Intellectual Property Organization passed a treaty in 1997 that has been ratified by a number of legislatures of signatory countries. However, like censorship, the protection of intellectual property may prove difficult—even impossible—to enforce. This book will consider these issues more fully in Chapter 14, "Cache Flow: From Bit-streams to Revenue Streams," and Chapter 15, "Brave New WWWorld."

Despite the barriers to a wholly integrated global communication system, I hope this chapter makes it clear that the world is well on the way to the establishment of an interconnected global network. An integrated infrastructure will certainly take some time, but satellite distribution is now everywhere. Some scholars believe that the media are a source of culture, in addition to interpersonal communication from family, friends, neighbors, and teachers. This opinion raises the question of the influence international programming might have on the development of culture around the world.

Does Global Television Presage a Global Culture?

A focus on the world as a single unit is a major idea of the twentieth century that permeates intellectual thought, historical action, and commercial development. Beginning in the 1980s, ubiquitous international communication, reliable transportation, international commercial outsourcing, the loss of jobs, and the massive movement of entire populations have driven home to many people that the geographical basis of culture may be changing. For example, the twelve to fifteen people sharing a meal in a bed-and-breakfast in Scotland may well have more in common with each other than they do with their immediate neighbors at home. This phenomenon occurs because the bed-and-breakfasters are likely to be a German computer programmer, a British professor, an American documentary film-maker, an Australian engineer, a French statistician, and an Italian architect—all members of the "Information Society," who read the same books, see the same movies, listen to the same music, and will exchange Internet E-mail addresses at the end of the meal.[77]

However, it is unlikely there will be a global culture that is like the geographically-situated culture we take for granted in our everyday lives. If one were to develop, it could take hundreds, perhaps thousands of years to evolve. If we broaden our definition of culture, then

> it may be possible to point to trans-societal cultural processes which take a variety of forms . . . processes which sustain the flow of goods, people, information, knowledge

and images which give rise to communication processes which gain some autonomy on a global level. Hence there may be emerging sets of "third culture," which themselves are conduits for all sorts of diverse cultural flows which cannot be merely understood as the product of bilateral exchanges between nation-states.[78]

This formulation of culture holds that global communication may lead people to joint memberships in overlapping spheres of cultural influence. They are born into their first culture, their ethnic culture. They are raised in their second culture, their civic culture. And they work and play in their third culture, the global communication culture.

Some postmodernist theorists concern themselves with these questions, and there is a significant research tradition that explores the dimensions of an emerging global culture. Since it is a growing and changing phenomenon, there is little agreement among scholars about the important variables, concepts, and processes that might be involved. Other scholars believe that the new communication technologies are destroying the "global village" recently created by the mass media:

While the new media communication system promises decentralization, individualism, and a break from passive viewing, it will end the global village as a dominant environment and the tribal culture which it has helped to recreate in the modern world. That culture bespeaks a commonality, a unity which derives from powerful bonding forces: a shared vision, shared values, a shared sense of security and comfort, much as America provided the many ethnic groups a new home in the New World and coalesced into the melting pot which was our societal core.[79]

In this view, the structure of the technology itself is the model for its effects; that is, the proliferation of digital delivery and distribution technologies, away from the centralized, unified analog systems, creates a society that loses their unifying features.

Localization: The Natural Limit to Globalization

There is considerable evidence that even when global communication is fully grown and available everywhere, some aspects of life and culture will remain local for decades (even centuries) to come. People respond to entertainment and information based upon their own experience. The stronger the material is connected to their experience, the more heightened the response.

It is true that great art transcends such parochial considerations, but a great deal of popular entertainment falls short of great art. More mundane fare requires the presence of people who are something like us to fascinate and involve us. The recorded music industry is seeing a rise in the popularity of local performers. A good example was provided by

NBC's foray into Asia, mentioned briefly earlier. This example is worth considering further in some detail.[80]

NBC Asia had distinct advantages over other would-be entrants. NBC was already a global concern, and the company could provide program packages, advertising discounts, and leverage its Internet presence to support its startup venture. At the same time, holding the line on operating costs was very important to NBC. In this spirit, the startup channel repurposed as much content as it could from material already acquired from its own information-gathering apparatus, rather than purchasing material from other companies or producing additional footage with local partners.

When the channel debuted in April, 1996, 11 hours of the 24-hour schedule were produced exclusively for the channel. By 1998, that total was only 3 hours per day, and most of the programs were shows acquired from NBC and MSNBC that had been produced for the U.S. audience. In addition, the small amount of material that was acquired was expected to appeal to audiences in both mainland China and Taiwan, despite the vast differences between these two audiences.

The key point here is that the demands of localization made it impossible for NBC Asia to operate profitably. Taiwan liberalized its media policies; China went the opposite direction and instituted more restrictive policies, necessitating separate satellite feeds. To make matters worse, technological systems imposed other restrictions on available bandwidth. The increase in the sheer amount of satellite time that had to be purchased effectively doomed the startup. NBC Asia was delivering programming via PanAmSat-2, AsiaSat-2, Palapa-C2, Intelsat, and Superbird C, as well as receiving programming from the United States and Europe.

This example demonstrates that a variety of cultural, political, economic, and technological conditions make the relationship between global ambition and local realities very complex.

A ubiquitous, inexpensive, high-speed global communication system would resolve some of these problems. But it will not address the most fundamental issues in media communications: the social and political context in which content is presented.

It is still too fragmentary to make accurate predictions about how this global network will evolve and affect everyone. However, we do know that one precondition of such a system is agreement upon a set of standards to guide the equipment, processes, and datastreams that will comprise a global infrastructure. The next chapter will consider the problem of global standards.

Notes

1. M. Featherstone, "An Introduction," in *Global Culture: Nationalism, globalization and modernity,* ed. M. Featherstone, London: SAGE Publications (1990):6.

2. P. Golding, "The communications paradox: inequality at the national and international levels," *Media Development* 41 (April 1994).
3. W. E. Kennard, "Vision to mission: A blueprint for architects of the global information infrastructure." Speech before the World Economic Development Congress, Washington, D.C., September 23, 1999.
4. "BT World Communications Report 1998/9," (May 1998). Available online at at http://www.bt.com/global_reports/1998-99/index.htm.
5. "Next generation broadband satellite networks: Executive summary." Pioneer Consulting (September 1999). Available online at http://www.pioneerconsulting.com.
6. Allied Business Intelligence, "Wireless systems outlook 2000: Markets, systems and technologies," Report Code:WSO00. Information about this report available at http://www.alliedworld.com/.
7. NASA/NSF panel and figures cited in "Satellite communications: The birth of a global footprint," *Voice and Data* (February 1997). Available online at: http://www.voicendata.com/feb97/4ib0081101.html.
8. U.S. Industry and Trade Outlook 1998, National Technical Information Service, Washington, D.C.: U.S. Department of Commerce (1998). Available online at http://www.ntis.gov/.
9. Pioneer Consulting, op. cit.
10. "I must go down to the seas again," *Public Network Europe* (December 1998/January 1999).
11. Pioneer Consulting, op. cit.
12. "G-7 nations hop on info superhighway," *Los Angeles Times* (February 27, 1995):D5.
13. "G7 countries agree to 11 'GII testbed networks,'" *NextNet* 4:5 (March 13, 1995):13.
14. J. E. Graf, "Global information infrastructure: First principles," *Telecommunications* (May 1994):72–73.
15. See the company's Web site at http://www.flag.bm/index_a1.htm. Be sure to have your PC speakers on—the audio is way cool.
16. The company's Web site is at http://www.globalcrossing.com/.
17. P. Golding, op. cit.
18. W. E. Kennard, "Unleashing the potential: Telecommunications development in Southern Africa." Keynote speech before the Annual General Meeting Telecommunications Regulators Association of Southern Africa (TRASA), Gaborone, Botswana, August 11, 1999.
19. D. Weld, "Africa and the global information wave," *The Journal of Public and International Affairs* (1997). Available online at http://www.wws.princeton.edu/~jpia/1997/chap2.html.
20. M. Jensen, "Where is Africa on the information highway? The status of Internet connectivity in Africa." Paper presented at RINAF Day/CARI 98, Dakar, Senegal October 16, 1998. Available online at http://www.unesco.int/webworld/build_info/rinaf/docs/cari98.html, or E-mail to mikej@sn.apc.org.
21. See the organization's Web site at http://www.global-learning.org/en/index.php3.
22. Cable television developments. Washington, DC: National Cable Television Association (Spring/Summer 2000):1.

23. As of June, 2000, the company's Web site was still up at http://www.metronet. ca. AT&T Canada is at: http://www.attcanada.com/.

24. The information about these trials came from the ADSL forum, at http:// www.adsl.com. There's another site about DSL at http://www.xdsl.com. I'm not sure you can still get data about worldwide trials. When I last checked, both these sites had removed much interesting material, or placed it in a password-protected area, rendering them far less useful. Too bad.

25. Look for the most up-to-date reports of cable systems to the CRTC at the Canadian Radio-Television and Telecommunications Commission Web site at http://www.crtc.gc.ca/welcome_e.htm.

26. These statistics came from a letter from BCE, filed with the CRTS, titled: "Licensing framework for new pay and specialty services a digital world—the need for new rules and immediate action for more consumer choice," addressing Public Notice CRTC 1999-19. Available online at http://www.crtc.gc.ca/ ENG/PROC_BR/NOTICES/1999/1999-19e/co032-2(submission).doc.

27. "Canadian Internet Survey Fall '97," A. C. Nielsen Company. Available online at http://www.acnielsen.ca/sect_studentcor/studtcor_en.htm.

28. P. Peters, "Going places: Emerging markets roundup," *Wireless Week International* (June 22, 1998).

29. British Telecom, "BT World Communications Report 1998/9." Available online at http://www.bt.com/global_reports/1998-99/regional.htm. The World Bank is another great source of information on teledensity and other national infrastructure data, at http://www.worldbank.org/html/fpd/telecoms/ subtelecom/ data_statistics.htm.

30. "Cisco finances FirstCom in Chile, Peru," *ATM Report* (November 22, 1998).

31. J. Dallas, "Digital DTH duels: Sharpening the aim in Latin America," *Multichannel News International* (1998):18, 20, 36–7.

32. S. Vonder Haar, "Portal player cultivates south-of-border market," *Interactive Week* 5:47 (November 30, 1998):48.

33. Nazca Saatchi, *Internet Survey* (1999). Download complete survey at http:// www.nazcasaatchi.com/.

34. "Star Media Annual Report," 10-K405 (3/30/2000). Available on company Web site at http://www.starmedia.com, or through the EDGAR database at http://www.sec.gov. 35.The European Union Information Society conducts surveys and reports information on communications infrastructure. Access data by region and country at http://www.ispo.cec.be/esis/default.htm.

35. "Screen digest," *European Cable and Satellite Economics*, research report (April 1999). Executive summary available at http://www.screendigest.com.

36. C. Forrester, "Europe's new programming paradigm: The continent's channel explosion is ushering in a host of new programming concepts," *Multichannel News International* (October 1999).

37. See the company's Web site at http://www.sohonet.co.uk/.

38. The Medienstadt Web site is at: http://www.babelsberg.de/.

39. There are so many projections and the information changes so quickly, so readers should access http://www.nua.ie for a "hype-o-rama" of the most up-to-date figures and estimates.

40. E. Neufeld and O. Beauvillain, "Online landscape: Vision report—France," Vol. 2. *Jupiter Communications* (December 19, 1999). For more information, contact Allison Tracy, (800) 481-1212 ext. 6177.

41. K. Jones, "Jolly old Cambridge targets high-tech success," *Interactive Week* 3:47 (November 30, 1998.).

42. I downloaded this information from China Research Corporation. The data has since been removed from the site. Contact information is: China Research Corporation, The New Century Group, No. L-N, 9/F Tower A, East Gate Plaza, No. 9 Dong Zhong Street, Dong Cheng District, Beijing 100027, PRC; telephone: (8610) 6418 1688, http://www.china-research.com/index.html.

43. K. C. Kwong, Secretary for Information Technology and Broadcasting, addressing the Asia Pacific Smart Card Forum and the Electronic-business Symposium, Melbourne, Australia, November 1998.

44. D. Lazarus, "Asia's racing toward wiredness," *Wired Online* (July 30, 1997). Available online at http://www.wired.com/news/business/0,1367,5622,00. html.

45. The APT Satellite Holdings Web site is at http://www.apstar.com.

46. N. Madhavan, "Asia lags in bumpy Internet revolution," Reuters (September 9, 1998).

47. M. Dorgan, "Malaysia: High tech utopia?" *San Jose Mercury News* (July 7, 1997). Online at http://www2.mercurycenter.com/archives/malaysia/stories/main.htm. Information about the University Tun Abdul Razak may be found at http://www.unitar.edu.my/.

48. The Australia Broadcasting Authority Web site is at http://www.aba.gov.au/.

49. The best compilation of Australian pay TV I could find on the Net is maintained by a senior analyst/programmer at the University of Western Sydney at http://www.nepean.uws.edu.au/vip/johnf/. There's also a pretty good Australian TV FAQ at http://www.su.swin.edu.au/~talor/ausfaq.txt.

50. See data at http://www.oecd.org/dsti/sti/it/cm/stats/newindicators.htm #chart1.

51. New Zealand's broadcasting regulatory agency's Web site is at http://www.nzonair.govt.nz/.

52. V. Crishna, N. Baqai, B. Pandey, and F. Rahman, "Telecommunications infrastructure: A long way to go," South Asia Networks Organization. Posted at http://www.sasianet.org/telecominfrastr.htm#hdg.2.

53. Reuters, "India on threshold of Internet age," *San Jose Mercury News* (online) (January 23, 1999). Online at http://spyglass1.sjmercury.com/breaking/docs/005829.htm.

54. M. Rao, "Indian ISPs form alliances to tackle market growth challenges," Indialine.com (March 4, 1999). Online at http://www.indialine.com/net.columns/column66.html.

55. An online resource page provided this information, at http://www.indo-net.com/indiainternet.html. Cited statistic from Net Wizards is at http://www.nw.com/zone/www/dist-byname.html.

56. D. McCullagh, "Report: Mideast misses the Net," *Wired Online* (July 8, 1999).

57. "The Internet in the Middle East and North Africa: Free expression and censorship." Compiled by Human Rights Watch (July 1999). Available online at http://www.hrw.org/reports98/publctns.htm.

58. D. Weld, op. cit.

59. M. Jensen, op. cit.

60. The RASCOM Web site is at http://www.esmt.sn/telrur/projet/rascomgb. htm. There are also some RASCOM documents at http://www.connect-world. co.uk/docs/articles/cwafrica99/rascomcwafrica99.asp.

61. The HealthNet satellite is operated by Satellife, whose Web site is at http:// www.healthnet.org/. See also P. Godard, *Africa and Sciences: The Availability of Computer Communications*, (1994). Available online at http://www.sas.upenn. edu/African_Studies/Padis/Godard.html.

62. M. Jensen, op. cit.

63. J. Emberg and J. Rose. Draft AISI/HITD Internet Connectivity Sub-Programme Framework. (April 1997). Posted online at http://www.bellanet. org/partners/aisi/proj/itufram.htm.

64. The Web site for the African Virtual University is at http://www.avumuk. ac.ug/.

65. M. Curtin, "Beyond the vast wasteland: The policy discourse of global television and the politics of American empire," *Journal of Broadcasting and Electronic Media* (Spring 1993):127.

66. B. Johansen, telephone interview, September 1999.

67. L. McElvogue, "Cannes now a prime-time player," *Los Angeles Times* (October 13, 1995):D4, D5.

68. J. Van Tassel, "World domination," *The Hollywood Reporter: Anniversary Issue* (December 1997):S-1, 5, 7.

69. Information posted on Viacom's corporate Web site at http://www.via-com.com/global.tin.

70. G. Castle, telephone interview, October 1999.

71. K. F. Kwong. "American Peacock on Chinese Soil—The Challenge Facing NBC Asia in Greater China." A Project Report Submitted in Partial Fulfillment of the Requirements for the Degree of Master of Arts in Communication, School of Communication, Hong Kong Baptist University (August 1998).

72. H. Mowlana and L. J. Wilson, *The Passing of Modernity: Communication and the Transformation of Society,* New York: Longman (1990):43–75.

73. T. Marshall, "EU panel urges tighter TV import quotas," *Los Angeles Times* (February 9, 1995):C-1.

74. Ibid.

75. B. Johansen, op. cit.

76. K. Gergen, *The Saturated Self,* New York: Harper Collins (1991).

77. M. Featherstone, op. cit.:1.

78. R. Weinman, "Anytime, anywhere communication," *New Telecom Quarterly* 4 (1994):18–22.

79. K. F. Kwong, op. cit.

9

Wired Bitpipes I

This chapter covers what is called the "plumbing" that will underlie the revolution in communications capacity. It is intended to provide more than a superficial glance at these wired bitpipes, yet each system, indeed each component of each system, could fill a book in itself. So the descriptions of individual bitpipes is necessarily curtailed, but they should contain enough information to enable you to fill in your knowledge of specific technologies through your own additional research.

The technical nature of the chapter should not disguise the enormous changes this infrastructure will bring, nor the thrilling sense of adventure its architects experience. The people who are engaged in the giant job of building the complex infrastructure required to bring broadband connectivity to homes and offices have a contagious sense of purpose, an almost romantic dedication to the mission. They attend numerous standards-setting meetings, conferences, and conventions, where they question, debate, and push the perspective of their industry, company, camp, or individual opinion. Twenty-five year old whizzes jump into the verbal fray, arguing vehemently with grizzled veterans over how best to build these advanced networks. People who are indifferent to or take this new connectivity for granted will miss one of the most exciting, passionate, and engrossing set of events occurring in their lifetime.

Little wonder! The communication revolution is a modern phenomenon. Beyond the beating of drums and signaling with flags or smoke, distance communication began in the nineteenth century. The telegraph, the telephone, and the light bulb all grew out of the understanding and commitment that scientists and inventors brought to the new technology of electricity. And despite the century and a half or so that separates us from them in time, many of the same issues that were involved in developing the earliest systems recur again and again, even as ever more sophisticated solutions are devised.

Spectrum and bandwidth, transmitters, receivers, radio waves, electrical pulses, light energy, relays, repeaters, wires, and switches—these were the components of the first wired systems, and they still constitute many of the elements of today's systems. Similarly, the design and development issues addressed by earlier developers are strikingly similar to the difficulties encountered today. Cost, capacity, configuration, the design of connection devices, consumer issues, compatibility and interoperability, economic benefits, and public policy—these considerations recur again and again in the creation of communications infrastructures generally, and with wired systems in particular.

In terms of the technology itself, though, tremendous advances have occurred in every area of endeavor. Engineers from the beginning of the twentieth century would find nearly all of today's communication infrastructure quite unrecognizable: components, configuration, and architecture. Yet, they would understand the issues quite readily.

So What's a Network to Do?

Defining network functionality—what the network should actually do—puts the end-point of the process first by specifying the features and capabilities the network must have after it is built and launched into service. Starting at the end is the only way to ensure that construction of the system will meet the goals and expectations that have been set for it. So what must the twenty-first-century broadband communication infrastructure do? Here goes. It must:

- Provide enormous capacity.
- Be flexible.
- Carry variable bandwidth traffic of every data type, encoded into any number of software formats.
- Deliver real-time interactivity.
- Provide universal and ubiquitous connectivity to anyplace in the world.

Let us look more closely at this list and flesh out these functionalities.

Function 1: High capacity—By any definition, even moderately good audio and video require substantial bandwidth. A worldwide network where most people are visually connected is currently beyond even theoretical possibility, but it won't always be.

Function 2: Variable bandwidth—The network must deliver efficiently everything from an HDTV movie or high-resolution medical graphics to an E-mail that says "Hi!"

Function 3: Universal translation—It must be able to accept, process, package, and deliver data from many different kinds of computers, which is encoded, compressed, modulated, and multiplexed in a gazillion formats. It must also be able to pass through analog signals and to convert analog signals to digital signals and vice versa.

Function 4: Real-time interactivity—It must allow for relatively symmetrical, two-way broadband communication in real time.

Function 5: Ubiquitous connectivity—It must provide for near-universal access, including remote access in public areas. Its addressing protocols, internal switching, and routing mechanisms must deliver messages to anyone on the network.

Function 6: Global reach—It must be worldwide in scope, capable of reaching anyone on the planet.

Well, we have a long way to go before this kind of global communications network is realized in its entirety. Make no mistake about it, communications infrastructures, especially of the wired variety, are monumental investments that take vision, billions of dollars, thousands of people, and a certain political wiliness. Let us now consider some characteristics all wired systems share. At minimum, the list includes:

- a physical wire or cable that carries information
- a source of power
- some way of making connections
- techniques to encode messages and signals efficiently
- methods of aggregating them and separating them again as they move through different sized channels, at various speeds, and over distances from a few feet to around the globe
- software for operation support, billing, and running applications
- reception devices

Network Commonalties

Since networks share functionalities, it comes as no surprise that they also need similar equipment. Each network type has its nuances, however, and employs alternative ways to accomplish the same result. For example, at the physical level, networks may use copper wire, coaxial cable, or fiber optic cable. The same network may use all of them, and many do. Power may be supplied by electricity off the grid, a generator, or batteries. And there are thousands of ways to encode messages.

Making Connections: Switches, Routers, Bridges, and Brouters

At minimum, a network links at least two reception devices. But if there are more than two, then there must be some way to insure that the intended receiver gets the message. Switches were the nineteenth-century solution to the problem, connecting people talking over the infant telephone systems of the time.

A switch is a device that opens or closes a circuit or electrical path between a sender and a receiver. The first-generation switch was human-mechanical: operators who, like Lily Tomlin's character Gertrude, sat at a switchboard, plugging incoming callers to the lines of the people they wanted to reach. The next generation switch was the widely used Strowger automatic, mechanical step-by-step telephone switch. (Armond Strowger was an undertaker. It is rumored he invented the switch because the local operator's boyfriend was a competitor, so she routed all the calls for undertaking services to her sweetheart!)[1] Mechanical switches became ever more complex until their replacement by modern electronic switches in the late 1960s and then optical switches in the 1990s.

Today's networks employ one of three ways to deliver messages over a network: circuit switching (CS), packet switching (PS), and asynchronous transfer mode (ATM), which combines elements of both circuit and packet switching. Table 9.1 compares them.

Circuit Switching   CS is the traditional way telephone companies connect calls. It gets its name from the fact that it connects circuits, or local loops, and all the network segments in between to establish a fixed, dedicated connection. CS is an excellent way of handling messages that are time-sensitive, or "isochronous," which require that the bandwidth be fixed at a certain rate, called "constant bit-rate" service.

Packet Switching   PS is effective for "bursty" traffic, like data transmission, where a computer may squirt out millions of bits per second for a few minutes, then stop for some period of time. Unlike a circuit-switched network, PS doesn't entail that a dedicated connection be kept open for the duration of the communication. Rather, messages are divided into packets, and each one carries the destination address with it as "overhead." The packet is divided into the "header" that has the address and the "payload" that is the actual message. Each packet is sent forward to the next network device forward without regard for its final destination except as a general direction. Some packets go one way, others go another. Eventually, the packets get closer and closer to the address and reach it, where they are reassembled back into the original message.

Packet switching is used for local area and metro area networks (LANs and MANs) or any IP (Internet Protocol) network. Routers, bridges, brouters, and switcher/routers perform the task of connection. All these devices take the place of traditional switches on packet-switched networks. Routers are computers that sit on a network, read the messages that pass through it, and figure out the most efficient way to route the bits

Table 9.1 Comparison of Circuit, Packet, and ATM Switching

	Circuit Switching (CS)	*Packet Switching (PS)*	*Asynchronous Transfer Mode (ATM)*
Hardware	Switches and switcher/routers	Routers, bridges, brouters,and switcher/routers	ATM switches and switcher/routers
Service	Constant bit rate (CBR)	Variable bit rate (VBR)	CBR and VBR
Data types	Time-sensitive audio and video	Time-independent, "bursty" traffic	All traffic
Native Network	Telephone networks	Computer networks	Fiber networks
Path	Dedicated	Multiple, opportunistic	Virtual
Noise handling	Assumes noise-redundant data	Assumes noise-redundant data	Assumes no noise–no redundant data
Busy signal	Caller call back	Network resends	Network resends
Addressing	No address	Address attached to message	Address attached to message
Bandwidth allocation	Fixed bandwidth	Dynamic bandwidth	Dynamic bandwidth

through the network. They have considerable network management capabilities, including responding to different levels of priorities of messages, statistical use monitoring and maintenance, load balancing across different network segments, and other troubleshooting tools.

A bridge is also a computer, but less intelligent than a router. It connects different networks to one another, passing messages back and forth between them. Brouters combine the network management functions of the router with the network connecting capabilities of the bridge.

ATM Switches   ATM switches incorporate both CS and PS in their operation. They are increasingly used for wide area networks (WANs) that distribute messages across some large, dispersed geographical area. For example, the Internet is a huge WAN. William Stallings, a noted expert on network design writes that ATM " . . . is in a sense a culmination of all the developments in circuit switching and packet switching over the past twenty years."[2]

ATM is potentially useful for systems that carry video because it handles different kinds of traffic simultaneously, and it is extremely fast and efficient. ATM design takes advantage of the low-noise characteristics of fiber optic transmission and doesn't add a great deal of redundant data, which ensures the information is received without error. CS and PS may add as much as 33% to the data just for taking care of error correction. In systems where the amount of traffic is already enormous, the reduction of overhead that ATM allows is highly valued.

Let's look more closely at the ATM datastream. The digital signals are repackaged into small, equal-sized "packets" called "cells." Each cell is made up of a byte. (Remember from Chapter 3, "Digitology 101," that a byte is 8 bits.) In ATM lingo, a byte is called an octet—"oct" means eight. There are 53 octets (or bytes) in a cell, which is the basic transport unit of ATM. Of the 53 octets, 5 are headers, which carry the unique identifier, the destination address, and error-correction information. The other 48 cells make up the payload.

It is the small, uniform cell size of ATM that allows the extraordinary speed of its bit rate—intended to operate in the range of up to several gigabits per second. Equipment that is now available permits data rates of 2.4 Gbps, with higher rates in the offing, carrying both constant bit rate (CBR) data and bursty (VBR) traffic. This rate compares to the speed of 140 Mbps of the fastest circuit-switching technique, and 2 megabits per second for the fastest packet-switching device.

In years past, there were questions as to whether ATM would be able to carry MPEG-2 compressed digital video. However, the International Standards Organization (ISO) approved a standard called the Delivery Multimedia Integration Framework Application Interface (DAI) that allows MPEG-2 to run over different kinds of transport, from the Internet to ATM networks.[3] In addition, in case of interactive operation across a network, it ensures interoperability between end-systems by mapping MPEG to the native network signaling messages. The speed of ATM and its flexibility in the types of data that it can carry make it ideal for the fiber optic networks under construction, as well as cable systems.

The last few years have seen the invention of a terabit switcher/ router from Nexabit, acquired in 1999 by LucentTechnologies.[4] The Nexabit product is being tested by Frontier Communications at 6.4 terabits per second. It will be used for next-generation networks, which will be covered in the next chapter.

Gateways   The last kind of connection device to consider is a gateway. These are devices that link networks with differing architectures that might otherwise be incompatible. For example, gateways link the circuit-switched public switched telephone network (PSTN) to the packet-switched Internet.

Modulation and Multiplexing

Recall from Chapter 3, "Digitology 101," that modulation refers to how a signal is encoded for transmission with variations that reflect the information that is conveyed by it. At the reception end, the signal is demodulated so it can be played on devices such as TVs, PCs, screen phones, and other equipment. All forms of modulation for communications involve one of three algorithms: frequency modulation (FM), amplitude modulation (AM), and phase modulation.

Every network design specifies its requirements for modulation. For example, cable systems use sophisticated variations of AM and phase modulation, QAM (quadrature amplification modulation) to carry downstream traffic, and QPSK (quadrature phase shift keying) for upstream traffic. QAM is the most efficient algorithm for high-quality, large bandwidth television signals, while QPSK is easy to implement and resistant to noise.

When telephone carriers must translate analog voice signals into digital data, they use a codec (coder/decoder) that incorporates one of two techniques: pulse code modulation (PCM) and delta modulation (DM). PCM is actually a complex sampling theorem for making the analog-to-digital (A-to-D) conversion. DM is an alternative to PCM that reduces the complexity of the A-to-D conversion by using a coarser quantizing algorithm. With the advent of high-speed data service over telephone lines, telephone companies also use a QAM variation called CAP (carrierless amplitude phase) to implement DSL (digital subscriber line) service for their customers.

Multiplexing techniques define how signals are combined into ever larger streams as they move from the originating source through the network, across the backbones, then are demultiplexed or decombined for carriage back down through lower network levels to its final destination. The equipment that handles these tasks on fiber networks are called "optical add and drop multiplexers," sometimes shortened to OA&DM. The two metaphors for multiplexing are the race track and the train. Frequency-division multiplexing (FDM) and wavelength-division multiplexing (WDM) aggregate the data that represents the signals in side-by-side streams, as shown in Figure 3.9 in Chapter 3, "Digitology 101." Time-division data handling is like a train, packing data any place in the stream where there is room to slot in another box car, as shown in Figure 3.10, also in Chapter 3. The messages have headers that allow them to be reassembled at drop points in the network or at the destination, the termination point.

A third type of multiplexing, code-division multiplexing (CDM), is used mostly for wireless, mobile telephone systems. It is a digital spread-spectrum modulation technique that assigns the digitized conversation a unique frequency code. The data is scattered randomly across the slice of spectrum that has been allotted to the service. The receiver will only pick up and decode the data that it recognizes from the assigned frequency code, reassembling it into an analog message for the human listener. Frequency-agile CDM changes frequencies from packet to packet, so the receiver must have information about the assignments used at encoding.

Software: The Brains behind the Network

The greatest complexity of software for wired broadband systems stems from the requirements for network management, operations support, billing, and applications. These tasks must be accomplished in addition to the information and signal processing I have already covered, such as analog/digital/analog and electrical/optical conversion, modulation, and multiplexing. There are several as yet unresolved questions regarding this additional layer of software, including:

1. Where should network resource allocation reside—in specialized management software or in the operating system?
2. How should resources be allocated: fixed versus dynamic, or simple versus complex?
3. Who owns and controls the resource control layer—network operators or software providers?
4. How are revenues from applications shared, and how will the accounting for them be accomplished?

One approach is that all these functions should be located in the operating system. The other notion is that layers of "middleware" should translate between multiple operating systems, handle specialized functions, and interface with the multifarious pieces of hardware in the network to the devices used by consumers.[5] Each industry approaches these questions differently, and there are many variations between individual companies as well, depending on their business plan and operating procedures.

Down-to-Earth Networking

There is a common repertoire of network designs to reach individual stations, nodes, households, or desktops. The most common designs are: tree and branch, bus, ring, and star, each with variations. The tree and branch is a type of bus network that is currently found in most cable systems. The star network is used by telephone companies, mandated by the FCC for incumbent local exchange carriers (ILECs). Both cable and telephone companies use ring designs when building synchronous optical networks (SONET). Finally, bus systems are often used for computer local area networks (LANs). As cable operators build advanced networks, extending fiber optic cable deeper into their systems, they are experimenting with mixed configurations: trunk/star, ring/star, and star/star/bus.

Figure 9.1 illustrates how the tree-and-branch design acquired its name. Typically used by cable companies, signals arrive at the headend— these are the roots. They are bundled into a single stream, the trunk, and transported to major regions of the service area where they are delivered to the feeders, the branches of the system. Feeders take the signals along streets and drop-cables tap into them to take signals to the home, which are the leaves.

Figure 9.2 shows a bus design, sometimes called a multipoint system. The stations are connected together in a linear transmission network. All messages flow up and down the length of the bus and can be received by any station.

Ring networks consist of stations joined together in a closed loop, as shown in Figure 9.3. Messages are addressed and sent in only one

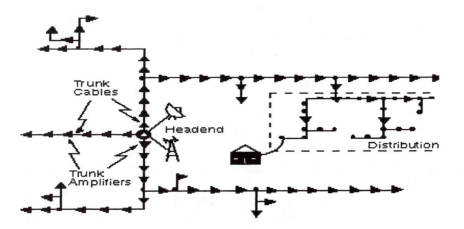

Figure 9.1 Network with tree-and-branch typology. Source: CableLabs.

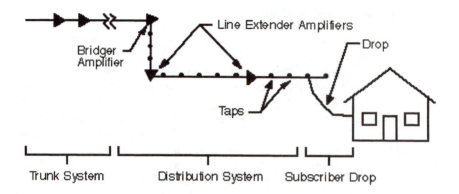

Figure 9.2 The bus network typology. Source: CableLabs.

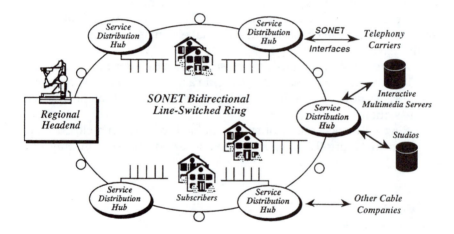

Figure 9.3 A SONET ring typology. Source: Nortel.

direction. When the message returns to its source station, it is removed from the ring.

SONET networks are built in the ring typology, in some cases with multiple rings. For example, the Cincinnati Bell Telephone MetroPLEX SONET network has a double ring with messages traveling in opposite directions. A variation is Cox Cable Communications' "ring-in-ring" design, developed to improve signal reliability and to reduce outages.

In the star and mini-star configurations, each station is connected to a switch at a central point—the typology of local exchange carriers. Even if a message goes to the station next to it, the message must pass through the switch. Now that cable companies are planning switched networks, some are adding mini-star networks of 150 to 500 homes at the neighborhood node of a tree-and-branch system. The mini-star configuration is a common architecture because it takes advantage of the existing fiber-to-feeder network, which offers cost-savings when operators decide to upgrade their systems. In 1995, the mini-star fiber-to-the-node design achieved cost parity with fiber-to-feeder installations because it requires fewer expensive "active" RF amplifiers.

Now this chapter will examine the three types of wired networks. Each was designed to deliver messages to a particular device that sits at the edge of their respective networks: the telephone, the television set, and the PC—tiny tails wagged by huge dogs.

Cable Systems

Cable systems began life as broadband one-way systems where signals ran over coaxial cable. Gradually, operators have introduced fiber optic cable into the backbone, extending it further into the "feeder" portions of the network. A particularly sophisticated and elegant design is the hybrid fiber/coax (HFC) network, conceptualized by Dave Pangrac of Pangrac and Associates on the back of a napkin during an airplane flight.[6] HFC was then further refined and implemented by a team led by James Chiddix of Time Warner Cable. By carefully thinking through the bandwidth usage, HFC allows every household in a 500-home neighborhood node to have an exclusive channel that will carry a unique broadband stream. The term "500 channel universe" came from the HFC network that could allocate a unique, dedicated channel to each household connected to the neighborhood node.

Almost all cable systems are somewhere in the process of converting to HFC. Many telephone companies are also adopting an HFC design when they build a network designed to deliver TV services. An overbuild is a separate network that follows the pathways of an existing network, usually to take advantage of right-of-ways. Some cable systems have fiber only in the backbone; others have pulled it to the feeders. Many operators are now installing fiber to the neighborhood node to enable an interactive digital TV tier and two-way high-speed Internet services that can grow as demand rises.

In an HFC cable system, the majority of the components are located in the headend, a transceiver node where television signals come in and are stored or immediately retransmitted. The rest of the hardware is distributed throughout the network, at nodes or reception sites. Table 9.2 shows what they are, where they are located, and what they do.

Table 9.2 Components in a Digital, Interactive Cable Television Network

Component	Location	Function
Satellite dishes: C-band, Ku-band	Headend	Signal reception
Microwave, FM, VHF, and UHF antennas	Headend	Signal reception
Fiber cable node	Headend	Signal reception
Router, bridge, or gateway	Headend	Connect signals from one network to other networks, such as telephone nets to cable or computer networks
A-D, D-A converters	Headend, Set-top box	Convert signals from analog to digital or vice versa
Descrambler	Headend, Set-top box	Specialized processor that can descramble incoming and upstream signals
Laser diode	Headend, Terminal node	Converts electrical pulses into on-off light signals (optical)
Modulator	Headend, Terminal node	Varies some characteristic of a signal as the information to be transmitted varies, so signal carries information
Processor	Headend, Set-top box	Converts bitstream into set-top format
Multiplexer/ demultiplexer/ mux/demux	Headend, Terminal node	Loads multiple streams of information onto the same channel, or transport medium, or reverse, disassembles a multiplexed bitstream
Fiber optic cable	Network transport	A transport medium, made of silicon, that carries light signals
Photodiode	Headend, Terminal node	Converts optical (light) on-off signals to electrical pulses
Switch	Headend	A device that opens and closes in order to complete or break a circuit
Router	Headend	Routes data streams to users on the network

Table 9.2 Components in a Digital, Interactive Cable Television Network (Continued)

Component	Location	Function
Combiners and splitters	Headend	Bundle separate signals together and separate bundled signals
Coaxial cable	Network transport	A conducting wire that carries great quantities of data (about 1 gigahertz) at very high speeds (about 200 megabits per second)
Amplifiers	Network	Amplify signals, every 1/2 mile or so
Tap signals	Receiver	A passive box (no electronics) that splits from coax and reduces their amplitude for delivery to customer.
Gateway	Receiver	The junction box where a cable enters a reception site, a household, or business.
Set-top box	Receiver	Essentially a computer that retrieves downstream signals and decompresses them. Addresses, scrambles, compresses, modulates, and sends signals upstream.
Software applications	Throughout network	Operating system software (OSS); software, server software; operation, administration and management (OA&M, or OSS- Operational Support System) software; transaction billing software, set-top box operating system and applications

The components are produced by many different manufacturers. Subcomponents are contracted out to many more companies. Standards, whether voluntary or enforced, are essential for all this equipment to interoperate.

System Capacity   Try reading this statement out loud: "this 750 MHz (megahertz) system passes 50,000 homes. To provide high-quality VOD [vee-oh-dee], assuming peak utilization of 25%, we'll need a server that can output 12,500 4-megabit per second MPEG-2 [em-peg two] video streams simultaneously. Network capacity on the backbone will have to be about 50 gigabits per second."

The above paragraph is fairly typical of the way an executive might describe a video-on-demand (VOD) system the company wants to test. It's easy to decode once you are familiar with the terms. The capacity of advanced cable television networks is measured in megahertz of bandwidth, such as 550 MHz or 750 MHz, or gigahertz for the largest systems, such as 1 GHz. When a system is described as a 750 MHz or 1 GHz system, the numbers refer to the capacity that can be delivered from the headend to each home or other receiver site, or its end-to-end throughput along the a downstream path. VOD stands for video on demand, a service that lets consumers request video material and receive it right away. Peak utilization is the highest usage the system must be able to accommodate under normal circumstances. The capacity of the backbone must be the biggest part of the system, or the sum of the bandwidth available for all the signals that are delivered to and from the neighborhood node, times the number of neighborhoods.

Figure 9.4 shows the spectrum allocation in an advanced full-service 1 GHz network. Over the past two decades, the average bandwidth of cable systems has been increasing, from about 200 MHz to 300 MHz and 450 MHz. Today, many systems offer 550 MHz capacity, and the current generation of upgrade is to provide 750 MHz or 1 GHz.

Until the transition to digital is complete, the system's capacity must accommodate both analog signals that every subscriber receives and digital signals that are addressed to the specific households that have paid for this premier tier of service. Typically, the top 200 MHz of the 750 MHz are reserved for digital services. This leaves the 54 MHz to 550 MHz for standard analog channels, which can continue to coexist with the new digital signals, resulting in no change for the subscriber on the day the system is installed.

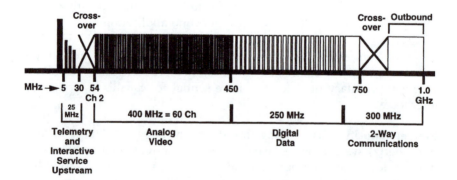

Figure 9.4 Spectrum allocation in a 1 GHz MHz system. Source: Carl Podlesney, Scientific-Atlanta

The Downstream Path   The process of delivering video channels over a cable system begins with bringing analog and digital signals coming into the headend via cable, satellites, and terrestrial broadcasting. In many systems, analog signals coexist with digital signals throughout the system, a situation that is likely to last for the next decade. The process of preparing the signals and transporting them to consumers is shown in Table 9.3.

Handling the analog signals is relatively simple, and most operations process the channels after they are multiplexed as a bulk operation. As they come in, they are demodulated, scrambled (optional), multiplexed, amplified, and demultiplexed for transportation to neighborhood nodes. If they are scrambled, the decoding will be done by the subscriber's set-top box.

Table 9.3 Moving Signals Downstream over a Digital, Two-Way Cable Network

COMBINED PATH

1ALL) Receive → 2ALL) demodulate and split analog (ANA) from digital (DIG) signals →

ANALOG PATH	DIGITAL PATH
→ 3ANA) scramble (optional)	→ 3DIG) digitize
→ 4ANA) combine all analog channels	→ 4DIG) compress
→ 5ANA) amplify	→ 5DIG) scramble (optional)
→ 6ANA) split to neighborhoods	→ 6DIG) store some signals on server
	→ 7DIG) break digital signals and assign addresses to each packet
	→ 8DIG) time-division multiplex packets into a single stream
	→ 9DIG) modulate for transport
	→ 10DIG) combine all multiplexed digital streams

COMBINED PATH: → 11ALL) combine analog and digital signals
→ 12ALL) modulate electrical-to-optical on neighborhood laser
→ 13ALL) send signals down fiber backbone and feeder pathways
→ 14ALL) at neighborhood node, demodulate optical-to-electrical for coax cable
→ 15ALL) add 90 volts of power
→ 16ALL) amplify signals (up to 4 times)
→ 17ALL) tap coax feeder and send signals down coax drop to home gateway
→ 18ALL) segregate telephone from television traffic
→ 19ALL)

Sending digital signals is new for cable operators, and it is much more complex than simply passing through analog channels, where the technology and processes are well known. To deliver colorful, rich digital TV pictures to their subscribers, a series of complex signal reprocessing steps must take place at the headend. Each digital channel is compressed and, in some systems, scrambled. Every channel will require between 3 Mbps and 5 Mbps. The higher rate is needed for a particularly fast-moving image, such as a basketball game. If it's an on-demand product (for example, a movie), the film will be sent to the media server, functioning as a kind of video jukebox, and stored until a subscriber calls it up. All digital signals pass through the server where they are tagged with a unique identifier, to ensure that subscribers get the right product.

The identifier is read by a switch or a router. Many VOD designs call for an ATM switch because of its speed, efficiency, and flexibility. It will also allow the cable network to handle Internet data and telephone calls as well as high-quality video. When the information stream passes out of the switch, it is time-division-multiplexed for further transport toward the consumer. Signals going to the same neighborhood are then all modulated and combined together and recombined with any analog signals. At this point, the analog/digital information stream is converted from electrical to optical signals for transport over the fiber portion of the system.

It is only now that the signals are actually leaving the headend! They travel down the fiber optic cable backbone, split off into feeder cables (either fiber or coax), and continue to the neighborhood node (usually coax). There, the signals are reconverted from optical to electrical, and 90 volts of power are added to them to power the amplifiers (and any telephone service that may be provided). As many as four amplifiers are "cascaded," or placed every few hundred yards to make sure that signal strength is maintained.

As the signals pass subscribers' homes, they are "tapped," that is, split off to the gateway outside the home. Data and telephone signals are segregated from the TV signals, which go into the set-top box (STB). The customer's PC and STB pick out the signals that are specifically addressed to them, descramble and decompress the information stream, and display any necessary graphics (such as navigation software) for subscribers to access the system. The set-top box displays the particular channel the consumer has selected and blocks the decryption of services the customer has not paid for so they cannot be selected at all. In households that do not have a PC, all the signals may go to the STB. If the signals are Internet IP data, it will display them on the TV via a navigation system.

The Upstream Path   In a 750 MHz system, all the upstream communication, including telephony, is usually assigned to the 5 MHz to 42 MHz low-end portion of the system's spectrum. However, not all that bandwidth is usable. There are problems below 10 MHz, but ignoring those, the 40 to 54 MHz portion must be set aside in order to avoid interference with Channel 2 at 54 MHz. Allowing for spectrum that cannot be used, some 25 MHz is left for system monitoring and all upstream services. The 54 MHz through 550 MHz accommodates analog programming, and 550 MHz to 750 MHz is reserved for a digital programming tier.

In a 1 GHz system, the 25 MHz at the low end of the spectrum can be used for upstream interactive data signals. At the upper end, setting aside about 150 MHz for guardbands (extra empty space) to avoid interference from the 700 MHz area where digital services stop, leaves the 850 MHz to 1.0 GHz area of the spectrum for a greater volume of telephony and high-speed two-way computer communication.[7] Since a compressed voice telephone call requires 4 kHz of spectrum, this 150 MHz would provide bandwidth for more than 37,000 simultaneous conversations.

The process for sending information upstream from the subscriber's home or business is quite similar to the downstream process, but it is not an exact mirror image, as Table 9.4 reveals. Some steps of the upstream process take place in different locations than occurred on the trip down. Also, the user's signal is never analog, so it doesn't have to be digitized. When subscribers want to send video upstream, they must predigitize it locally before transmitting it.

In the past, cable systems left the 5 MHz to 40 MHz portion of that bandwidth empty, even when they never planned to offer interactive services. Sometimes operators didn't use it at all. Others used it for network management or for shuttling video from local sites upstream to the head-

Table 9.4 The Interactive Hybrid Fiber/Coax (HFC) Network— Moving Signals Upstream

DIGITAL UPSTREAM PATH

1) Send signal via TV remote control, computer (modem), or telephone to set-top box or gateway → **2)** set-top box or gateway digitizes, scrambles, compresses, encodes signal into ATM cells, and modulates it into code-division multiplex format → **3)** transmit signal through site gateway over coax drop cable to neighborhood node → **4)** strengthen it with amplifiers every ∫ mile or so → **5)** at neighborhood node, remodulate signal → **6)** convert electrical RF into optical signals → **7)** send signal over fiber to headend → **8)** convert optical to electrical RF signal, demodulate, and descramble for switch or router → **9)** switch or router sorts signals for transport to server or to router into public switched telephone network (PSTN) → **10)** when appropriate, split off billing and account management information → **11)** convert telephone traffic from electrical to optical signals and multiplex for transport to PSTN.

end. For example, when the local Little League team plays it may be carried on the public access channel and the cable system may have a hookup so the video can be sent upstream to the headend from the remote location at the park.

Now some cable operators have adapted the 5 MHz to 40 MHz part of their system to offer upstream services to subscribers, for such activities as high-speed Internet access, telephone service, and responses to interactive advertising and programming. As mentioned, there is less bandwidth here than meets the eye, as frequencies below 10 MHz are difficult to use and those above 20 MHz are susceptible to interference. To get around these problems, operators with 1 GHz capacity networks have allocated the top 900–1,000 MHz band for the return loop.

Allocating the spectrum on a cable system is primarily an economic decision. If operators think that interactive services such as high-speed Internet access will attract subscribers, then using the high end of the spectrum will allow for expansion as more customers sign on. However, if it would be more profitable to add additional programming services, then they should put interactive services at the low end to leave room at the top for the one-way services to grow.[8]

Still another solution is called overbuilding. In some cases, a system might have two separate cables from the home to the fiber feeder, one carries downstream services and the other transports upstream communication. In the case of telephone companies, overbuilding usually means constructing a complete HFC system on top of their copper wire network because they can get all the added capacity for little more than it might cost them to upgrade their copper network. This low-maintenance technique eliminates the need for expensive electronic spectrum management and filters to separate the two streams, although it costs more for cable and associated amplifiers. An example of a telephone company overbuilding is a GTE-built HFC network in Thousand Oaks, CA.[9]

Another approach to solving the problem of return bandwidth is to use "spectrum management," or operating support system (OSS) software. Currently, cable systems allocate specific spectrum assignments to specific channels. However, network management software can allow operators to use dynamic spectrum allocation and management, assigning bandwidth to a service as needed and only when needed. Network management requires a complex set-up that includes a processor, detecting sensors, agile modulation, and automatic adjustment of frequencies and power.

Digital Services over Cable OpenCable, DOCSIS, and PacketCable comprise the legs of cable's digital triangle. The process of providing high-speed Internet access and broadband services over cable networks really began with the OpenCable initiative to insure standardized digital set-top boxes (STBs). The FCC has mandated that subscribers must be able to buy their STBs at retail and that consumer electronics manufacturers needed to know how to build them. Since the production of such an item requires a long lead time, it was essential to start early.[10]

OpenCable was launched in July 1997, and a request for information (RFI) was issued to leading consumer electronics and computer industry companies. The RFI described devices that would deliver digital video, data, and interactive services to a television set, with several components: (1) define a group of advanced digital devices, including STBs; (2) create a blueprint for delivering advanced services to consumers; and (3) establish a consumer brand. The specification is open rather than proprietary, in order to create competition in pricing and features among many vendors. It is also flexible so it can interoperate with VCRs, television sets, DVD players, and personal computer cards, and accommodate the introduction of new services and revenue streams. The CableLabs Executive Committee reached consensus on key OpenCable elements in November 1997:

- formats for digital cable television signals (MPEG-2)
- a consumer privacy system
- DOCSIS (Data-Over-Cable Service Interface Specification) interfaces for high-speed connections
- authoring interfaces to create interactive applications
- a copyright protection system

DOCSIS 1.0 was ratified by the International Telecommunication Union (ITU) in March of 1998. Compliant cable modems are integrated into STBs for TV sets that follow the OpenCable standards. They will also support the high-definition television (HDTV) standard. DOCSIS also specifies modulation schemes and the protocol for exchanging bidirectional signals over cable. It supports data rates up to 30 Mbps (megabits per second). An important element of DOCSIS is its security components that provide secure data transport services across the shared cable network. It is designed to give modem users data privacy and to prevent unauthorized access to data on the network by encrypting the stream from the cable modem to the cable modem termination equipment in the headend.

The third leg of cable's digital triangle is PacketCable. The initiative defines the specifications for an interoperable Internet Protocol (IP) network that will enable sophisticated interactive services. PacketCable rides on top of the cable modem infrastructure, using IP to deliver multimedia services, such as video conferencing, carrier-quality IP telephony, interactive gaming, and general multimedia applications. In September, 1999, CableLabs released seven draft specifications for PacketCable 1.0, sending them to 240 participating industry vendors for review and comment, and adding them to the previously released network-based call signaling spec (a way to manage telephony services).

The PSTN

The public switched telephone network is an astonishing achievement that has taken more than 100 years to construct. It is a highly centralized, very reliable network, with a rigid operational structure. Its design and its importance to the national economy mean that innovation and change come regularly but carefully, requiring considerable coordination. The infrastructure, including its construction and operation, is regulated by federal and state agencies.[11]

The company that provides telephone service to most residences is called a local exchange carrier (LEC) or incumbent local exchange carrier (ILEC). If the LEC used to be part of the old national AT&T telephone system, it might be called an RBOC, standing for regional Bell operating company.

In many cities, large companies have the option of receiving telephone service from a competitive access provider (CAP), which are private companies that do not have a governmentally-established geographical service area. There are a few private companies that do provide residential service; they are called CLECs, competitive local exchange carriers.

The basic telephone line is composed of two wires that run between the telephone and the telephone company's central office (CO), sometimes called a branch exchange. Carrying signals and electricity in both directions between the home and the central office, this pair of wires is variously called a drop, a loop, a pair, twisted pair, copper pair, or a circuit. The electricity is provided by loading coils if the loop is analog. If it is digital, the electricity must be carried on a separate line to avoid interference. The local loop is the physical layer of the interface between customers and the PSTN, providing the familiar dial tone, touch tones or DTMF (Dual Tone Multi-Frequency), and busy signals.

The twisted pair come in and out of the home or commercial building through a box called the network interface device (NID). Leaving the NID, the wires are tied into a neighborhood wiring pedestal or telephone pole, where it meets other circuits from the same area. If the neighborhood is further than 18,000 feet, or 3.5 miles, away from the CO, then the bundle of messages from that area is converted to digital signals, and carried by a digital loop carrier (DLC), normally fiber optic cable. DLC systems consist of physical pedestals containing line cards that concentrate residential links onto digital circuits. The signals will probably need to be strengthened because they lose power as they travel; the equipment that amplifies them is called a repeater. If the local loop is still all analog and within 18,000 feet of the CO, the signal is converted to digital by a channel bank in the central office. A channel bank is digital-to-analog (D-A) and analog-to-digital (A-D) conversion equipment that can also mix together (multiplex) or separate out (demultiplex) many different telephone calls into a single signal for forwarding.

The central office also houses one or more switches that transmit each signal towards its final destination. The path from origination to termination will stay open throughout the length of the communication, a dedicated circuit. Central office (CO) is the designation in the United States for a telephone company facility, although the correct network term is "transceiver node," any point on the network where signals are both sent and received. Central offices vary in their size and sophistication of equipment. Some COs use new digital technologies, while others are still analog. The class of the CO is ranked by its switching capabilities, interrelationships with other COs, and the transmission requirements placed upon it.

If the call is a neighborhood call, it might be switched directly to the terminated line; if it is within the local area, it will be forwarded to the switch in the central office where the terminated line is located. If it is a long distance call, outside the area of the telephone company's service area, then it goes to the switch of an interexchange carrier (IXC), a long distance company. The location at the IXC is called a point of presence or POP.

Long distance companies either own or lease a network that is called a backbone network. A long distance call goes from the originator to the LEC central office, to the nearest IXC point of presence, through the IXC backbone network, to a remote POP nearest the dialed number, through a LEC switch, to the called party. Telephone switches respond to touch tones or dial pulses from the telephone set. They send a request over a trunk line, a high-speed digital circuit that links components within the telephone company network.

Long distance messages are sent to a remote computer called a signaling control point (SCP), which directs the switch where it should route the call. SCPs come in pairs, one primary and one backup. The switch follows the SCP's direction and sends the call ahead to a signaling transfer point (STP), which establishes a route between the SCPs and switches of all the different telephone companies. The signaling structure between SCPs and STPs is called Signaling System 7 or SS7.[12]

Signaling refers to the information used to establish, maintain, and end a telephone call. Traditionally, the signaling took place in the same circuit as the telephone call, a technique known as in-band signaling. Today, most of the world's telephone networks communicate with each other and accommodate signaling using SS7, a separate fault-tolerant network that operates parallel to the telephone call.

When someone makes a telephone call, the information to set up, route, and end the call travels on the SS7 network, not in the same circuit; this technique is called out-of-band signaling. The overall technology is called common channel interoffice signaling (CCIS), introduced in 1976 by AT&T. SS7 performs many functions and gives telephone companies opportunities for revenue-generating services. It provides routing and billing information for telephone services such as 800 numbers, 900 numbers, 911, and custom features like caller ID.

DSL: Digital Subscriber Line   Digitization has proceeded throughout the PSTN backbone, and now it is moving into the local loop. Figure 9.5 shows an early view of how telephone companies would provision a broadband network. ISDN, or integrated services digital network, as a way of delivering digital services to consumers is being overtaken by a newer technology called DSL. DSL comes in many flavors; xDSL refers to any or all of them. The various types are ADSL, ADSL Lite (G.lite), RDSL, HDSL, SDSL, and VDSL. Asymmetric digital subscriber line (ADSL) is the most common installation in homes, as shown in Figure 9.6. In the past few years, it was trialed by more than forty companies, mostly in North America and northern Europe. Commercial service began in 1997, and now it is rolling out in many markets.[13]

ADSL transmits 1.5 to 8 Mbps downstream and from 16 to 640 Kbps upstream over twisted-pair copper wires. It is essentially a local area Ethernet network, which we will discuss later in this chapter. It supports up to 8 Mbps but the residence must be located within 18,000 feet of copper wire, which excludes as many as 20% of U.S. loops, and another percentage are equipped with loading coils that keep ADSL from running.

ADSL requires some equipment, including a PC, special modem, and a device called a splitter, that separates the voice from the data communications. In the central office, the provider must put in a DSLAM, a digital subscriber line access multiplexer to take in the signals, and to route the voice calls to the PSTN and the data to a high-speed digital line to an Internet point of presence (POP).

The DSLAM in the CO itself is composed of CO modems and a service access multiplexer (SAM) that acts as an interface to the data network. At the other end of the fiber line carrying the data traffic, a remote DSLAM interfaces the data onto the backbone network.

Telephone companies make ADSL work by coding the signals onto the copper wire using a modulation technique called DMT, discrete multitone. Here the question of standards comes up again, because the American National Standards Institute has specified another modulation scheme called carrierless amplitude modulation (CAP). But local telcos are actually using DMT because they say it is more reliable.

DMT divides the available frequencies into 256 discrete subchannels or tones to avoid high-frequency signal loss caused by noise on copper lines. A more advanced version is DWMT, discrete wavelet multitone, that creates even more space between the subchannels. CAP is a version of quadrature amplitude modulation (QAM), where part of the modulated

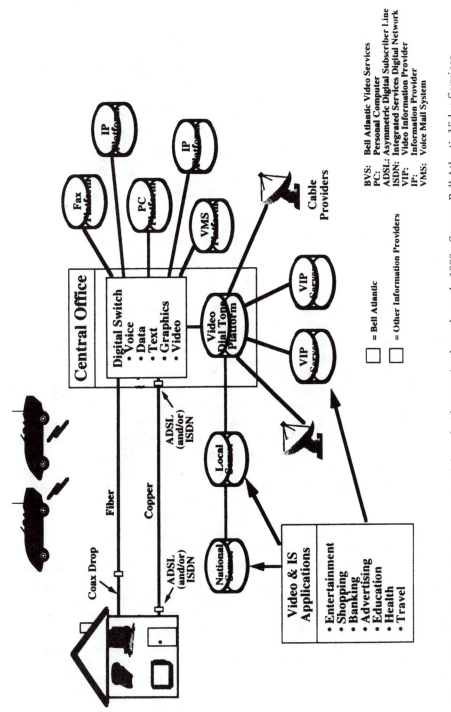

Figure 9.5 A vision of a telephone network-based video service from the early 1990s. Source: Bell Atlantic Video Services.

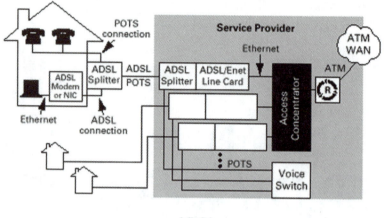

ADSL

Figure 9.6 Diagram of ADSL implementation.

message signal is stored in memory and then reassembled in the modulated wave. The carrier signal is eliminated from transmission because it contains no information and is added again at the receiving modem, giving this modulation scheme its name.

G.lite is a variant of ADSL that was approved by the ITU in June, 1999. It provides an inexpensive version of DSL technology, because it works with existing equipment and does not require any additional gear at the customer's premises. It is a little slower than DSL, providing a maximum downstream bit rate of 1.5 Mbps and an upstream bit rate of 512 kbps. The importance of no additional equipment at the customer site is that it would make it possible for the consumer to install DSL themselves. The modems are standardized as well, so people could buy them at a consumer electronics outlet, hook up their phone lines and computers, and call the telephone company for service. The telco would be able to remotely configure the DSLAM and switch or router in the central office.

R-ADSL, rate-adaptive digital subscriber line, utilizes a new variation on CAP that splits the transmission spectrum into discrete subchannels and adjusts each signal transmission, depending on the quality of the local loop. HDSL stands for high bit rate digital subscriber line, a symmetrical service that delivers 1.54 Mbps both downstream and upstream. And SDSL is single-line digital subscriber line, essentially HDSL over a single twisted pair.

VDSL is a form of DSL that carriers are thinking of using to provide multichannel digital television service to customers. It stands for very high bit-rate digital subscriber line, a technique that uses an advanced type of DMT modulation called "zipper" to enable a data rate from 1.54

Mbps to 52 Mbps. One limitation is that VDSL has a maximum range of 1,000 feet to 4,500 feet on 24-gauge wire.

Two trial installations of VDSL are in Canada and Phoenix, Arizona in the United States. In New Brunswick, Canada, NBTel, the area's major telephone company, is using equipment from iMagicTV. NBTel tested the system with 200 viewers and launched commercially in late 1999 with more than 100 channels of digital TV, pay-per-view movies, music, high-speed Internet access and telephone service on the same line. In the United States, US West has been testing VDSL service with about 5,000 households in a suburb of Phoenix, offering 170 channels of TV, 40 channels of digital audio, and high-speed Internet access, in addition to the phone. It is not known how the merger of US West with Qwest will affect these plans.

Computer Networks

Computer networks, sometimes called Ethernets, come in small, medium, large, and extra large sizes known respectively as LANs (local area networks), MANs (metropolitan area networks), WANs (wide area networks), and GANs (global area networks).[14] (See Figure 9.7.)

LANs networks are geographically contiguous. MANs are metropolitan area networks that may be dispersed over 50 miles or more from one end to the other. And WANs are dispersed over national, even international territories, although a new acronym has been recruited to describe global dispersed networks—GANs.

Dispersion means that parts of the networks are geographically separated from each other. They are connected by lines leased from local exchange or long distance carriers, sometimes called network service providers (NSPs). The speed of the network depends on how many users

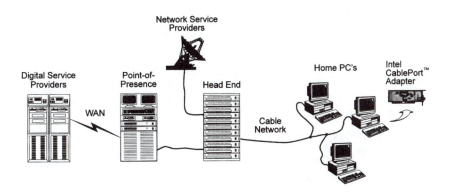

Figure 9.7 Network architecture for a LAN over a cable system, connected to a WAN. Source: Intel.

there are, the size of the files and messages moving between them, and the capacity of the leased lines.

Most Ethernets usually have a server sitting on the network, and they are frequently called "client-server networks." The server receives E-mail and holds it until requested by the user it is addressed to. It also stores material that users need to share so that everyone authorized to access it can do so. Often complex software will keep track of the various versions of project materials, making a record of all changes that have been made.

Ethernets also run at different speeds by design. 10Base2 Ethernets and the variants that begin with 10Base carry information at a rate of 10 Mbps. Fast Ethernets transmit data at 100 Mbps and Gigabit Ethernets at 1 Gbps.

Local area network (LAN) locations are usually less than a mile or two apart at their furthest reaches. A LAN links workstations and computers together via copper twisted pair, coax, or fiber optic cable, with as few as two or three PCs, or as many as thousands. The computers are loaded with software and a network interface card (NIC) that provides an interface between the computer and the physical media of the network and software.

LANs are packet-switched networks that transport information on shared media. In its simplest form, LANs carry unstructured streams of messages, and the edge devices (PCs and workstations) pick them out of the bitstream. For a protocol, many LANs use the IEEE 802.3 carrier sense multiple access with collision detection (CSMA/CD).

Since the rise of the Internet, however, many computer networks have adopted TCP/IP, or transmission control protocol/Internet protocol. Considering how warp-speed of Internet change, it is amazing how little TCP/IP has changed since the 1960s. The protocols were developed to survive military engagements—war. The idea of TCI/IP was to create a decentralized and redundant network where intelligence was distributed throughout it so that in case of a breakdown of one part of the network, information could be rerouted to another node and get to its destination via some other pathway.

TCP/IP operates at Open Systems Interconnection (OSI) layers 2, 3, 4, and 7. It runs over any physical network, so it does not address the physical layer of the OSI model. It does not allow differences in classes of connectivity, so it does not reference the session layer either. Finally, it moves data without regard to type or format, so it does not involve layer 6. These are the 7 OSI layers:

Layer 1: Physical network

Layer 2: Data link (transfers information across physical network)

Layer 3: Network (synchronization, error and flow control, connection set-up, maintenance, and termination)

Layer 4: Transport (end-to-end recovery and flow control)

Layer 5: Session (establishes, manages, and terminates sessions)

Layer 6: Presentation (allows interoperability between data and syntax)

Layer 7: Application (access for users)

Here's what TCP/IP does do: Internet Protocol covers how Internet addresses are handled on a network; each address is a unique, 32-bit set of numbers. Network routers read them to direct packets (also called datagrams) to their destination. TCP defines the transport layer and how the network carries the unstructured bitstream, and it guarantees delivery.

Each computer on an Ethernet has a network interface card that is assigned a unique 32-bit number called its media access control (MAC) address, which is known to the physical layer network. The MAC address is wrapped in a datagram, like a letter in an envelope, that remains sealed across the network until it is recognized by the machine it is addressed to. The MAC address isn't even read by the routers across the network or Internet; "packet sniffers" or "network analyzers" just read the address of the destination host computer, not the unique addressee.

The connectivity between multiple computers and workstations can be provided by one of several devices, such as hubs, fan-out boxes (concentrators), repeaters, multistation access units (MAUs or tranceivers), and Ethernet switches. A hub is a central point where multiple cables come together. A hub usually allows 8, 16, or 64 node connections to communicate. If any single connection disconnects or is having problems, the hub removes it from the network and allows all the other nodes to continue operating without interruption.

Transceivers, also called media attachment units (MAUs), connect computers and workstations to the various Ethernet media, wires, and cables. Repeaters connect two or more Ethernet segments of any media type to extend them further. Repeaters also link segments of different media.

Bridges connect separate Ethernets together by mapping the addresses of the devices residing on each segment and then allow only properly addressed traffic to go through. Bridges read the address and direct packets to their destinations. Bridges are store-and-forward devices because they halt the packet stream before they forward them.

An Ethernet switch is a specialized bridge that connects more than two segments together. It forwards only messages addressed to the linked segment. In this way, the switch component improves network performance by eliminating inappropriate traffic from each segment.

Routers work in a manner similar to switches and bridges in that they filter out network traffic. However, they do it by established network policies and protocols, rather than by deciding on individual packet addresses. Routers divide a network into subnets to limit traffic to appropriate segments. It makes them a little slower than bridges and switches because protocol filtering takes more time. Table 9.5 shows how routers and bridges differ.

Table 9.5 Comparing Routers and Bridges

Routers	Bridges
Have a node address	Invisible on the network
Treat packets according to priority	All packets are equal
Send packets to a specific destination	Send packets ahead to any forward destination
Use multiple information sources to forward packets	Use only addresses to forward packets
Can read packets and payload	Just forward the packets
Return error messages	Usually do not provide error messages

More sophisticated routers also perform the functions of gateways. A gateway links different kinds of networks, not just LAN segments. These new routers provide internetworking services—that is, tasks involving integrating the functions of disparate networks. They examine a packet and then forward it to any network that will provide the most effective route to the destination, whether it is an Ethernet network or not.

The Internet

Begin by considering how Internet service, based on IP networks, differs from telephone service, delivered over the PSTN, as shown in Table 9.6.

The Internet began as a project of the U.S. Defense Advanced Research Project Agency (ARPA), originating as a theoretical network architecture on paper, in 1962.[15] The Net's military background is extremely important to its current operation, because it established from the outset a reliance on multiple routes to destinations that makes it resilient to breakdowns. If there is a problem, the Net routes around it. By 1969, it was a real network, linking four U.S. universities: the Stanford Research Institute, U.C.L.A., U.C. Santa Barbara, and the University of Utah.

The network achieved a rarefied popularity quickly, with E-mail emerging as the first "killer app." A year later, ALOHANET went online at the University of Hawaii. By 1971, there were twenty-three universities and government research units connecting to ARPANET. In 1973, University College in the United Kingdom and the Norwegian Royal Radar Establishment made the fledgling network an international phenomenon. In the mid-1970s, ARPANET began moving away from its military and research origin, and in 1975, the network's operations were transferred to

Table 9.6 Comparing the Internet Network to the PSTN

PSTN Service	Internet Service
Circuit-switched	Packet-switched
Dedicated physical connection	Connectionless
Static switching	Dynamic routing
Geographic numbering	Nongeographic addressing
Intelligence at the center of the network	Intelligence distributed throughout network and at the edge
Transport, plus value-added applications	Transport, plus free applications (This is changing.)
Settlements	Peering agreements
Metered usage-based pricing	Flat-rate pricing
Reliability before innovation	Innovation before reliability
Regulated	Unregulated

the Defense Communications Agency. By 1981, there were 213 hosts, with a new one added every 20 days. The academic segment called BITNET connected most universities. In the early 1980s, Bob Kahn and Vint Cerf headed up a team that created the common language and protocols for the Internet, TCP/IP. This development made the Internet into what could be recognized as an operating internetwork, coherent across its various segments. The mid-80s saw the introduction of PCs, and many corporations began participating on the network. By 1987, there were more than 10,000 Internet hosts, then 100,000 in 1989, and 300,000 in 1990!

By this time, the decision had already been made to get the government out of the network business. In January, 1989, a Merit/IBM/MCI team put together a plan to increase Internet capacity and speed, using 45 Mbps T-3 telephone lines in the backbone network. IBM developed the first router capable of operating at those speeds, routing 100,000 packets per second.

In 1990, the three companies created an independent nonprofit organization, Advanced Network and Services, Inc. (ANS) to operate this backbone, called the NSFNET (National Science Foundation Network).

Three events came together in 1991 that led to the present-day popularity of the Internet. First, it became a graphic medium when Tim Berners-Lee introduced the hypertext transfer protocol (HTTP) and related hypertext markup language (HTML)—creating the World Wide Web. About the same time, the University of Minnesota released the first "gopher," providing an Internet navigation system. Its name was a pun on the school's athletic teams, the Golden Gophers, and the program's ability to "go-fer" material on the Internet. Finally, the National Science Foundation lifted restrictions against commercial use of the Net.

In 1993, the National Science Foundation privatized the Internet backbone entirely. It let out contracts for the construction of the four original network access points (NAPs) in Washington, D.C., New York, Chicago, and San Francisco. The design of the NAPs made it possible for any provider with a high-speed network to connect to it, not just the NAP contractor. No restrictions were placed on usage or content. One company, Merit (now owned by MCI/WorldCom), received the contract to be the routing arbiter, maintaining a database of routing and interconnection information. In 1995, the NSFNET backbone shut down.

The ability of all network operators to interconnect with the Internet is extremely important, since operators are not inclined to share their interconnection points. But the NSF did not specify how these operators would come to an agreement about payments for interconnection and reciprocal carriage. This continues to be a contentious problem as the Internet has grown.

Basically, there are three ways traffic is carried on the Internet. The first is peering, which is an "I'll carry yours if you'll carry mine" type of agreement. Although all operators originally peered with one another, in the mid 1990s, UUNet dropped small networks from its agreements and since then, only the largest carriers peer with one another. The other agreements are called "peering payment" and "transit payment," both of which involve some kind of fee charged by the larger operator for carrying the traffic of the smaller operator. The fee contracts can become very expensive for small operators; however, today's rapid expansion of bandwidth is working in their favor.

The actual architectures of the major carriers is proprietary information—no one knows how the Internet actually works in complete detail. In a router-based network, there is no dedicated connection open for the duration of the communication. This "connectionless" type of network is often represented in graphic depictions of networks as a "cloud," and a puffy curved entity often sits in the middle of diagrams of network architecture. It really misrepresents the complexity of network connectivity but satisfies corporate marketing departments' communication requirements.

Internet domain names are another occasion for strife. Corporations and "cybersquatters" fight over them, and even the authority to assign them has been the topic of much dispute. For many years, Internet domain names were provided by InterNIC, but in recent years, that function has been given to other agencies. The overall authority is exercised by the Internet Corporation for Assigned Names and Numbers (ICANN), a private, nonprofit corporation that is responsible for Internet address space allocation, protocol parameter assignment, domain name system management, and root server system management. Previously, these functions were performed by the Internet Assigned Numbers Authority (IANA).

Routing information is similarly dispersed. Each router makes a decision about how to forward a packet by referring to a lookup table that identifies the next node to which the packet should be forwarded to move it toward its destination. Lookup tables are loaded by the router itself, based on routing information supplied by other routers or on some kind of human input.

The Internet is the world's biggest WAN—a GAN (global area network), if you will. It is composed of thousands of networks internetworked together. No one owns the Internet. The system of addressing is based on "domain names" that are held in specialized computers on the network called nameservers. They translate domain names, such as www.bhusa.com, into its numeric IP address, such as 193.0.0.195. The numeric address is the one used by Internet protocols to deliver packets.

Root nameservers handle the first part of the translation from domain name to IP address by locating the nameservers for the top level domains (TLDs): NET, COM, ORG, EDU, GOV, FR (France), JP (Japan), AU (Australia), PE (Peru), and others. The root nameservers are dispersed across territories and the Internet itself, so that failure in one root nameserver will not affect the operation of others. The other nameservers that participate in the Domain Name Service (DNS) are dispersed among hundreds of thousands of other nameservers for each domain. Each TLD has at least two independent nameservers and often more.

If this arrangement sounds somewhat anarchic, it is. And these principles were established as integral to the Internet from the beginning. The network was designed to have an open architecture that would allow for virtually unlimited growth. The four principles that guided its foundation included:

- Each distinct network had to be independent, and Internet connection would not require any internal change to it.
- Message delivery would be made on a best-effort basis. When packets were not delivered, a feedback message would notify the source to retransmit it.
- Network connection devices were not specified. However, information about messages that passed through them could not be retained.
- No overall control and operations mechanisms were specified.

The Internet does have a hierarchy and some geographically situated elements: NAPs, IXs, MAEs, POPs, ISPs, and users. The top of the hierarchy begins with:

- Level 1: Network access points (NAPs), large interconnection facilities where the largest Internet service providers (ISPs) and network service providers (NSPs) all mount racks of equipment in the same location.
- Level 2: Data emanates in and out of the NAPs along large-size lines that form the backbone of the network.
- Level 3: The information lands at the interconnection point of a regional network.
- Level 4: Data is routed to a local Internet service provider.
- Level 5: Customer premises—businesses and homes.

The biggest telecommunications companies like MCI/Worldcom/Sprint and Cable and Wireless co-locate at the network access points at the highest level of the Internet. There are four top-level NAPs: in Chicago, operated by Ameritech; in New York, operated by Sprint; in San Francisco, operated by PacBell; and in Washington, D.C. (MAE East), operated by WorldCom. A great deal of Internet traffic no longer goes through these exchange points or NAPs; rather, it is moving through the private interconnects of the backbone providers.

One reason for using private interconnects is that it is easier to build big fiber backbones than it is to build scalable interconnection points. Switching equipment is expensive and doesn't always have the capacity the manufacturers claim for it. It's not always a company's fault. The Internet is so large, it cannot be simulated in a laboratory; so engineers have no way to test interconnection devices before they are put into use.

The large co-located telecom companies are the Internet's backbone carriers, providing the lines that make up the second level. The regional networks are the Internet's third level, provided by telecom companies that operate in the area. They provide backbones within a state or multi-state area, similar to the way national carriers do at the top level. They connect to the national backbone operators, often more than one, at metropolitan area exchanges (MAEs), which exist at most major markets in the country. The MAEs provide connections to the Net for the Internet service providers and company point-of-presence (POP) sites at the fourth level.

Internet service providers range from very small operations to huge businesses with hundreds of thousands of subscribers like Mindspring and Earthlink. ISPs do not operate networks; rather, they focus on customer service and education. However, increasingly, the largest ISPs are owned by network operators. For example, Earthlink was purchased by Sprint in 1999, and the LECs often offer ISP service.

At just about every node on the Internet sits a server. A Web site is really a software construction that is stored in a server, so when a user "visits" a Web site, it really means that the graphics, text, video, and audio—the content—is transferred to the user's local machine by a server.

Of course, in reality, the user isn't going any place; the bits that comprise the Web site move across the Internet to the user's machine.

Servers may be huge pieces of equipment—"Big Iron"—operated by ISPs or large private companies. Or they may be merely user PCs that are software-enabled to dish out data. Whether they are businesses or people in their homes, customers are at the fifth level of the Internet. If the subscriber is a business, then the organization often has a LAN that supports any number of employees, extending the Internet yet further.

When a computer is connected to a network, data flows in and out. For many, this is quite worrisome. Companies are in a position to act on their concerns, and security is manifested in hardware as "proxy servers" or "edge servers." This specialized server maintains a "firewall" that protects unidentified data from entering the company's LAN or WAN, although it can authenticate authorized users and give them the appropriate level of access. One problem with proxy servers is that companies sometimes assign them addresses that do not conform to Internet standards, violating the basic assumption of an open system upon which the Internet is constructed.[16]

On-Demand Network Components

As broadband capacity to consumers becomes more widespread, many providers will want to offer material on-demand, especially movies. But other media forms will also be important, including video and audio of all kinds, such as movies, TV shows, music concerts and performances, shorts, and even commercials. Providing on-demand media requires more than high-speed, high-capacity networks. The next section looks at media servers and storage requirements for this type of service.

Media Servers

In order to provide a movie, TV show, video, or audio stream when a customer orders it, program distributors they must have a system for sending out the ordered program to the right place and a place to store it. The equipment that takes the order and routes the show to the viewer is called a "media server," a kind of video jukebox. The equipment where programming is stored is called "mass storage," and increasingly access to it is provided by a "storage area network," or SAN.

The issues of the design of media servers is the same whether the provider is a cable system, Internet service provider, or telephone company. When video leaves the media server, it is transported to the address from where the request was made—and only that address. The ability to send material to a specific household is accomplished through switching.

Essentially, a server is a switch with storage, or mass storage linked to a switch, because the switching function is inherently incorporated into a media server.

When a server stores video material for access by users, it is called a "video server." Video servers are already penetrating the professional broadcasting market where they are used for news editing and for ad commercial playback and insertion. Increasingly, they are part of a media asset management system that encompasses the creation, repurposing, and distribution of a company's content. When a server is part of a commercial operation and billing records are part of its function, it is called a "media server."

We need to begin by comprehending the scale of on-demand programming. Steve Rose writes:

> the entire Internet ran successfully for years on a backbone of 45 Mb/second for the entire United States (and much of the rest of the world). Video-on-demand for a small town of 10,000 customers, assuming 20% online (2,000 simultaneous digital, compressed video streams), requires a throughput of 8 Gb/second, or about 200 times the Internet backbone for much of its life![17]

Media servers are part of the center-versus-edge revolution in computing, where the relationship has changed between the central site (where stored material resides) and the terminal that requests the material. In the previous era of mainframe computing, the term that characterized the relationship was "master-slave" because no intelligence resided in the terminal. Today, the term "client-server" describes how users on a network with intelligent terminals can call for information stored at the central site and use it locally.

From a hardware standpoint, a media server is actually a special-purpose computer that makes possible some of the most potentially profitable services that providers anticipate offering.[18] A customer order comes into the media server; the server locates the desired information in storage, retrieves it, and sends it downstream to the viewer. It then sends a message to the billing software to charge the consumer. The ordered material might include video of movies, television programs, and direct-to-home programs. It may be entertainment-oriented or informational in nature. Other content could be music, games, catalogs, lists, and announcements.

Media servers use all-digital processing and storage techniques. Current questions about media server design are: Is it possible to build a cost-effective server? What type of processing should be used? Where in the network should the server be located? What kind of storage should be used?

The cost-effectiveness of media servers is described in terms of "cost-per-bitstream" (CPB) or "cost-per-user," with CPB being the preferred term, because not all users access bitstreams. Cost-per-bitstream calculations are based on cost of storage, plus the cost of the special- purpose computer. In the spring of 1995, the cost per bitstream dropped from an average of $1,200 to $440 (including storage costs) when nCube came out with a new-generation server. According to Steve Rose, in 1999, that figured has not changed, even though there are quotes of $150.19. However, by the time ancillary equipment for modulation or output is factored in, the cost rises to $350 to $500 over a three- or four-year period, a realistic price that includes maintenance as well as delivering a modulated stream to real users.

The question of the cost per bitstream is intimately tied to the underlying technology for media servers. Several types of computers can be made to operate as media servers, ranging from single PCs and workstations to mainframes. Even the new generation of switcher/routers could lower the cost per bitstream, but a great deal of design work must be accomplished before it could happen. There are three favored computer architectures for media servers: "loosely coupled computers" (LCC); "symmetric multiprocessing" (SMP); and "massively parallel processing" (MPP).

Servers based on the LCC model are big—too enormous for the average headend to accommodate because they are composed of so many machines operating together. The LCC model is best exemplified by the Microsoft approach. The company's Tiger software, based on the Windows NT operating system, coordinates several hundred 486- or Pentium-level PCs on a bus, switch, or mesh. The linking method slows processing and limits the number of processors that can be interconnected.

An example of the SMP model is Silicon Graphic's Challenge design that was used in the Time Warner test trial in Orlando, Florida. This design calls for many processors, all linked to a single disk controller and shared memory. Unfortunately, although this media server can be made to operate for the smaller number of users in a test site, it will not scale up to meet the requirements of a 50,000-subscriber cable system—or even a 10,000-subscriber cable system.

The MPP design is similar to the LCC server, except that the processors are linked in such a way that they allow thousands of powerful chips to be assembled in a relatively small space. There are many advantages to the MPP design. It is the only server that is truly scalable up to the requirements of a 10,000- to 50,000-subscriber cable system. Its enormous processing capacity allows it to perform the switching function, eliminating specialized (and expensive) switching equipment. Finally, only MPP allows an unlimited number of simultaneous downstream flows of a single on-demand product from one storage cache.

This issue of simultaneous flows of one product has been a thorny problem for would-be providers of on-demand video. When a new movie comes out, everybody wants to see the new hit at once. Upwards of 90% of the traffic could be generated by five or six current hit films. At the video store, all the copies of the popular film are gone and customers have to be placed on a waiting list—by contrast, an interactive system just crashes.

The MPP server is composed of dozens of processors. It breaks a movie (or any video) into tiny pieces of just a few seconds and shunts them into a fast-access buffer. Each piece is handled by a different processor, then replaced in cache memory. The information is reassembled into a complete bitstream as it leaves the server. Since only a few seconds at a time are being worked on, when another subscriber calls for that same movie ten seconds later, the material is still available. Combined with sophisticated mass storage equipment, the MPP can produce almost as many bitstreams as there are processors. Media servers by nCube, IBM, Silicon Graphics (which bought the Cray MPP technology), and Intel all use the MPP model.

Once a media server is selected, the question of where to locate it arises. For a time CableLabs proposed locating giant servers in regional hubs, where system operators would access little-used material. Similarly, AT&T originally thought to archive its material in Manhattan, storing only a small portion of a program or film locally. Arguing against this model, Rose makes the following pragmatic observation:

> the network operator owns the local network and must pay to transport material accessed outside the network. The most popular movies and programs will reside in memory locally. This means that the least popular products will be the furthest away, cost the most to transport—and the operator must charge less for this seldom-called upon material![19]

At the same time, locating the media server in a neighborhood node means there must be a server for every 500 subscribers, a very expensive approach. Thus, although there is continuing experimentation to find the best location for the media server, chances are the cable system headend, the telephone company's nearest central office, and a geographically nearby Internet mirror site will be the appropriate locations.

Fundamentally, where servers should be located is an economic question. They should be placed as close to the final consumer as the cost of facilities and storage will warrant. Densely populated, heavy usage areas will support the closer servers in greater numbers.

Storing the Movie Store

Another important type of capacity to consider for an on-demand service is how much material will be stored, available for subscribers to choose from. Although there are subtly different storage requirements for different programs, the following figures represent an average. One very popular type of material is motion pictures. They are also the most challenging because of their quality, length, and desirability. A movie service also needs to offer a fairly diverse catalog, especially to a large audience. Probably no service would need to store more than 1,000 movies at any given time. They would probably want to rotate both popular and historical material, working from a base of popular older films for which they have made a one-time payment.

People want reasonably high quality to view a 90-minute feature, so storing the average movie compressed in MPEG-2 requires about 3 GB (gigabytes). Three hundred films would require about 1 TB (terabyte) of storage, 1,000 movies would take up about 3.3 TB, and 3,000 movies need 10 TB. An hour of television would need about 1.5 GB and a half-hour program about 750 MB.

When the numbers get that big, storage gets expensive. If an operator decides she needs more mass storage, the cost increases linearly. In late 1995, one gigabyte of storage cost about $200. Add 100 gigabytes, and the cost was $20,000. Today, that same gigabyte is around $5.00, and the operator can add a terabyte of storage (300 movies) for $5,000! A professional robotic system will certainly cost more—but the lowering prices are still dramatic.

A service will probably use more than one type of storage. The advantages of different types of memory are shown in Table 9.7. Often-used programming will reside on an array of magnetic disks, called a RAID array, standing for Redundant Array of Inexpensive Disks. This ingenious storage solution allows operators to back up programming using only 25% more disk space. If there is a malfunction, the system has enough redundancy built into it that the flaw will not be apparent to the customer.

When a customer orders a program from the media server, the server looks first in the RAID array where the last-ordered material is still stored. As new material is called up, the oldest videos are pushed back into longer-term storage such as tape or optical disc. When a customer requests rarely-requested video, it will be fed into the disk array. The most

Table 9.7 Comparing Mass Storage Media

Medium	Cost	Speed	Capacity	Maintenance
Random access cache memory (digital RAM)	high	fast	small	low
Magnetic JBOD (just a bunch of disks)	moderate	moderate	large	low
CD-ROM (digital)	moderate	slow	moderate	moderate
Videodisc (analog)	moderate	fast	moderate	moderate
Tape (digital)	low	fast	large	high

requested material stays in the RAID, until its appeal declines. The advantage of the RAID over long-term storage media is that it allows access of multiple requests separated only by a few seconds. This use of RAID storage is called "caching."

In the past few years, the path between media server or the server farm and mass storage has itself become a network, called a Storage Area Network, or SAN. A decade ago, a simple interface sufficed, but today's higher capacity, data rates, and performance demands require a more responsive, robust architecture. A common architecture for this function is Fibre Channel (this is the spelling designated for the technology), a fiber-based gigabit speed network that includes a physical network and topology, switching or shared network protocols, addressing, data transport, and network management capabilities. The FOX Broadcasting Network Operations Center uses Tektronix servers on a Fibre Channel network, accessing stored programming for playout to a satellite uplink for national distribution.[20]

This gigabit-per-second network is a harbinger of the terabit backbone network. It's already been demonstrated, and it's coming soon to a network near you. The next chapter looks at next-generation networks.

Notes

1. H. Newton and R. Horak, *Newton's telecom dictionary*, 16th ed., Telecom Books/Miller-Freeman (2000).
2. W. Stallings, *Data and computer communications*, 4th ed., New York: Macmillan (1992):803.
3. At the ISO, DAI is covered under the nomenclature JTC 1/SC 29—Coding of audio, picture, multimedia, and hypermedia information. Until 2001, the committee chair is Dr. Hiroshi Watanabe (Japan). More information is at http://www.iso.ch/projects/tcinfoFrame.html?6JTC,1,29%138EN.
4. The company's Web site is at http://www.lucent.com.
5. M. Magel, "The box that will open up interactive TV," *Multimedia Producer* (April 1995):30–36.
6. S. Rose, consultant, Viaduct Corp. Personal interview, Malibu, CA, May 1995.
7. C. Podlesny, "Hybrid fiber coax: A solution for broadband information services," *New Telecom Quarterly* 1 (1995):16–25.
8. R. Brown, "The return path: Open for business?" *CED Communications Engineering and Design* (December 1994):40–43.
9. R. Kietzman, Director of Product Marketing, GTE. Telephone interview, November 1998.
10. Up-to-date information about these topics is available at the CableLabs Web site at http://www.cablelabs.com.
11. R. Hodges, analyst, Technology Futures, Inc. Telephone interview, October 1999.

12. ADS, a company that designs, manufactures, and markets SS7 equipment maintains a Web site with information about SS7 at http://www.ss7.com/.
13. There is a lot of information about DSL on the Internet. Two sites dedicated to DSL technologies are at http://www.adsl.com and http://www.xdsl.com.
14. A good source of information about Ethernets is at http://www.techfest.com/networking/lan/Ethernet.htm.
15. Information about the history of the Internet is available at http://www.isoc.org/internet-history/.
16. The problem of proxy servers is much-discussed among Internet engineers. In March 2000, a one-day seminar—a Special workshop—on intelligence at the edge of the network, was held by USENIX, the Advanced Computing Systems Association, to consider the issues. USENIX is at http://www.usenix.org/.
17. This comment was written by Steve Rose in a proprietary report, "Media server system overview." For a copy of the full report, contact Steve Rose at P.O. Box 100, Haiku, HI, 96708-0100.
18. J. Van Tassel and S. Rose, "The evolution of the interactive broadband server," *New Telecom Quarterly* 1, 2 (1996). Information about this publication at http://www.ntq.com/TOC/NTQV4I1TOC.html.
19. S. Rose. Personal interview, Las Vegas, NV, April 2000.
20. A. Setos, Vice President, Technical Operations, FOX Broadcasting. Telephone interview, October 1998.

Wired Bitpipes II: Next-Generation Networks

10

Convergence: Next-Generation Fiber Optic Networks

Chapter 1 of this book, "The Digital Destiny," defined convergence as the coming together of telecommunications, computers, and broadcasting, at the levels of data, devices, networks, and organizations. But no matter what kind of network it is, converged or not, the functionalities are the same for all twenty-first-century networks, or next-generation networks (NGNs), that are to be the communication infrastructure of a wired world. And they are coming more quickly than could possibly have been imagined just a decade ago.

In 1993, communication networks of 10 Mbps speed/capacity were the norm. In 1995, the standards for 100 Mbps networks were set and implementations carrying gigabits of information were common within three years. This represents a ten-fold increase in network capacity in five years. Now at the beginning of the new millennium, we are seeing all the standards and equipment becoming available for 1 Gbps and 10 Gbps networks and beyond, capable of delivering terabits of information—a ten- and hundred-fold exponential leap in capacity just another five years later.

This growth holds true whether one considers leased lines, frame relay for routers, ATM core for routers, or packet over SONET. In the router market, the bar for companies making these devices to play in the high-speed network space has been raised to an astonishing 10 gigabits per second. Table 10.1 compares the growth of basic communication technologies and the demand for the Internet and data communication.

Most network operators are retrofitting their existing systems rather than building new ones because it is less expensive. The cost of plowing in new fiber optic networks ranges from about $35,000 per mile in rural areas to about $100,000/mile in urban areas, more expensive than redesigning

296

Table 10.1 Communication Supply and Services Demand

BASIC TECHNOLOGY	PERFORMANCE DOUBLING TIME (months)	GROWTH RATE (per year)
Capacity of chips (Moore's Law)	18	59%
Optical Fiber—bits per second, per fiber	12	100%
Packet switching—bits per second, $	12	100%
BASIC DEMAND		
Internet users	12	100%
Data bits	7.5	300%
Internet core	4	1,000%

Source: Broadband World.

the legacy networks. The cost of right-of-way and design add considerably to the overall costs, according to a recent report of one large company that builds networks:

> *Fluor is installing hundreds of miles of conduit, between 15 and 100 miles per city, and it is about 20% complete, says Gregory J. Amparano, Fluor project director. He estimates at least one more year before completion. Depending on quantity, method of installation, labor costs and restoration requirements, installation costs are currently running about $35 to $100 per ft.[1]*

At these prices, it's no wonder that dark fiber that was previously installed along with lighted fiber, perhaps many years ago, is the new gold in the ground.[2] It's a treasure that many are planning to mine now that the technology to build all fiber networks is becoming so sophisticated. Optical modulation and multiplexing have attained a new level of maturity, and tunable lasers now operate over an enormously wider range, with innovations on the front burner that will make even greater bandwidth available. The power of output lasers continues to grow, and the fiber itself has become capable of carrying more information along narrower channels with less noise.

The remaining frontier for multi-gigabit fiber optic networks is in fiber switching and high-speed terabit switches and routers from a few startup firms satisfy the market until optical switches become available. These switches are on the way. In late 1999, Cambridge University in the United Kingdom demonstrated the first optically transparent fiber-to-fiber switch that combined a hologram with ferrous liquid crystal, similar to a liquid crystal display (LCD). Built by Nortel and British Aerospace, the device is a holograph beam steering switch with a reconfigurable refractive element that directs light from fiber-to-fiber.[3]

The emergence of dense wavelength-division multiplexing (DWDM) is allowing operators to upgrade their single fiber mode networks to multi-wavelength systems, at least in the backbone. The number of channels that can be feasibly multiplexed continues to grow. Just a few years ago, 4-channel WDM was considered high-capacity; today, it is possible to multiplex upwards of 80 channels onto a single fiber.

One byproduct of DWDM technology is that it will allow some layers of the seven-layer OSI model to be collapsed, making optical networks even faster and more efficient.[4] Major efforts are underway to develop standards and products that will eliminate one or more of the intermediate OSI layers, allowing terabit switches and routers to integrate an optical WDM layer and optimize network performance. For example, the ITU recently standardized a DWDM channel to provide interoperability between DWDM and existing fiber architectures. This will allow carriers to migrate their existing network communications into the optical layer.

The typical components of WDM systems are optical combiners and splitters, wavelength selective filters, optical amplifiers, and specific wavelength sources. More advanced systems incorporate wavelength add/drop multiplexers (ADMs) at intermediate sites so that wavelengths can be picked up and dropped off as required by customers. However, fixed systems require physical changes to add and remove channels, so network management is difficult. This means that the most common network applications are point- to-point with all wavelengths accessed at the terminal locations.

One barrier to the implementation of advanced WDM systems is the immaturity of network management. Few standards exist, and not many engineers and managers have the needed know-how and skills. Legacy systems were not designed to accommodate wavelength management, and existing network elements must be redesigned or replaced to handle it. As a result, most analysts believe that newly constructed systems, such as those being built by CLECs (competitive local exchange carriers) and CAPs (competitive access providers) will be the first to market with DWDM technologies.

All the legacy systems use fiber optic technology in some part of their networks. Whether it is a cable, telephone, or computer network, there is usually a fiber backbone because this is where all the upstream and downstream signals come together and require the most capacity. Many systems have also replaced coaxial cable and even copper wire with fiber further down into the network, which carry traffic to some major portion of the network.

One attraction of fiber optic systems is the extraordinary reduction of costs, which accounts for the growth of these networks, as shown in Table 10.2. For example, before 1994, a 2.5 Gbps circuit on OC-48 cost $750. By 1997, that amount had dropped to $95. And telephone companies are gradually replacing copper wire with fiber optic cable as part of their regular plant maintenance because it has become cheaper to lay fiber than copper.

When operators upgrade existing systems, they simply replace existing wires with fiber, leaving the existing "tree and branch" structure in place. When networks are constructed from scratch, some builders choose SONET ring designs.[5] SONET stands for Synchronous Optical NETworks. This technology has already been quite successful, and beginning in 1996 long distance and local telephone companies began quadrupling the speed of their SONET networks by replacing OC-48 (OC stands for Optical Cable) cable with the larger capacity OC-192.[6] A more detailed discussion of ring network structure appears later in the chapter. Table 11.2 shows how legacy networks will migrate to the newer high-capacity networks.

We have already covered the equipment and processes that compose a wired network in Chapter 9, "Wired Bitpipes I." The components unique to a fiber network include the fiber optic cable itself, the transmitting and receiving devices—laser diodes and photodiodes—and regenerators and amplifiers, which strengthen the decaying light signals. (The decay of light signals is called "attenuation.")

The optical signals carried by a fiber optic network are light waves. They may also be called optical or photonic signals. Normal light waves spread out and dissipate as they travel. However, a special kind of wave, solitons, do not attenuate. This desirable feature makes it likely that when it is feasible at some time in the future, optical networks are likely to carry solitons.[7]

The first scientific observation of solitons was made by Scott Russell in 1834 when he was experimenting with water wave propagation along a canal in Scotland. He chased these slow-moving surface waves for several miles. In 1965, they were christened "solitons" because they can collide without being deformed, as if they had followed their own solitary path. It was understood that if solitons could be generated as light waves and they

Table 10.2 Miles of Fiber Optic Cable in the United States, 1994–2003

1994	1999 (estimate)	2001 (estimate)	2003 (estimate)
105,000	190,000	230,000	288,000

Source: Pioneer Consulting.

**Table 10.3 Optical Networking Application
in Metro Network Segments**

	Metro	Core Metro	Access Enterprise
Typical Distances	25–100 km	5–25 km	Less than 10 km
Typical applications	Point-to-point trunking, mesh connectivity	Optical add-drop bandwidth management, ring networks	Data center interconnect, storage area networking
Systems in place today	SONET OC-12/48	SONET OC-3/12 ATM VP Rings	Dedicated Fiber
Upgrade Strategies	OC-192 or DWDM Terminals for 40 Gbps capacity, Optical switch/ crossconnect	Hybrid SONET-ATM, ATM service access multiplexers (SAMs), optical ADMs	DWDM point-to-point or managed ring for virtual dark fiber
Key Technologies	Low-cost optical amps, optical crossconnects and switches	Programmable/ Dynamic ADMs, Optical switching, Optical performance monitoring	Low-cost DWDM filters and transponders
Key Benefits of DWDM	Fiber facility expansion, savings scalability, rapid restoration	Flexibility, rapid service provision, scalability of fiber	Efficient use capacity. Cost over dark fiber

Source: Pioneer Consulting.

could be controlled over a fiber, that the resulting system would not need to have a string of expensive signal regenerators, repeaters, and amplifiers.

In 1993, AT&T Bell Laboratories researchers transmitted solitons over 13,000 kilometers at 20 Gbps per second, double the previous world record.[8] Now techniques are extending this range even further. In 1999, scientists at Nippon Telephone and Telegraph (NTT) successfully sent 40-Gb/s single-channel soliton data over 70,000 kilometers of fiber optic cable.[9] The NTT team put together a 250-km dispersion-shifted optical fiber loop for the experiment. They superimposed four 10 Gbps soliton units at slightly different amplitudes, also reducing soliton-soliton interaction. The 40-GHz clock signal was made with an ultrahigh-speed photodetector, then used to drive a lithium-niobate modulator that performed soliton control. Without such control, noise and timing jitter increase the bit-error rate rapidly beyond 4500 km.

Whether the light waves are solitons or simply laser-generated light pulses, the components of fiber optic networks are the same. Fiber optic cable is basically a waveguide—a container composed of long strands of glass that carry light waves, surrounded by lightproof, reflective cladding. It offers many advantages, such as high bandwidth, relatively low cost, low power consumption, small space needs and total insensitivity to electromagnetic interference.

The first suggestion for silica fiber as a transmission medium was made in 1966 by researchers K. C. Kao and G. A. Hockam, working for Standard Telephone Laboratories in London.[10] The first AT&T field trial took place in 1977 in Chicago, and development of the technology has been rapid since that time. The production of purer silica strands and reflective cladding material decreased the attenuation of the light signals, resulting in increased bandwidth over ever-longer distances without having to regenerate the signal. The bandwidth of fiber optic cable was originally 50 MHz. Today, with DWDM, it is 400 GHz, that is 40 channels of 10 Gbps. The spacing of regenerators was 50 kilometers in 1986, 70 km in 1988, 130 km in 1991, 400 km in 1998, and 480 km in 1999.

To provide a sense of scale for just how much capacity fiber can grow to accommodate communication—or its "head room"—Russell Dewitt of Contel is quoted in Harry Newton's *Telecom Dictionary*: "For the future, the ultimate potential of a single mode fiber has been estimated. It is about 25,000 Gbps (25,000,000,000,000,000 bits per second). At that rate, you could transmit all the knowledge recorded since the beginning of time in 20 seconds."[11]

Tom Bowling argues that fiber optic technology should be regarded as an entirely new medium. Figure 1.5 (in Chapter 1) shows how Bowling plotted the growth of bandwidth since communication by semaphore in the early 1800s. His point is that historically, the world's demand for bandwidth is voracious, will likely continue, and will be made possible by fiber. Writes Bowling: "It is all new and different. The technologies, the content, the selection of programming, the need for sophisticated search capabilities, and the methods of marketing, billing, and research will be entirely alien to the older television and cable industries."[12]

Improvements continue. In June, 1995, Boston Optical Fiber, Inc., announced that they had produced a clear cable capable of carrying nearly as much high-speed communication traffic as glass—but made of plastic. In 1997, the ATM Forum established a standard for plastic optical fiber (POF) as the preferred medium for high-speed networks.[13] Now several companies are producing GIPOF, graded-index plastic optical fiber, at a fraction of the cost of similar cable made of silicon glass. Another advance has been the development of a piece of equipment called the erbium-doped fiberoptical amplifier, the EDFA.[14]

Regenerators, repeaters, and amplifiers all work by intercepting light signals, converting them to electrical signals, amplifying them, then converting them back to optical signals to continue on their path to the ultimate receiver. By adding the rare earth element erbium to a stretch of optical fiber, the fiber itself becomes an amplifier that boosts the signal. Thus, the EDFA amplifies optical signals directly, without conversion, making the process faster and eliminating a possible source of noise and breakdown.

As mentioned earlier, an important technology undergoing rapid development is that of laser emitters, which can be tuned electrically to produce light waves of different wavelengths. These "wavelength tunable lasers" or "wavelength agile lasers" are important, because the coding techniques of DWDM and wavelength-division multiple access (WDMA) allow a large number of parallel channels of information to be sent down the same fiber at the same time.[15] At the receiving end, a photodiode is tuned to the desired reception frequencies and retrieves only the messages directed to it.

Three problems limit the acceptance of all-fiber systems.

1. It is difficult to connect two fibers together in the field, whether because of a break or a planned expansion.
2. The cost of photonic switches and amplifiers is still very high.
3. Optical transmission equipment is complex and precise. For example, a laser optical termination unit requires the alignment of the laser diode and the optical fiber within microscopic tolerances. More flexible systems will someday reduce the cost considerably.

Carriers and companies who get the rights-of-way to plow in fiber optic cable usually lay dark fiber, extra fibers, and fiber bundles. Often they don't "light it up" because the capacity isn't needed at the moment. Or they may want to wait for lower equipment costs—although the expense for fiber optic cable has been dropping rapidly over the past fifteen years. In addition, the standards for interconnecting optically-enabled devices are delayed. The standardization of IEEE 1394 in this arena will speed up the use of fiber optical to both the desktop and the TV top.

A regular alphabet soup of acronyms describes bringing fiber closer and closer to the consumer. "Fiber-To-The-Node" or FTTN means that the fiber comes to the neighborhood and some other kind of cable goes from the neighborhood service area to the customer premises. More general terms for this advanced architecture is FSA, or Fiber-to-the-Serving-Area, or Fiber Deep. "Fiber-To-The-Curb" (FTTC) means that only the last few hundred feet would be nonfiber-type cable, although telephone companies sometimes use the term FITL (Fiber-In-The-Loop) to designate networks that extend fiber to a service area that includes no more than a few hundred customers. Finally, "Fiber-To-The-Home" (FTTH) envisions an all optical network. An all-fiber network is also sometimes called a Passive Cable Network (PCN), or Passive Optical Network (PON).

Network Configurations

In the community that follows the current enthusiasm for network expansion, there are vibrant arguments about the best network architecture. One particularly vociferous discussion revolves around smart versus dumb networks. It was all started by David Isenberg, who in 1998 worked for AT&T. That year, he wrote a brilliant paper, "The Rise of the Stupid Network: Why the Intelligent Network Was Once a Good Idea, But Isn't Anymore."[16] Although he was not fired for writing it, his well-argued points so cut across the grain of the thinking of major carriers, it became uncomfortable for Isenberg to continue working at AT&T, and he has since made his living consulting, writing, and speaking.

Isenberg argued that carriers built networks based on four assumptions:

1. Bandwidth is scarce
2. Voice messages dominate
3. Circuit switching is superior
4. Control is in telephone company hands

The paper proceeded to shred each of these assumptions. However, based on their network and business models, carriers had long promoted the "Intelligent Network." The architecture to implement this concept put computers in the center of the network, and rolled out Signalling System 7 (SS7) across the PSTN. The carriers marketed intelligent services, executed them via their computer-controlled network, and charged accordingly.

But something happened on the way, wrote Isenberg—stupidity:

The Internet breaks the telephone company model by passing control to the end user. It does this by taking the underlying network details out of the picture. Let's look at how this works in the case of voice. To the telephone company, there is one main way of transmitting voice—sampled in 8-bit bytes, 8000 times a second, for an aggregate rate of 64 kbit/s. The entire telephone network is designed around this rate. But if you want to send voice on the Internet, you can encode it at any rate you want, and send it at any rate up to the one that the slowest underlying network link supports. The recipient must have the right decoder running in her intelligent terminal, too.

Another way of framing this issue is to examine the relationship between bandwidth and processing. In the PSTN, intelligence resides in the computers in the center of the network. In a computer network, intelligence resides in computers at the edge of the network. Consider how the two designs make connections. In the PSTN, the network opens a path and holds it open throughout the call. All the intelligence to do it is in the network. In a computer Ethernet, there is some intelligence to forward messages, but fundamentally, the PC at the edge "reads" the messages coming across and picks out the ones addressed to it. Thus, intelligence is distributed throughout the network—but the most intelligence is at the periphery, as shown in Figure 10.1.

It is this command over processing that gives users communicative power. However, the condition prevails only as long as processing is cheaper than bandwidth, argue some companies, notably Sun Microcomputers, Oracle, and Netscape (before browser pioneer was digested by AOL). According to their philosophy, immortalized in the slogan "The network *is* the computer," when bandwidth becomes cheaper than processing, then intelligence will move back into the network and be on-call to dumber edge devices.

The argument rages. The eminent technologist George Gilder, a proponent of the stupid network, waxes eloquently about the democratic triumph of user control at the edge, but seems unaware that the Sun/ Netscape/Oracle position undermines that set of assumptions. Marc

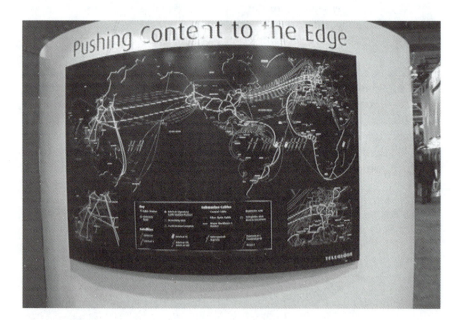

Figure 10.1 Teleglobe is part of the movement to move content distribution to the edge of the network.

Andreeson (Netscape, now AOL), Bill Joy (Sun), and Larry Ellison (Oracle) support it enthusiastically. The rival Microsoft camp has jumped on board, promoting the ASP (Advanced Services Platform) concept that would standardize services provided over the Internet.

The way the new smart network would work is that since bandwidth is cheap, Internet users will increasingly "rent" software, accessing it on Web sites when they need it and leaving it there, instead of buying and installing expensive programs. Consumers would stop buying storage; they could just save their data to a service provider on the Internet and access their files whenever they need them—from anywhere on the Net. Indeed, we now see all these services emerging.

This is the type of argument that will not go away any time soon, if ever. The relationship between bandwidth and processing is likely to sway back and forth like a pendulum . . . now processing is cheaper, now bandwidth is cheaper. Users may like having processing power and control over computing so that if the cost of bandwidth is not substantially lower than processing, they may opt for their own processing power anyway. They are certainly concerned over privacy, and it is difficult to see people willingly turning their most private data over to Web sites where security must always be a concern. (It's also a problem on a networked computer, but data feels more secure on your own hard drive, even if it isn't.)

An unanticipated development in this design battle is Gnutella, a computer program that lets people check other people's hard drives to find and download software. When a user launches Gnutella, the Web site searches the hard drives of other users of the program to see if they have the desired content. If they do, the originating user can download it. Talk about security issues—paranoids need not apply.

But Gnutella is just the beginning. Peer-to-peer or P2P networking can be used for all kinds of parallel processes. A group of PC users can agree to share computer processing cycles they are not using to work on a common problem. In short, there is no telling how those innovative Internet companies are going to figure out new ways to harness users together in all kinds of fascinating patterns.

The question of intelligence versus stupidity is one overarching philosophical question of current network architecture. However, a handful of specific designs dominate the actual construction of networks. In the next section, we examine a number of ways the components can be assembled into advanced networks.

HFC Moves up to Ultraband

We have already considered hybrid fiber/coax (HFC) cable systems at some length. However, there is a new super-HFC design that will leapfrog today's designs called Ultraband. It was developed by a group led by Dave Pangrac, the engineer who envisioned HFC. It is an open-architecture design over a hybrid-fiber coax network that provides a 40 megabits per second (or even higher rate) connection to each household.

Ultraband is an OSI Layer 3 (network layer) technology. Steve Rose designed the architecture of the equipment located at the point-of-distribution, the POD. It supports content management, user profiling, caching, and network management. A consumer can download a DVD movie in thirteen minutes over Ultraband; the same movie would take a week over dialup and eighteen hours over a cable or DSL broadband network.[17] (See Figure 10.2.)

The Grand Designs: SONET/SDH and ATM

SONET, the Synchronous Optical Network, is a physical network with standards for the other OSI layers that are designed to provide a universal transmission and multiplexing scheme. It is a very high-speed network with transmission rates in the gigabit per second range, and has sophisticated operations and management software as well. The standards have been set by the American National Standards Institute (ANSI) T1 committee.

Often SONET is referred to as SONET/SDH. SDH stands for synchronous digital hierarchy, a European version that was approved by the International Telecommunications Union (ITU). It is very similar to

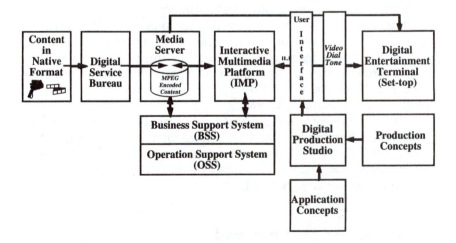

Figure 10.2 Ultraband will enable services like those envisioned by Bell Atlantic. Source: Bell Atlantic Video Services.

SONET, except that its multiplexing hierarchy is a subset of SONET. The differences between the two networks are not great, and global messages flow across them seamlessly. SONET networks are usually dual-ring structures that uses a synchronous transmission scheme to send different-sized frames (similar to cells or packets), depending on the channel rate, every 125 microseconds. SONET runs at various speeds, as shown in Table 10.4.

Notice that in the chart, speeds are given for both optical signals and electrical signals. The two measures appear because while SONET offers network management capabilities, it does so only with electrical signals. In this sense, these networks are not truly all-optical. In addition, the signals travel in optical form only from one point to another. Every time the traffic must be managed, the signals need to be converted from optical to electrical.

Optical signals leave a router and go into a SONET terminal where they are converted to electrical pulses and managed. The ongoing traffic is reconverted to optical, forwarded to a WDM (wave-division multiplexing) transponder, then to a WDM multiplexer. Every time the traffic hits another management point, it must go through this process.

SONET architecture is most cost-effective for operators who plan to connect their networks to the public switched telephone system, particularly heavily-trafficked point-to-point runs. For cable operators, SONET would be more costly than current hybrid fiber/coax (HFC) architectures. Moreover, when operators add switching capacity to their systems, they will want to run cable television, high-speed data, and telephony over their network, traffic that is composed of many different data types, better handled by an ATM or IP network

Asynchronous Transfer Mode (ATM) is a very fast general-purpose, connection-oriented transport mechanism that can be implemented on any number of physical networks, including SONET. Its cell-based switching and multiplexing technology carries different data types such as text, video, and audio. The cells are fixed in length and the switching is hardware-based, enabling this network to be extremely fast. One of the

Table 10.4 SONET Network Speeds

Electrical Signal	Optical Signal	Gross Rate (Mbps)	Payload Rate (Mbps)
STS-1	OC-1	51.84	49.536
STS-3	OC-3	155.52	149.460
STS-12	OC-12	622.08	594.432
STS-48	OC-48	2488.32	2377.728

advantages of ATM technology is that it provides a wide range of messaging and network management features, including network-based routing, addressing, flow control, and quality of service guarantees.

To transport ATM traffic over a SONET physical layer network, the ATM layer continuously maps the ATM cells into the SONET frame. Payload IP signals can also be formatted into ATM cells, allowing content from IP networks to migrate to ATM as the underlying data transport technology, while still using the existing applications of their legacy IP systems. To complete the conversion, the operator replaces the wire used in the LAN with an ATM switch and software.

However, IP over ATM is not very efficient. To start, ATM requires more overhead for the information in its header—that's what gives it so many useful network management parameters. When IP datagrams are mapped to ATM cells, they are sometimes bigger than the cells, so many of them are not completely filled up. This means that IP-to-ATM inefficiency must be added to the already-high ATM overhead. IP content will also run directly over SONET, and many Internet service providers are thinking about implementing this type of network because of its greater efficiency. They are considering many of the issues involved, as shown in Table 10.5.

When ISPs and enterprises consider whether to implement IP over SONET they are attracted to the higher speeds of running IP directly on SONET. But if they have a need for network management, they are likely

Table 10.5 Comparing IP over SONET with IP over ATM

Property	IP over SONET	IP over ATM
Available Bandwidth	95%	80%
Bandwidth Management	No	Yes
Quality of Service	Only point-to-point	End-to-end
Addressing and routing	Yes, but requires extensive provisioning	Yes, simple, fast provisioning
Flow control	No	Yes
Multi-protocol encapsulation	Yes	Yes
Fault tolerance	Yes, via dual SONET ring structure	Yes, via dynamic routing protocol

to decide on the ATM configuration. Telephone carriers, on the other hand, now favor the SONET design in their backbones because they need to move such massive amounts of information around the world.

There is also a great deal of talk about eliminating SONET and ATM software altogether and running IP over raw fiber. This approach would be the fastest solution possible, and Nortel Networks announced that it was moving to introduce such a network configuration. At the Telecom Business 1999 conference, the company announced it had broken the network speed/capacity record with an 80 Gbps/80 wavelengths platform that will carry 6.4 terabits per second (Tbps) of Internet and other traffic over a single, hair-thin strand of fiber, using DWDM.[18]

The all-optical network remains a Holy Grail to many networkers. "When you are working with 32 channels [wavelengths] of 10 gigabits each, from an efficiency point of view, it makes a lot of sense to do the restoration of service and the rerouting at the optical layer, and not convert those high-speed signals to electrical pulses that can be managed by SONET systems," says Pawan Jaggi, manager of the optical networking group at Fujitsu Network Communications.[19]

In the last decade, network design has become exceptionally challenging as architects try to put together efficient, cost-effective systems using circuit and packet switches, routers, crossconnects, and multiplexers of different granularities, capabilities, and capacities. "Each element has its own set of capabilities, constraints, cost structure, and interoperability requirements. When viewed in its entirety, even in the backbone, a network with these elements has many layers of hierarchy that interact with one another and change as new layers are added and old ones removed," notes one white paper.[20]

So where will the next-generation network (NGN) come from, and how will network designers learn what works and what doesn't? How does the development of innovation occur so that they can be brought into the PSTN and other important networks where reliability is at a premium? Just as the U.S. government played the key role in enabling the Internet, it is now fostering innovation in more advanced networks. Other governments around the world are partners in one part of the U.S. initiative, Internet 2, making high-speed networking a global effort.

Testbeds for the NGN

The point agency for U.S. government networking is the National Science Foundation (NSF). The agency funds high-level research in many different scientific domains that are important to policy-makers. Its success with the Internet gives it a sterling track record for heading one of the more admirable federal efforts in the past two decades.

Very High-Speed Backbone Network Service (vBNS)

The vBNS began in 1993 when the NSF realized that high-speed network-ing would make a significant contribution to further advances in the research areas the agency supported. Moreover, the federal government had made its interest in computers and high-speed communication mani-fest in the High Performance Computing and Communications (HPCC) program. To support the aims of the HPPC as well as its own agenda, the NSF budgeted $50 million, solicited proposals for a very-high-perfor-mance network backbone, and created network access points (NAPs) to support it.

MCI won the contract to build the very-high-speed backbone net-work service (vBNS) in 1995. The company built an OC12 SONET net-work that used the IP protocol over ATM transport. It originally ran at 622 Mbps, with a charter to always have greater capability than any commer-cially available telecommunications network. Designed to allow funda-mental research into advanced networking technology, vBNS links two NSF supercomputing sites and will eventually connect about 150 U.S. research institutes.

In 1996, then-President Clinton pledged $100 million per year for three years to foster partnerships among academia, industry, and govern-ment to connect about 100 research universities and national laboratories

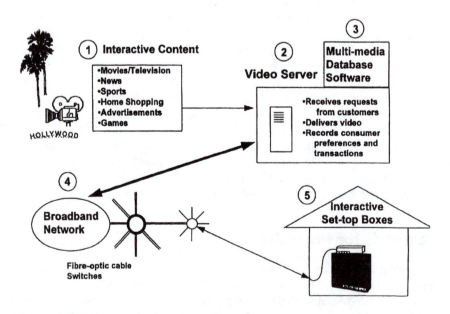

Figure 10.3 Advanced networks will let entertainment companies move their content around the world. Source: Hewlett-Packard.

with high-performance networks. Called the Next-Generation Internet (NGI) initiative, it would also promote experimentation with next-generation networking technologies and demonstration of new applications. In 1998, the NSF included NGI-related activities in its budget. In fiscal year 1998, the NSF budget request for NGI-related activities is $10 million. Essentially, the NGI initiative has been merged with the earlier NSF effort.

One important benefit that government support of the vBNS brings is open architecture. The agreement with MCI and other technology partners is a full and open disclosure of all network and engineering plans and activities in both online and hard copy formats. In addition, there are publicly available reports and findings covering service objectives and network traffic and performance, made on a monthly, quarterly, and annual basis.

The vBNS serves as a testbed for innovations designed to improve the Internet. Some of the improvements in trial are IP multicast, IPv6, MPLS, and DiffServ. As of late 1999, vBNS had 104 connections, 97 high-performance connection awardees, 4 supercomputing centers, and 3 collaborative institutions. An OC-48 link ran between Los Angeles and San Francisco, with several more links planned. The latest broadband broadcast software was in place, and the newest version of Internet Protocol, IPv6, continued to be implemented. Finally, there was progress towards developing guaranteed quality of service on the network.

These implementations of the vBNS point up several perceived shortcomings of today's Internet. They include:

- A need for more advanced network protocols
- Insufficient bandwidth for multipoint-to-multipoint broadband (video) distribution
- Inefficient use of bandwidth through unicast audio/video streams
- Provision for differentiated classes of service

IP Multicast and the MBone   A key advance in using the Internet for audio/visual streams would be to improve the efficiency of the process. Currently, when users want to see or listen to audio/visual (AV) material over the Net, they point their browser to a Web site and request the stream. It is a one-to-one connection—one request, one stream. And if there are 500 requests, then there are 500 streams. Limitations on servers, ports, and Internet bandwidth limit the number of simultaneous viewers that are possible, preventing the Net from assembling audiences in the millions.

This method of delivering AV is wildly inefficient and highly problematic for the PSTN as a whole. If several million users happened to request AV streams at the same time, it is possible that the entire U.S. telephone system would crash. To improve the efficiency and to increase the potential audience for webcasts, most computer companies support the IP Multicast Initiative. It was developed more than thirty years ago to send a single multimedia stream of each "program" over a dedicated backbone infrastructure called an Mbone (a contraction of "multimedia backbones") to the lowest possible level of server, where the stream is then replicated to users.

The IPMI system means that for most of its path, a given program is transmitted only once, until it reaches the local level—typically the ISP. This solution would eliminate duplicated streams and greatly reduce the overall level of AV traffic over the Internet, at least over the PSTN backbone portion of it. Supporters hoped for rapid diffusion of IPMC (IP multicast) technology, but there has been little actual progress.

The biggest problem is that when the current Internet was privatized, the old Mbone was dismantled. So the first action item to implement IPMI would be to rebuild some kind of media backbone. Another barrier to acceptance is the cost to ISPs for new routers that can interpret Mbone addresses. There is little incentive for ISPs to make the investment, as users are generally unwilling to pay for AV service (except for "adult" market aficionados). Operators are in a commodity business with narrow profit margins, and they are generally unwilling to buy new equipment that will not pay for itself. Issues of archival storage of multicast material by the ISP also remain to be resolved. It will probably take a critical, "not to be missed event," for IPMC to be seen as an essential part of an ISP's service repertoire.

IPv6 Throughout the Net, the current version of Internet Protocol is IPv4. IPv6 is the new version, designed to coexist with IPv4, so chances are that users will experience little or no inconvenience as IPv6 is implemented. They probably won't even notice it.

Indeed, IPv6 is already in use on parts of the vBNS that are known as the 6Bone. For example, some v6 packets already move over the public Internet. They are encapsulated inside v4 packets so Internet routers don't have to handle them in any special way. It will take a long time before all Internet routers can handle the new IPv6 addresses, which are much longer than those in IPv4. The extraordinary growth of the global Internet makes it essential to have more addresses (or "eddresses") available to accommodate everyone who will be on the Internet in just a few years. IPv4 allowed for 4 billion eddresses—but there are 6 billion people, and many more computers, appliances, vehicles, and other devices that will soon be assigned IP eddresses as well!

IPv6 has new security built into it and provides for some quality of service that goes beyond the current "best effort delivery" that is given to all IP messages. It also includes a protocol called MPLS, multiprotocol label switching. MLPS performs as its name implies. We have talked about addresses that are embedded in the headers of IP packets, as well as addresses in the headers of ATM and SONET cells. Each of these "addresses" is different and requires some kind of mapping as the message moves from one network to another. Naturally, this constant readdressing adds considerable overhead to the processing load of network routers and switches, and may cause bottlenecks and delays.

MPLS requires some changes in the topology of Ethernets, including the Internet. It places routers at the edge of the network, where the messages enters. It is immediately assigned an MPLS label in place of the destination address. From that time on until it reaches an edge router near its destination, the message proceeds under this new label. Hence the name, multiprotocol label switching.

MLPS does a bit more. In its header, it also assigns a priority and a particular path for the message, all the way to the destination. Thus, it introduces quality of service (priority), security, authentication, reliability, and switching into the Internet. These are deemed important so that the Internet can be used for "mission-critical" business processes that were not included in the original Internet protocols.

MLPS is controversial because it replaces connectionless routing protocols with connection-oriented switching technology. While the Internet Engineering Task Force is working to standardize MLPS, as of late 1999, there were a number of alternatives and all were linked to some specific company's products. In short, MLPS is chum, shark-bait, in the competitive waters of the profitable world of telecommunications networking equipment.

The proponents of MLPS say that routing is too slow; opponents point to the new terabit switcher/routers and the explosion of bandwidth. Recall the argument about smart versus stupid networks. Here it is again. The smart network folks want MLPS. The stupid network people want to keep the intelligence distributed and see the new protocol as an insidious recentralization of the network.

One implementation of network quality of service guarantees is called DiffServ. A group of the Internet Engineering Task Force (IETF) is working to define how to mark packets to indicate the priority of service they should receive, called Class of Service or COS. Not only will it introduce different ways of handling packets, it will also bring in problems of differential pricing, billing, and resource allocation that have been heretofore avoided.

The testbed for DiffServ is a part of Internet 2 called the Qbone. DiffServ works a little differently than some service protocols in that it defines a cloud-to-cloud path, rather than a router-to-router path, ensuring that it will be faster than the typical QOS differentiated service. At the top level is a QBone Premium Service (QPS) that provides one-way assurance, independent of other traffic on the network—that is, whether it is heavy or light. There should be no packet loss due to congestion, low latency, or low packet-delay variation.

In addition to MPLS and DiffServ, there are a number of other schemes that bring these features to the Internet. Explanations of these alternative proposals are beyond the scope of this chapter but readers can find more technical information about additional protocols through an Internet search engine:

- LAN Emulation (LANE)
- Classical Model IP over ATM (RFC1577)
- Next-Hop Resolution Protocol (NHRP)
- Multi-Protocol Over ATM (MPOA)
- Integrated PNNI (I-PNNI)
- Multicast over ATM
- IP Integrated Services over ATM
- IPv6 over ATM

The familiar set of smart-versus-dumb network arguments arise over quality of service or differentiated classes of service. On the current Internet, the Net makes a "best effort" to deliver the message. The original message was broken into datagrams and reassembled at the destination. If some packets were missing, then a message is sent back to the originating server, "send this message again." Those who argue for a stupid network believe that introducing different classes of service, each one assigned a certain level of priority, also introduces a new and unwelcome level of complexity and management to the Internet that they reject on both philosophical and practical grounds.

By contrast, telephone companies have long had guaranteed quality of service. Network failure is regarded with shame and abject apologies. This means that they accept the business obligation to guarantee that the call (or message) will go through. The more customers pay, the higher priorities their messages receive, and the more guarantees they are granted. And the more complex the network, the more network providers can charge.

Internet 2

A closely related project to the vBNS is Internet 2 (I2), one of the informational tributaries flowing into the mighty vBNS. I2 connects more than 150 U.S. universities that collaborate with one another and government and industry researchers.[21] Emphasis is placed on uses for higher education, telemedicine, digital libraries, and virtual laboratories that are not possible with the technology underlying today's Internet.

I2 is not a single network; it connects member networks of campus, regional, and national networks to carry applications and to foster engineering development efforts. The infrastructure is mostly provided by a Qwest experimental network called Abilene. (See Figure 10.4.) The

Internet2 GigaPoPs *(as of February 1999)*

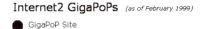

● GigaPoP Site

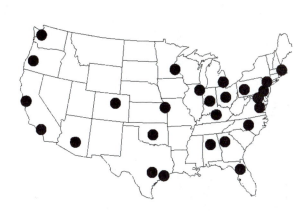

Figure 10.4 Internet2 "gigaPOPs," gigabit per second points-of-presence that route high speed traffic.

administration of I2 falls under the University Corporation for Advanced Internet Development (UCAID).

The decision to locate an untried advanced network with opportunities to innovate new structures, processes, and applications in universities is not surprising, considering the success of the first-generation Internet. The nature of university research requires collaboration among colleagues who work at institutions around the country. And many of the nation's foremost networking experts teach and work in them.

I2 has become an international phenomenon, at least partly through the efforts of the U.S. government. Table 10.6 shows networks affiliated with I2, and provides their Internet URLs for readers who would like to learn more about them.

Other U.S. Advanced Networking Initiatives: NREN, DREN, and ESNET

These projects by U.S. federal agencies all use the vBNS, at least in part, as their network platform. They were funded as part of the Clinton-Gore administration's NGI (Next-Generation Internet) program.

NASA used the Research and Educational Network (NREN) to direct the rover on Mars, assist surgery on the other side of the world, control airframe design in a distant wind tunnel, and operate virtual laboratories. It is a small-scale testbed for the NGI technologies of the future,

Table 10.6 Networks Affiliated with Internet 2

Network/Country	Description	URL
CANARIE/ Canada	Established in 1993, it is Canada's advanced Internet development organization, a coalition of 120 members and 500 project partners, including government, industry, research, and educational communities.	http://www.canarie.ca/
CUDI/ Mexico	Promotes and coordinates the development of telecommunications and computer networks, focused on educational and scientific development.	http://www.cudi.edu.mx
Deutsche Forschungsnetz (DFN)/ Germany	A high-speed broadband network that links universities and institutions throughout the country.	http://www.dfn.de/
INFNet/ Italy	The Istituto Nazionale di Fisica Nucleare, INFN, coordinates subnuclear and nuclear research in Italian universities, 4 laboratories, the national networking center, and 6 associated sites. INFNet is affiliated with the Italian academic network, GARR, and supports scientists who work in remote areas to facilitate distributed collaborations.	http://www.garr.it/
Internet-2/ Israel	Eight Israeli universities are part of their country's participation in Internet2. It enables real-time multimedia applications such as videoconferencing, multimedia broadcast, and other applications that require 2–30 Mbps of dedicated bandwidth.	www.internet-2.org.il/

Table 10.6 Networks Affiliated with Internet 2 (Continued)

Network/Country	Description	URL

WIDE Project
Japan Gigabit
    Network
Genesis
IMnet
Cyber Kansai
    Project
CRLnet
ETL/
Japan
	Japan's participation includes 7 different groups. WIDE stand stands for Widely Integrated Distributed Environment, which puts together a wide area network with telecom infrastructure.
	The Kansai Project is an advanced, high-speed Internet system, a joint project between business, government and academic entities. It is a demonstration network and is used to encourage participation on the Internet in Japan.
		http://www.internet2.edu
		    /international/html
		    /jairc.html
		http://www.wide.ad.jp
		    /backbone/index.html

NorduNet/ Nordic countries	NORDUnet interconnects the Nordic national networks for research and education and connects these networks to the rest of the world. It offers network services that are based on the Internet Protocol only. In addition to basic Internet service NORDUnet operates a DNS root server, information services, USENET NetNews and MBONE connectivity to the Nordic national networks.	http://www.nordu.net /basics/
Renater/ France	National Network for Technology, Education, and Research: a 155 Mbps network that connects universities and research institutes throughout France.	www.renater.fr/
SingAREN/ Singapore	A high-speed broadband network that provides video, audio, and MBone connectivity.	http://www.sren.net/

Table 10.6 Networks Affiliated with Internet 2 (Continued)

Network/Country	Description	URL
SURFnet/ Netherlands	High-speed network to promote innovation in infrastructure, information handling, and network management. Focus on directory services (electronic addressing), standardization of E-mail, real-time multimedia types, and e-commerce.	http://www.surfbureau.nl
TERENA Trans-European Research & Education Networking Association	Formed in 1994 "to promote and participate in the development of a high-quality international information and telecommunications infrastructure for the benefit of research and education" (TERENA Statutes). Carries out technical activities and provides a platform for discussion to encourage the development of a high-quality computer networking infrastructure for the European research community.	http://www.terena.com/ info/
UKERNA UK	UKERNA manages the operation and development of the JANET networks under agreement from the Joint Information Systems Committee (JISC) of the U.K. Higher Education Funding Councils. In 1995, SuperJANET II increased the size of the original network and helped to create a number of Metropolitan Area Networks. It is IP over ATM using a combination of 155 Mbit/s and 34 Mbit/s curcuits. SUPERJANET III: Consolidating network with a 155 Mbit/s ATM backbone between central ring of switches at London, Bristol, Manchester, and Leeds. Extended network using 34 Mbit/s and 155 Mbit/s links to backbone edge nodes.	http://www.ja.net/ http://www.superjanet4. net/

because its missions require the agency to have networking capabilities of two to three orders of magnitude improvement over today's high-performance networking. For example, NASA's Mission to Planet Earth, for example, will have to move petabytes of information (trillions of bits) in the coming years. Other projects involve advanced aerospace design, telemedicine, astrobiology, astrophysics, remote operations, and simulations, and many other activities that depend on high-speed networking.

The Defense Research and Engineering Network (DREN) supports high-performance computing to provide the U.S. military with a technological advantage, particularly in weapons systems design. According to the Department of Defense, the use of high-performance computing in the early stages of the system acquisition process aids in decreasing the total life-cycle costs of fielding new combat support systems.

ESNET is Energy Sciences Network, the underlying activities of the Department of Energy (DOE). The program supports advanced network and distributed computing capabilities needed for DOE scientific research and other programs such as energy resources, environmental quality, and national security. The network links the DOE national laboratories and facilities into a cohesive, integrated research environment. The agency plans to add shared data services and a collaborative environment for distributed research teams.

The Wired World Meets the Wireless World

The last decade has seen significant advances in the technology of both wired and wireless systems. (See Figure 10.5.) There is general agreement that the twenty-first-century network will incorporate many different kinds of networks. Rather than an entirely wired network, wireless technologies will be partners in the enterprise, and a seamless union of the two will form a single interoperable, interconnected, seamless wired-plus-wireless infrastructure. The next chapter covers the fascinating, new wireless world.

Notes

1. D.K. Rubin and W.J. Angelo, "Level 3 grows on high fiber diet," *Telecommunications* (November 1, 1999). Available online at http://enr.com/new/C1101.asp.
2. T. Nolle, "Take a look at the dark side of fiber," *LANTimes* (October 1997). Available online at http://www.lantimes.com/97/97oct/710a063b.html.
3. G. Middleton, "University unplugs last electronic bottleneck," *TechWeb* (September 27, 1999). Available online at http://www.techweb.com/wire/story/TWB19990927S0003.

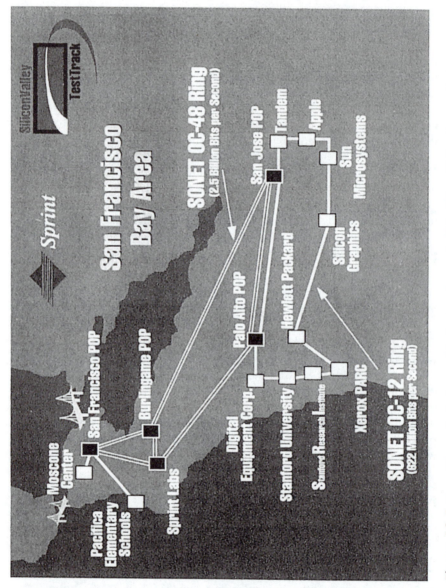

Figure 10.5 One of the earliest advanced test networks was built in Silicon Valley. Source: Sprint.

4. B. Hawe, C.T.O., Nortel Networks. Speaking at Telecom Business 1999, Anaheim, CA, March 1999.

5. Information about SONET is provided at http://www.sonet.com. It's populated and maintained by Light River Technologies and has many good links.

6. C. Wilson, "Telco networks take the fast lane," *Inter@ctive* 2:9 (May 8, 1995):35.

7. Check out the solitons home page at http://www.ma.hw.ac.uk/solitons/. They have all kinds of great information and wonderful Net movies of solitons in action.

8. "Soliton waves double fiber-optic capacity," *New Telecom Quarterly* (2nd quarter, 1993):6.

9. "40-Gb/s soliton transmission travels 70,000 km," *Photonics Technology News* (March 1999). Available online at http://www.laurin.com/Content/Mar99/techNippon.html.

10. K.C. Kao and G.A. Hockham, "Dielectric-fibre surface waveguide for optical frequencies," *Proceedings of the IEEE* 133:7 (July 1966):1151–1158.

11. H. Newton and R. Horak, *Newton's telecom dictionary*, 16th ed., Telecom Books/Miller-Freeman (2000).

12. Tom Bowling, "A new utility," *New Telecom Quarterly* (4th Quarter, 1994):14–17.

13. This information is from a company press release dated September 17, 1997. Available online at http://www.bostonoptical.com/9-17-97.htm. The site has information about plastic optical cable at http://www.bostonoptical.com/WhatisPOF.htm.

14. Information on EDFA available at http://wwwtios.cs.utwente.nl/Docs/tios/dg/tst/Assignments/avail/D-opdracht.html.

15. B. Kim, Y. Chung, and S. Kim, "Dynamic analysis of widely tunable laser diodes integrated with sampled- and chirped-grating distributed Bragg reflectors and an electroabsorption modulator," *Institute of Electronics, Information, and Communication Engineers* E81-C:8 (August 1998):1342–1349. Download from http://search.ieice.or.jp/1998/files/e000c08.htm.

16. D. Isenberg, "The rise of the stupid network." Available at http://www.isen.com (1998).

17. D. Pangrac and S. Rose. Personal interview, New Orleans, April 2000.

18. B. Hawe. Op. cit.

19. P. Jaggi and L. Steinhorst, "IP over photonics: Making it work," *Lightwave* 16:1 (January 1999).

20. C. Wilson, "Network architectures of the future," *Interactive Week White Paper* (May 4, 1998).

21. Find a lot of information about Internet 2 at http://www.internet2.edu.

11

Wireless Bitpipes:
The Invisible Infrastructure

The frenzy to build wired networks often distracts observers from the advances made in wireless technologies. New satellites, orbits, receiving dishes, modulation schemes, compression, and transmission techniques, in concert with the digital revolution, increase the likelihood that broadband communication and digital television may well find their way to consumers from the airwaves, rather than the wires. While over-the-air spectrum is finite, it is not static. In addition to compression, new equipment makes ever-higher frequencies in the gigahertz range usable for wireless communication. While the airwaves may never equal the extraordinary bandwidth of fiber optic cable, there nevertheless will be more spectrum than anyone would have predicted as little as a decade ago.

This chapter will start where the process of communication ends—with consumers and some of the new ways they will receive television and other moving images over wireless devices. We begin with the new handheld devices, including third-generation wireless telephones, because they are the key to understanding why there is such an explosion of growth in wireless connectivity. Indeed, other than the Internet, there are few consumer electronics products that have spread as quickly and decisively around the world as wireless telephones. It is nothing short of a commercial avalanche. And given a similar dramatic growth rate with the Internet, it is a small leap of faith to imagine that there is significant potential in bringing the Internet to mobile telephones, allowing people to be online wherever they want to be—no strings attached.

Research by Scottsdale, Arizona-based Cahner's In-Stat Group projects that the wireless data market will grow to 25 million subscribers by 2003, up from 1.7 million users today.[1] The company's findings indicate that users want access to data on wireless devices. In-Stat thinks that mobile workers will be the largest market, followed by corporate usage to give employees access to E-mail, fax, and Internet access. The consumer market comes last as the service becomes more mainstream and affordable.

3G Handheld Device

Broadband over your mobile phone and other handhelds. That is the promise of third-generation (3G) technologies. From the PalmPilot to the Nokia MediaScreen—these are just the harbinger products of the wave of the future that will arrive in stages over the next decade.

A great deal of effort has been expended on 3G, well before the networks could even be designed, let alone built. Groups have met for several years to set a multitude of standards for everything from networks to receiver devices, to markup languages that automate the conversion of standard Web pages for the small screens of mobile devices. Probably the biggest remaining hurdle is actually batteries—figuring out how to power the increasingly sophisticated and therefore power-hungry combination of telephone+computer+display.

There are three types of emerging products available to consumers, each one reflecting a particular component more strongly than another. The PalmPilot is a personal digital assistant, a computer, that will also provide wireless connectivity, including telephone calls and Internet access. In the next few years, it will be capable of receiving and displaying rich media from the Internet. So will the 3G mobile telephone, which will have powerful processing chips to carry out these tasks. And portable e-pliances like the Nokia MediaScreen, as shown in Figure 11.1, will offer mobile telephone and Internet access. On its twelve-inch color LCD screen, the battery-operated MediaScreen features a digital TV receiver that decodes TV transmissions and Web pages encoded in the TV signal. It also lets users connect to the Internet, send and receive electronic mail, and listen to the radio.

All these devices offer similar functionalities, packaged in different form factors. They face nearly the same challenges and must proceed through approximately the same evolutionary development. The work underway to accomplish the goal of seamless, mobile broadband wireless connectivity is occurring under the rubric of the 3G mobile phone. Before getting to 3G, we must start with 1G and 2G.

The first generation of mobile telephones were analog. From the get-go, the world was divided into the U.S. Advanced Mobile Phone Service (AMPS) and the European Groupe Speciale Mobile (GSM) camps—it was no great surprise that everyone wanted their own way. Two-G was supposed to pull the two standards together. Instead, it drove them ever wider apart, hardening the leading players into heavyweight opponents, intent on dominating the world in their respective domains.

Figure 11.1 Nokia MediaScreen is one of the new wireless broadband reception devices. Source: Nokia.

Three-G is called IMT-2000, or International Mobile Telephone-2000, by the International Telecommunications Union (ITU). Three-G started off with the same idealism and soon degenerated into an unusually brutal, rancorous fight to commercial death, with Ericsson and Qualcomm as the corporate contenders. All 2G digital telephones used time division multiple access (TDMA), but each trading block had its own variation—GSM (Europe), D-AMPS (United States, Latin America, Eastern Europe, Asia Pacific), and PDC (Japan).[2] But Qualcomm had developed code division multiple access (CDMA) and unabashedly evangelized on behalf of spread spectrum technology, which added the claimed virtues of more efficient operation and better clarity to the incentive of ownership—the company had a patent on it. Spread spectrum schemes sprinkle a message's bits across a slice of spectrum and transmit the key to reassembling the bits at the receiver end.

No manufacturer was thrilled about paying the Qualcomm tax (license) on CDMA, but Ericsson seemed absolutely determined not to do so. The Swedish giant developed a competing technology called Wide-CDMA, and quite naturally Qualcomm was incensed over what it considered a bastardization (and commodification) of its modulation scheme, as well as a violation of its patent. Representatives from the two companies brawled publicly in an unseemly (if entertaining) fashion throughout the late 1990s. Lending the normally sedate trade show environment a bit of excitement was of some value, but the battle brought further development of 3G telephony to a halt.

There was wide recognition that failing to create global interoperability could deal a crippling blow to the wireless telephony industry as a whole, stalling its progress for years into the future. Consumers could not be expected to carry around multiple telephones. A confusing marketplace would turn them off, potentially establishing an adverse relationship between providers and users that would not be easy to overcome.

Finally, in 1999, Ericsson and Qualcomm came to an agreement, ending their patent dispute. As part of the deal, Ericsson acquired Qualcomm's infrastructure unit and agreed to put an office in San Diego near Qualcomm. The two companies decided to work on a common 3G solution, uniting their divergent W-CDMA and CDMA2000 proposals.

Just twenty-four hours after the two opponents settled their differences, the International Telecommunications Union meeting in Brazil approved key characteristics of 3G networks. The ITU will support combination TDMA/CDMA protocols, so that 3G networks will support multimode, multiband handsets, allowing 3G network users to roam to existing CDMA and TDMA networks. The agency believes this approach will give consumers maximum flexibility.

As a result of the agreement between Ericsson and Qualcomm, the ITU now expects 3G networks to appear sometime next year in some regions around the world, depending on market conditions.[3] Each 2G standards camp will have to move through an evolution of gradually increasing capacity in their networks and their telephone products.

Perhaps it is merely a cynical view to suggest that this will allow consumers to pass through eight years of mobile phone limbo, purchasing a new model every two years before getting to 3G heaven. A research study from Ovum, "Third Generation Mobile: Market Strategies," reported that significant investments are now being made to upgrade existing second-generation (2G) networks.[4] However, consumer adoption will depend on the availability of inexpensive, standardized networks, services, and telephones.

Wireless Application Protocol: Standards for Wireless Apps

The goal of WAP (Wireless Application Protocol) standards is to let developers plan services for consumers that ensure fast, secure, reliable, and flexible interactive services, regardless of which wireless device they happen to use. WAP is a network-agnostic set of communication protocols that will work with many different kinds of wireless networks, including

CDPD, CDMA, GSM, PDC, PHS, TDMA, FLEX, ReFLEX, iDEN, TETRA, DECT, DataTAC, and Mobitex.[5] It is also a communications protocol and application environment that will ride on top of most operating systems, including PalmOS, Symbian's EPOC, Microsoft's Windows CE, FLEXOS, OS/9, and Sun's JavaOS. WAP provides service interoperability between different device families, such as telephones and pagers.

Applications for WAP-enabled devices include online access to make reservations and purchase tickets for every sort of travel, service, and event imaginable. A host of information services, from banking to shopping to looking at a map of an unfamiliar area could be accessed. Communication services include E-mail, telephony, unified messaging, paging, and various kinds of alerts. All the activities people do now with wireless phones, with the added convenience of a more convenient and easy-to-use screen, will be online.

A key element to making it possible to use a portable handheld appliance for visually-oriented applications is a micro-browser. These devices have a small screen. They may have limited processing power. And they certainly have limited battery power. Therefore, they need an online browser that is fast and compact, uses little processing or memory, and offers good visibility on a small screen area. The major player in this arena is a company called Phone.com, which was one of the founding WAP members.

Wireless Internet Service Providers

In February 1999, Cisco and Motorola made a deal to invest $1 billion to set up a wireless Internet service provider network.[6] They were just keeping pace with Lucent Technologies, which bought WaveAccess in November of 1998 for $50 million and has its own plans for supplying the radio technology for wireless Internet services.[7] In another agreement, British Telecommunications and Microsoft are developing a worldwide Internet and corporate data services business. Trials started in the United Kingdom in the spring of 1999, testing applications such as personalized Web content, E-mail, and online information services on digital mobile phones, pagers, or handheld and laptop computers.[8]

The new 3G wireless telephone systems won't be your parents' old, slow mobile networks either. For example, in February of 1999, Qualcomm partnered with U.S. West to test high-speed wireless Internet access at data rates of 1.8 Mbps, based on 3G standards. It may not be coming soon—but it's coming.[9]

Overview: Wireless Networks

For the most part, for wireless communication to occur, the transmitters must be within line-of-sight of the receiving antennas, whether the signals travel across the room or across the sky from a satellite in geosynchronous orbit 22,247 miles away. Wireless systems fall into two categories: earth-based (terrestrial) systems, which this chapter will cover first, and space-based satellite systems. Terrestrial systems include a variety of infrastructures: the television stations we have watched since the 1950s; fixed broadband wireless, including multichannel, multipoint distribution systems (MMDS) and local multipoint distribution systems (LMDS); interactive video and digital services (IVDS); and low-power television (LPTV). The airborne and space-based delivery systems include everything from unmanned vehicles to giant satellites to . . . balloons!

Before looking at each kind of system, this section presents some basic information about wireless signal transmission. Post-Newtonian modern physics has revealed a world that, at its most basic level, has both a material aspect and a nonmaterial aspect. As material, reality is made up of atoms—but atoms themselves are composed of subatomic particles whose properties must be described as "energy" as much as "matter." Early in the twentieth century the famous scientist Neils Bohr proved that this energy results from the movement of electrons "jumping" from one atomic shell to another.[10] Similarly, the radio waves that travel through the atmosphere are caused by the purposive excitation and control of electrons, an expression of energy.

The energy that emanates from the waves is called electromagnetic radiation—hence the term "radio." The energy that causes the electrons to jump is provided by electricity, and all wireless transmission requires some amount of power. That power ranges in strength from the 50,000 watts used by commercial television stations, to 100 watts for LPTV, low-power television. Systems operating in the gigahertz part of the spectrum use even less power, from 10 to 20 watts for satellite signals, to fewer than 5 watts for LMDS, or local multipoint distribution services.

Radio waves travel through the air, a phenomenon known as "wave propagation." Wireless signals are waves that have been modulated to carry information. Electromagnetic signals can be modulated in one of three ways: amplitude modulation (AM), frequency modulation (FM), or phase-shift keying (PSK). AM changes the height of the wave. FM changes how often the wave oscillates per second. PSK changes the timing of successive waves in relation to previous ones.

The modulated waves are then sent out, or propagated, through an antenna that repels electrons—this force generates the electromagnetic waves traveling through the air. Sound (and pictures) are converted to electrical energy and connected to a transmitter. The electrical signals are amplified and converted into waves, composed of a carrier signal that is modulated to carry information by its associated sideband frequencies that also travel with the carrier. Think of how a pebble generates waves when tossed into a pond. The movement of the water results from the energy of the traveling pebble impacting on the stationary liquid. Similarly, the repelled electrons generate the modulated, traveling sine wave, which oscillate at an assigned frequency within the electromagnetic spectrum, as was shown in Figure 3.6 in Chapter 3, "Digitology 101." Table 11.1 shows the nomenclature for ranges of frequencies.

The FCC allocates spectrum for commercial broadcast of radio waves in five frequency ranges. All these transmission frequencies are much higher than the human ear can hear or the eye can see. In order for our perceptual system to process the information that radio waves carry, they must be down-converted to our perceptual frequencies.

- Low and medium frequency range from 10 to 3,000 kilohertz for AM radio
- High frequencies of 3–30 MHz for FM radio
- Very high (VHF) and ultra high (UHF) broadcast frequencies at 30–100 MHz for broadcast television
- Ultra (UHF) and super high (SHF) frequencies of 1–12 GHz for microwaves
- Extremely high frequencies (EHF) at 30–34 and 38–39 GHz for cellular television.

As we learned in Chapter 3, "Digitology 101," radio, or sine, waves have height (amplitude) and length (wavelength, which varies inversely

Table 11.1 Radio Wave Frequency Ranges

Radio Wave Frequency Ranges		
Abbr.	*Name*	*Frequencies*
ELF	Extremely Low Frequencies	30–300 Hz
VF	Voice Frequencies	0.3–3 kHz
VLF	Very Low Frequencies	3–30 kHz
LF	Low Frequencies	30–300 kHz
MF	Medium Frequencies	0.3–3 MHz
HF	High Frequencies	3–30 MHz
VHF	Very High Frequencies	30–300 MHz
UHF	Ultra High Frequencies	0.3–3 GHz
SHF	Super High Frequencies	3–30 GHz
EHF	Extremely High Frequencies	30–300 GHz

with frequency.). However, they also have rotation (left or right), orientation (horizontal or vertical), and velocity (186,000 miles per second). Each of these properties is important in one or more of the new, advanced wireless technologies.

In terms of transmission, radio and television are similar. Depending on the type of system and the environment in which it operates, the transmitted signals may need to be strengthened with repeaters. Or their direction may need to be deflected to serve a difficult-to-reach area with reflector equipment called a "beam bender."

Wireless transmissions bring some disadvantages. Since they need to be within line of sight of receivers, they may not work well in heavily forested areas, hilly regions, or urban centers with tall buildings. Wireless signals are subject to interference from other radio waves, reflections, and even rain. As a result, in a cable system, each 6 MHz TV channel will support 27 Mbps of downstream data using 64 QAM modulation, but a wireless system can only transmit 19.2 Mbps, due to the greater interference. Sometimes these problems can be overcome with additional transmitters aimed at problem reception areas.

Radio waves are received everywhere, in some surprising places. They cause the free electrons in all nearby metal objects to vibrate in response—lamp posts, chain fences, railroad tracks, perhaps even some people's metal dental work![11] But the idea is that there is a metal antenna at the receiving end, and free electrons move down the antenna creating voltage. Inside the receiver, the voltage is amplified and filtered with a tuner that selects a particular frequency. The electrical signals are downconverted to frequencies that can be heard by the ear and amplified to the desired volume through loudspeakers.[12] Add a camera at the transmitting end, some processing complexity, and a television at the receiving end, and the same model will suffice for understanding how television is sent over the air.

Terrestrially-Based Broadband Wireless Delivery Systems

One way to distinguish between the wireless modes of delivering television is the frequencies they occupy. The next section will begin with the familiar delivery system of over-the-air broadcast television, and how its technology for distribution industry will be affected by the transition to digital.

Free TV: Over-the-Air Broadcast Television

As with all complex innovations, television is the result of the work of many different contributors. The first suggestion that pictures could be sent via telegraphy was in 1842 by a Scottish engineer, Alexander Bain, who also anticipated the scanning process of today's system. Sometimes credited as television's inventor, Philo Farnsworth patented a television system in 1927. The other individual often cited as a creator of TV was Vladymir Zworykin who developed the iconoscope (a camera) in 1923. Subsequent lawsuits over the patent kept TV held up in court until 1934. Finally, Zworykin's version was exhibited at the 1939 World's Fair, but went virtually unnoticed.[13] In 1941, the National Television Standards Committee, under the direction of the FCC, adopted the NTSC's standards for black-and-white television.

World War II intervened to put television on the back burner. In the meantime, radio had become extremely popular. Its commercial nature was well established and a contractual network-affiliate distribution system was in place. It seemed natural that television would follow the same development.

Starting in 1946, early television was distributed over coaxial cable. New York and Philadelphia were linked in 1946, Boston in 1947, cities in the Midwest hooked up in 1948, and the West Coast connected in 1950. By that time, there were 94 television stations and by 1951, there were 17 million television sets. Today, there are about 1,600 TV stations in the United States. About 98% of the U.S. population has a television set and, on the average, sets are on about seven hours a day.[14]

In January of 1999, there were 1,216 commercial TV stations, 562 on VHF frequencies and 654 on UHF frequencies. The number of commercial networks had grown from four to seven: ABC, CBS, FOX, NBC, UPN, WBT, and PAX. The result is that there are almost no entirely independent stations, although UPN is not an "on-pattern" network, so its affiliates can integrate network programming into their programming with some flexibility. On-pattern network contracts require their affiliates to play programming at designated time slots. UPN, WBT, and PAX do not yet provide shows to fill all primetime hours, seven days a week, so their stations fill more of their schedule with syndicated programming.

In retrospect, the TV industry has always had to evolve, in both technology and business practices—from black and white to color; from advertiser to studio to independent production; from sponsorship to spot buys; and from over-the-air broadcasting to multichannel cablecasting. However, in recent years, network and station executives have been wrestling with a truly disturbing trend that threatens their business model: the declining network share of the audience. About twenty years ago, the audience share for broadcast television during primetime hours began declining. In the 1978–79 season, ABC, CBS, and NBC had a 90% share or more. By the 1986–87 season, that figure had dropped to 75%, then to 47% in 1997–98.

It would be natural to think that as the audience declined, the value of the networks would decline, but that is not the case. Rather, networks have increased in value and their commercial time is ever more expensive. The reason for this surprising consequence is simply that for advertisers who need to address a national audience, the networks are still the only game in town. At any given time of day, basic cable networks account for about 20% of the viewing audience fragmented across nearly 100 networks. For advertisers, premium cable networks don't matter because they offer no advertising opportunities. Radio stations don't pull a national audience; neither do newspapers, except for *USA Today*, but few read it compared to viewers of network television. Thus, the remaining audience has become more valuable than ever—so networks charge even more for it.[15]

Broadcast networks have been helped by the 1994 action of the FCC to change the "financial interest and syndication rule." Previously, networks had been prohibited from any ownership position in programming, making independent studios and production companies very profitable. The rule change meant that networks can now produce and own their programming, generating revenue from the syndication market as well as the first-run sales. Ironically, the proliferation of new networks—PAX, UPN, and WBT—has filled the programming pipeline for previously independent stations, closing off many time slots for syndicated material. At the same time, networks' new ownership rights have deprived independents of some primetime first-run opportunities for their wares.

At the local station level, anther regulatory action resulted in significant consolidation within the broadcast industry. The FCC rule change allows networks and station groups to own up to twelve stations. Station groups are companies that own multiple local television stations. The larger groups are better able to finance the transition to digital, and it offers them some additional opportunities that will ultimately offset the cost.

DTV allows them to broadcast multiple channels in their digital spectrum, developing additional revenue streams in addition to their income from the advertising-supported local station operation. They will be able to offer additional programming, programming to PCs, and digital data transmission and information services. They will develop whole new program genres and venues, and they will be able to participate in e-commerce. From a business standpoint, digital technologies allow them to improve the efficiency of their holdings by centralizing operations. Enterprise resource planning (ERP) and media asset management (MAM) systems give groups greater control over employee work flow and use of their content. Activities within a broadcast facility will be connected to a central video server through high-speed data networks, as shown in Figure 11.2.

Broadcasters' Digital Spectrum   The new digital bandwidth and the realities of the multichannel world where broadcasters have to compete

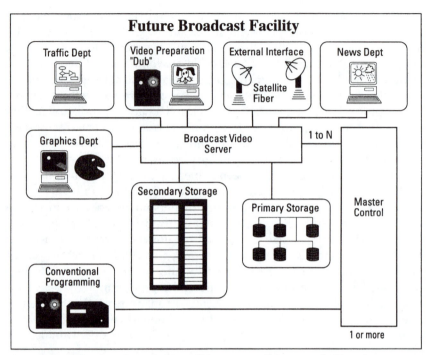

Activities within a broadcast facility will be connected to a central video server through high speed data networks.

Figure 11.2 The advanced broadcasting facility. Source: Hewlett-Packard.

against cable and DBS will fundamentally alter the business of local TV stations. Many will exploit their digital capabilities in a variety of ways. Recall that digital throughput on a 6MHz channel is 19.39 Mbps. An ATSC 1080i signal takes up 17.56 Mbps, or nearly all of it. A 730p signal requires 10.2 Mbps, and a standard definition channel (SDTV) needs 3.96 Mbps to reliably replicate the quality of an NTSC analog signal. These numbers mean that if stations use 720p HDTV, they can run the main HD signal plus two channels of SDTV. Or they could run four SDTV channels. As earlier chapters indicated, digital compression is elusive, so if a given video stream did not have a great deal of motion, or if lower quality images were acceptable, then stations could raise the compression rate (lower the bit rate) of SDTV channels and squeeze more of them into their 6 MHz allotment.

The ability to transmit more than one programming channel is called "multicasting" in the broadcast world. It's an unfortunate usage of the term because in the Internet world, "multicasting" means using an Mbone to stream audio/video content once over the backbone down to servers as close as possible to consumers. Where are the terminology police when you need them?

It appears that many stations will adopt a multicasting strategy. They are also likely to use their digital bandwidth for datacasting, or transmitting a digital bitstream to PCs and TV set-top boxes. It would let them offer services like providing statistics about sports events, giving supplementary data about an advertised product such as features, prices, and store listings, distributing computer software and video games, sending out stock quotes, and providing audio services and other digital information services.

A PricewaterhouseCoopers report on digital television identified three technical approaches that allow local stations to datacast auxiliary data services over the air: analog, in-stream, and parallel datacasting.[16] A number of proprietary schemes for analog datacasting already allow broadcasters to send out a limited amount of digital data on unused portions of their current signal. These include the vertical blanking interval (VBI), the vestigial sideband (VSB), and the horizontal overscan (HOS). The low data rate has restricted these services pretty much to text-based services, such as WavePhore. However, one innovative use came from Microsoft for its Actimate line of toys. In one instance, data is transmitted to a radio-receiver-equipped Barney toy, allowing it to "talk" to kids who are watching the TV show at the same time.

In-stream datacasting occurs when data packets are inserted into unused portions of the DTV data stream like the Wink System, shown in Figure 11.3. You may remember that the bit rate for a compressed video stream varies, depending on how much change is occurring in the picture. In-stream techniques take advantage of the times when there is less TV data being sent to add data packets. Because this method is very "bursty," it will generally be used for software downloads and other applications where a constant bit rate is not important.

Parallel datacasting allows stations to send out time-coherent "channels" of information. The bandwidth can be assigned dynamically or guaranteed, allowing assured delivery of richer content forms such as low-res video, audio, and graphics. This type of datacasting will be used by iBlast, a consortium of large broadcasting companies and content providers who plan to transmit rich Internet content, including video streams, over their DTV spectrum. GeoCast is another company using this technology to provide datacast service to local broadcasters.

As networks move toward the multichannel, multimedia world they will need to consider what their affiliates want to do. It is surprising for viewers to realize that there is a finely-tuned power balance between networks and their affiliates that has moved first in one direction and then another. Today, the power is more centered in stations than ever before because their ownership of 6 MHz of local bandwidth gives them enormous independence from network programming services.

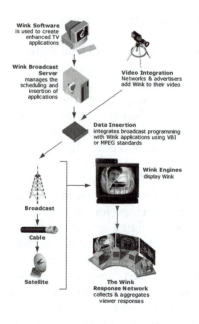

Figure 11.3 Wink's technology is a good example of in-stream datacasting. Source: Wink.

Each broadcast network is developing a strategy to deal with the changes in the media world that it sees coming. Singly and collectively, they are actively engaged in defensive moves to shore up their existing businesses. They work effectively through trade associations, the National Association of Broadcasters and the Television Bureau of Advertising, and they are quite protective of their resources. For example, they have fought DBS services vigorously to prevent them from deploying "spot beam" technology that would allow them to transmit local signals into local areas, called "local in local," a battle they lost in 1999. However, DBS is making some headway by showing broadcasters how they can use an alternate technology to keep the upper hand in their negotiations with local cable operators.

Chapter 15, "Brave New WWWorld," will look at the business strategies for the new communications landscape. For now, it is enough to say that all the networks have constructed digital network operating centers and invested in cable networks that have global distribution. ABC owns ESPN1 and ESPN2 and part of A&E Network. CBS owns the Nashville Network (TNN) and Country Music Television (CMT). NBC has CNBC and part of MSNBC. FOX packages different versions of its holdings of regional sports networks.

CBS excels in improving its internal functioning through advanced technology. LIDIA, Local Identification Inserted Automatically, is a digital system that lets affiliates display their call letters and logos during network promotion. CBS also has its own computer network, CBS Net-Q, which allows affiliates to log on and view the New York Master Control schedule, allowing them to schedule insertions or to turn them over to remote network control. In 1998, CBS News announced it would install the IBM Digital Library and asset management system. FOX Broadcasting lit up its Network Center and plays out all its programming, commercials, promotions, and long-form programs, from networked video servers. And NBC is in the process of building the Genesis digital server system.

The outlook for over-the-air, free local TV is unclear. Since the widespread deployment of cable and DBS services, it has become obvious that broadcasters will continue to lose audience share as the number of delivery platforms increases and other media formats proliferate. For the present it is still a profitable business. The bandwidth local stations control is a substantial asset. They may not make the same profits as in the past from their television operations, but they will have opportunities to replace it with profits from their bandwidth capacity.

Wireless Cable: Fixed Wireless Broadband Systems

Wireless cable is sometimes called MMDS, which stands for multichannel, multipoint distribution services. Currently, 250 MMDS systems serve 1.1 million subscribers in the United States, and about 4 million households in 90 countries around the world.[17] Wireless cable offers significant benefits. It can be deployed quickly and less expensively than wired systems. The signal can be received for about 35 miles from the transmitter in all directions, although operators may need to place additional transmitters in the area in order to extend the range into particular communities. Viewers use a special antenna, a down-converter, and a set-top box to receive the signals.

The first fixed wireless broadband system began in 1973 as a multipoint distribution service (MDS) that transmitted a single channel over the air. Some readers may remember SELEC-TV, a now defunct pay-TV service that utilized MDS technology. The more widely known premium service HBO used MDS in the early 1980s, eventually reaching 500,000 subscribers. However, multichannel cable systems ultimately put most MDS companies out of business.[18]

The technology then languished until passage of the 1992 Cable Act, which for the first time guaranteed that cable programming had to be made available at parity prices to any buyer. Before then, cable systems received favorable rates from suppliers such as Ted Turner, who offers such popular networks as CNN, Headline News, TBS, and TNT. In 1994, wireless cable operators cracked open the champagne bottles at their annual convention in Las Vegas to celebrate the anticipated growth of fixed broadband wireless as a vehicle for delivering multichannel TV. Even though they had only 700,000 subscribers nationwide who generated $117 million annually, they believed that wireless cable was on the verge of taking off.[19]

In the early 1990s, the FCC's more relaxed regulatory climate allowed operators to assemble several blocks of spectrum in the 2.1 to 2.7 GHz band to make up a multichannel service, as shown in Table 11.2. The blocks include allocations for multipoint distribution service (MDS), multichannel multipoint distribution service (MMDS), instructional television fixed service (ITFS), and wireless communications service (WCS), adding up to about 200 megahertz that carry more than 30 analog channels and, of course, many more digital channels.

The ability to deliver between 150 and 300 channels makes fixed broadband "wireless cable" competitive with wired cable systems, at least as they are presently configured. In mid-1995, financing for digital conversion came from telephone companies that sought to enter the video delivery market at that time.[20] The value of MMDS channels to telephone services continues, but now it is more likely to be used to provide high-speed broadband wireless access to the Internet.

Wireless cable was originally intended as a broadcast platform, but operators are now testing and deploying high-speed Internet services

Table 11.2 Wireless Cable Spectrum Usage

Frequency Range (GHz)	Service Allocation	Number of Channels	Channel Width
2.150–2.162 GHz	Multipoint Distribution Service (MDS)	2	6 MHz
2.305–2.320 GHz	Wireless Communication Service (WCS)	2	5 & 10 MHz
2.345–2.360 GHz	WCS	2	5 & 10 MHz
2.500–2.596 GHz	Instructional TV Fixed Service (ITFS)	16	6 MHz
2.596–2.644 GHz	Multipoint Multichannel Distribution Service (MMDS)	8	6 MHz
2.644–2.686 GHz	ITFS	4	6 MHz
2.686–2.689 GHz	MMDS	31	125 MHz

over their systems. Wireless cable systems vary in their ability to support two-way service. Many wireless cable trials began by using a telephone return for the upstream path, but increasingly operators are deploying full two-way technologies. Retrofitting legacy one-way wireless cable systems to support two-way communication is technically challenging, requiring operators to convert their broadcast systems into networks that more closely resemble a cellular telecommunications platform.

In October 1999, one consequence of the proposed $115-billion merger between MCI/WorldCom and Sprint would be to create a wireless cable powerhouse. Just prior to the announcement, Sprint had spent about $1.3 billion to acquire MMDS systems that could reach 30 million people, and MCI/WorldCom bought CAI Wireless Systems for $350 million to serve 24 million homes. "Fixed wireless, to us, looks like it's the most economically feasible because you can build a couple of towers and hit 80% of the market," said Sprint Chief Executive William Esrey.[21] The consolidated company would have the capability of providing broadband wireless access to more than 54 million U.S. households at a competitive price, more than AT&T could reach even with its Telecommunications, Inc. (TCI) and MediaOne holdings.

The technology to deliver broadband wireless access (BWA) is similar to outfitting wired cable systems to carry data services, less expensive only because it does not entail the labor to dig trenches. The wireless operator must install a cable modem router and networking equipment in the headend. Digital data such as Internet content is modulated onto radio frequency (RF) channels for over-the-air broadcast transmission to rooftop antennas at subscriber homes and offices. Coaxial cable runs from the antenna to a down-converter that transforms the microwave signals into cable television frequencies and then carries them to the cable modem.

The cable modem demodulates the high-speed data signal and sends it to the subscriber's PC or local area network (LAN) though a 10 Base-T Ethernet network interface card or cards. The customer must also have a "transverter" device to transmit data upstream. According to wireless cable equipment provider Hybrid Networks, a 6 MHz wireless cable channel can support 650 to 1,500 simultaneous active high-seed data users, providing peak downstream burst rates up to 1.5 Mbps or more to each one. Typically, operators assume that about 20% of users are online at any given time, so each 6 MHz channel serves up to 5,000 home-based and business subscribers. Table 11.3 shows broadband wireless access trials in 1999, and Table 11.4 shows the companies supplying technology for them.

Wireless cable isn't the only fixed wireless technology that provides broadband service. Local multipoint distribution service (LMDS) is an alternative platform that has a similar structure to cellular telephone systems.

Table 11.3 Wireless cable modem trials and commercial deployments in United States and Canada.

Wireless Operator or ISP	Description	Location	Equipment
American telecasting	Commercial launch of WANTWeb high-speed MMDS Internet service with Online System Services (OSS)	Colorado Spr., CO Denver, CO Portland, OR Seattle, WA	Hybrid Networks
CAI Wireless	Commercial launch of high-speed Internet MMDS service for $50 (residential) with unlimited access	Rochester, NY	Hybrid Networks
CAI Wireless	Same as above	New York, NY	General Instrument
CAI Wireless	High-speed Internet MMDS trial	Boston, MA	General Instrument
CAI Wireless	Same as above	Washington, DC	Hybrid Networks
CS Wireless	Commercial launch of "The Beam" high-speed Internet over MMDS	Dallas, TX	Hybrid Networks
DirectNET	Commercial high-speed Internet service offered in MDS spectrum band; SOHO package is $195/month plus $395 installation	Ft. Lauderdale, FL	Hybrid Networks
GoFast	High-speed Internet trial using ITFS spectrum	Minneapolis, MN	Hybrid Networks
Heartland Wireless	Commercial launch of HeartNet Internet service over MMDS	Sherman, TX	Hybrid Networks
IJNT	Commercial launch of UrJet wireless Internet service	Salt Lake City, UT Orange County, CA San Francisco, CA Houston, TX Beaumont, TX	Hybrid Networks

Table 11.3 Wireless cable modem trials and commercial deployments in United States and Canada.

Wireless Operator or ISP	Description	Location	Equipment
Look Comm.	Wireless cable Internet trial	Milton, Ont. (Can.)	Hybrid Networks
Metro.Net	Commercial high-speed Internet service using MDS spectrum	Las Vegas, NV	Hybrid Networks
QuadraVision	Commercial launch of Wireless Express Internet Services with Online System Services; basic service $49.95 per month	Reno, NV Sparks, NV Carson City, NV	Hybrid Networks
People's Choice TV	Commercial launch of Speed-Choice Internet service; for$49/month	Phoenix, AZ	Hybrid Networks
People's Choice TV	High-speed Internet trial over MMDS	Detroit, MI	Hybrid Networks
Third Rail Comm.	Deployment of two-way MMDS data services	Nashua, NH	Integrity
Wavepath	Two-way 384-Kbps Internet service from Concentric Networks, $150/month	San Francisco, CA San Jose, CA	Hybrid Networks
WBS Cable TVM	Commercial launch of XspeedI Internet service offered with Micron Services; Starts at$49.95/mo for 256-Kbps plus $600 for the modem and installation	Boise, ID	Hybrid Networks
Wireless One	Commercial launch of two-way WarpOne Internet Service; $450/mo.for 256 Kbps and $1,250 per month for 1.54 Mbps	Baton Rouge, LA Jackson, MS Memphis, TN	Hybrid Networks

Source: Cable Datacom News.[22]

Table 11.4 Wireless cable equipment manufacturers.

Supplier	Product	Downstream	Upstream	Trials/ Deployments
General Instrument	SURFboard	30 Mbps	Telco return (56 kbps)	CAI Wireless, CS Wireless
Hybrid Networks Wireless	Series 2000	30 Mbps	Up to 5 Mbps RF, or telco return	American Telecasting, CAIWireless, CS DirectNET, Metro.Net, PCTV
Integrity Comm. Nashua, NH	PTP-8 Modem	667 Kbps to 8 Mbps	667 Kbps to 8 Mbps RF	Two-Way MMDS with Third Rail/
New Media Comm.	CyberCity	5.5 Mbps or 30 Mbps	Telco return	Not disclosed

Source: Kinetic Strategies, 1999.[23]

Fixed Wireless Broadband Systems: LMDS

". . . the higher the frequency, the shorter the wavelength, the wider the bandwidth, the smaller the antenna, the slimmer the cell and, ultimately, the cheaper and better the communication,"[24] writes George Gilder about LMDS technology. Local multipoint distribution service (LMDS) is a method for distributing many channels of television, using the super-high-frequency portion of the electromagnetic spectrum, at 27.5–29.5 GHz. Recall that "giga" means billion and that frequency is the inverse of wave length; thus, waves in the gigahertz range are so small, they are called millimeter waves. As Gilder correctly observes, the use of such a high frequency means that LMDS signals are very small (and therefore weak), transmitted by only 10 milliwatts of power.

LMDS was born when engineer Bernard Bossard was working at SpeedUS.com and decided to develop a working point-to-multipoint video system in the 28 GHz band. Rather than send high-powered, low-frequency TV signals over long distances, he took on the challenge of transmitting low-powered, high-frequency signals for only short hops. And he elected to encode them with frequency modulation (FM) rather than amplitude modulation (AM), as do traditional TV broadcasters. A side benefit of this choice is that not only does FM tolerate rain interference better, it also delivers a far higher quality image than does AM.

If weak signals are to be reliable, they must be transmitted over a substantial swathe of bandwidth. LMDS channels take up 20 MHz, in comparison to the 6 MHz of over-the-air or cable channels.

However, with transmission in the gigahertz frequencies, the size of the signal doesn't really matter because there is so much bandwidth in that part of the spectrum. Texas Instruments estimates that the LMDS spectrum will support 224 digital video channels and 16,000 telephone channels per residential node, or 192 T-1 (1.5 Mbps) circuits and up to 4,608 standard voice circuits per business node. Most potential operators will offer a mix of services including digital TV channels, telephone services, personal communication services (PCS), and data transmissions. In addition, there is sufficient capacity to allow for a return path from subscribers, enough even for videoconferencing. Given the popularity of the Internet, most LMDS operators are looking to provide a suite of services that will combine voice, video, and data. Both ATM and IP architectures work well with LMDS systems that carry these different types of traffic.

An LMDS network has four parts: (1) a network operations center (NOC), (2) a fiber optic connections, (3) a base station, and (4) equipment at the customer premise.[25] The NOC holds network management equipment, and operators can link to the PSTN and interconnect their NOCs from different regions via fiber optic terrestrial networks, such as SONET, ATM, or IP networks. The NOC is usually hard-wired to base stations, which is typically where the wireless infrastructure begins. Base stations, or network nodes, have a network interface for fiber termination, modulation/demodulation, and microwave transmission and reception equipment. In some systems, billing, channel access management, registration, and authentication take place within the base station as well. The radio frequency (RF) gear includes transmitters and receivers as well as transceivers and the antennas they feed. Individually modulated signals, such as TV channels, are combined and sent to the broadband transmitter. Within the transmitter, VHF signals are converted up to the desired carrier frequency, amplified, and applied to the antenna for transmission.

The small transmitter cells that transmit the signals from the base stations are theoretically placed three to ten miles apart, broadcasting to other tiny "cells"—the reason LMDS is sometimes called "cellular TV." Cell sizes are strongly affected by the overall environment. Dense foliage, rainfall rates, height of the transmitter antenna, height of the customer's antenna, and the modulation scheme used can all affect how well and how far the signals travel. LMDS pioneer SpeedUS.com actually deployed systems, and their experience is that the optimum distance between cells is actually about 1.25 miles, even smaller. The signal carries more than a gigabit of traffic on a beam no bigger than a length of common kitchen string.

Customer premise equipment varies from one installation to another because subscribers and their premises differ so greatly—from large office buildings, hospitals, and entire campuses, where the system serves microwave equipment that is shared between many users, to single-family residences that have only some TV channels, a 10 Base-T high-speed Internet connection, and one or two plain old telephone service (POTS) lines. However, all systems will include microwave equipment for modulation/demodulation, control, and subscriber interface, either inside or outside the premises. The LMDS antenna is only about four-by-four-inches square, and it receives signals that produce exceptionally high-quality TV pictures. The broadband receiver gets the signals at carrier frequency and converts them as needed for routing to the network interface unit (NIU), the gateway between the RF components to the various user appliances—TV, PC, and telephone. NIUs are managed by the network management system provided in the NOC, and can be scaled to meet subscriber requirements.

The system is capable of rapid two-way transmission with sufficient speed and bandwidth for videoconferencing. Two-way data network applications require a transceiver for the return path, and uses the same four-by-four-inch antenna to send data back to the base station. It can be this size because LMDS cells are only a short distance from subscribers and the transmission frequency is high, so it take little power for subscribers to send signals back to the base station cell with a correspondingly small transmitting antenna.[26]

There are widely different estimates for the cost of building LMDS infrastructure because per subscriber estimates vary according to the level of market penetration that are assumed in making the projection. For example, according to the National Institute of Standards and Technology, a reasonable LMDS network can be built for between $700 and $1,000 per subscriber. SpeedUS.com claims that it spent around $300 per subscriber for infrastructure and startup operating costs for a one-way analog network, compared to an estimated $700-per-subscriber cost for MMDS 200 MHz band service, and $800-per-subscriber cost for DBS. However, Hewlett-Packard believes that once there is higher demand, LMDS infrastructure will cost about $150 per customer covered, as opposed to $1,000 for fiber to the home. At 25% market penetration, fiber would cost $4,000 per customer, and LMDS would cost $600 per customer. Texas Instruments says LMDS will be about half the cost of hybrid fiber/coax (HFC) at 30% market penetration.[27]

LMDS systems can thrive even when they start with only a few customers. Operators must install such elements as hubs, with trunking and cell sites at the outset, but once these are deployed, additional costs are incurred only as additional customers sign on. Ironically, the largest fixed expense to build out an LMDS system is likely to be the cost of wiring the hub to the transmitters via fiber optic cable, rather than the transmission equipment itself. By contrast, wired network operators incur the vast majority of their costs before the first paying customer gets service, so they need a fairly high penetration rate to survive. Another advantage of LMDS systems is that they can be deployed quickly. Once the LMDS equipment is assembled, hub installation takes only a few days, and it takes only a few hours to add new customers. For example, after the massive earthquake in 1995 in Kobe, Japan, authorities constructed a 23 GHz broadband network to replicate the previously wired cable television system in just a few weeks.

As LMDS wireless access systems evolve, standards for the technology will become increasingly important. Standards activities currently underway include activities by the ATM Forum, DAVIC, ETSI, and ITU. The majority of these methods use ATM cells as the primary transport mechanism. De facto industry standards are also important in this arena. In late October, 1999, Cisco Systems established a group of networking and communications companies—Bechtel Telecommunications, Broadcom, EDS, KPMG, Motorola, Pace, Samsung, Texas Instruments, and Toshiba—to set a new standard for broadband wireless communication called Vector Orthogonal Frequency-Division Multiplexing (VOFDM).[28] The goal is to drive the cost of manufacturing and customer prices down, encouraging greater construction of both MMDS and LMDS networks through the VOFDM open standard that transcends any particular company's proprietary products.

This technology has been crying wolf for a decade, casting doubt on whether it will ever actually come on the market. The FCC controls LMDS spectrum licenses for carriers, and in 1998, the agency finally awarded licenses for most major metropolitan areas, chopping up a huge chunk of spectrum into multiple slices. Each metropolitan area was awarded two LMDS licenses, one massive 1,150 MHz block broken into three equal-sized chunks, and a 150 MHz block divided into two pieces. Only the larger license holder has sufficient bandwidth to enter the broadband market effectively. Moreover, the LMDS spectrum holder can sit on the license for ten years or subcontract its usage to other companies rather than provide direct-to-customer service.

One interesting license holder is Virginia Tech. A research center, the Center for Wireless Telecommunications (CWT), bid and acquired LMDS spectrum for the greater Roanoke, Danville, Martinsville, and Kingsport-Johnson City market areas in Virginia. In early 1999, the university began limited deployment of LMDS in Blacksburg, using equipment from Wavtrace. The CWT sponsors a variety of research projects on wireless broadband technology. One of the most rewarding studies was conducted by Katina Reece, "Modeling the Propagation and Latency Effects of the Integration of ATM in a Seamless Terrestrial and Satellite-Based Wireless Network," which will help foster the integration of LMDS into other networks. According to information posted on the Web site:

The ultimate goal of this work is to develop an ATM-based protocol that would work in a network whose nodes were LMDS sites, LEOSATS, and GEOSATS. A critical examination of the propagation and latency effects associated with the satellites and the effects of the data rate mismatches that may occur between fiber, LMDS, and satellite links will be carefully observed.[29]

Another broadband research center is the National Wireless Electronic Systems Testbed (N-WEST), a project of two agencies of the U.S. Department of Commerce, the National Institute of Standards and Technology (NIST), and the National Telecommunications and Information Administration (NTIA).[30] N-WEST is a measurements and standards resource for the broadband wireless access industry. It promotes the development of the industry by creating and carrying out tests and measurements at the system and component levels and promotes sound operational standards and specifications based on open technical results. N-WEST invites the participation of service providers, system integrators, and component manufacturers. The agency contributes findings to the IEEE 802.16 Working Group on Broadband Wireless Access of the IEEE 802 LAN/MAN Standards Committee.

There are a number of trials and deployments of gigahertz spectrum services in the United States offering broadband wireless service, not all in the LMDS spectrum, but very close to it. The actual rollout of service has occurred at a glacial pace. SpeedUS.com began transmitting multichannel cable TV service in Brooklyn and Brighton Beach, New York in the early 1990s. The company now uses its LMDS network to provide Internet access, delivering 1.5 Mbps to the customer for $50 a month. SpeedUS.com has important partners, including regional Bell operating company (RBOC) Bell Atlantic, Philips Electronics, and the David Sarnoff Research Center, but they haven't done much to put LMDS in actual use. In the late 1990s, a few other systems came up. In December, 1998, Liberty Cellular of Kansas announced it would trial LMDS across the state, with full commercial deployment to follow in 1999 or 2000.[31] Similarly, Home Telephone, Inc., a telecom service provider that holds LMDS spectrum, said it would build a network and offer commercial service in the Charleston, South Caroline market.[32]

The biggest market for LMDS may prove to be businesses, and there are a number of companies developing fixed broadband wireless services to them. One company, ART, uses a 39 GHz spectrum to connect 126 buildings to its network with service orders from 250 customers. ART offers commercial service in Portland, Oregon, Phoenix, Arizona, and Seattle, Washington, via point-to-point technology. In addition, it is conducting point-to-multipoint throughout the U.S. Denver-based Formus Communications is carrying out point-to-point testing and trials in Denver, Budapest, and Strasbourg, France. The company introduced live point-to-multipoint service in 1999.

In 1998, Nextlink acquired WNP, a company that had bought up LMDS licenses auctioned by the FCC earlier in the year. The purchase gave NextLink 95% of the top 30 U.S. service areas. It is testing and making trials for true LMDS point-to-point and point-to-multipoint technology in its Plano, Texas, lab facilities and commercial environments in the United States. It will offer commercial broadband wireless service following the trial.[33]

Teligent uses the 24 GHz band to offer commercial service in 27 markets comprising 430 cities and towns. It has more than 3,100 buildings under lease or option, and has already installed 444 buildings with fixed wireless equipment, using point-to-point and point-to-multipoint technology. By the end of 1999, Teligent expected to provide service in more than 40 markets.[34]

Another operator, WinStar, operates in the 39 GHz portion of the spectrum, with access rights to 4,800 buildings and service to 17,000 customers. Currently, it offers commercial service in more than 30 markets using point-to-point and point-to-multipoint technology. It plans to reach 45 U.S. markets as well as 6 international markets in 2000.[35]

Prospects are even brighter outside the United States. In Canada, LMDS spectrum is allocated at 28 GHz, and MaxLink holds the licenses to serve 207 communities.[36] The company is now operating a test network in Ottawa and built out that city, as well as Montreal and Calgary, Alberta, in 2000, ultimately serving Canadian cities from coast-to-coast. Australia has just auctioned its LMDS spectrum at 28 and 31 GHz, but no network construction has yet been announced. Similarly, in March, 1999, the Israeli government formed a committee to begin the spectrum allocation process for LMDS.

There are operational LMDS networks in South America and Asia as well. In the Philippines, Pilipino Telephone (PILTEL) Corp. LMDS network, dubbed the Wytec Causeway2, has demonstrated that the technology can deliver high-speed data service reliably, even in a high-rainfall region. In Manila, where torrential storms can deluge the city with five inches of rain per hour, the PILTEL system has proved that it can work. Wytec Causeway2 delivers up to eight OC-3s per cell, and consumer network interface units provide 8 POTS lines, a T1/E1 link, and a 10/100 Ethernet interface.

Other operational LMDS Systems include:

- Hong Kong: A two-way digital system that includes high-speed data access
- Korea: A two-way digital system that includes high-speed data access
- Baku, Azerbaijan: One-way service for TV TelCom

- Valencia, Venezuela: One-way television service for TV-Cellular
- Caracas, Venezuela: One-way television service for Viva-Vision
- North Island, New Zealand: Trial of two-way network for voice and data for carrier Clear Communications prior to national rollout

The LMDS Market   The first LMDS trials and deployments were aimed at bringing multichannel television service to consumers. However, the transition to fully two-way data-capable systems has moved the center of the target towards the business market. Douglas Lockie, Vice President at millimeter wave subsystem manufacturer Endgate Corp., once joked: "It will be easier selling broadband wireless services to companies trying to interconnect LANs at reasonable prices than selling yet another cable TV service to holders of overdrawn credit cards."[37]

According to Pioneer Consulting, the total worldwide market for LMDS business services will grow from $157 million in 1998 to just over $11.5 billion in the year 2007, as shown in Table 11.5.[38] The size of the increase suggests that LMDS is beginning to carve a sustainable global niche for itself among network architectures. The study reports that there is growing recognition that LMDS will be able to give business customers with high bandwidth service for 50–75% less than the cost of traditional private line services.

Table 11.5 Estimated Total Revenues from LMDS Services to Businesses ($ Millions)

	Small Business	Medium Business	Large Business	Total Business Market
1998	$ 56.70	$ 12.10	$ 27.30	$ 96.10
1999	$ 151.07	$ 32.23	$ 58.19	$ 241.49
2000	$ 523.26	$ 137.39	$133.33	$ 793.98
2001	$ 752.85	$ 219.63	$213.15	$1,185.63
2002	$1,083.17	$ 304.30	$272.60	$1,660.06
2003	$1,558.42	$ 421.60	$334.10	$2,314.12
2004	$2,242.20	$ 584.11	$391.68	$3,217.98
2005	$3,225.99	$ 809.27	$459.17	$4,494.43
2006	$3,953.81	$ 991.85	$513.83	$5,459.49
2007	$4,845.84	$1,162.77	$574.99	$6,583.61

Source: Pioneer Consulting, 1999.

Low-Power Television (LPTV): Home-Grown TV

Would you rather watch your child play Little League baseball in the local park or a mid-season game between the Braves and the Giants? A few minutes before you leave for work, are you more interested in the national weather picture or the chance of rain in your area? Since it was established in 1982, low-power television (LPTV) was a wireless television delivery system that transmitted at a maximum of 10 watts VHF or 1,000 UHF. The signals could travel as far as 35 miles, but the effective reach was more like 10 to 15 miles, and even fewer miles in areas with tall buildings.[39]

Currently, there are about 2,200 licensed LPTV stations in approximately 1,000 communities, transmitting in all 50 states. They serve both rural and urban audiences. In some areas that are not large enough to support their own broadcast license-holder, or the allocations are taken by nearby large markets, LPTV stations may be the only TV station that provides truly local news, weather, and public affairs programming. Stations are operated by such diverse entities as community groups, schools and colleges, religious organizations, radio and TV broadcasters, and a wide variety of small businesses. The service has provided first-time ownership opportunities for minorities and women.

Even in well-served markets, LPTV stations serve small communities within a larger area, broadcasting to viewers that belong to specific ethnic, racial, and interest groups, including foreign language enclaves. They usually offer local news, sports, weather, talk, children's programming, religious programming, and civic services such as council and local committee meetings. Most offer occasional live programming, but this is usually only a small part of an LPTV's overall programming. Despite complaints from both broadcasters and cablers, LPTV stations fall under the cable "must-carry" rules, which means that local cable operators must carry them like any other local station.[40]

The only available survey of LPTV stations is rather dated, with data reported in 1990. It indicated that 46% of the stations covered a rural area, 22% broadcast in an urban neighborhood, and 17% reached a suburban population. The remaining 5% covered a combination of markets.[41] Of those LPTV stations operating at that time, religious organizations had about 200 facilities and commercial interests accounted for about 350 more. The rest either delivered over-the-air broadcast signals to difficult-to-reach areas, functioning as what are called "translators," or offered some combination of educational or community service telecasting.

The most recent trend in LPTV is multichannel subscription service. Operators put together several FCC licenses in the same area and offer 10 to 15 channels as a low-cost alternative to cable service. Subscribers receive a set-top box to descramble the signals but no special conversion is needed, as LPTV operates in the VHF or UHF frequencies. Although the technology itself is relatively simple, LPTV is an advanced system because it makes broadcasting personal by moving it to the neighborhood. There is as yet no provision for interactive services, but the FCC is considering how operators can move to a digital platform.

LPTV has been gravely affected by the transition of the TV industry to digital, because the huge allocation of UHF bandwidth often displaced their spectrum. The FCC made provisions for them to apply for other spectrum and to receive priority treatment. At the end of September, 1999, the agency opened up a proposed rulemaking to finalize the status of LPTV.[42] The agency asked for comments on a wide variety of issues: power, transmitter height, transmission range, spectrum allocation, service class, commercial status, equipment requirements, programming restrictions, ownership requirements and limits, and a variety of other potential provisions. Given the 90-day comment period, the FCC will probably make its decisions sometime in the first half of 2000.

Over the air, LPTV will never compete with commercial television services. However, the Internet gives LPTV operators new opportunities. Ron Nutt was the first broadcaster to put TV signals on the Net.[43] The LPTV owner was the first to realize that he could obtain a certain parity with even the largest stations by webcasting his signal over the Internet. His success has enabled him to get a 150,000-watt license, and he has now applied for a Class A 500,000-watt license that will give his station the same power to transmit locally as a network affiliate.

LPTV is one more source of terrestrial television programming that results in the further fragmentation of the traditional mass television audience. And in the mid-1990s, there was even more splintering of TV viewers with the inauguration of popular satellite services. Advances in processing made it possible for new satellite design and placement. In the next section, we will look at wireless broadband from space.

Space-Based Delivery of Television and Broadband Services

According to an April 1999 survey by *Satellite Today* magazine, at that time, there were 199 Western-built, geostationary, commercial communications satellites in orbit equipped with about 4,500 C-band and Ku-band transponders. Fifty-six more satellites are on order, which will increase the number by another 1,800 C- and Ku-band transponders. The world's first two commercial LEO systems began operating, and a third was nearly completed, altogether adding up to 134 LEO satellites. In 1998, 31 commercial geostationary satellites were ordered and 22 of them were launched. Total 1998 industry revenues were nearly $66 billion, up 15% from 1997.[44]

In 1994, the world of satellite television changed forever when space-based direct broadcast satellite (DBS) service moved in, the new kid on the block bringing TV programming to viewers.[45] Full-page ads and well-produced commercials extolling the benefits of 18-inch pizza-pan-sized satellite dishes appeared in newspapers and on television sets across the United States. Sure, satellite dishes had been available for twenty years. But the older C-band dishes were 10 to 15 feet across, required motors to aim them in the proper direction, received signals in the 3.7 to 4.7 GHz range, and cost thousands of dollars. By contrast, as shown in Figure 11.4, the new RCA dishes were small, stationary, high-powered (11.2 to 12.7 GHz range), made possible by a higher-powered satellite that transmitted a stronger, more focused beam than that sent by previous "birds" (a term for a satellite). Initially, the small dish cost $700; within four years, they were advertised at prices ranging from $99 to free, as part of promotional packages.

Figure 11.4 The RCA digital satellite dish, manufactured for DIRECTV. Source: DIRECTV.

There are three different types of satellites, identified by where they orbit in space. Geostationary earth-orbiting satellites (GEOS) are at 22,247 miles (36,000 km) altitude, requiring 3 to 6 satellites to cover the earth fully. Middle-earth orbiting satellites (MEOS) travel between 8,000 to 20,000 km above earth, and it takes a fleet of 6 to 20 satellites to cover the world. Low-earth orbiting satellites (LEOS) track at 300 km to 2,000 km and need 40 to 70 birds to cover the globe. GEOS satellites have a life expectancy of about fifteen years, while the smaller LEOS should perform well for about seven years.

The first satellite dish available to consumers was TVRO, standing for TV-receive-only. The 6 GHz (uplink) and 4 GHz (downlink) signals were handled by GEOS, and the large 10- to 15-foot-diameter dishes were called C-band earth stations. DBS is also a GEOS service. In Europe, DBS service is often called DTH (direct-to-home). The newer services are handled by high-power GEOS that uplink at 14 GHz and downlink at 11 or 12 GHz, and they are received on the smaller Ku-band and Ka-band earth stations (dishes). Marketers call the complete equipment package located at the consumer premises DSS, or digital satellite system.

Satellite systems are sometimes known by the frequencies over which they transmit and receive. For instance, C-band is often called "6/4" for the 6 GHz uplink and 4 GHz downlink frequencies; Ku-band is "14/12" for the uplink and downlink frequencies of 14 GHz and 12 GHz, respectively. The newest type of service is Ka-band, or "30/20," as its uplink is at 30 GHz and its downlink is at 20 GHz. The uplink is always stated first, and it is nearly always the higher of the two figures because the high frequency requires a bigger antenna; the larger one stays on the ground and the smaller one is sent into space.

The number of U.S. subscribers to the first type of consumer service, TVRO, is declining, while the number of DBS customers is growing. The main reason is that the new dishes are so much cheaper and easier to install and use. By 1992, there were about 3.5 million TVRO satellite dish owners, but this figure has declined steadily with the competition from DBS, as shown in Table 11.6.

Table 11.6 Number of U.S. Satellite TV Households

Service	August, 1997	August, 1998	August, 1999
EchoStar	717,601	1,537,227	2,845,352
Primestar	1,790,010	2,133,301	1,900,000
DIRECTV/USSB	2,804,769	3,953,037	5,559,000
C-Band	2,167,051	1,992,667	1,748,958
TOTAL	7,481,430	9,617,232	12,053,310

Source: *Satellite Business News.*[46]

How Satellite Communication Systems Work

Satellites vehicles vary considerably in size and cost. For the government's space program, they can weigh as much as 1,000 pounds up to a ton, and carry payloads up to 10,000 pounds, and cost upwards of $50 or $60 million. Since 1985, however, there has been a trend towards. smaller, less expensive space vehicles called "smallsats." These tiny vehicles weigh as little as 100 pounds. The launch price runs between $500,000 to $15 million, with costs moving down towards $100,000.

Think of a satellite as a cable headend in the sky. Signals are collected at a broadcast center from satellite, fiber optic land lines, and videotape. The programming is digitized, encrypted, and transmitted up (uplinked) to the satellite, then transferred to a transponder and downlinked. An uplinked signal comes from one site; the downlink goes to a potentially infinite number of receiving dishes; thus, this type of system is called point-to-multipoint. If the signals are encrypted and can be decrypted only at one site, then satellites can also function as point-to-point systems.

The area where the signals can be received is called the satellite's "footprint." The footprint from a satellite 22,247 miles away covers about one-third of the entire world (except for the poles), a hemisphere, a region, or even a fairly small area, depending on how strong the signal is. The stronger the signal is, the more tightly focused it can be. Signals transmitted from satellites are simply microwaves, located in the super-high-frequency (SHF) portion of the electromagnetic spectrum, between 3 and 30 gigahertz. The signals received by earth stations are measured in EIRP, "effective isotropic radiated power," and expressed in dBw, the number of decibels per watt.

No matter how strong the traveling signal is initially, it spreads out and weakens as it moves, so that by the time it reaches earth it is usually very weak indeed. "Detecting a transmission from a 100-watt transponder is like clearly seeing a typical light bulb from a distance of 22,247 miles," writes one author about the challenge of receiving a signal from a geostationary satellite.[47] Signals from MEOS or LEOS do not attenuate as much as those from GEOS because they are closer to earth. Satellite transmissions are powered by solar panels, which constitute much of the weight of the vehicle. The information processing units aboard are called "transponders." Satellites now have as many as 48 transponders with several 36-MHz-wide video channels transmitted on each transponder.

The Receive Site   The earth station or receive system on the ground, consists of a dish, an amplifier, a feed, and a down-converter, as shown in Figure 11.4. The dish is actually an antenna that collects and concentrates the signals at the "focal point." Some dishes are equipped with an "actuator," a motor that moves the dish so that it can point at and receive signals from different satellites.

The feed collects the signals reflected off the dish it faces, a process called "illuminating" the dish. The feed automatically detects how the signals are polarized: horizontally, vertically, or circularly. Attached to the feed, a small processor called the low-noise block (LNB) down-converter transforms the signals to an electrical current, amplifies them, and down-converts them to a lower frequency for transport to the indoor satellite receiver. The new Ku-band digital services require special LNBs that are "phase-lock looped" (PLL) to receive a precise frequency without drifting. From outside the home, the down-converted electrical signals go to the inside receiver via coaxial cable. The receiver includes a tuner to separate the channels and modulation/demodulation equipment to process the signal so it can be displayed on the owner's television set or PC.

Limited interactivity is possible if all that is needed are a few pulses, such as those needed to order a pay-per-view (PPV) movie. The return path is usually a low-power FM signal that goes to a local receiver. The local receiver sends back a message to the set-top converter, directing it to open a path to receive the channel carrying the requested material. Some systems use the telephone for a return path. The incentive to provide on-demand or near-demand service is to allow impulse buying, which greatly increases the "take rate" for pay-per-view.[48]

However, for a satellite service provider to offer two-way service, whether it is narrowband for telephone calls or broadband for TV and Internet traffic, requires considerably more power and costly equipment. One way to add interactivity is to link the satellite system by terrestrial microwave, local cells, or even telephones. However, if the signal must ultimately reach a GEOS, one difficulty is "latency," the length of time it takes for a signal to travel 22,247 miles. For interactive-rich media, video, and voice data, this delay can be a problem.

The business problem for GEOS systems handling interactive traffic is backhaul: the cost of getting traffic to and from satellite uplink hubs. When there is substantial multipoint-to-multipoint broadband traffic, there is no cost-efficient way to deliver service. Moreover, while DBS downstream bandwidth is enormous, it is not infinite, particularly compared to the new optical technologies coming on the market. However, as we will see later in this chapter, closer-to-earth MEOS and LEOS do not have this disadvantage, and they offer the opportunity for interactivity.

Administering the Satellite System

The regulation of slot assignment in the Clarke orbit, 22,247 miles in space, is carried out by the International Telecommunications Union, through its control of the assignment of spectrum for transmission. The ITU sponsors administrative conferences, called World Administrative Radio Conferences (WARCs), to examine spectrum allocation. In the

United States, the FCC establishes the rules for applications, evaluates requests, and forwards their recommendations to the ITU.[49]

The first global satellite organization was INTELSAT, a not-for-profit cooperative headquartered in Washington that administers and managed the first satellite network. INTELSAT puts up satellites, then makes service available to 170 countries, both members and nonmembers. It is still one of the largest networks, and many of the members of INTERSPUT-NIK, organized by the former Soviet Union during the Cold War, have now joined. There are a number of regionally-based systems: ARABSAT, EUTELSAT, CONDOR (Latin America), and RASCOM (Africa). In addition, a number of countries have launched their own satellites, including India, Indonesia, Brazil, Mexico, France, and China.

GEOS: The "Deathstar"

"Deathstar" is what cable operators call GEOS, because the new DBS programming services confront them with their first truly substantial competition. Interestingly, the ability of geostationary satellites (GEOS) to reach a national audience also makes them competitive with broadcast networks because they are the only other potential packagers of a national audience. The term "geostationary" means that the satellite appears to stay still when it is positioned in an orbit 22,247 above the equator. This precise distance, the geostationary arc, is named the "Clarke orbit" because it was first proposed by science-fiction writer Arthur C. Clarke in 1945.

The Clarke orbit allows line-of-sight transmission to any place on earth, except the poles, using only three satellites. This means that the earth stations can be fixed, and that they do not need a motor to move them around. Often the transmission is focused to send a stronger signal to an even smaller area than one-third of the globe. In fact, downlink antennas can focus transmission to global, hemispherical, zone, or relatively small areas through the use of spot beams.

Getting a GEOS into orbit isn't easy. Nevertheless, commercial launches have become more common from locations not associated with high technology like French Guiana, China, and Russia. A measure of the technical difficulty, however, is that in 1994, three of five Chinese launches exploded, destroying the satellites they were carrying as well as the rockets. A limited number of liftoffs also occur from Vandenberg and Cape Canaveral in Florida. There have also been an increasing number of in-orbit satellite failures. In 1998, 24 satellites experienced total or partial failures, costing the satellite services industry more than $1.9 billion in losses. Some experts believe that these failures are the result of the rapid growth in the satellite industry, with manufacturers hurrying up production to meet client demands for more birds in shorter time spans.[50]

The space available for geostationary satellites is very limited. Ostensibly, 2 degrees (916 miles) separates each satellite in the Clarke orbit, limiting the number of "parking spaces" to 180. However, there are actually 199 operating satellites in the geosynchronous orbit, and there is little chance of the vehicles colliding. Nevertheless, competition to claim a spot is fierce as demand for international communication grows. The allocation of satellite slots has become increasingly difficult and rancorous, despite ITU regulation, and the picture is even more complicated by the presence of orbiting "space junk"—the 500 or so "dead" satellites whose useful life of 10 to 25 years has ended but which are still in place.[51]

However, digital compression, an increase in the number of transponders per satellite, and the use of middle- and low-earth orbits has brought much new satellite capacity, resulting in dropping prices. The cost of renting a satellite transponder has decreased considerably in the last five years, from $5 million per month for transmitting television signals 24/7, to $1 million per month, to as little as $500,000. Most satellite traffic in the developed countries is video, and telephone calls are routed over fiber optic cable. However, in regions where there is little installed fiber, satellites still handle telephony. Increasingly, satellites carry significant amounts of Internet traffic, particularly between continents.

C-Band TVRO: The First Service from GEOS   Within the United States, satellites have been widely used for the past decade to distribute television programs from production centers to television stations and cable headends. Credit for the first use of satellites for non-live program distribution to cable headends goes to Home Box Office, which adopted the technology to reduce the costs of getting their shows to the 1,800 cable systems that carried the HBO service. Previously, the channel made expensive copies of the shows, called "dubs," and sent them by courier or mail. Twenty-four hour distribution, while expensive, proved to be much cheaper than "bicycling" the tapes, as the earlier process was called.

By the mid-1980s, many programming services followed HBO's lead and sent channels and programs to cable headends and television stations over satellite. Program distribution is still an extremely important and growing aspect of the GEOS business. In early 1995, two telephone companies won a court victory, allowing them to receive video programming from satellite feeds.[52]

Before long, technically clever entrepreneurs figured out how to downlink the satellite signals for private use. As the number of people with TVRO dishes grew to 1 and then 2 million, programmers realized they were losing substantial subscription fees. As we learned in Chapter 4, "High-Definition Television: Or How DTV Came into Being Because HBO Needed Cash," within a few years programmers began to scramble their signals, initiating an ongoing "infowar" between program suppliers and pirates. "The airwaves should belong to the people. If a TV signal comes trespassing onto my property, I should be free to do any damn thing I want with it, and it's none of the government's business," said one pirate.[53]

Today, there are about 1.75 million TVRO owners in the United States alone and many more in foreign markets. These large, motorized dishes can receive 350 channels, many of them still unscrambled, from dozens of satellites. However, when TVRO owners want to receive such popular programming services as CNN, HBO, and Showtime, they must pay for service from a channel-bundler and a decoding set-top box from General Instruments. (See Chapter 2, "From Show Biz to Show Bits: Digital Technology in the Media and Entertainment Industry," for a description of compression and scrambling.)

VSATs: Very Small Aperture Technology Networks   VSAT stands for Very Small Aperture Technology. They are a network of microwave terminals that converge on uplink dishes at a "hub" to send up signals to a satellite. VSAT systems transmit data in the 14/12 GHz Ku-band of the spectrum. Some VSAT systems are entirely wireless, where multiple VSATs "squirt" data over a microwave transmitter to a hub, while others are hybrid networks that use landlines to the hub. The hub bundles messages and uplinks them to a satellite, which then beams the bundle back down to the VSATs. This means that within a few seconds, geographically dispersed sites all have the same information—allowing inventory and sales records to be updated in near real time.

The bird then retransmits the bundle so they can be picked up by many small earth stations.

Many VSAT networks are designed for point-to-multipoint communication, such as the Fordstar network. The company's headquarters can send training videos, graphics, voice, and data communications simultaneously to all their dealerships, worldwide. Auto dealers are major VSAT networkers; 49,000 VSATs have been sold to automotive companies of one sort or another around the world, and dealers make up about 15% of the world's users, and gas stations comprise another 15%.[54]

The Fordstar VSAT network has been an enormous success. It links 6,800 sites throughout North America, carrying 1,250 hours of live programming per month over 8 channels of compressed video to dealers. The primary application is training for company employees, and it has enabled the company to move 92% of its training activities to on-site classrooms, rather than paying to bring employees to a training center.

Retail outlets such as Wal-Mart, Toys "R" Us, and K-Mart also use VSAT networks for credit checks and inventory control. In these applications, the store contacts the headquarters by landline, where the request for information is processed by a computer. The computer sends the answer (also by landline) to the VSAT, which uplinks the answer to the satellite, and it is instantly beamed back down to the retail outlet—all in a few seconds.[55]

The number of VSATs is growing, largely because of the explosion of Internet traffic.[56] For example, NSI Communications' top-selling model is now its TDMA-based VSAT Plus 2. The product provides ISP connectivity at 2 Mbps to multiple distribution points with low-speed return channels. Another product introduced in 1999 by Matra Marconi Space was the FreedomIP VSAT. This product allowed dynamic allocation of bandwidth and the capacity to scale in tens of megabits to optimize load and reduce communication costs. These products demonstrate that VSATs are quickly becoming capable of handling higher-bandwidth traffic. Data rates began in the tens of kpbs and now operate bidirectionally at sustained rates of up to 8 Mbps. They are being redesigned to support 155 Mbps and above once broadband satellite networks come online.

The latency from the ground to the satellite and back to multiple VSATs on the ground has been reduced to about 1/3 of a second, so VSATs are now able to support a wide range of broadband applications, including IP multicasting, LAN/WAN interconnection, voice telephony, and videoconferencing. Moreover, the technology has proven that it can support ATM, frame relay, and IP protocols—essential in the new all-digital world, which is gradually growing from one that could support data rates in the tens of kbps to one that can support data rates at T1/E1 and above.

The VSAT marketplace has been burgeoning in the broadband world, and it has proven that it can efficiently and cost-effectively support popular protocols including ATM, frame relay, and especially IP. However, if VSATs are to remain competitive, they must become less expensive to match the falling prices of wired bandwidth. Lower-cost modems and tranceivers are an area where many manufacturers hope to cut costs.

Ku-Band DBS: Digital Broadcast Satellite   It took twelve years to launch high-power 14/12 GHz satellite transmission services. The FCC authorized eight orbital slots back in 1982, but substantial technical development was required before services could actually launch. When DBS arrived in June of 1994, it was an instant hit. The 18-inch RCA-brand DSS dish was shipped to retail outlets in 150 markets, and every unit was sold out within the first 72 hours. Within six months there were 500,000 subscribers. This growth is expected to continue for the next decade.

The attraction of the new high-power satellite technology is its simple installation, smaller size, and lower price. The dish can be smaller because of the higher signal power of 100 to 400 watts, compared to the less than 60 watt 6 GHz/4 GHz C-band signal. DIRECTV, the first digital DBS service, used MPEG-1 digital compression, because MPEG-2 was not yet ready. Shortly after launch, the service went to MPEG-2, dramatically improving the image quality, and giving retail sales outlets a powerful new argument in favor of DBS.

DBS is particularly welcome in areas where multichannel TV is not otherwise available. Twenty-one percent of rural households subscribe to it, compared to 4% of urban households and 9% of suburban households. The majority of new DBS subscribers are interested in DBS for the programming options and the benefits that it provides, not because they are anti-cable. Indeed, 17% of DBS households also subscribe to cable.

The cost per subscriber to deploy high-power DBS systems is $275 per household if there are 1 million subscribers; with 10 million customers, the price tag drops to $25. Even at the higher rate, DBS is cheaper than many forms of TV delivery. One reason is because the consumer pays for the receiving dish, effectively transferring the cost of equipment from the provider to the subscriber. In 1998, the average of cost of reception equipment (dish, LNB, and home satellite receiver) was $429. By 1999, the average price dropped to $260, and further price drops are expected. The lower prices now allow satellite operators to promote free dishes if consumers sign up for service.[57]

Despite the excellent business prospects, only two operators do business in the United States. In 1996, there were four DBS services: DIRECTV network packaged with USSB programming; PrimeStar; Echo-Star; and AlphaStar. By 1999, there were only two operators. AlphaStar went into receivership in 1997; PrimeStar lost its orbiting slot and associated frequencies, and, along with USSB, was acquired by Hughes, the corporate parent of DIRECTV. Now only that company and EchoStar are still standing to fight it out in the consumer marketplace against cable companies and one another. Figure 11.5 shows the DIRECTV network.[58]

One disadvantage faced by DBS operators during their first few years in business was their inability to give their subscribers local television stations. If viewers wanted to receive them, they needed a two-input A-B switch to bring in signals from rabbit ears or a cable service, allowing them to switch back to the satellite feed when they wanted DBS programming. Until 1999, the technology to push tightly focused spot beams into a local market did not exist; moreover, satellite operators were restricted from doing so by regulation.

Then in 1999, Capitol Broadcasting Company (CBC) announced plans to market a configuration package to let satellite operators deliver local signals via tightly focused spot beams. And the U.S. House of Representatives passed legislation that would permit DBS operators to offer "local in local," as the delivery of local stations is called. A few months later, President Clinton signed the bill into law. This change will help DBS compete with cable much more successfully, and give broadcasters considerable bargaining power in their negotiations with cable operators.

Figure 11.5 Overview of the direct broadcast satellite (DBS) service of DIRECTV.

However, the rise of the Internet tilts the balance back towards cable companies because they can provide high-speed Internet service. For DBS companies, the Internet presents a short-term opportunity and a long-term problem. The opportunity comes in the form of downstream Internet service, linked to a telephone return path. For example, DIRECPC delivers datastreams to subscribers via satellite. Customers install an add-in card into their PCs that lets them take in information from coaxial cable. The cable is attached to a 24-inch satellite receiver dish that receives the over-the-air signals.

The most important advantage the system offers is a download speed of 400 kbps. The least expensive residential plan offers service, including Internet access, for $29.99 per month for 25 hours per month;

the next tier is 100 hours for $49.99; and the highest tier is $129 for 200 hours. These prices are not competitive with either cable or telephone company DSL packages, but in areas where it is the only one available, DBS serves a purpose.. The long-term problem is that a GEOS network simply cannot compete in a world that is wired for broadband because of the cost of backhaul.

New Broadband GEOS Systems[59]   Going forward, there are a number of new broadband GEOS coming online. They range from single birds to multisatellite constellations, but most are making some plans to carry the growing broadband Internet traffic. Astra-Net is operated by European Satellite Multimedia (ESM) Services SA, an entity jointly owned by Intel Corporation and Societe Europeenne des Satellite (SES). The multimedia platform transmits DVB material over the Astra satellites, using Ku-band to achieve 38 Mbps data rates. After 2000, Ka band 30/20 GHz transponders offered two-way interactive service and bandwidth on demand, directly to PC users in Europe. Both business and residential markets are targeted.

Astrolink is a $4 billion Ka-band 29/19 GHz project from Lockheed Martin Telecommunications scheduled for launch in 2003. The company hopes to launch the first global, wireless broadband telecommunications service. Plans call for global coverage by nine GEOS in five orbital slots to provide global coverage with data rates between 16 kbps and 9.6 Mbps. Astrolink will provide voice, data, video, and multimedia services. Their Web site is at http://www.astrolink.com.

CyberStar is a Loral Space and Communications project that was licensed in May of 1997.

The system originally entailed plans to utilize 3 Ka-band GEOS, but as of late 1999, CyberStar was essentially a service packager that uses the Loral TelStar bird to reach the United States. The company offers six types of service: eCinema, enhanced business television, digital delivery, narrowcast TV, distance learning, and high-speed Internet browsing. Coverage for Europe, Middle East, and Asia will be available in the near future. Alcatel is a major investor in CyberStar. The two companies will jointly market their services after Alcatel rolls out its $3.5 billion Skybridge LEOS fleet.

EAST stands for Euro African Satellite Telecommunications, a terrestrial network extender. The partners in EAST are Matra Marconi Space, Digimed, Matra Hautes Technologies (France), Nera (Norway), and Aon Space. The $1 billion project is expected to launch in 2001 and begin operation in 2002. It will use a Eurostar GEOS to reach Africa, the Middle East, and parts of southern Europe. EAST will provide low-cost rural voice and data services, plus Internet access at 57.6, 115.2, and 384 kbps.

EuroSkyWay was introduced by Alenia Aerospazio Space Division, a subsidiary of the Finmeccanica Group. It has a license for two orbital slots and transmission in the Ka-band at 30/20 GHz frequencies. EuroSkyWay intends to be the first European satellite network to offer high-speed two-way information capacity to mass market users and broadband bandwidth on-demand to service providers and carriers. The cluster of five GEOS will support ISDN, ATM, IP, DVB, and MPEG standards, with a 45-Gbps capacity.

Plans call for the system to rollout in two phases, beginning with two co-located satellites set to launch in 2001, providing service to Europe, the Middle East, Africa, and some former Soviet block countries. In phase 2, an additional three satellites will extend coverage to Africa and Asia. Services will include Internet and intranet, videoconference, and webcasting. It is envisioned as providing bandwidth for carriers, service providers, and large enterprises. The EuroSkyWay Web site is at http://www.euroskyway.alespazio.it/.

Eutelsat, a Paris-based satellite company with a fleet of fifteen birds, began developing a multimedia digital platform to deliver Internet services to PCs and to allow their customers to engage in data broadcasting. Eutelsat has sophisticated onboard processing capability that allows it to deliver what it terms "microbroadcasting" to reach small markets and still maintain bandwidth efficiency. The company will dedicate a four-transponder Ka-band payload to serve small offices and businesses with interactive IP and DVB service, with capacity available in 2 to 6 Mbps slots and up to 18 carriers per transponder.

KaStar is a $520 million Ka-band 30/20 GHz, 2-GEOS project planned by Lockheed Martin. It will go operational in 2001 to provide broadband ATM data, digital video, and voice at data rates from 64 kbps to 5 Mbps to markets in the United States, Mexico, the Caribbean, South America, and parts of western Europe. Two new GEOS will be built by Space Systems/Loral with sophisticated spot-beam technology, and launched by Arianspace into its two FCC-assigned orbits before 2003, providing 10 Gbps capacity. This will be a giant step towards the company's goal of building a global Ka-band satellite system that delivers low-cost, interactive, on-demand multimedia services to DBS, cable, and other terrestrial network providers. The Web site is at http://www.kastar.net.

Expressway is the name of a Hughes Communication fourteen-bird GEOS fleet that will launch in 2000 at a cost of $3.2 billion, to be in service by 2002. It is designed to provide global high-speed data services over Ka-band at 30/20 GHz. A constellation of twenty LEOS birds will also be launched to complete the network, which will provide global broadband high-speed data services, such as high-speed Internet access, corporate intranet communication, virtual private networks, and multimedia broadcasting. The initial launch will be a North American constellation of two geosynchronous satellites, plus an in-orbit spare. The communications capabilities will include onboard digital processors, packet switching, and spot-beam technology to provide point-to-point communications, direct connectivity without routing through a hub, in addition to broadcasting throughout the service area. The project's Web site is at http://www.spaceway.com.

WEST (Wideband European Satellite Telecommunications) is a $2 billion initiative from Matra Marconi Space, collaborating with Nortel Networks, to develop a satellite-based platform for multimedia services by 2003. It is a broadband interactive communications network that will debut with a Ka-band GEOS, called the WeB for WEST Early Bird, covering Europe and adjacent regions. A complementary constellation of up to nine medium earth orbit satellites (MEOS) will provide additional service and extend coverage as demand justifies network expansion. Broadband applications planned for the network include interactive TV, telemedicine, telebanking and investment services, and interactive gaming and gambling. The first bird will allow onboard crossbeam connection to provide least-cost routing. After 2004, later satellites will have onboard ATM switches to address and rout packets to any location within the constellation's footprint and interconnect seamlessly with terrestrial networks.

Finally, GE Americom has filed with the FCC for 5 Ka-band slots in a $4 billion project using nine GEO satellites from Alcatel. The company is one of the largest global operators, with a fleet of thirteen satellites. In 1999, it acquired Columbia Communications, which operated three birds, and GE Americom will launch three additional GEOS in 2000.

MEOS and LEOS: New Satellites for a New Century

As mentioned earlier, the Clarke orbit, 22,247 miles above the earth, is the distance from which a satellite is stationary over the same place at the equator. This positioning allows global coverage with only three satellites using inexpensive fixed antennas on receiving earth stations. Advances in computer processing power now make it possible to locate satellites closer to earth and to send and receive signals from a much smaller area. When the message is addressed to a location outside the range of one satellite, it is switched to another satellite in the fleet, one that can reach the designated address.

Low-earth orbits extend from 644 km (402 miles) to 1,600 km (1,000 miles); and medium-earth orbits start after the 1,600 km limit. MEOS and LEOS have advantages and disadvantages, and in recent years serious disagreement has arisen over which approach produces better results. For example, the higher the satellite, the larger the area covered by the satellite footprint, so the fewer the number of satellites are needed in the fleet—this favors the MEOS. However, a larger distance also means longer latency for signal traveling time, so for very fast data exchanges, the LEOS are better.

The European Space Agency commissioned the Mitre Corporation to study LEOS systems.[60] They found one technical problem, in that to achieve global coverage some satellites would have to be placed at very low elevation levels, making it difficult for users to link with them during certain times of the day.[61] MEOS and LEOS technology are well-understood; the former U.S.S.R. launched a number of both types of satellites. Two MEOS systems are in the works, as well as several LEOS systems that will deploy many small satellites (smallsats) for global coverage—the so-called Big LEOS systems. The Little LEOS fleets have only narrow bandwidth capability. Most plans for such fleets will offer global service and plan to tap the potentially huge market of people unable to get wired telephone service, or who prefer mobile service.

The cost of a LEOS or MEOS system ranges from $200 million to several billion dollars. Such a wide disparity in cost exists because each system is quite different. The total startup expense depends on how many satellites are used, the orbit (lower orbits require less power), the number of transponders, and the number of links to other satellites.[62]

The Federal Communications Commission began accepting applications for multisatellite systems in January, 1995. The agency issued the first four licenses for service links (downlinks, satellite-to-earth) in March of 1995. Licenses for feeder links (uplinks, earth-to-satellite) and intersatellite links were issued after the ITU's World Administrative Radio Conference, held in October, 1995. International approval was given for nonexclusive use of the Ku-band (10–18 GHz) for fixed satellite services by nongeostationary (NGSO) systems at WRC-97, as long as they do not interfere with GEOS service and terrestrial fixed service systems (FSS). In 1999, Ku-band spectrum sharing was accepted by the FCC, despite objection by GEOS operators.[63]

In the next few years, satellites will be extending their signals into the V-Band and Extremely-High-Frequency range of the spectrum. These frequencies will be used primarily for bandwidth-intensive services, like aggregated network trunking lines, because they afford such enormous capacity. Allocation is already underway, but there will be little more than regulatory activity for the near future of two or three years. For example, it took three years for the mobile satellite industry to move Ka-band spectrum from the regulatory domain into the world of commercial service and products.

Broadband MEOS Systems   The earliest MEOS systems were Orbcomm and Odyssey, both designed as multiple-bird communication systems, supporting telephone, paging, and other narrowband services. Orbcomm is still in operation, but Odyssey closed down in December, 1999. The backers transferred their holdings and assets to ICO, another fleet operator providing narrowband services.

ICO began its existence in 1995 as a plan for a high-orbit MEOS fleet that would market global telephone service, a competing scheme to OrbComm, Odyssey, and the LEOS system, Iridium. The ICO scheme called for satellite placement in the high portion of the MEO, allowing the company to cover the earth with only ten relatively slow-moving satellites. This plan meant that there wouldn't need to be as many satellites or so many handoffs between them as other fleets demanded.

But narrowband services have not proved to be viable in the marketplace. Terrestrial networks have flourished, bringing the cost of bandwidth down. So, like its competitor Iridium, ICO defaulted on the interest payments to service its $43 million bond debt, and filed for bankruptcy protection in August of 1999. While in "bankruptcy hell," ICO was able to convince its backers to put up another $225 million, allowing the company to reformulate its business plan, retrofit its satellites, and offer broadband service. In May, 2000, Craig McCaw and Teledesic made a $1.2 billion investment and changed the name to the reconstituted company, New ICO.[64] This financial rescue ensures that the company's twelve satellites will be launched in 2000. ICO's Web site is at http://www.ico.com.

LEOS Systems   Just launching a Big LEOS fleet was a technical triumph for the satellite industry, despite some spectacular launch explosions and rather poor performance on the business side. When Iridium declared bankruptcy in August of 1999, it led some investors to question the viability of Big LEOS fleets. Iridium was licensed by the FCC in March, 1995, a creation of an impressive corporate consortium led by Motorola and McDonnell-Douglas, that also included Sony, Mitsubishi, Lockheed, and Sprint. The $3.4 billion Iridium system put 66 satellites 420 miles up into space to offer satellite phone and pager services around the world. Ultimately, the project cost $4.4 billion outright, and half again that much was needed as the company foundered financially.

Iridium began operation in November, 1998, but it never really took off. Initially, there was a shortage of handsets. Then there were complaints that the cost was too high. The first phones cost $7,000 and per-minute charges ran as high as $9.00. The phones, when they finally arrived, were too heavy and looked clunky. In an eerie parallel to its competitor ICO, Iridium filed for bankruptcy protection in August of 1999, owing $90 million dollars.[65] Unlike ICO, however, investors did not come to the rescue. For one thing, the satellites were in the air—too late to modify them for broadband service or any other kind of service. Currently, Iridium is restructuring under Chapter 11 protection. Their Web site is at http://www.iridium.com.

A second LEOS system is GlobalStar, whose main proponents are Loral/Qualcomm, joined by Hyundai of South Korea. Service began in October of 1999. Before it is completed, the network will cost $1.8 billion and place 48 LEOS 750 miles in space. It will start out providing only narrowband telephone services, using CDMA modulation, with the capacity to support 7.5 million of the potential 40 million subscribers. The amazing part is that GlobalStar is up and running at all. It was delayed by a launch disaster when twelve satellites were lost, with the failure and explosion of the Zenit-2 rocket in September 1998. A few months later, the next launch generated some frictions between the United States and Russia over inspection of the military base launch pad, but these were resolved in January of 1999 and launches continued.

At the commencement of service in late October, 1999, 44 satellites had been launched into orbit. The least expensive phone cost $880 and airtime was set at $1.50 per minute for national calling. Service will rollout throughout 2000 to about 54 nations on a country-by-country basis as all the system's birds came online in late 1999. Ultimately the system will be able to carry any sort of data transfer, including video. From the outset, this system will be interactive.

The GlobalStar satellite is simple and proven. An antenna, two solar arrays, and a magnetometer rest on a trapezoidal-shaped body. The satellite orbits at 1414 km or 876 miles. The network architecture is a "bent-pipe" design that requires little processing onboard the satellite. Sometimes this design is called "the big repeater in the sky," where a satellite merely forwards a call to a satellite dish located at a gateway, which then routes it through the PSTN.

The GlobalStar system is designed to provide affordable satellite-based digital voice services to a broad range of subscribers and users. GlobalStar will meet the needs of cellular users and global travelers who roam outside of cellular coverage areas, with a particular focus on residents of underserved markets who will use GlobalStar's fixed-site phones to satisfy their needs for basic telephony. But like its predecessors, Iridium and ICO, the company is besieged by financial difficulties. GlobalStar's Web site is at www.globalstar.com.

Constellation, formerly known as ECCO, is another narrowband telephone service that plans to provide satellite telephone service to underserved regions of the world. The major investors are Orbital Sciences Corporation, Bell Atlantic Global Wireless, Inc., Raytheon Systems Company, and SpaceVest Fund. The $450 million system consists of twleve LEO satellites at an altitude of 2000 km. Brazil and South America are seen as the focus of the coverage. The Web site for Constellation is at http://www.cciglobal.com/.

Skybridge is Alcatel's $3.5 billion entrant into the Big LEOS market. It calls for a constellation of 80 Ku-band birds at 1,469 km (913 miles), a low-earth orbit that holds latency at 30 milliseconds, permitting rapid interactivity. The system will incorporate a sophisticated handover procedure between satellites, with traffic running transparently from user terminals through the satellites, without any on-board processing. Approximately 200 gateway stations with a 234-mile radius (350 km) are planned for worldwide coverage. The gateways will interface seamlessly with the terrestrial networks via an ATM switch. Residential and enterprise users—companies, factories, hospitals, or schools—will have small, low-cost terminals, priced at about $700 for a residential terminal.

The SkyBridge system is designed primarily to address the broadband local access market, specifically "last mile" connectivity to the home. The company cites estimates of 250 to 500 million Internet users by 2005, resulting in 72 million broadband residential users and a $100 billion business market for such services. The applications for broadband would be as high-speed Internet, corporate intranets/extranets, LAN/WAN remote access, e-commerce, videoconferencing, and interactive entertainment. The Web site is at http://www.skybridgesatellite.com/.

Hughes' Spaceway system will also have a LEOS component. Spaceway NGSO (nongeostationary orbit) is a 20-satellite constellation that will operate in the 30/20 GHz Ka-band frequency range. The satellites are intended to incorporate multiple-beam antennas, and they have digital processors to switch traffic among beams and intersatellite links (ISLs) that connect the satellites. This means that a signal received by one satellite can be relayed directly back to the same beam, switched to another beam, or relayed by ISLs to other satellites.

Spaceway will offer nearly global service in North and South America, Europe, Africa, the Middle East, and the Asia Pacific region. It will offer customers bandwidth on demand to give businesses and consumers fast access to the Internet, intranets, and local area networks. Its 66-centimeter-diameter transceiver antennas uplink at a data rate as high as 6 Mbps. The Hughes board of directors approved a $1.35-billion-dollar investment to build the North American piece of the network. Two satellites will launch so the company can begin service in the Ka-band in 2002.

Teledesic, the brainchild of Bill Gates and Craig McCaw, is the most ambitious of all the Big LEOS systems. By 1998, it has evolved to a $9 billion undertaking that would orbit 288 birds about 438 miles above the earth to offer 24-hour seamless coverage to over 95% of the Earth's surface, reaching almost 100% of the Earth's population. However, developments have conspired to make it more likely that Teledesic would go forward, but less certain as to its ultimate configuration. For one thing, several other projects have been merged into Teledesic, beginning with Motorola's Celestri system. As a result, Motorola became the prime contractor and has a 26% stake in Teledesic. This project changes as it is developed. Check out the most recent configuration at the company's Web site at http://www.teledesic.com.

Celestri was itself a $4 billion plan that called for 9 Ka-band GEOS vehicles. It absorbed two other Motorola space ventures, the Millennium and M-Star projects. Millennium was a $2.3 billion network of four GEOS backed by Motorola and Vebacom to provide interactive video and data broadcasting. M-Star was Motorola's $6.1 billion 72-LEOS constellation. More information about Celestri is at http://www.celestri.com.

There are a number of sophisticated elements to Teledesic, and they were spelled out in unusual detail from the beginning. It was conceptualized as an interlinked LEOS constellation that would provide global access to a broad range of voice, data and video communication capabilities. The boomer billionaires put together an array of global partnerships, so seamless integration with many other networks was envisioned at the outset. Initially, Teledesic was designed to support as many as 1,000,000 two-way high-speed connections, with the capability of handling millions of simultaneous users. Moreover, the architecture was deliberately scaleable, allowing it to grow to carry much higher capacity without fundamentally changing the architecture, spectrum usage, or user terminals.

The LEOS strategy was indicated because of its low latency, the reliability of its connections through coverage redundancy, the low power requirements to transmit both uplink and downlink signals, and the small size of its antennas. The Ka-band was chosen because of its high bandwidth capacity. Each satellite was conceptualized as a node in a fast packet-switched network, and plans called for each one to be connected to eight adjacent birds via intersatellite links, forming a nonhierarchical geodesic, or mesh network, tolerant to faults and local congestion. The overall network would use ATM switching technology.

The Teledesic system software was designed to maintain a database of the earth's surface mapped to a fixed grid of approximately 20,000 "supercells" composed of 9 cells each. Each supercell is a square, with 160-km sides arranged in bands that are parallel to the Equator. There are approximately 250 supercells in the band at the Equator, and the number per band decreases with increasing latitude. Frequencies and time slots are associated with each cell and managed by the current-serving satellite. As long as the receiving terminal remains in the same cell, it is supported by the same channel assignment for the duration of the call—no matter how many satellites and beams are involved. This handling reduces channel reassignments and processing overhead for frequency management and handoff.

A variety of multiple access methods are used to aggregate the traffic, depending on the link. This combination of earth-fixed cells and multiple access methods results in very efficient use of spectrum. The Teledesic system will reuse its requested spectrum over 350 times in the continental United States and 20,000 times across the Earth's surface.

The Future of Broadband Satellite

The bankruptcy of Iridium and ICO, the financial problems of GlobalStar, the ubiquity of terrestrial networks, and the fierce competition from other satellite companies cast doubt on the viability of this and other planned satellite fleets. And they are very expensive. Nevertheless, the vision of a globally ubiquitous two-way system of broadband communication rests on this technology.

In the 1990s, the worldwide wave of deregulation affected the satellite industry in many ways. One of the most significant effects was the launch of many privately-owned satellites—by the end of the decade, the industry was largely commercially rather than governmentally operated. The growth of the market, private investment, and consolidation resulted in the emergence of several "mega-operators."[66] For example, in 1997 Hughes Galaxy and Panamsat combined to become one of the largest companies involved in distributing cable programming, providing private satellite communication services, and offering programming packages directly to consumers. Other major powerhouses are Intelsat, Loral, GE Americom, New Skies, SES Astra, and Eutelsat.

Deregulation in many areas of the world also led to an increase in the number of international TV channels. Throughout much of the 1990s, deregulation created a boom for satellite companies, but by 1998, there was excess transponder capacity, particularly in Asia and Africa, in spite of the increase in Internet traffic. Demand for transponders is still high in the United States and Europe, however.

More birds will be coming online. Euroconsult, a research firm that specializes in the satellite industry, estimates that between February, 1998, and the end of 2007, about 276 to 348 satellites will come on the market, worth approximately $29.7 to $38.6 billion.[67] Eighty-six satellites, or 25% to 31% of the total market, have already been ordered. The continuing expansion in the number of transponders leads some analysts to predict flat revenues for the satellite industry until the middle of the next decade. Indeed, the emergence of untried Ka-band technology poses a risk to several operators. Moreover, the high-profile failures of several satellite fleet systems, such as Iridium and ICO, and the ongoing weakness of Global-Star demonstrate that there is uncertainty about the marketability of services. The considerable financial risk has made some investors wary of backing such ambitious undertakings, putting it in doubt as to whether some of them will even get off the ground.

However, Pioneer Consulting, a research firm that specializes in satellite communications has a much more optimistic view. In their 1999 report, "The Broadband Satellite Market, 1999–2008,"

these analysts write that the future for broadband satellite services through the next decade will be strong and vibrant, especially after the launch of two-way, high-bandwidth services in the Ka-band.[68] Pioneer's projections for global business and residential subscribers are presented in Tables 11.7 and 11.8. The global business subscriber base for broadband satellite services will increase from 30,000 businesses in 1999 to almost 7 million in 2008.

Table 11.7 Estimates of Global Business Demand for Broadband Satellite Services, 1999–2008 (Millions)

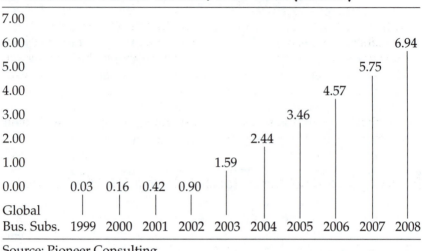

Global Bus. Subs.	1999	2000	2001	2002	2003	2004	2005	2006	2007	2008
	0.03	0.16	0.42	0.90	1.59	2.44	3.46	4.57	5.75	6.94

Source: Pioneer Consulting.

Table 11.8 Estimates of Global Residential Demand for Broadband Satellite Services, 1999–2008 (Millions)

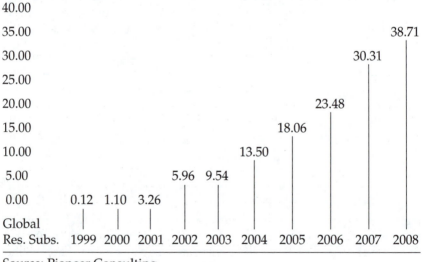

Global Res. Subs.	1999	2000	2001	2002	2003	2004	2005	2006	2007	2008
	0.12	1.10	3.26	5.96	9.54	13.50	18.06	23.48	30.31	38.71

Source: Pioneer Consulting.

According to Pioneer, global residential demand shows the same trends as business demand.

They project that global residential subscribers will increase from close to 100,000 in 1999 to over 39 million in 2008. Both Table 11.7 and Table 11.8 show slow growth for satellite access services over the next two to three years, but after the launch of Ka-band service in 2003, subscriber totals increase dramatically. While much of the demand will come from North America, the demand from Europe will also grow substantially. The Asia Pacific region will be a distant third-largest market in terms of the number of subscribers, and the most significant growth will come sometime beyond the forecast period.

Proto-Space and Commercial Air-Space Platforms

Proto-space is the area in the sky above commercial air space. It starts at 13 miles above earth until outer space is reached at 70 miles out. The projects planned to deliver communications from above always sound like science fiction. Perhaps they are, but their proponents seem quite serious.

It's a Bird, It's a Plane . . .
No, It's a Communication Network!

Angel Technologies has put forward an intriguing, innovative idea, although the company has not published a rollout schedule.[69] Angel's HALO Network is essentially LMDS in the sky. It is planned as a wireless broadband metropolitan area network that would interconnect as many as hundreds of thousands of users, giving each one multi-Mbps data rates for multimedia file transfers and interactive applications.

A piloted, fixed-wing, high-altitude long-operation (HALO) aircraft will circle at 52,000 feet, above commercial airline traffic and terrestrial weather conditions. It will carry equipment that allows it to act as the hub, or base station, of a wireless broadband communications network. An aircraft provides an excellent line-of-sight 75-mile footprint to offer ubiquitous high-speed, broadband access throughout an entire metropolitan center, including its suburbs, and beyond, as shown in Figure 11.6.

The HALO Network can complement satellite systems by concentrating local traffic for uplink and serve as a spot beam to transmit data locally on the downlink. Since the HALO aircraft is so much closer to earth, it requires less power and operates with far less latency. In addition, the network can be extended on a market-by-market basis. Finally, the aircraft and equipment can be maintained, serviced, repaired, and updated on a regular basis.

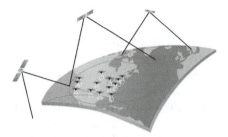

Figure 11.6 Angel Technologies' HALO aircraft is an LMDS network in the sky.

HALE UAV Platforms: GEOS at 14 Miles Up

This chapter ends with even more speculative technologies: HALE, UAV, RPV . . . and balloons.[70] These acronyms stand for "high-altitude long-endurance," "unattended autonomous vehicle," and "remotely piloted vehicle," respectively—and balloons (we all know what they are).[71] The HALE, UAV, and RPVs are similar in concept to Angel Technologies' HALO Network, except that the aircraft would not be piloted. Strong, light, solar-powered HALE UAVs (or RPVs) would fly for several days at a time using their power to stay above a fixed location on earth. They would be controlled from earth and destroyed in the event of malfunction. Transponders on these craft would support between 10 and 100 beams, with the capacity for hundreds of thousands of telephone lines serving millions of subscribers. The allowable operational coverage would be a diameter of 500 kilometers, or about 300 miles. (See Figure 11.7.)

Two other high-altitude systems are just coming out of research labs. In Japan, a study by the national telephone company confirmed the feasibility of a 270-meter airship running on solar and fuel cells that could carry a ton of communications equipment. The study suggests that a prototype could be readied sometime between 2001 and 2002. RotoStar is a solar-powered unmanned aircraft from Silver Arrow. Designed for the Israeli government for military use, it would hover at 70,000 feet for as long as six months. Solar panels would heat argon gas and provide fuel for a thermodynamic propulsion engine. More research would be required to develop the materials for solar storage.

Finally, former Secretary of State Alexander Haig is a founder of Sky-Station, which is making helium-filled, lighter-than-air vehicles (balloons!) to float above metropolitan areas at altitudes between 22 and 250 kilometers. This is an enormous version of existing vehicles that currently measure ozone-layer depletion. Although somewhat vulnerable to solar winds at 70,000 feet, the SkyStation would incorporate some kind of propulsive

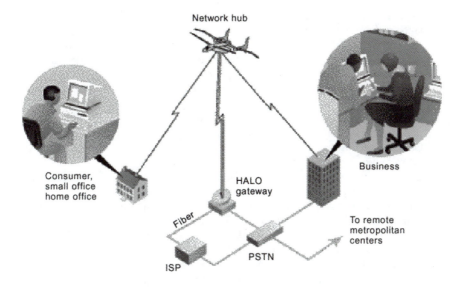

Figure 11.7 The network-in-the-sky footprint. Source: Angel Technologies.

power to stay on course. The company has secured FCC approval for V-band spectrum and hopes to deploy its first system in 2000. The communication systems onboard the balloons will provide 2 Mbps on the uplink and 10 Mbps on the downlink, with estimated costs of a few cents per minute. SkyStation's strategy is to ally with existing national providers worldwide for interconnection, and it has already penned fifteen joint-venture agreements. Their Web site is at http://www.skystation.com/.

All these platforms are new and experimental. However, they demonstrate the recognition that there is ongoing demand for last-mile connectivity. It may well come from the widening world of wireless communication, as ever-higher portions of the electromagnetic spectrum become available for communication uses.

So far, we have examined the features of wireless technologies, and seen how they fit together in complex wired and wireless systems. In the next chapter, we will look at the information that flows through these advanced systems, and how the adoption of technologies and content will affect people's lifestyles. Welcome to the digital living room.

Notes

1. Cahners In-Stat, "Wireless cable report," *Cable Datacom News* (April 30, 1999). See http://www.instat.com.

2. A good site for global listings of GSM networks is at http://www.cellu-lar.co.za/gsm-worldnetworks.htm.

3. P.D. Henig, "Qualcomm, Ericcson unite over new wireless standard," *Red Herring Online* (March 29, 1999). Available online at http://www.redherring.com/insider/1999/0329/news-qualcomm.html.

4. D. Gardiner, et. al. "Third-generation mobile: Market strategies," Ovum (October, 1999). To get more information about this proprietary market research report, contact Ovum in the United States at 781-246-3773, or in London: +44 (0) 20 7551 9000. The Web site is at http://www.ovum.com.

5. Get current information on WAP and activities concerning the standard at http://www.wap.com.

6. Considerable information about this partnership is posted at http://www.cisco.com/warp/public/756/partnership/motorola/.

7. Press release is posted at http://www.waveaccess.com/news/lucentac.htm.

8. "Microsoft and British telecom join forces against symbian alliance," *EPOC Times* (February 10, 1999). Available online at http://pdacentral.flash-net.it/5alive/Archives99/Feb10_613.htm.

9. E. Clampett, "Qualcomm, U.S. West test high-speed wireless solution," *Internetnews.com* (February 1999). Available online at http://www.internetnews.com/bus-news/print/0%2C1089%2C3_65391%2C00.html.

10. G. Zukav, *The dancing Wu Li masters*, New York: William Morrow (1979):12–13.

11. J.G. Truxal, *The age of electronic messages*, New York: McGraw-Hill (1990):309.

12. J. Hanson, *Connections: Technologies of communication*, New York: HarperCollins College Publishers (1994):100.

13. Ibid.:127.

14. The Television Advertising Bureau publishes "Television Facts" on its Web site at http://www.tvb.org/.

15. J. Flanigan, "TV networks evolve from dinosaurs to darlings," *Los Angeles Times* (October 5, 1994):D1, 2.

16. PricewaterhouseCoopers, "Digital television '99: Navigating the transition in the United States." Position paper from the PricewaterhouseCoopers Entertainment/Media Group (1999):35–6.

17. "U.S. Wireless Broadband: LMDS, MMDS, and Unlicensed Spectrum." Strategis Group proprietary report (December 6, 1999). Press release available at http://www.strategisgroup.com/press/pubs/wbb99.html.

18. C. Stern, "FCC moves to strengthen wireless cable," *Broadcasting and Cable* (June 13, 1994):11.

19. D. Blankenhorn, "Wireless cable operators form alliance," *Newsbytes* (June 22, 1994).

20. C. Stern, "Telcos hedge bets with wireless wagers," *Broadcasting and Cable* (May 1, 1995):14. And M Berniker, "PacTel joins wireless migration," *Broadcasting and Cable* (April 10, 1995):35.

21. E. Douglass, "Wireless cable may prove golden alternative to copper wires," *Los Angeles Times* (October 11, 1999):C4.

22. Trials listed by Cable Datacom News at http://www.cabledatacomnews.com/wireless/cmic11.html.

23. Ibid.

24. G. Gilder, "The new rule of the wireless," *Forbes ASAP* (April 11, 1994):99–110.

25. A resource page for LMDS technology is at http://www.lmdswireless.com/.
26. J. Skoros, "LMDS: Broadband wireless access," *Scientific American* (October 1999).
27. E. Steckley, "Broadband wireless access: Dawn of a new era," *Telecommunications Online* (February 1998).
28. "Cisco drives industry standards for broadband wireless internet services," Cisco press release (October 26, 1999). Archived at http://www.cisco.com /warp/public/146/october99/26.html.
29. K. Reece, "Modeling the Propagation and Latency Effects of the Integration of ATM in a Seamless Terrestrial and Satellite-Based Wireless Network." Project at the Center for Wireless Telecommunications, Virginia Tech, described at http://www.cwt.vt.edu/.
30. The Web site for N-WEST is at http://www.nwest.nist.gov/.
31. Liberty Cellular will be using Nortel Networks equipment. The press release is at http://www.nortelnetworks.com/.
32. "Home Telephone selects Newbridge for LMDS trial." Posted at http://www.mainstreetexpress.com/doctypes/customerstory/hometel.jhtml.
33. "Ericsson, Nextlink launch wireless broadband trial," *IDG News* (September 22, 1999). Posted online at http://www.idg.net/crd__85607.html.
34. See the company's Web site at http://www.idg.net/crd__85607.html.
35. WinStar's home page is at http://www.winstar.com/home.asp.
36. Press release posted on Canada News Wire at http://www.newswire.ca /releases/August1998/19/c3402.html. The company's Web site is at http://www.maxlink.ca/.
37. I. Brodsky, "What will LMDS hatch?" *Telephony* (September 1996).
38. "LMDS and the Buildout of Broadband Wireless Networks: Worldwide Market Opportunities for LMDS, MMDS and MVDS Technologies." Pioneer Consulting proprietary research (August 1998). For more information, contact company at http://www.pioneerconsulting.com.
39. S.E. Coran, "Low-power subscription television," *Wireless Broadcasting Magazine* (April 1995):14–16.
40. M.J. Banks, "Low-power television," in A.E. Grant, ed., *Communication technology update,* 3rd ed., Boston, MA: Butterworth-Heinemann (1994):107–115.
41. M.J. Banks and M. Harice, "Low-power television 1990 industry survey." Unpublished report of the Community Broadcasters Association (December 14, 1990).
42. Federal Communications Commission, MM Docket No. 99-292 Establishment of a Class A RM-9260 Television Service NOTICE OF PROPOSED RULE MAKING. Washington, D.C.: Federal Communications Commission (September 29, 1999).
43. Ron Nutt. Telephone interview, May 2000.
44. C. Boeke and R. Fernandez, "Via satellite's global satellite survey satellite industry trends and statistics," *Satellite Today* (July 1999). Available online at http://www.satellitetoday.com/viaonline/backissues/1999/0799cov.htm.
45. E. Hartenstein, president and CEO, DIRECTV. Personal interview, El Segundo, CA, April 1999.
46. The Satellite Business News Web site has a lot of information. It's at http://www.satbiznews.com.

47. F. Baylin, *Miniature satellite dishes: The new digital television*, Boulder, CO: Baylin Publications (1994).
48. "Buy rates skyrocket for DBS pay-per-view," *Interactive Video News* (April 3, 1995):1–2.
49. R. Akwule, *Global telecommunications*, Newton, MA: Focal Press (1992):57–87.
50. K. McConnell, "The future of the earth station industry: a billion dollar bet," *Satellite Today* (May 1999). Available at http://www.satellitetoday.com/viaonline/backissues/1999/0599cov.htm.
51. M. Stein, "Satellites: Companies, nations fight for spots in space," *Los Angeles Times* (September 20, 1993):A1, A16.
52. J. Sanchez, "Two phone giants open cable TV's door," *Los Angeles Times* (February 12, 1995):D1, 2.
53. C. Platt, "Satellite pirates," *WiReD* (August 1994):8, 122.
54. J. Careless, "Pumping up sales: VSATs and the automotive industry," *Satellite Today* (July 1998).
55. G. Lawton, "Deploying VSATs for specialized applications," *Telecommunications* 28:6 (June 1994):27–30.
56. M. Crossman, "Dollars and aense: Cyclical growth ahead for satellite manufacturing," *Satellite Today* (June 1999).
57. "Direct Broadcast Satellite: 10 Million and Still Growing." Yankee Group study (August 1999). The company's Web site is at http://www.yankeegroup.com/.
58. Theresa Foley, "Mega operators: Big fish in a very big pond," *Satellite Today* (November 1999). Available at http://www.satellitetoday.com/viaonline/issue/1099cov.htm.
59. "Next Generation Broadband Satellite Networks." A proprietary study by Pioneer Consulting (October 1999).
60. See company site at http://www.mitre.org.
61. R. Tadjer, "Low-orbit satellites to fill the skies by 1997," *Computer Shopper* 15:3 (March 1995):49.
62. J. Krauss, "NGSO satellites: Creating new spectrum capacity," *CED* (March 1999).
63. Read the FCC order denying a petition filed by Hughes and GE Americom to turn back spectrum sharing with the new LEOS systems at http://www.fcc.gov/Bureaus/International/Orders/1999/fcc99375.txt.
64. See news release on the Teledesic site at http://www.teledesic.com/newsroom/articles/2000-05-17%20ico.htm.
65. "Iridium files Chapter 11," *CNNfn Online* (August 13, 1999). Available online at http://cnnfn.com/1999/08/13/companies/iridium/.
66. T. Foley, op. cit.
67. Study cited in M. Crossman, op. cit.
68. Ibid.
69. The company's Web site is at http://www.angeltechnologies.com/.
70. Network Computing published an excellent series of articles that cover GEOS, MEOS, LEOS, and these newer platforms. They are online at http://www.networkcomputing.com/905/905f2side9.html.
71. J.N. Pelton, "Geosynchronous satellites at 14 miles altitude?" *New Telecom Quarterly* (2nd Quarter, 1995):11.

Living Digitally

12

As new technologies emerge, the enthusiasm of consumers in accepting them seems endless—radio, films, television, portable radios and TVs, home theaters, sound systems, cable, VCRs, camcorders, CD-ROM players, PCs, pagers, wireless and mobile telephones, DVDs, MP3 players, and digital still and video cameras. In the past two decades, there has even been an explosion of all kinds of traditional print titles—books, magazines, and journals. Media and technology aren't just something we live with, they are threads in the fabric of life itself.

Home Is Where You Hang Your @

The last few chapters dealt with the implementations of entire communication systems. We have now reached the edge of the network and beyond—the last mile to the home and, increasingly, the home office. Large organizations, businesses, schools, and government agencies have always been able to pay for fast, high-quality communication facilities when they were required. But households and even small businesses have had to settle for one, or perhaps a few, 4-kHz telephone lines.

Now that's changing. Higher speed communication is beginning to extend to the residential market. Cable systems are finally reaching large numbers of people. According to Pioneer Consulting, in 1998 the total subscriber base of broadband access systems was infinitesimal, just 750,000 subscribers. In August, 1999, the cable industry installed its one-millionth cable modem in North America, about 70% of them in the United States. High-speed access was available to about 32 million households, or 30% of all cable homes passed. On average, more than 2,500 new cable modem customers were installed every day, so that 500,000 went into service in the first half of 1999. By November, 1999, there were 1,403,333 subscribers, so cablers will have gone into nearly three-quarters of a million homes in just one year. Two companies provided access to

90% of cable customers, Excite@Home claiming a 59% market share, and RoadRunner serving about 32% of subscribers.[1]

Broadband access from telephone companies is also rising rapidly. Pioneer Consulting projects that the number of DSL subscribers in the United States will grow from 760,000 in 1999 to more than 12 million in 2003.[2] Local phone company Bell Atlantic adopted a plan to make high-speed Internet access available to 21 million customers by early 2000, requiring the company to equip more than 1,000 central offices with DSL technology. Telephone company SBC Communications established Project Pronto in October of 1999, allocating $6 billion to deliver broadband to an estimated 77 million Americans, about 80% of its Ameritech, Nevada Bell, Pacific Bell, SNET, and Southwestern Bell customers by the end of 2002.

By 2007, Pioneer forecasts that the worldwide number of subscribers to broadband access will reach 136.05 million, with 48.33 million of them being located in North America. Online access is increasing even faster. In September, 1999, Nua Internet Surveys estimated there were 201 million worldwide Internet users, increasing to about 350 million by 2005.[3] Many of the new subscribers will live in Europe, says research firm IDC (International Data Corporation), forecasting that 35% of Western Europeans will be using the Web by 2002.[4] In the United States, PC ownership and online access continues to grow, as shown in Table 12.1.

Indeed, access to the Internet is considered so important that a number of cities and towns plan to extend service to all residents, including Lynchburg, Virginia; Harlen, Iowa; Glascow, Kentucky; Newnan, Georgia; and Tacoma, Washington. In Pelham, Minnesota, the East Ottertail Telephone Company has wired all the homes in town. And in Sweden, the city-owned communications company Stokab is installing over 1100 km of optical fiber around the city.

There's No Place Like http://www.home.net

The world is getting wired, the local area is getting wired—and so is the home. Extending a communication network into the home, whether it is

Table 12.1 Market Penetration of Home Computers and Related Products, U.S. Households

	Jan. 1999	July 1998	Jan. 1998	July 1997	Jan. 1997	July 1996	Jan. 1996	July 1995	Jan. 1995	July 1994
Computers	50%	45%	42%	39%	37%	36%	35%	32%	31%	27%
Online	33%	27%	23%	19%	17%	14%	11%	9%	7%	6%

Source: Odyssey Home Front survey.

wired or wireless, begins with a network interface unit (NIU), the central point where information enters and exits. In the near future, many homes will be wired with home networks. And as the home gets more and more networked, it may require many of the same services and equipment that business establishments do. For example, some users will want to utilize a gateway, which will perform security functions such as locking electronic intruders out of their computers, routing messages around the home network, and serving content out to the Internet. An example of this kind of technology is made by 2Wire at http://www.2wire.com.

So why would anybody want a network in their home? According to IDC research, about 17 million U.S. households own more than one computer, and that number will rise to nearly 23 million by 2002.[5] A home network allows multiple computers to share access to peripherals like printers, scanners, fax machines, and storage units, to leverage the parallel processing power of their machines, and to transfer files back and forth between them.

If the home has broadband service, all the networked machines can draw on the bandwidth resources to let more than one user surf the Web simultaneously. It also makes it possible to network entertainment and communication systems, as well as other consumer appliances. For example, the time is not far off when people will be able to turn on their ovens from work via remote access to their computer if the stove is networked to it. These functionalities led IDC research to estimate that while only 1.9 million homes had actually installed a home network by the end of 1999, they will exist in almost 12 million homes, or 12% of all U.S. households over the next few years.

There are dozens of proprietary systems offered by individual companies, as well as a confusing array of standards under development. Wireless technologies exert considerable allure because of the "no strings attached" untethered mobility they offer. Proposals for wireless standards include HomeRF, Bluetooth, W-Ethernet, and Wi-LAN. On the wired side, HomePNA uses existing telephone wiring for data signals, standard Ethernet over copper wire or coaxial cable, and networking over power lines in the home.

Other pieces of the puzzle are also falling into place. An IBM-led coalition of companies is collaborating on wiring standards for newly built homes. Another group is working on a common set of software interfaces to insure that products will work regardless of what kind of network they are attached to. Hewlett-Packard has developed a communications protocol that will work across all of the various systems, and Honeywell plans to introduce an Internet-based applications package to let users control household appliances. Table 12.2 shows the various schemes for home networking.

An important group for making this product category consumer-ready is the Wireless LAN Interoperability Forum. The WLIF is a world-

Table 12.2 Home Networking Systems

Standard	Bandwidth (bit rate)	Web site address
Wireless		
Bluetooth	1 Mbps	http://www.bluetooth.com/
HiperLAN	2 Mbps	http://www.wlif.com
HomeRF	1.6 Mbps	http://www.homerf.org/
(SWAP)	10 Mbps (in development)	
W-Ethernet	11 Mbps	http://computer.org/student/looking/summer97/ieee802.htm
Wi-LAN	46 Mbps[a]	http://www.wi-lan.com
Wired		
HomePNA	1–10Mbps	http://www.homepna.org
Ethernet (IEEE802.3)	1 Mbps	http://www.ieee.org
Powerline	—	
Independent Platform		
Home Application Programming Interface Working Group	NA	http://www.hapi.org

[a]Demonstrated by Philips sending two live MPEG-2 streams to separate receivers, this technique actually delivers 23 Mbps to the user.

wide not-for-profit organization that was established in 1996 to promote wireless LAN technology and to ensure interoperability of equipment from multiple manufacturers. Its members are companies that develop, produce, and market wireless LAN products or services. It tests products and certifies that they meet overall standards.

Schemes to wire homes are in the works as well. The Home Phone-line Networking Association is a group of 97 companies that developed a 1-Mbps solution to send data over telephone lines in 1998. The home owner plugs a device into a telephone jack, the computer into the device, and uses similar equipment to recover the data in another room. As long as the home is well supplied with jacks, the technique will work well. HPNA is now working on a second-generation 10-Mbps solution.

Of course, it would be much easier to network homes if they were wired for data as they were being built. A consortium of 15 companies, including IBM, Lucent, and Intel, have created a program called "Wiring America's Homes." The group has developed a standard for what they say is an easy and affordable system for unified wiring in new homes that will allow computer control of heating, lighting, and other appliances, as well as provide Internet access. (See Figure 12.1.) These companies believe that by 2004, almost half of all home owners buying a new home will want these $750–$2,000 wiring systems put in as a needed feature.

Home networking software is coming along, too. Hewlett-Packard has designed a protocol package called HP JetSend Communications Pro-tocol. The popularity of the company's printers has led it to develop the software that uses wireless infrared technology to send data to devices that are near one another. There are already more than 3 million JetSend-enabled products in the marketplace.

Another software initiative is the Home Application Programming Interface (HAPI) Working Group, involving such companies as Honey-well, Compaq, Intel, Microsoft, and Mitsubishi, which are currently pre-paring a common set of "application programming interfaces," or APIs, for networked home products that will work regardless of the type of net-work or protocol that they might run over. The interface standards will let manufacturers of TVs, VCRs, lights, security systems, and common household appliances be controlled remotely, either by a home computer or by a user accessing their home computer via the Internet. Honeywell is developing a product for this market, the Home Controller, a HAPI-com-pliant offering that lets people control home appliances remotely over the Internet. Using the Home Controller, a person at work could contact their home computer and tell it to turn on a furnace or air conditioner, lights, and other household devices.

Even before families put in home networks, they are using their communication devices together in novel ways that manufacturers never predicted or intended. For a long time, viewers who craved better audio have hooked up their TV sets to play its sound over their stereos. Now they are using their PCs in conjunction with their TVs to get more infor-mation about what appears on their television set. In some ways, consum-ers are converging technologies well ahead of the companies that market to them.

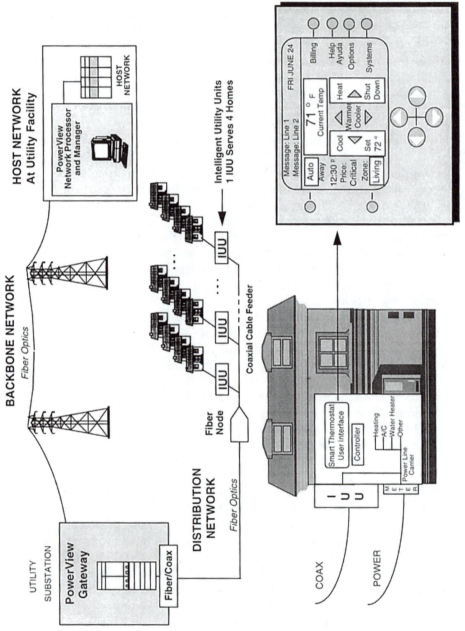

Figure 12.1 The networked home has computer monitored energy usage. Source: First Pacific Network.

The Converged, Digital Family

In 1997, Showtime Networks showed a boxing match between Mike Tyson and Evander Holyfield. As an added promotion, they came up with the idea of letting TV viewers who had PCs vote along with the judges. The viewers could go up on the Internet, sign onto Showtime's Web site, and vote. The network could collect all the numbers, average them, report the results to the viewers live over the channel, comparing them to the actual judges.

At a point in the match, the live announcer told the viewers to go to their computers and vote. Within a few seconds, Showtime's Web site servers crashed. This occurred twice more during the evening. Frantic emergency surgery by techs allowed Showtime to collect and report back the viewer information that night for the broadcast. But the network's researchers were intrigued. How could the servers have crashed so quickly if viewers had to go down the hall to a PC, boot it up, log on to their ISP, surf over to Showtime, and register their vote? Impossible!

So they embarked on a study and were amazed to find that 18% of their viewers kept their PCs in the same room as the TV, and many of the computers were on at the same time. Some of their most-devoted viewers (young men ages 18 to 35) were watching TV and surfing the Net simultaneously. The phenomenon is called "co-location," and the finding has been replicated by several different research teams.

In July of 1999, Media Metrix found that the simultaneous use of PC and TV was on the increase. According to MM, simultaneous use increased by 1.8 million households in the first six months of that year alone. Over half (52%) of PC-owning households, co-located their TVs and PCs: That's 18.8 million homes, up from 16.4 million just six months previously, as shown in Table 12.3. Nearly all of them (18.1 million) reported that they have both devices turned on at the same time.

Table 12.3 Television and PCs in the Same Room in U.S. PC-Owning Households

(Millions of Households)	Mar-1999	Sept-1998	Growth
U.S. Households Owning PCs	36.6	35.0	1.6
TV in Same Room as PC (Millions of Households)	19.0	17.3	1.7
TV in use "Some or All of the Time" in Same Room as PC	18.1	16.4	1.7

Source: *Media Metrix 1999 HardScan Report Volume I.*

Typically, people use the two appliances independently; in other words, they watch a show on the TV while working on finances, checking E-mail, or surfing the Web during commercial breaks. For producers, this means that people are not staying on a site or in a chat room that is related to the program. By 2000, the number of "viewers" had risen to 44 million, about 12% of them broadband service subscribers. Gartner Group research also found that 60% of them had gone online to get more information about a product advertised on TV, and 32% had actually made an online purchase.[6]

To target these involved "viewers" or "telewebbers" as these convergent consumers are sometimes called, several companies produced "syncTV" applications. These are online applications and content that are designed to synchronize with a scheduled TV program. In November and December of 1999, ACTV HyperTV demonstrated "15 Days of Bond," a James Bond marathon that they developed online companion programming for with Turner Network Television. Spiderdance created a syncTV application with MTV called "webRIOT." Softcom designed a broadband-only site for an HBO boxing series.

Another study conducted in August of 1999 by SRI/SMART, the "Home Technology and Television Ownership" report, suggests that while consumers co-locate the two appliances in the same room and sometimes use them at the same time, they still see them as quite distinct. SRI conducted 6,032 interviews between January and April of 1999 and found that about 47% of home computers were co-located—a bit lower than the Media Metrix finding, but in the same ballpark. The SRI results indicated that more than half of used the computer at school or work, where the PC does not compete with the TV at all. Moreover, 74% of the online users think that the primary benefit of the Internet is information, research, and E-mail, and only 3% cite entertainment as key.

The significant aspect of SRI's study is that households with premium PC technologies like high-speed Internet service are much more likely to have advanced television equipment and services as well. They are twice as likely to own two or more VCRs (56%) as the average U.S. household (25%), twice as likely to have bought a pay-per-view movie in the last thirty days (10%) as the average home (5%), and almost twice as likely to have rented a videocassette in the last month (70% versus 40%). The advanced PC homes also increased subscriptions to multichannel wired cable and satellite TV services, compared to non-PC homes that showed a decline.

The PC isn't the only computer that is in the same room as the TV. The innocuous-looking set-top box is really a computer, too, and it is becoming more and more like it's bigger cousin—so much so that computer companies are jumping into the lucrative set-top box market. What's more, there will be more kinds of set-top boxes that perform specialized functions than are commonly found in today's living rooms.

The Top of the TV Set:
The Most Valuable Real Estate in the World

For most of television history, there was no set-top box. It came into homes along with cable as the device that turned on and off the spigot of channels for which the consumer had paid. Now with DTV, everyone will have a set-top box (STB). And as the new generation of TV signals pump out from the heart of Washington, D.C., through the arteries and veins of networks, station groups, and local stations, they will soon fill the capillaries of the American living room.

The top of the television set may be today's most valuable piece of real estate because it controls the way consumers access the world's most successful and lucrative medium. The next-generation set-top boxes can keep track of viewers' choices and offer them new untried ideas for shows to watch. It can replace commercials for existing commercials in the bit stream with different ones, or eliminate commercials altogether. The STB software interface can show banner ads and promote one TV show over another.

The current generation of STBs is analog and addressable, which means that the cable operator can send a signal down the cable to open and close premium and pay-per-view channels remotely. (In the first generation, a cable technician had to come to the box outside the house and manually change the settings.) The guts of some of the next generation of STBs will be analog, but they will also have the ability to do limited digital processing, so they are called hybrid analog/digital units. An example is the Scientific-Atlanta 8600X, which displays graphic electronic program guides, accepts different software downloads to alter STB functions, and writes information to the TV screen much more quickly than the purely analog boxes did. Similarly, the CFT 2000 from General Instruments (GI, now owned by Motorola) is a modular, upgradeable STB that will deliver these same features. In addition, the GI model offers greater data security, expandable memory, and a signal generator that allows consumers to order pay-per-view movies with their remote control device.

Even the analog/digital boxes are starting to look more like computers, with many different input and output ports. There are variations by manufacturers, but they will all offer some combination of infrared, telephone jacks, PC connection, and high-quality VCR like S-VHS and composite RCA plugs. In addition, they will have an internal high-speed modem to provide Internet access through the same device.

The all-digital units may be a few years away yet. But when they come on the market, many manufacturers are planning to put Motorola RISC (reduced instruction set chip) processors inside, the same ones that now run the PowerPC. The digital STB will have two modules: the network interface module (NIM), which will vary depending on the network to which it is attached, and the processing module. Digital STBs will send and receive broadband information, compressing and decompressing it, scrambling and descrambling it, and either multiplexing it directly or addressing it for the multiplexer at the curb or neighborhood node upstream.

The New STBs—Not All of Them Are STBs

One of the most advanced STBs being ordered today is the Sony Aperios, which Cablevision Systems has commissioned Sony to design, build, and deploy. The billion-dollar baby will be an operating-system-driven interactive set-top box and system that will handle Cablevision's 2.7+ million subscribers on their fiber optic network in New York City. Cablevision plans to offer video-on-demand (VOD) services, access to the Internet, and an electronic program guide (EPG). The Aperios will compete with Motorola's digital generation STBs, and both will strive to fend off possible STB replacements, including WebTV-type devices and personal video recorders (PVRs) like TiVo and Replay.

WebTV is a set-top device that allows consumers to see E-mail and Web pages on their TV sets. The company invested in the development of technology to display text clearly on a screen that was that not designed for it. WebTV contains a modem and uses a telephone line return path; the company provides the ISP service for $24.99 per month. Subscribers to the Dish Network satellite service order WebTV as part of their package, integrated into the set-top box. Similarly, AOL-TV, designed to compete with WebTV will become part of the DIRECTV lineup.

DIRECTV has adopted an interesting position vis-à-vis its STB. The company has made a range of products available to its consumers and will let them choose the services they want to receive. The choices include TiVo, AOL-TV, Internet access with DirecDuo, and the Wink interactive service, each one enabled on a different STB.

In Europe, the STB market is becoming more competitive with the introduction of a product from NetGem, the NetBox. Like WebTV, it offers Internet access over the television set. It is currently marketed in fifteen countries, coming with a high-quality video and graphics display. This allows Web pages to be shown on the TV screen with the same clarity and shape as on a PC. Unlike WebTV, NetBox can be used to surf the Internet freely; it connects to any ISP, not just a proprietary service.

Another harbinger of the future comes from Mark Cuban, founder of broadcast.com (now part of the Yahoo empire). He created an HD-enabled set-top box that really is a PC, as shown in Figure 12.2. The $1,899 "Cuban box" is a stealth version of a full-bore, RAM-busting, mouse-to-the-metal screamer of a computer. Add a wireless keyboard and this monster can be used to run a company from the living room while the CEO watches HDTV. As the dynamic, outspoken Cuban put it in a recent

Figure 12.2 Mark Cuban's design of the future set-top box on the TV set will be a powerful computer. Source: CompUSA.

speech: "A set-top box is a stupid computer that isn't upgradeable, has no hard drive, and can't be returned to the retailer. What good is that?"[7]

The Cuban box is a combination computer/home entertainment center that will be offered commercially by CompUSA. The HD-ready device will have an HDTV decoder tuner card, supported by a 600 MHz Pentium III processor, DVD-ROM drive, 27 GB hard drive, Ethernet card, AM/FM tuner, and enough video in and out ports to run a digital studio. "It's in a black, smoked-glass, sexy-looking container," says CompUSA PC general manager Vice President, Rob Howe. "It is really very pretty and elegant, with an enormous amount of utility designed into a small, living-room-ready package, about the size of a multi-CD changer."[8]

It's an idea that seems to be in circulation. "CompUSA isn't the only company who wants to introduce something like the Cuban box. We've been talking to all the major PC manufacturers, and they are watching this DTV thing very closely," says Ken Aupperle, president of Hauppauge Digital, whose card is inside Cuban's creation.[9]

Another possible platform for the STBs of the future are game-players like the Sega DreamCast or the Sony PlayStation II. Both current versions support 56k modems and can be upgraded to send and receive broadband signals well. The players have real processing muscle that delivers spectacular graphics and lightning speed. The PlayStation II that will be introduced in the fall of 2000 has five times the graphics-processing power of today's most powerful graphics workstation, generating 75 million polygons per second. Polygons are the basic unit used to generate the surface of 3-D models. By comparison, today's top 3-D cards pump out between 1 and 5 million polygons per second. This means that the player could easily decode and play high-quality Hollywood movies, adding interactive sequences if the market demand justified them.

Hauppauge would like to bypass the STB altogether and out and out replace it with a PC. The company is the leading producer of TV tuner cards for PCs, and its executives believe there's a big market for HD viewing on the PC because they've watched their analog TV card sales double every year for the past four years. According to Hauppauge research, about 80% of their customers use them to watch TV, but they don't watch it full screen. The TV picture is relegated to a corner while they are surfing the Web.

Hauppauge will market a high-end DTV tuner card for the PC that will hit retail outlets in late 2000 or early 2001, costing about $299. A true HD card will come out about the same time for about $500 that puts 720 lines of progressive HD video on most 17-inch computer screens. Not everybody agrees. HD on PC is only a hobbyist market, says DTV guru Mark Schubin: "From a technological standpoint, TVs and PCs are converging, but I don't see usage changing. I don't think people will watch TV on computers or compute on TVs."[10]

For many analysts, the more likely scenario is the rise of what Forrester Research calls "e-ppliances." These are a new class of more capable consumer products that carry out specific functions and can better satisfy consumer needs and desires. The personal video recorder, or PVR, is a good example of an e-ppliance.

The Personal Video Recorder (PVR)

The PVR is an e-ppliance and service that digitizes incoming TV signals, then uses a hard drive in a set-top box to store them. Dedicated software lets viewers manipulate the video as if it were a videotape, but with all the advantages of instant random access. Thus, viewers can pause, rewind, and fast forward (if the video has been captured earlier), but they don't have to wait for these functions to be carried out. If they know where the piece of video is stored, they can just push a button and the material will come up on the screen instantly. PVRs let consumers time-shift just as they would with a VCR by recording video for later viewing, but the makers of these services have gone to great lengths to make it easier to use than programming recording on a VCR.

There are some other terms for the PVR, such as the Digital VCR (DVCR), the digital video recorder (DVR), and the digital personal recorder (DPR). Essentially computers in the form factor of an STB—complete with high-capacity hard drives—these will also sit on top of the TV set. They are an extension of the set-top box (STB), but are a little more than an STB, offering a way for consumers to exercise more control over their viewing than existing STBs, as well as a range of other functionalities and services. The two most well-known PVR makers are TiVo and Replay.

TiVo is a subscription service that costs about $10 per month. The customer buys the box and pays $9.99 per month to receive a guide and other services. The TiVo technology provides a continuously updated program guide, and automatically records designated programs by name, category, time, or channel. The system builds a viewer profile and, based on past choices, will offer up similar programming for consideration. TiVo receives analog and MPEG-2 streams, decodes digital material for the analog TV set, and stores either 14 or 24 hours of programming, depending on which STB the consumer purchased.[11]

The set-top PVR connects to a telephone for backchannel communication to TiVo, as shown in Figure 12.3. It makes a short call late at night to retrieve programming-guide data and downloadable promotions from a TiVo server. The company has made deals with more than ten programming services including Time Warner and Sony. The arrangement with Time Warner alerts TiVo subscribers to HBO programming and prompts them to record the offerings, including movies, series, documentaries, sports, family shows, and comedy and music specials.

Sony invested in the PVR company and will manufacture the devices, which are also produced by Philips and sell for $499. In addition, Sony entertainment will use the platform to promote programming from its TV and movie studios. Sony has also been developing interactive versions of its valuable properties, *Wheel of Fortune* and *Jeopardy*, that will run on the TiVo box.

Figure 12.3 Personal video recorders give consumers control over their television viewing.

The entertainment giant could also use the platform as a way to download entire movies and to offer them as a pay-per-view attraction that is featured when the consumer next turns on their TV.

Replay sells a box as well, but it gets its money from advertisers not subscribers. The ReplayTV set-top box takes in MPEG-2 digital and analog television programming, which it converts to a digital stream. The device records up to 28 hours of MPEG-2 recording, depending on the quality level the user selects—2 Mbps for VHS quality, 5 Mbps for DVD quality. Replay Networks has licensed its technology to Panasonic, which manufacturers a $699 model of the device. Replay also plans to incorporate on-screen sponsorships and has already signed deals to promote the Showtime and E! Networks.

There are many potential services a PVR can offer, depending on the specific hardware configuration and software platform. VCR functionality—record, pause, fast forward, rewind—is just the entry-level benefit, with digital technology conferring other richer features. For example, PVRs can identify and eliminate commercials. Or they can receive commercials addressed specifically to a particular subscriber or to a class of viewers, replacing the commercial that was originally inserted by a programmer. PVRs can receive and store detailed programming information and allow users to browse it by program, name, time, network, subject matter, genre, or any other variable the software will recognize. And the software can be used to track viewers' favorites, then seek out other similar programs that the same individual might like to see.

At the high-end of PVR technologies is a Philips product, still in development. It will have storage capacities of 9 GB—four hours of digital video and audio on single-layer optical disks. The high-bit densities and bit rate are achieved by using a high-numerical-aperture doublet lens instead of the standard single lens in CD and DVD, and by using new phase-change materials. The high data throughput of 35-MHz user bit rate and the short access times enable a new "read-while-write" feature. In the near future, when blue lasers become available, capacities of 20 GB will be achievable.

An interesting transition product is the $699 Hughes HDR-205 DVR. It is really a digital version of a VHS machine—it records the data that makes up digital video, along with the service's CD-quality audio, onto standard VHS tape. It is connected to the satellite dish that decodes the digital video and sends the signal to the analog TV set, as if it were coming directly from the satellite. The dish serves as the decoder for both the recording and the playback.

Another entry into the PVR market comes from Europe, the OpenTV, Télévision par Satellite (TPS), and Thomson device. OpenTV is a set-top box and interactive television services provider, TPS is a French satellite service, and Thomson Multimedia is a French hardware manufacturer.

The three companies will join to install a digital personal recorder into TPS boxes, primarily for European distribution. This new box will provide interactive services that Replay and TiVo boxes don't have because of the inclusion of OpenTV's proprietary operating system.

Finally, the Dazzle Digital Video Recorder is designed to turn a PC into a personal TV receiver, using it as an analog VCR. It connects to the TV and the PC, capturing both analog and digital signals and recording them in MPEG-2 video.

These early products will pave the way for the introduction of even more advanced equipment that will let consumers manage their TV viewing. Ultimately, there will be purely software applications that allow people to hook up their TVs to the hard drives on their PC, with the computer doing the processing instead of a separate appliance. Joe Butt, head of the Forrester Research entertainment practice, thinks that the new generation set-top technology such as TiVo and Replay are the wave of the future. "We see the PC moving to CE (consumer electronics), rather than the other way around, CE to PC," explains Butt.[12]

More E-ppliances: Internet Access Devices and Digital Audio Players

The 1999 COMDEX (Computer Dealers Exposition) convention featured a host of portable Internet access appliances for the home. One example is the Cendis 2000, a system that includes a base unit that sends Web pages to as many as fourteen handheld devices over the 900-MHz portion of the spectrum. The color touch screen is 6 1/4 inches wide. Another product was a stationary consumer appliance designed for Internet access by Compaq that is about the size of a fat notebook computer, with an LCD screen and a pull-out keyboard. It doesn't have disk drives; instead information is stored in flash memory.

Web tablets will also come on the market in the next few years. They are small info appliances that use flat screen technology and are usually wireless. A Web tablet has important advantages for Web access—they are mobile, easy to set up and use, and very reliable. National Semiconductor has a reference design for its WebPad that will probably be manufactured by Acer in 2000, and Samsung showed the Izzi that it expects to bring to market soon. Qubit has a prototype wireless Web tablet that it claims will be the most full-featured product. All these products have a way to go before they will be successful in the marketplace. Currently, their prices are high, partly because there are no economies of scale, but also due to expensive components like flat panel displays, batteries, and wireless technologies.

Even webcasting is getting portable. ZuluTV has developed a portable handheld device that broadcasts live, high-quality video and audio to the Internet. Called Video Vamoose, the battery-operated technology works with any camcorder and lets people schedule and webcast their own events through the ZuluTV service.

Another appliance that is coming on the market is the personal digital audio player, capitalizing on the popularity of MP3 music downloads from the Web. The Rio player from Diamond Multimedia was the first player, which now faces some stiff competition from major consumer electronics manufacturers. It's easy to see that 3G cell phones will make it easy for users to download music from the Web to their MP3 players.

The RCA Lyra product decodes both MP3 and G2 (Real Networks) formats. The 3.3 ounce player holds more than two hours of music on a compact flash memory card. It costs between $200 and $250, depending on the amount of memory included. In addition, the new Sony Walkman will also support the MP3 format used to download music from the Internet. The product records on the company's proprietary Memory Stick that competes with flash card memory technology. Owners of Sony Vaio computers will be able to download digital music to their PCs and transfer it to the Memory Stick for the player.

Blue-Skying in the Bitstream

At the MIT Media Lab, researchers and designers have put together a home that's wired down to the kitchen counters. Just about everything in the home is tagged with a radio ID that transmits to a computer. For example, if someone sets down food products taken from the cupboards on the counter, a computer keeps track of them and begins looking for recipes of dishes that can be made from the assembled ingredients. The Kitchen Sync will preheat your oven, check your refrigerator and cupboards to make sure you have all the ingredients you need, keep track of what you've measured, and make sure the meal doesn't burn in the oven.

It's all part of the Things That Think consortium at the Media Lab.[13] Another innovation is the travelling welcome mat, a video display that follows a person from room to room, offering them relevant information about where they are at the moment. The computers that manage this home environment are nowhere to be seen. This reflects the belief held by many designers that computers must become invisible to realize their full measure of influence. Other elements of MIT's home of the future are Anchored Displays, inexpensive battery-operated display screens that can be affixed to walls, doors, and desks. The displays can present information such as weather, traffic, stock quotes, sports scores, and other information taken from the Internet.

Architectural firm Hariri and Hariri, which specializes in experimental designs, is also looking at ubiquitous displays inside and outside of the home. In one recent conceptual design, large LCD screens replace the glass in windows. When they are turned on, the screens display images, video, text, or any data in the windows. When they are turned off, they are transparent, appearing to be ordinary glass. The designers believe that the LCD panels will be prefabricated off-the-shelf building blocks of the home of the future.

In a video shown at the Museum of Modern Art in New York, someone cooking is watching a chef demonstrating a recipe on an LCD panel in front of the kitchen counter. In other rooms, entertainment pours forth from the walls. Similarly, video telecommunications are also displayed on the screen. Even if cost were not an issue, the technology does not actually exist. LCD displays can become clear but when they show images, the material must be projected onto them. Direct view displays require tubes or grids for color and definition, so they cannot be transparent.

However, the point is not so much that this or that technology or function is possible or available. Rather, it is that innovative thinking is being applied to everyday living in what may some day be ordinary homes.

The Connected Family and Environs

There is some research that is looking at the way families use communications technologies today, and there is some cause for concern. Jan A. English-Lueck, Associate Professor of Anthropology at San Jose State University, has been working as part of a team of anthropologists that includes Charles Darrah and James M. Freeman. The Silicon Valley Cultures Project is a ten-year research program to conduct interviews and carry out observations, studying technology and community in Silicon Valley.[14] According to Dr. English-Lueck, "We discovered that people don't just own or use individual devices, but ecosystems of technologies at home. Pagers, faxes, cell phones, telephone answering systems, and computers are used together to serve the goals of individuals and families."

The researchers call "infomated households" those that have at least five information devices, including VCRs, CDs, laser discs, fax machines, answering machines, voice mail service, computers, and cell phones. The anthropologists wanted to understand how people used these devices and how they were affected by them. They looked for answers through 450 interviews, conducted at work and at home.

They were surprised to find that the infomated household, as they called it, revolved around work. "Work" did not always mean paid employment; these families saw their relationships among one another as projects that needed to be "worked on" as much as formal occupational activities. Nevertheless, work often does mean work, and there was a marked intrusion on family time by work time, requiring parents to spend considerable time dealing with job-related tasks at home.

People use their communication devices to coordinate the interactions of family members. They use pagers, cell phones, answering machines, PCs, and palm pilots to manage the work, school, and recreational activities that the individuals in the modern family take for granted. They require complex, sequenced patterns of behavior if the needed transportation and other support mechanisms are to unfold in the proper time and place. As English-Lueck puts it, "Message machines and pagers allow plans to be created, shifted, and coordinated in the space of a single afternoon."

The research suggests that time is in very short supply for the infomated family. Each member has a full schedule that allows for very little unprogrammed leisure time. The infomated family defines "doing things together" as being in the same room when members are involved in their individual interests. Thus, if the parents are working and the children are doing homework, then the family is considered to be doing things together. Communication technologies have enabled people to plan ever more activities with ever greater speed. There's a sense that these families are enjoying connectivity rather than connectedness.

Converged families aren't just networked at home, as shown in Figure 12.4. The environment that surrounds them will also become an extension of the Net. Welcome to the digital neighborhood.

Connectivity in the Public Sphere

The wiring of the public sphere is just beginning. It is already occurring in one common application—ATM banking, where machines routinely

Figure 12.4 Education is a major reason for high-speed networking. Source: Sprint.

connect to mainframe database records.[15] But Internet access for individuals to connect to their own accounts will occur in the future as well. One example is the partnership of GTE and U.S. West, with their plans to bring the Internet to public venues such as airports and malls. U.S. West put in its first public Internet terminal in the Seattle-Tacoma International Airport and now has 100 such kiosks in airports and shopping malls.

The terminals have ISDN or xDSL connections and let customers access airport and local information and a few major Web sites, such as CNN and ESPN, for free. It costs $3.95 for 15 minutes of access to E-mail and $1.95 for each subsequent 15-minute block of access time. GTE has 26 kiosks in the United States, for which it charges $3.75 for 15 minutes of time. Many analysts believe that this service will eventually become the norm, freeing users from having to lug their laptops everywhere.

Advantageously placed webcams may also be in the offing. Le Printemps is a well-known Paris department store that has equipped four salespeople with rollerblades and lightweight video cameras.[16] The skating clerks, called Webcamers, assist online shoppers, broadcasting live onto a small screen on the store's Web site, www.webcamer.com. As the Webcamers skate around the aisles, they chat online with shoppers, who make their requests for merchandise. The clerks use their laptops in shoulder harnesses to query shoppers as to color, size, brands, and other details. The products are sent to customers, arriving a few days after purchase. (See Figure 12.5.)

Le Printemps is getting 30 to 100 shoppers per day on their Web site, who make an online appointment to engage a Webcamer's services. Although much of the innovative way of helping customers is for show, the store's management points out that it obviates the necessity of putting the retailer's 1.5 million inventory of merchandise into a catalog. Just updating that gargantuan number of ever-changing products would be difficult and very expensive. In addition, the program helps Le Printemps reach customers who would never actually visit the store—about 25% of the e-shoppers come from outside France.

There are plenty of technical problems with the system. And even at its best, the images are updated only about once every three to four minutes, rather than the vision of high-speed video that the initial description brings to mind, because the cameras are connected via slow wireless technologies. The store is also pursuing more conventional online shopping services, one modeled after the Home Shopping Network.

e-Cinema and Digital Theaters

In the near future, theatrical motion pictures will be released and distributed to theaters digitally.[17] Although people have varied reactions to the quality of the images, digital distribution would save money for studios and distributors and make money for theaters. For studios, the cost of

Figure 12.5 Fry's Electronics is such a fixture in Silicon Valley, that employees of high-tech companies don't need to clock out from work to go to the store to shop.

making release prints for distribution is a huge expense, averaging $3 million per picture in 1997. Naturally, they try to minimize the number of prints they must make by forecasting the demand for them as closely as possible. Indeed, the entire pattern of the early release window—which theaters, how much revenue they will bring in, where they are located—is tied to the cost of prints and estimated revenues.

Digital distribution changes all this. It reduces distribution costs significantly and makes them virtually the same for narrow or wide releases. It also alleviates the recurring issues distributors have with the physical handling of film, the reduction of piracy, and the cost-effective management of distribution. Digital technology provides potential answers to many of these issues through the use of known encryption techniques and distribution via satellite or optical disks.

For theater owners, the availability of as many copies of the film as they may want to show simultaneously and the high-quality of each and every copy are powerful incentives in themselves. But there are other advantages for them to put in the needed reception equipment. The ability to feature live events opens up immense new revenue streams. They can use their exhibition space to host large crowds for daytime functions, including fashion shows, business meetings, and other such occasions.[18] In the past five years, theater operators have spent millions of dollars putting in stadium seating, digital sound systems, and other amenities designed to appeal to consumers. Some see the ability to receive live digital feeds as a way to recoup some of this spending.

Estimates for when digital cinema will become widespread in major markets hover around 2004 or 2005. Many analysts believe that the economics of the film distribution and exhibition process must eventually catch up to the investment costs of changing equipment.

The two best-known manufacturers of equipment for electronic cinema are Hughes/JVC and Texas Instruments (TI). TI's product is called Digital Light Processing, or DLP. At the heart of the system is a digital micromirror device that is essentially a light switch. Thousands of tiny, square, 16-by-16mm mirrors are fabricated on hinges atop a static random access memory array. Each mirror is capable of switching a pixel of light. The hinges allow the mirrors to tilt between two states: +10 degrees for "on" or –10 degrees for "off." When the mirrors are not operating, they sit in a "parked" state at 0 degrees.

The Hughes/JVC product is based on Image Light Amplification technology, or ILA. A projector separates the projected film image into its red, green, and blue components. It then passes each one through a reflective LCD panel, converts it photoelectronically, and illuminates it by a beam of light generated by a high-intensity arc lamp. By aiming the beam at the side of the stacked light layers, the entire image is reflected back through the optical system of the projector and digitally converged for registration on the screen. The companies' claims of brightness, brilliant color, contrast, and resolution are confirmed by large numbers of professional viewers. The first public demonstration of digital cinema was a showing of *Star Wars* in the Loews Cineplex Odeon Tenplex in Paramus, New Jersey, on June 18, 1999. Before then, there had been a number of private screenings for film professionals, such as at ShoWest in January of 1999. Since the first exhibition, a number of films have been shown digitally, including *Star Wars*, *An Ideal Husband*, *Arlington Road*, and *Tarzan*.

This chapter has now looked at the last extensions of high-speed networks into the home and beyond. But how does it affect the people who receive the information? Now we turn to the consumers—and producers—of digital messages.

The Digital Consumer: Teaching Old Media New Clicks

The rise in popularity of television in the 1950s and 1960s led to the burning question in communications research of the time: "Why do people use the media?" One of the most elegant answers was provided by Katz and Blumler in a seminal 1974 article that suggested that there are two over-arching reasons: uses and gratifications.[19] They argued that people went through a process in which (1) their social and psychologically based needs (2) generated expectations of (3) the various media, (4) leading to differential patterns of exposure and engagement, and resulting in (5) the gratification of their needs. They defined these needs as:

- cognitive: information, knowledge, and understanding of the environment
- affective: aesthetic, pleasure, and emotional experience
- personal integrative: credibility, confidence, stability, and individual status
- social integrative: contact with family, friends, and the world at large
- escapist: escape, tension release, diversion.

Within this framework, we can consider the ways that people will make use of and receive gratifications from broadband connectivity. It is easy to see that the hype over the Internet will soon attach itself to high-speed access. Perhaps the media has already done so, given the following headlines and slogans:

"Need for Speed" *Wired*

"Wired for Speed" *Los Angeles Times*

"@Home—the cable.internet.revolution" Company slogan

"Give computer users a taste of the high-speed Internet, and they'll take up arms if they are forced to go back to dial-up speeds . . . " *CNET News*

Media reports generate public expectations about what high-speed Internet and broadband access will do for subscribers, leading the people who have read about them to sample these new services. After they have tried broadband service, they use it to fulfill their needs and wants. If they have already found satisfaction with the Internet, they are likely to find broadband an improvement. A number of research studies conducted in the uses and gratifications tradition employed surveys, asking people which media they used for what purposes, and how well the media met their needs.

Marshall McLuhan also considered how people became involved with a range of media. He observed that the emergence of a new medium did not drive existing media to extinction; rather, it creates a new balance in the minds of the public who re-arrange their usage patterns to fit the new array of available media.[20]

Connectivity and Media Use

People in the United States use the media a great deal. Statistics from Simmons Market Research indicate that on an average day, 94% read magazines, 77% listen to the radio, 72% watch TV, 66% read newspapers, and only 14% go online.[21] It may be surprising that online usage is so much lower than other uses of media, but it is still a new medium that is growing quickly. In 1998, a spate of studies from Nielsen Media, Strategis Group, and True North indicated that Internet usage was causing television viewing to decline. The Nielsen study of 5,000 homes said time spent on the Internet reduced time spent watching television by 15%, although only by 6% during primetime.[22] The Strategis research indicated that 64% of the respondents said they were spending less time with TV because of the Internet.[23] The True North effort found that AOL users watched 15% less television, except during primetime hours when they were more likely to watch TV.[24]

A review of data over time suggests people who sign up for Internet access watched less television before they began surfing. Research from Burke Information, Communications and Entertainment Research for MTV, and Turner Entertainment Networks sheds additional light on the relationship between the two media.[25] According to Burke, less than 2% of people with Internet access traded time spent watching TV solely for Internet usage. The study drew on the firm's entertainment industry and online survey database, compiled between December 1998 and August 1999, which asked if Internet usage was the primary reason why those respondents who reported lower TV viewership were watching less television. Only 1.5% of the 17,000+ randomly intercepted visitors from nine television network sites surveyed said they were watching less television only because they were online more.

Burke found that more people are multitasking, however. The survey found that 14.7% of households are participating in activities other than going online in place of watching television. More than 5% of the site visitors explained that TV viewing levels declined because they were spending more time with family and friends, engaging in athletics, working longer hours, and spending more time online.

The Burke survey uncovered other interesting aspects of the TV-Internet relationship. About 66% of the TV viewers who had Internet access had visited a network's Web site for more information about a program after watching it. Moreover, 42% of those questioned reported they had watched a TV program based on information they found on the Web site before the program appeared. Finally, 39% said they had watched TV and surfed the Internet at the same time.

As a result of these and other studies, by 1999, the idea that TV and the Internet were in a struggle to the death of one or the other seemed naïve. Perhaps, advanced some analysts, the two media actually complement one another. From phenomena like PC/TV co-location to using the

Internet for e-commerce extensions of traditional media, many studies found that Internet users consume more media than nonusers generally; they don't see media so much as either/or as both/and.

A 1999 Arbitron NewMedia Pathfinder study corroborated this hypothesis, as shown in Table 12.4.[26] The study found that people who used the Net more were more likely to consider television very important to them. Of the low-Net-users, 23% rated TV as very important to them; 34% of users who spent a medium amount of time on the Net thought TV was important, and 39% of those who spent the most time on the Net rated TV highly.

The Web users under study did watch slightly less TV than nonusers. But the researchers believe that their viewing habits are more related to the income, education, and age profile of Web users rather than to displacement of TV viewing by Internet use. During the peak hour of Internet use at home, 8:00 to 9:00 P.M., 52% of the heavy online users said they watched television, a level only slightly lower level than that of all Web users (55%) or the total population (59%). Depending on how much time they spend on the Net, users have distinct preferences for broadcast programs. Heavy users like classic rock on the radio; on TV, *The Simpsons* is their favorite show; on cable, they watch the Discovery Channel, Learning Channel, and HBO; their most-read section of the newspaper is business news.

None of these findings should lead to an underestimation of how important the Internet is to the people who use it. In 1998, Roper Starch Worldwide surveyed 1,001 adults with access to online services randomly by telephone. Sixty-seven percent said they would rather have a computer connected to the Internet than a working telephone (desired by 23%) if they were stranded on a desert island for a lengthy period of time.[27]

Internet Use—How Much, Who, and Why

At any given time, various statistics are released about the Internet and its users, and the figures change so rapidly and come from so many different

Table 12.4 TV Viewing by Net Users

Hours per Week	
Internet	*TV Viewing*
Internet	TV Viewing
1 hour or less	18 hours
2-4 hours	15 hours
5+ hours	20 hours

Source: Arbitron NewMedia.

research groups using different sampling procedures and analysis methodologies that certainty about the actual state of affairs is virtually hopeless. The numbers cited here are taken from late 1999. Two Web sites that maintain a wealth of information about the current crop of reports at any given time are: http://www.emarketer.com and http://www.nua.ie/survey. Nua is an Ireland-based research firm that has been tracking the Internet for several years. Figure 12.6 shows the results of their surveys of global Net access since 1995, projected out to 2005, independent of access method.[28]

One fairly long-term trend is the upscale skew of online users. According to the eMarketer Usage Report, the median income for online households is 57% higher than the average American household, or $58,000 versus $37,005. However, as Net use edges up towards 50% and beyond, the demographics will come to resemble the general population.

The fact that Internet access occurs with, by, and through computers makes this form of communication the most measured media in existence, particularly unusual so early in its life cycle. Media Metrix regularly releases a report on the "Top 50 at Home and at Work Combined Digital Media and Web" audience ratings.[29] In the United States for October 1999, the Web tracking company announced that the U.S. Internet crossed an important milestone that October: Users accessed an average of more than 1 billion Web pages per day, a 49% increase over October of the previous year. The overall growth in the number of users continued to rise at a steady rate of 12.5%, as it had done for the past three years, reaching 63.9 million unique visitors during October, 1999. Media Metrix measured the most popular sites as AOL, Yahoo, Microsoft, Lycos, Go Network (Disney),

Figure 12.6 Growth of Net Users, 1995–2005. Source: NUA Internet Surveys.

Excite@Home, Amazon, Time Warner Online, Go2Net, and Bluemountain-arts.com.

European Internet penetration was somewhat lower in 1999, with about 15 million users, said Jupiter Communications and forecasted that this number would rise to 49 million by 2003, raising the penetration rate from 10% to 32%. The research company further estimated that 4.2 million Europeans shopped on the Internet in 1998, about 13.5% of continental online users.[30]

There are reasons why the number of people going online is growing rapidly. The gratifications that people get from it can be summarized as the 5 Cs: convenience, choice, control, community, and commerce. All of these are magnified by broadband connectivity when compared to narrowband access—broadband offers more, better, faster:

- Convenience: always on
- Choice: smooth, detailed video, high-resolution graphics, high-quality audio, and text—from millions of Web sites
- Control: speedy downloads allow fast movement from one page or site to another
- Community: live audio and visual chat and multimedia E-mail
- Commerce: shopping for products with video and 3-D surround technology and simultaneous telephone assistance from a salesperson

The last two features are much-studied phenomena and deserve some special comment. Community on the Internet refers to social interaction that can be conducted in chat rooms or via E-mail. Many observers have called E-mail the "killer app" of the Internet, meaning that this is the most popular activity and the one that leads people to open online accounts. No one would argue with the assertion that community is one of the most significant reasons people get online and spend time there.

According to Forrester Research, about 55 million American homes are now able to send and receive E-mail communication, and the number of messages now totals more than 150 million messages per day. A 1998 Activmedia study concluded that Internet users both strengthen existing relationships and add to their social networks online, disputing the earlier finding of a Carnegie Mellon survey that claimed that the Internet reduced people's contact with friends and family.[31]

According to the PricewaterhouseCoopers (PwC) 1999 Consumer Technology Survey, 48% of U.S. users reported that they go online for E-mail, while 28% said they do it for research. Last year, PwC said the figures were exactly reversed.[32] eMarketer studied the use of E-mail and came up with the following statistics:[33]

- 3.4 trillion E-mail messages delivered in 1998
- 9.4 billion messages exchanged every day of the year in the United States
- 81 million Americans use E-mail , at least occasionally
- The average American sends or receives 26.4 E-mail messages every day
- On a daily basis, the total number of E-mail messages sent by U.S. internet users is 2.1 billion (81 million X 26.4 messages/day>2.1 billion)

Indeed, an American Psychological Association convention survey stated that Internet users were more likely to go on-line in search of social entertainment as opposed to looking for information.[34] The study focused on international Internet users and found that 35% used chat rooms and 28% participated in interactive online fantasy games. The report states: "pathological Internet users were logging on in a bid to create another persona." (Oh, goodness! Clearly these researchers fail to grasp the unique pleasures that text-based interaction offers to the more identity-adventurous among us, condemning all to the tyranny of physicality.)

Nevertheless, it is also true that for some people, the Internet is a venue for escape that knows no bounds. It can compete with the real world for attention, insulating them from reality. Internet addiction has been cited as a contributing factor in the disintegration of marriages and families, school dismissal, and the collapse of careers. "Internet Addiction Disorder" is now a recognized behavioral problem that describes clinically out-of-control behavior that can have disastrous consequences for the person who suffers from it.

Aside from community, no other aspect of the Internet has brought about so much discussion as e-commerce. A 1998 CommerceNet/Nielsen Internet demographic survey noted that the number of people purchasing products and services via the Internet hit 20 million that year, double what it was nine months previously. And as consumer trust grows, the amount of money being spent online will increase.[35]

To some extent, the reasons people use the Internet differ, depending on the demographic niche to which they belong. Historically, there have been more men than women on the Net in the United States, but this gender disparity is changing. There are now 27 million women online in the United States says eMarketer, and they account for 46% of all U.S. Internet use. A 1999 report by NetSmart America found that 58% of the new Internet users in the United States are women (up from 44% in 1998), and predicts that at current growth levels women will constitute 60% of Net users by 2002.[36] A large proportion of women, about 73%, use online access to obtain product and service information—more than any other kind of information. However, most online shoppers (64%) and purchasers (71%) are men.

Age matters as well, and the distribution of people online by their age is shown in Table 12.5. The number of people over 50 who access the Internet may also surprise some observers. More than 13 million U.S. adults in the 50+ age group go online, and this number is growing rapidly, according to a 1999 study conducted by SeniorNet and Charles Schwab, Inc. They constitute about 16.5% of the total U.S. online population. About 40% of them have a computer, up from 29% in 1995, and 70% of these computer owners surf the Web. This age group reports that they use the Internet for E-mail (72%), to research issues of interest (59%), obtain news and current affairs information (53%), conduct travel research (47%), and get weather updates (43%), and 40% said the Internet helped them make investment decisions. Moreover, Internet users over the age of 55 are highly educated, affluent, and have a higher tendency to purchase online than younger users. The report noted that the travel industry is expected to represent 35% of all online sales by the year 2002.[37]

Young people may not be the largest group online but they are the heaviest Internet users, according to eMarketer. Currently, adolescents average 8.5 hours online per week, about 27% more time than average Net users. A huge percentage of college students are currently online, 87%, and they are the most active single group on the Net. They are heavily targeted by marketers who acknowledged the potential spending power of this group. Moreover, their younger siblings are also busy online. According to the research, the number of teens online will grow 38 percent, from 11.1 million in 1999 to 15.3 million by 2002.

The new America Online and Roper Starch Youth Cyberstudy sampled approximately 500 youths between the ages of 9 and 17 over the telephone.[38] They found that youngsters between 9 and 11 use the Internet

Table 12.5 U.S. Internet User Population by Age Group, 1999

Age	Number Online (in millions)	% of Total People Online
1–12	11.7	14.5
13–17	11.1	13.7
18–34	23.0	28.5
35–54	25.2	31.2
55+	9.8	12.1
Total:	80.8	

Source: eMarketer, 1999.

three times a week on average, going online for real-time chat, games, writing letters, and downloading music. Those aged between 15 and 17 go online on average five times a week, and what they report they would most like to do is:

- Send and receive pictures online—78%
- Download music—76%
- Live videoconference with friends—70%
- Watch short cartoons or video clips—63%

While middle-aged adults use the Internet mainly for research and surveillance of the environment, people between ages 18 and 24 use the Internet as much for entertainment as for research. Moreover, younger users access the Internet for a wider variety of activities than their elders, and they incorporate the Internet into just about every phase of their lives.

Research by advertising agency Saatchi and Saatchi coined the term "connexity" to describe the way that Generation-Y young people (ages 12–24) relate to the world.[39] They take being connected to the world for granted and place a value on the importance of growth acquired by staying connected. Knowledge is important to them, and they understand and acknowledge the importance of self-reliance.

The agency came to these conclusions after hiring child psychologists and cultural anthropologists to interview and observe this group to understand how they are affected by digital media. Based on the findings, Saatchi and Saatchi advises marketers about the changes they must make to take this influence on Gen-Y attitudes into account. The researchers found that for members of Gen Y, digital media is an extension of their social, intellectual, and emotional selves. This group has considerable personal confidence gained through their greater access to information. As a result, marketers cannot talk down to the Gen Y cohort, lest the youngsters tune them out.

There isn't much information about the youngest children online, but at the Digital Coast '99 conference in Los Angeles, the founder of The Palace (http://www.thepalace.com) Mark Jeffrey mentioned how surprised observers were at their behavior. The Palace is a collection of user-created rooms; using proprietary software on the Web site, it takes users about an hour to put their personal room together. Visitors on the site check out one another's rooms as well as hang out in public spaces. At any point, users may engage in chat with others who are visiting the same space.

What has surprised Jeffrey and others at The Palace is that kids use the chat function very differently than older children and adults: "Young children engage more in collaboration than chat. They draw for one another, and examine and exchange one another's creations. We've been surprised that they end up spending much of their time engaged in such collaborative activities. You can't really call it chat at all. It's communication, but it isn't chat," he reported.[40]

Broadband Users   How can we expect media consumers to change as bandwidth increases? A great many companies have been conducting research to find answers to these questions. Jupiter Communications examined what people do when they are online, as shown in Figure 12.7. Most analysts believe that as broadband connections become more pervasive, usage will move towards the left side of the array—that is, more towards entertainment and less towards information gathering. The reason? Broadband connection provides a much more satisfying entertainment experience than the narrowband dialup Internet is able to do.

Several studies cluster online and broadband users into "psychographic" or lifestyle categories. Jupiter's analysis of early broadband consumers showed two distinct customer segments: the Hipsters and the Sophisticates. The Hipsters are younger, more diverse entertainment-oriented consumers who are most likely to lean towards cable modem service. They want fast downloads for their entertainment, and cable marketing has been effective in reaching them. The Sophisticates are older

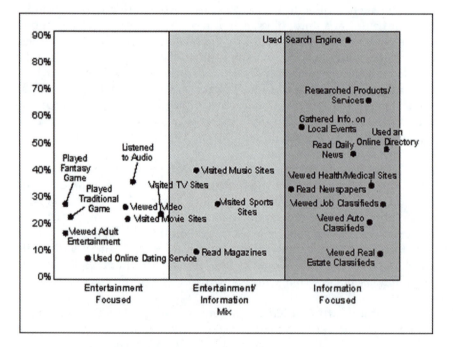

Figure 12.7 What users do on the Net. Source: Jupiter Communications.

high-income people, attracted by the "always on" feature. They are information-oriented and tend to subscribe to DSL service because they are more likely to be aware of it through their research of broadband options.

Another study by Media Metrix and McKinsey divided online users into six behavioral groups, based on how much time they spend online, the pages and domains they access, and the amount of time they spend on Web pages: Simplifiers, Surfers, Connectors, Bargainers, Routiners, and Sportsters, as summarized in Table 12.6.[41]

And research from Arbitron NewMedia/Edison Media looked at an important category of Internet user that they call "streamies."[42] These are Internet users who watch streaming audio/video content. They average 11 hours per week on the Net, almost 50% more time than the average online user, and, according to the research, 60% of them have made an online purchase, making them extremely attractive to Internet marketers and advertisers.

The New Consumer: Desperately Seeking Experiences

Some analysts say that underlying all these studies is a larger trend in consumer motivation and behavior. They believe that people want to purchase experiences rather than merely goods and services.[43] In a recent

Table 12.6 Types of Online Users

Who They Are	What They Want	What They Do
Simplifiers	End-to-end convenience	Spend little time online, but account for 50% of transactions Log on with specific purpose
Surfers	What's new?	8% of users, account for 32% of time Use Web for many activities
Connectors	"Newbies" looking for a reason to be online	E-mail, chat, send greeting cards
Bargainers	Shopping and more shopping	8% of all users . . . but 52% of eBay customers
Routiners	Valuable information content	Visit fewer domains but spend twice as much time per page as average users
Sportsters	Exciting, fun entertainment	4% of active users Spend few hours online

book, *The Experience Economy,* an important point is that a key difference in providing experiences, as contrasted to selling traditional goods and services, is that the locus of the definition of an experience is in the customer, not the provider.

When considering the consumer experience of software content, Pine, Gilmore, and Pine analyze material along two dimensions: whether the audience is active or passive, and whether they become absorbed or immersed in the material. Table 12.7 shows the results of their analysis.

Experience is a psychological state, although the creation of an experience probably involves extensive use of hardware, real-world locations, and objects. These kinds of products might include theme parks or the new themed theater complexes. There's a reason for this emphasis on theme. The authors define five design principles for creating experience products: (1) theme; (2) impressions and cues that support the theme and create the experience; (3) elimination of negative cues; (4) availability of takeaway memorabilia of the experience; and (5) the engagement of all five senses.

Film exhibition is a good example because the industry has been moving in this direction for some time with the construction of megaplexes. Here's how Hamid Hashemi, president of Muvico, an innovative regional theater circuit based in the South, describes the design of his theaters:

> Each megaplex is highly themed and offers a different environment carried through the entire facility—we don't take a cookie-cutter approach. For example, one has a Fifties theme, another has an Egyptian theme—and it's probably the best theater that's been built in 40 years. It has old-fashioned balconies, a 1200-square foot bar/restaurant, and child care facilities run by certified teachers for kids 3 to 8. They can play with educational computer programming and arts and crafts.
>
> In addition to the environment, there's also the presentation and service. Patrons are greeted by ambassadors, dressed in red uniforms, little hats, and white gloves. Their job is to make people feel like a star going to a premiere. We serve a variety of foods besides traditional concession items, like chicken wings and tenders and popcorn shrimp.
>
> We have stadium seating, which means each row is on a slightly higher riser than the row in front of it so the view of the screen is never blocked. We have wide seat spacing and reclining chairs.[44]

Table 12.7 The Experience of Content

	Active	*Passive*
Absorption	Education	Entertainment
Immersion	Escapist	Esthetic

Summary

As we have seen, digital technology is moving out of the workplace and into the community, the home—and the heart. The future holds media that combine the visual involvement of television with the choice and depth of the Internet. This convergence augurs a next generation of digital technology and content that will be far more immersive than today's TV programming or Internet have ever provided. The next chapter will look at the content side of this equation—what it will be like and how it will be developed.

Notes

1. Kinetic Strategies monitors the cable modem market. Get the most up-to-date information from the company's Web site at http://www.kineticstrategies.com.
2. "Data CLECs: xDSL Markets and Opportunities for Small and Medium-sized Businesses." Study by Pioneer Consulting (1999). Information available at http://www.pioneerconsulting.com.
3. Consulting firm Nua tracks online usage, complete with disclaimers about how difficult it is to compile reliable figures. However, they've been doing it longer than most. Their Web site is at http://www.nua.ie/surveys/how_many_online/index.html.
4. S. Elmer, "Internet Usage and Commerce in Western Europe 1997–2002," International Data Corporation (IDC) research (February 2000). The press release is at http://www.idc.com/Press/default.htm.
5. K. House, "To Connect or Not to Connect: Home-Networking Market Review." IDC research Study and Forecast, 1998–2002, Report #W18220 (March 1999).
6. S. Ramnarayan, "Telewebbers on the rise." Research publication of the Gartner Group (June 5, 2000). Summary available at https://gartner4.gartner-web.com/ggbin/ggpurchase?prod=ehmeusdp0002.
7. M. Cuban. Personal interview, Manhattan Beach, CA, (May 1999).
8. R. Howe, Vice President, CompUSA. Telephone interview, May 1999.
9. K. Aupperle, president, Hauppauge Digital. Telephone interview, May 1999.
10. M. Schubin, HDTV analyst. Telephone interview, September 1999.
11. S. Jolna, Vice President, TiVo. Telephone interview, November 1999.
12. J. Butt, media analyst, Forrester Research. Telephone interview, November 1999.
13. Check out MIT's "Things That Think" projects at http://www.media.mit.edu/ttt/.
14. J.A. English-Lueck, C. Darrah, and J.M. Freeman, "Technology and Social Change: The Effects on Family and Community" (June 19, 1998). Information about this research is available at http://www.sjsu.edu/depts/anthropology/svcp/.
15. A good source of information about public kiosks is at http://www.kiosks.org. A press release from GTE about airport kiosks is at http://www.gte.com/AboutGTE/NewsCenter/News/Releases/022497.html.

16. See the department store's Web site at http://www.printemps.fr/.

17. Some Web sites for further information about digital cinema and digital distribution and exhibition of motion pictures are at: http://www.hjt.com/products/ila12k.html; http://www.cinedc.com; http://www.hjt.com/technology/technology.html; and http://www.ti.com.

  Good photos of the Cinecomm technical set-up are at: http://members.aol.com/dennyd1/DigitalCinemaJPG1 and http://members.aol.com/dddegan/DigitalCinemaJPG2.

  Information about the Texas Instrument DLP system is at: http://www.ti.com/dlp/resources/whitepapers/overview/state.shtml.

18. M. Moritz, president and CEO, National Theater Owners, California and Nevada. Telephone interview, January 2000.

19. E. Katz and J.G. Blumler, *Uses of mass communications: Current perspectives on gratifications research,* Beverly Hills, CA: Sage (1974).

20. M. McLuhan, *Understanding Media: The Extensions of Man,* New York: McGraw Hill (1964).

21. Simmons conducts annual surveys of media usage. The site is at http://www.smrb.com/.

22. Nielsen Media press release, dated November 11, 1998, available at http://www.nielsenmedia.com/newsreleases/releases/1998/HHtuning.html.

23. "Internet usage threatens TV viewing," *Inside Cable* (August 16, 1998). Available online at http://www.inside-cable.co.uk/n98q3alt.htm.

24. "Net usage versus offline media," *eMarketer* (November 27, 1999). Available online at http://e-land.com/estats/usage_net_vs.html.

25. "Internet surfing not impacting TV viewing," summary of Burke study, available at CyberAtlas at http://cyberatlas.internet.com/big_picture/traffic_patterns/print/0,1323,5931_214791,00.html.

26. B. Cox, "Study: Web no threat to traditional media," *Internetnews.com* (October 6, 1999). Available online at http://dev-www.internetnews.com/IAR/print/0%2C1089%2C12_212391%2C00.html.

27. M.L. D'Amico, "Internet has become a necessity, U.S. poll shows," *CNN.com* (December 7, 1998). Available online at http://www.cnn.com/TECH/computing/9812/07/neednet.idg/.

28. This graph is available online at http://www.nua.ie/surveys/analysis/graphs_charts/comparisons/how_many_online.html.

29. Find regular reports at the company's Web site at http://www.mediametrix.com/landing.jsp.

30. F. Swerdlow, "Growth Abroad Requires Change by U.S. Sites Wishing to Compete." Concept report from Jupiter Communications (October 25, 1999). Figures available in press release at http://www.jup.com/jupiter/press/releases/1999/0922.html.

31. "Activmedia: Web improves relationships." Activmedia research. Archived at http://www.glreach.com/eng/ed/it/150998.html. See also: A. Harmon, "Sad, lonely world discovered in cyberspace," *New York Times* (October 30, 1998).

32. PricewaterhouseCoopers research cited by a Reuters report dated October 1, 1999, posted by Nua Surveys at http://www.nua.ie/surveys/index.cgi?f=VS&art_id=905354444&rel=true.

33. Information about the eUser and Usage Report is at http://www.emarketer. com/.

34. "Webaholics Anonymous." Article posted online at http://www.concen-tric.net/~Astorm/iad.html. This site also has a good article by S.A. King, "Is the Internet addictive or are addicts using the Internet?" (December 1996). Available online at http://www.concentric.net/~Astorm/iad.html.

35. The CommerceNet/Nielsen Internet Demographic Survey. Research by Com-merceNet/Nielsen (June 1998). Press release and summary statistics posted at http://www.commerce.net/news/press/19980824b.html.

36. A summary of demographic information about women online as of November, 1999, is online at http://cyberatlas.internet.com/big_picture /demographics /article/0%2C1323%2C5901_221541%2C00.htm. The NetSmart America Web site is at http://www.netsmartamerica.com/.

37. Summary of this study, conducted by SeniorNet and Charles Schwab, Inc., October, 1998, is available at http://www.nua.ie/surveys/index.cgi? f=VS&art_id=905354444&rel=true.

38. A PDF file of the entire Roper-Starch Youth Cyberstudy can be downloaded from http://www.corp.aol.com/press/roper.html?.

39. R.X. Weissman, "Connecting with digital kids," *American Demographics* (April 1999). Available online at http://www.demographics.com/publications/ad /99_ad/9904_ad/ad990405d.htm.

40. M. Jeffrey. Personal interview, Los Angeles, September 1999.

41. "Net consultants identify six types of surfers." (April 20, 2000). Available online at http://canadacomputes.com/CCP/Print/1,1040,3310,00.html. For more information, contact Media Metrix at http://www.mediametrix.com.

42. Online Radio Listeners are the New Breed of Internet Consumers. Posted online at http://www.newmediamusic.com/ps/arbitron_net_services.html.

43. B.J. Pine, J.H. Gilmore, and B.J. Pine, II, *The Experience Economy*, Boston: Har-vard Business School Press (1999).

44. H. Hashemi. Telephone interview, January 11, 2000.

13

From Reel (Movies) to Real (Networks)

The last chapter explored some of the ways people will actually use the communication systems covered in this book. Now we look at the digital content—communication, information, transactions, and entertainment—that flows through the pipes, with a focus on some of the content projects underway as of early 2000. This chapter will cover how content is being developed for broadband and narrowband, wired and wireless platforms, an environment where consumers will access content via an array of mobile, fixed, and embedded reception devices and appliances.

It then turns to the production, packaging, marketing, and distribution of content as a coordinated business effort, examining how digital video becomes entertainment and information products. In many cases, content is developed within large media organizations, but there is also a strong tradition of independent creative work. The last section will examine the differences between these two creative environments in new media companies.

There are some areas of content creation this chapter does not cover. The first is linear media. Virtually every university and college in the nation offers courses in media writing and television and radio production. Thousands of books explicate the creative problems and solutions in traditional media and channels. Here, we are concerned with the potential of the new networks, ranging from the Internet to interactive TV.

A second domain largely beyond the scope of this chapter is that of personal creative projects; this book concerns itself with professional production. Nevertheless, it must be understood that individual communications are a significant part of the context of the networked environment, cumulatively posing a formidable challenge to the efforts of global entertainment companies. The Internet is probably the most accessible medium in history for people to share their ideas, interests, and talents. In short, everyone is potentially a content producer. Creating material for an audience can be a business enterprise or a labor of love, and anyone who invests their personal time into producing a personal Web site can do so in a few hours, although more elaborate creations can become a fulfilling hobby that absorbs many hours. Information about how to create personal Web sites is readily available both offline and online.

Nor will this chapter cover videoconferencing and other forms of user-created and -supplied visual material over telephone lines. Like telephone calls, the business model for such activities falls under the provision of bandwidth as a commodity telecommunications service, although there will be an ancillary market for software to enable the applications, including independent programs, Internet service applications, and browser extensions and helper apps. We will, however, consider professionally produced material designed for display on wired and wireless screen-enabled telephones.

Finally, content for HDTV was already covered in Chapter 4, "High-Definition Television: Or How DTV Came into Being Because HBO Needed Cash." SDTV digital programming provided by satellite systems consists of digitized versions of popular networks and programming services. Since one-way television cannot provide interactivity to profit from e-commerce, these suppliers are looking to the Internet, enhanced TV, broadband TV, and interactive TV to open up this potentially enormous revenue stream for them.

Content—Oh, Those Beautiful Bitstreams

An examination of media introduced in the past shows that perceptions and concerns about them move through similar stages as the technologies and content spread throughout society. These stages can be summarized as:

- Technological concerns
- Aesthetic, production, uses, and other content issues
- Study of unique social and psychological effects
- Near invisibility

Consider the adoption of the film medium. From the end of the nineteenth century until well into the second decade of the twentieth century, the literature about film is concerned with film catching on fire, breaking, and tearing; the safety of theaters; and the difficulties of distributing the reels. Through the 1920s, attention turns to the rapidly maturing artistry of films, and people create and worship the stars that fascinate them. In the 1930s, there begins to be disapproval of sexuality, smoking, drinking, and other behaviors taken to be undesirable by a given critic. Indeed, during World War II, the belief that movies exert some profound social influence is shown by the recruitment of movie stars to appear at events and appear in films, encouraging people to support the war effort, and by the banishment of the unfaithful (gasp) Ingrid Bergman from Hollywood. Today, studios are satisfied if their stars are not indicted for heinous felonies, and the content of films is largely taken for granted—although from time to time a film will raise the ire of certain groups and create considerable controversy, as in the case of *The Last Temptation of Christ.*

The focus of digital technologies is now shifting from the first stage to the second stage—from the technology itself to its content. For more than two decades, the center of concern has been on the hardware almost to the exclusion of all else. Fueled by the rapid adoption of the Internet, there is a great deal of activity to develop and create digital, interactive material—information, entertainment, games, and applications. As understanding of the new formats on the Web grows, many of the same skills apply to broadband TV and handheld mobile broadband appliances, which are also digital and interactive, as well as to traditional television where a given platform permits a return path for upstream communication.

The perspective of content creators is much different than that of the researchers and engineers who design and build the underlying technological platforms. The technical people are likely to say, "bits are bits." And as far as transport is concerned, this observation is quite true.

Nevertheless, it is also true that all bits are not created equal. From the creative and business points of view, some bits are worth more than other bits. Disney bits, Viacom Bits, *Titanic* bits, Placido Domingo bits, Ricky Martin bits, *Dawson Creek* bits—all these configurations of bits bring in a lot more eyeballs and revenue than random bits, or even carefully organized bits created by artists or offered by brands that are less well known. However, it is growing more and more difficult to establish brands in a media and entertainment world of expanded choices. "Mindshare" and "attention economy" are both terms that have come into common parlance to describe the necessity of capturing consumers long enough to engage the attention, awareness, and interest that building a strong brand requires.

Why There Are No Cyberhits

One reason it's so hard to create a hit on the Internet, or to establish a brand for that matter, is that there are so many roads to the consumer. The gazillion-channel universe makes assembling an audience something like herding bees that fly from flower to flower in a meadow of infinite choice. The hit business is now extremely difficult, even in the traditional media environment. The number of distribution outlets for "content," the entertainment and information products, has steadily increased over the past two decades.

In 1970, there were 3 broadcast networks and a handful of independents in larger markets. In the 1980s, there were 4 broadcast networks, some independent stations, and 12 to 30 channels on cable systems. Today, there are 7 networks, 54 to more than 100 channels on cable systems, up to 300 channels on DBS systems and C-band dishes, plus growing numbers of broadband services over cable and telephone lines, wireless cable, LMDS, and the Internet. And the Internet expands the number of available channels to infinity.

This proliferation of distribution outlets has changed the marketplace for content in several important ways. It has resulted in the production of vastly more material. No wonder! In 1979, there were 14 cable networks, 94 in 1994, 139 in 1998, and 234 national and regional cable programming services, including audio and interactive channels in 2000— with 62 planned services in the wings.[1] The 1990s also saw a dramatic increase in prices for content, as broadcast network expenditures for programming rose from about 64% of their gross revenues in the 1980s, to 80% in the early 1990s.[2] The inevitable result is a splintering of the audience across channels and across programs.

The kinds of entertainment and information products available to consumers have also changed. Broadcasting became "narrowcasting" under the cable regime, then morphed into "slivercasting," niche casting, and microniche-casting on cable, and mass customization of content to the individual on the Internet. Broadcast networks always point to the "broad" aspects of their industry. Cable systems and now DBS offer ever-greater multichannel capacity, enabling them to present niche content, designed to appeal to relatively small segments of the audience with special interests. Sports, religion, entertainment genres (comedy, science fiction, action, mystery, etc.), news, public affairs, culture, documentary, and business are all examples of narrow subject matter that make up entire cable channels.

Now on the Internet, pick your interest—dog-sled racing, environmental activism to save the Hudson River, Assyrian jewelry, and matrix organization—and you will find the all-whatever channel, whenever you want it.

The Internet is the most gargantuan repository of every kind of content imaginable, one that far exceeds any other collection of information ever compiled. Whether an individual seeks to learn about issues of universal concern, such as parenting, housing, cars, and leisure activities, or investigating the most narrowly-focused topic, like "perturbed sine-Gordon equations in the presence of periodic point-like weak inhomogeneities," it's somewhere out on the Net.

Content in Context

Chapter 5, "Digital Shape-Shifting: The Many Forms of DTV," introduced ten kinds of digital video or television. They resolve into two categories, narrowband and broadband:

BROADBAND	NARROWBAND
HDTV—high-definition television	eTV—enhanced television, data with TV
SDTV—standard-definition television	PConTV—Net content seen on the TV
DVD—digital versatile or video disc	TVonPC—TV content seen on the PC
BBTV—broadband television, transported over cable and telephone networks	CD-ROM—compact disc-read only memory
ITV—Internet television, streamed over computer and telephone networks	
TTV—telephone television, where data is displayed on screen-enabled phones	

To some extent, the amount of bandwidth each type of digital content uses is determined by how it was produced. But the platform it runs on can make a difference. Some content will become "midband" when it is distributed via LAN, cable modem, or xDSL:

BBTV	Bandwidth 300 kbps and 1.5 Mbps, arguably a midband type;
eTV	When broadcast with SDTV, eTV can use more bandwidth and edge towards midband, or BBTV;
PConTV	When sent over a company LAN, xDSL line, or cable, can become midband, or BBTV;
CD-ROM	Running at 1.5 Mbps, arguably midband, or BBTV;
ITV	As more consumers get high-speed Internet access, streamed material runs at 300 kbps to 1.5 Mbps and becomes midband BBTV;
TTV	As 3G phones and other handheld devices accept video, TTV edges towards the midband range, or BBTV.

Many producers that create content for a narrowband environment are developing it with plans to migrate it to a broadband environment, inspiring one conference audience member to proclaim that: "Narrowband programming is just broadband programming waiting to be born!"

However, keep in mind that broadband actually refers to midband programming, in terms of the bandwidth it uses. The truly broadband programming types are HDTV and SDTV, but these will be referred to specifically, leaving the term "broadband" free to describe midband programming—as it is employed in actual usage.

In the digital world, "content" refers to any material that appears on viewer screens. Writers, directors, producers, and programmers become "content providers," a much-disliked term among traditional practitioners in these creative fields. Content that is stored in quantity that can be accessed and repurposed for digital playback is sometimes called a "bucket" or "buckets of content."

Linguistic compromise is sorely needed in the arena of content creation and production. In the television industry, the images and audio presented to viewers are called "programming," referring to programs and shows. In the CD-ROM and DVD industry, the images, audio, and text that users get on their screens are collectively called "software." In computer companies, programming is a document coded in a special language, specifying the rules and variables that structure the operations of the computer; software doesn't mean what's on the screen—it's the computer programming.

In the entertainment business, the script is a blueprint composed of the scenes and dialogue that will make up the finished product. In multimedia, the script is software programming in an authoring language that defines what is going to happen as the user progresses through the material. In television and film, the person who oversees a project is called a producer; in the multimedia and video game fields, that individual is a developer. In show business, the finished product is a program, a show, or a film; in multimedia, it's called a title; for online projects, it's an application.

The terminology problem is not simply that the creative groups speak different languages; the bigger problem is that they often use the same words to express different underlying objects and processes. Beyond language, the computer-based creative companies (video games, Internet dot-coms) have a different culture than the traditional creative companies (TV, film). The disparate cultures lead to profound disagreements over what content should look like, where it should come from, and how it relates to its audience. Even here, the defined reality is contentious: in the computer industry, audience members are users; in the entertainment industry, they are viewers; in the gaming industry, they are players. A compromise of sorts is reached with the terms "viewser" and "telewebber" to describe people who converge their usage of TV and the Internet.

No matter how the work is labeled, or how difficult it is to break into the field, the fact remains that many people dream of doing creative work. A perennial romantic aura surrounds content. At the many conventions, demonstrations, expositions, and shows that feature new communication technologies, the phrase "content is king" is a cliché. The lure of content origination blinds many companies to its costs, both financial and organizational, at least where visual material is concerned. Although sometimes it may be cheaper to make original content, there are many advantages to licensing already-created visual intellectual property, rather than funding the development and production of new material, as well as providing the venue for its distribution. Licensing can be a less expensive, less risky way to acquire visual content, but it lacks the mythic power conferred by the creative process.[3]

The Search for the "Killer App"

When a new communications technology is first introduced, much time is spent looking for the "killer app." The term is short for "killer application," and it refers to content that is so appealing that it entices consumers to buy the whole system in order to get the one service or program.

A good fairly recent example of just how powerful a killer app can be was shown by *Star Wars Episode 1: The Phantom Menace*. It put the Apple QuickTime media player on 10 million desktops in four weeks in 1999. The killer app for TV is well-produced free programming. For online access, it has been E-mail, followed by information access and entertainment. Popular games are the killer app for proprietary game consoles like Sega DreamCast and Sony PlayStation II. Untethered telephone communication is the killer app for mobile phones. For cable, movies were the first killer app, selling subscriptions to the cable itself and premium channels; CNN's coverage of the Gulf War was the second.

In the early 1990s, many people believed that video-on-demand (VOD) or even near-video-on-demand (NVOD), would be the killer app to pay for the enormous cost of building wired interactive systems. However, the interactive TV trials of that time demonstrated that on-demand services would not bring in sufficient revenue. Ultimately, most companies concluded that the killer app wasn't a single service; rather, it was a bundle of interactive services, taken together.

Vincent Grosso, who worked on the early AT&T interactive TV trial, argues that there aren't so much killer apps as there are killer attributes. That is, what will make interactive systems successful is the way they are designed, rather than their subject matter, service area, or packaging.[4] AT&T continues to use this philosophy as it prepares its interactive TV services over the cable systems it operates after acquiring cable giant TCI.

Designing for Interactive Platforms

Developers recognize four types of content: (1) prepackaged transient, (2) prepackaged evergreen, (3) user-created added data, and (4) user-created content. Prepackaged content is prepared offline and uploaded to a Web site where consumers can access it. Much of this material is transient, replaced within an hour, a day, a week, or a month. If it is relatively permanent it is called "evergreen," meaning it will be of interest over some length of time. Prepackaged content may be structured so that users can add to it, gradually building a user-created database. Or the content may be entirely created by users, such as an online chat room or videoconferencing, or video telephone conversations, or rich collaboration environments like ThePalace.com.

Interactivity is an add-on to television, but it is native to computers. For that reason, much of the terminology and activity involved in developing and creating interactive material comes from the computer industry. For example, one of the first elements that must be defined in a project is the software and how it launches the various components of the interaction. Typically, the instructions are described by an API (application programming interface), which defines how an application or interactive program hooks into the system software and triggers services that are transported across the network. "Middleware," translation or gatekeeper software, stands between the specific application, the set-top box, and the operating system, making sure all the software components can talk to one another when an application is run.

At the moment, there are several incompatible hardware and software schemes to integrate interactive elements into a content stream, requiring content providers to write interactive sequences for each implementation. Multiple formats make the process expensive and time-consuming, so creative companies would much prefer standard APIs across platforms. The ATVEF (Advanced Television Enhancement Forum) is performing this function for enhanced television; HTML and related languages such as SMIL (Synchronized Multimedia Integration Language) and XML (eXtensible Markup Language), both under development by the World Wide Web Consortium, and WML (Wireless Markup Language) are all efforts to standardize and automatize the conversion between platforms and formats.

Standardization needs to address three levels of the content creation chain, or the 3 Ds:

- Design content once, using the creators, designers, and tools;
- Distribute content across multiple formats and channels: analog/digital, cable, telephone, computer, terrestrial broadcast, satellite, and mobile wireless networks;
- Display universally: digital/analog TV sets, set-top boxes, PCs, cell phones, and PDAs.

Other difficulties specific to interactive media also exist. For example, there is what is often referred to as the "chicken and egg problem":

- There is no consumer demand;
- Because there is no consumer demand, there is no content;
- Because there is no content, there are no transport services;
- Because there is no content or services, people can't see programming to know whether they want it, so there is no consumer demand.

The Internet broke that cycle. It demonstrated, for the first time in the 30 years of evolution of interactive platforms, that there is consumer demand for interactive content and services. The Internet has been a catalyst for releasing investment money both to build broadband cable and telephone networks and to develop and create programming.

Coming up with appealing interactive material is a daunting task that does not have much of a record of success. In fact, it doesn't have much in the way of a track record at all, and that is one of the most difficult aspects of the process. Every creative effort has a history to draw upon; but for the content providers of these new technologies, it's a very short one. Much is unknown: formats, appeals, flow, styles, uses, abuses, and limits.

The technology itself is a frustrating blend of opportunity and obstruction that changes constantly. Producing interactive programming requires a much greater coordination between creative and technical people than either group has experienced in other media. Nor is it likely that this collaborative overhead will decrease any time soon, even as tools and platforms become more sophisticated, because the ambition and complexity of interactive projects continues to grow.

Tom Connor, who heads marketing and brand strategy at creative development company 3 Ring Circus commented on the conceptual work that underlies a content project:

> We have to know how people think and how they want to be approached by convergent media. We have to know how to organize the information, how to layer it, how to signify it, and how to communicate with that box. For example, I sit almost on top of my computer, but I watch my television from the couch seven feet across the room. It is a different experience, and we have to understand that. We're starting to introduce the viewer to the idea of looking at television as an active experience rather than a passive one. The remote control itself is a very interactive device. It's probably what's going to make people more comfortable with the idea of interactive services being available through their television.[5]

Creating Digital Content

Although creating digital and interactive programming covers a remarkably wide range of devices, formats, and genres, there are some similarities that apply across them all. For example, it's more time-consuming (and therefore more expensive) to develop and produce interactive material than linear content. Even Web sites take considerable effort and teamwork.

The required investment means that preparing the business plan is typically the first step in a creative endeavor. (Some people would say that in the new media area, the business plan itself is a creative work.) The initial piece of documentation is a management overview and statement of business requirements that incorporates two key sections: the market analysis statement and the intended product statement. The market analysis statement includes market research and competitive analysis. The intended product statement gives a description of the intended service or program and lays out the resources that will be needed to get it up and running. It states the title of the project or program and its prospective launch date. The technical and performance requirements will be specified. Finally, it will address the programming strategy, outlining the intended audience, how the work will appeal to the audience, and ancillary selling points.

So what is appealing in the interactive domain? Listening to executives from the Internet, broadband, and interactive TV companies as they make speeches and appear on industry panels, one hears the same few concepts over and over. They might be called the 6 Cs of interactive programming:

- Choice: many different kinds of programming that are easily accessed;
- Convenience: always on 24/7, easy to use, easy to modify;
- Control: the ability to navigate and personalize access to content as a way of exercising choice in a megachannel environment. In addition, people want to be able to output material as they wish for E-mail, local storage, reuse, and repurposing;
- Community: enabling people to communicate with one another, and offering content that has contextual relevance to the group so they want to talk about it;
- Customization: enabling people to personalize content for themselves, their systems, and their circumstances;
- Cool: there have to be elements that make customers feel good about themselves or believe that others will think well of them. The Cool Factor will increasingly be defined by the generation that is growing up digital, but while we may not be able to define it, we know it when we see it.

The Web is the petri dish of content creation. Increasingly, it is the design environment for the development of interactive programming, no matter what medium or format it will eventually take, such as a game or broadband platform. The lower cost of Internet production is one incentive.

Although it is by no means inexpensive to produce a critical mass of sophisticated content for a Web site, it is considerably cheaper than doing so for an interactive television system. The Internet also encourages innovation because risk-takers are rewarded over risk-avoiders.

The fast and furious pace of the Internet plays a role as well—people sometimes say that Internet years are shorter than dog years, or that the events of a month take place in a week in the Internet space. One real-world month equals 1 webweek, and 1 real-world week = 1 webday. News travels fast on the Net, so what doesn't work has to be reworked right away—or at least rebuzzed—until it does. Producers have more control over their success, as they are able to match content with platform and audience in rapid innovation cycles. A Web site offers them the opportunity to look at consumer-usage patterns, bottlenecks, and other factors, and then make changes, all at modest costs compared to other interactive media.

These characteristics lead to a Net culture that supports creative endeavor. Innovation and experimentation are the norm. People who work in Internet time quickly learn not to underestimate the rapidity of change. Moreover, more than with any other media environment, creatives must respect the consumer's control—one click, and they're gone, outta there.

Despite the support, not everyone agrees that content is king in the digital world. Some claim that context is king, especially in a megachannel environment where it may be the only way people can understand what to expect from a particular program. For example, on ESPN, the audience knows to expect sports; on CNNfn, they get financial news, and on A&E they'll find cultural programming. Producers can't just put material out there; they have to engage people through its context. Another point of view is that the customer is king (or queen), directing content providers to study, understand, and serve their targeted audience segments.

Content may not be royalty, but it still exercises significant influence over the diffusion of communications platforms. Companies that will benefit from the spread of broadband networks like Intel, Microsoft, RoadRunner, Excite@Home, cable, and telephone companies are major sources of money for creating the new interactive content. For example, through the American Film Institute, Intel funded the AFI-Intel Enhanced Television Workshop program, which provides money, equipment, and mentoring to creatives for proof-of-concept materials that they can use to pitch entertainment and media companies to produce their ideas.

Now in its second year, the latest projects show a new integration of material and interactivity. As one of 1998's grant recipients, John Niemack, observed: "The ideas [this year] are much more attached to the content than they were last year. We struggled as documentary filmmakers to justify enhanced component and we tried to answer, 'What do people want to do? What do we want to lay over our own work that we spent months working on?' Today the enhancements are much more integrated with the content."[6]

There's a lot of interactive content creation going on. Table 13.1 lists just a few of the digital content projects that incorporate moving images that are in development or have been produced in the past couple of years.

Production Floor Content Management

Producing content in the digital world requires specialized systems to track work in progress. Since everything resides on hard drives in many different pieces and stages of development, a coherent way of searching, locating, retrieving, and tracking changes is essential. (Sometimes people call content management "media asset management" or "digital asset management," but these terms are more likely to refer to systems that personalize and distribute content to end users.) The elements of the project—text, graphics, audio, and video—are called the "assets," and the overall technology itself is called the "production floor content management system."

Originally, media companies installed proprietary systems that linked centralized storage, databases, and work stations. Disney pioneered production floor content management of feature animation assets with its DALS (Disney animation library system). Dreamworks built Nile, and Sony developed a pilot system. Pacific Data Images, the company that actually created *AntZ* found that their existing system that they used to track commercial production was inadequate for a feature, and they had to build a production floor capability as they also created the film.[7]

Dylan Kohler, who helped design Dreamworks' Nile system for *The Prince of Egypt* and other animated features, notes that it is a delicate business that must take into account how creative people actually work:

> *Just finding material isn't all that straightforward. When artists are working on a cell, there's a kind of shorthand. For example, there's the cell of Moses in the bulrushes that Bob liked, the one that Tom liked, the one that management approved, the one that the key chain manufacturer wanted to license, the early draft nobody liked, the revision Ann drew that had some interesting colors—and none of these terms are searchable by someone looking for the cell two years later to find a graphic for the cover of an educational DVD.[8]*

Despite the challenges, some kind of production floor asset management is necessary for companies involved in digital production, whether it is high-end postproduction and special effects for broadcast television commercials or an even moderately large Web site. Tracking may be achieved by a simple database created in-house or a multimillion dollar client-server network, but every enterprise that produces content ignores some

Table 13.1 Digital Content Projects

NAME	CONTENT	VENUE
ENHANCED TV		
3 Ring Circus	Interactive services	Multichoice (Satellite service, South Africa)
RespondTV	Domino's Pizza TV spot Discovery Channel	Multiple platforms
Caroline May Productions	Ken Burns, *Frank Lloyd Wright*	PBS
GTEMainStreet	Virtuality	GTEMainStreet
Intel-PBS	Zoboomafoo	PBS
National Geographic	Documentary	Intertainer
NBC, Interactive	Saturday Night Live	NBC
Programming	1998 U.S. Open Golf Championship	NBC
Pittard-Sullivan	TV spots for NTL and	NTL (UK cable system)
Pittard-Sullivan	Discovery Shop	Discovery Channel
Pittard-Sullivan	Rose Parade: Enhanced HDTV broadcast	Syndicated
Comspan/ Steeplechase	Judge Judy	Syndicated TV show
BROADBAND TV		
AFI-Intel projects	Expedition 360; *From a Whisper to a Scream;* Interactive *Academy Award Show; The Eddie Files; Liquid Stage*	In development
Big Band Media	Interactive StarGate SG1 (MGM property)	Showtime
BSkyB	Interactive sports channel	BSkyB satellite service (U.K.)

Table 13.1 Digital Content Projects (Continued)

NAME	CONTENT	VENUE
First Virtual Corporation (www.fvc.com)	Live content for telecom companies	Many partners and clients, including Bell-Attl.
FOX Sports	News and events	RoadRunner
Game Show Network	Jeopardy, Wheel of Fortune	25 million cable households
ImagicTV	Interactive services	6 telco customers in North America and Europe (including NBTel, Canada)
Innovation.com (www.innovatv.com)	Cooking show Sports show Financial	RoadRunner and others
Karaoke on Demand	Singalong service	ISDN service in Japan
NBC, Interactive Programming	TeenNBC	TBA
NetGateway	e-commerce apps	6 million CableOne and MediaOne cable customers
Next Level	Interactive services, Net access, 170 TV channels, digital audio	More than 12 tests and deployments in North America—phone companies (including U.S. West)
OpenTV	13 digital networks; one show: *Open Roads*	4 million viewers, including BskyB subscribers
Quokka Sports[a]	Several sports-related shows	RoadRunner
SourceNet	Interactive services: Movies on Demand, Net access	15 phone companies
VPRO	User-supplied content	Dutch cable channel, www.vpro.nl

Table 13.1 Digital Content Projects (Continued)

NAME	CONTENT	VENUE
INTERACTIVE TV		
Intertainer	Movies, music, TV, and information programming on demand; shopping	Partners: Comcast, U.S. West, NBC
Xing	Movies on demand	Dutch cable systems
INTERNET + TV		
ACTV	HyperTV: Web material synchronized with TV program	Showtime Network (Viacom) TBS, *Cyberbond* (Time Warner)
Spiderdance	*WebRIOT*, cable TV show with synchronized Web site for online play	MTV Network (Viacom)
INTERNET TV		
AtomFilms	Films, esp. shorts and animation	www.atomfilms.com
BitMagic	Animated clips, interactive games	www.bitmagic.com
Entertaindom	Warner Bros. Online entertainment portal site	www.entertaindom.com
Gamesville	New site from Hasbro Toys	www.gamesville.com
Go Network	Disney's entertainment portal site	www.go.com
CinemaNow	750 feature-length films	www.cinemanow.com
Shockwave.com	Flash-enabled games	www.shockwave.com

Table 13.1 Digital Content Projects (Continued)

NAME	CONTENT	VENUE
The Spot (and East Village)	Famous firsts— Webisodics that put entertainment on the Net	defunct
WireBreak	Several shows, including *It's Saul Good* and *Girls Locker Room*	www.wirebreak.com
WIRELESS HANDHELD CONTENT		
AOL	AOL Anywhere	Mobile devices
AvantGo	Formats *Wall Street Journal*, *New York Times*, and *USA Today*	Palm devices
C-SPAN	Election news	Palm devices
Puma	Net content	Palm devices
Riverbed	ScoutWeb—Net content	Mobile devices
Yahoo	Acquired Online Anywhere to develop programming: News, sports, weather, and E-mail	Sprint's WAP-enabled PCS phone
MISCELLANEOUS AND MIXED FORMATS		
Warner Bros. Online	WebDVD: Distribute weekly TV shows, including *Drive-On*, other popular WBTV shows, and original shows	Content on DVD, authorized when user goes to Web site

[a]Also creates Internet programming.

form of content management at their own peril. The next sections will explore the experiences of people who have been working to create interactive content for the new digital services.

The First Content: Interfaces and Navigation

Interfaces and navigation begin with the device needed to access them—some kind of remote. It is arguable that in all its 50-year history, nothing changed television viewing so much as the remote control device (RCD). The new television systems call for a new generation of RCDs that are much more complicated to use than the current, familiar remote. For one thing, interactive programming requires more buttons than people are used to. Viewers must be able to browse menu pages, highlight the available choices one at a time, make a specific choice, and sometimes send their choice upstream. For video, the RCD needs VCR-type buttons (rewind, fast forward, and pause) as well. Finally, the RCD must be comfortable to hold and the buttons need to give a tactile response to the viewer's actions—a sense of a "click" or snap. It often takes several years to develop an RCD that will handle the functionalities required as well as incorporate people's physical requirements for ease and convenience of use (a field sometimes called "ergonomics").

The interface and navigation system software may seem negligible. Not so—they occupy a powerful and sensitive position in the overall success of the system and specific programming carried on it because they control screen real estate and are the gateway into programming.[9]

The interface is the visual and functional field that lets the subscriber access the available programming and services. Navigation is the set of procedures embedded within the interface that the subscriber uses to move around the system. Usually, interfaces and navigation systems are based on an operating metaphor, such as a mall, a neighborhood, or a dial/tuner. As they become more sophisticated, interfaces are beginning to become metaphors in their own right. Ideally, the graphics and text that make up the interface and navigation software are consistent across the applications. The Intertainer interface is a good example of an elegant contemporary interface/navigation design, as shown in Figure 13.1. Compare it to an interface designed for the Internet in Figure 13.2 for the same service.

A key component of navigation is the electronic programming guide (EPG). Satellite companies' adoption of interactive EPGs has been important to the growth of their subscriber base, according to satco (satellite company) marketing executives. EPG leader TV Guide, Inc. reports that about 40% of viewers say that the guide influences their decisions about what to watch. Subscribers recognize the importance of EPGs to their

Figure 13.1 An interface for interactive television. Source: Intellocity.

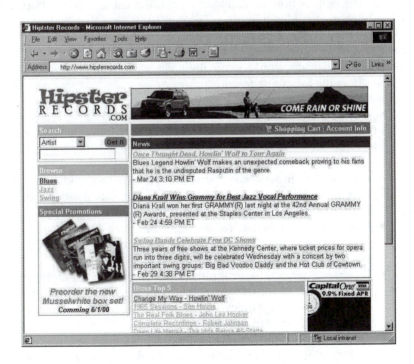

Figure 13.2 An interface for the Internet. Source: Intellocity.

viewing experience—some customers even pay monthly fees for the most sophisticated versions.

In a megachannel environment, an EPG is the only way viewers can figure out the programming and services that are available to them. A system with 70 to 80 channels can deliver about 10,000 different programs across a week, an unwieldy number for printed guides. Further, an EPG offers many features that the paper version cannot, such as point and record, point and order, sorting functions, genre listings, program reminders, bill viewing, favorite channel selection, and channel lockout to let parents prevent children from accessing certain channels.

The most common EPG is the familiar TV Guide Channel that is carried on 2,200 cable systems, and has 50 million viewers that tune in more than 100 million times per month. It is more popular than HBO, E! and Headline News in terms of viewership. The TVGC is a scrolling service that, given the growth in the number of channels, has become somewhat unsatisfactory. In a large capacity system, it can take more than 6 minutes to scroll through all the channels. In addition, scrolling can only show the next few hours of viewing.

The TV Guide Channel Interactive replaces the scroll with user-controlled interaction. TVGC-I is the leading digital cable interactive program guide in the United States, distributed in more than 800 systems by over 25 cable MSOs. It is now available in 34 million cable homes; about 1.5 million of them actually subscribe. The company says that their interactive subscribers report a high degree of satisfaction. More than 76% consider it one of the most valuable components of their digital package, and 40% report that they would drop their cable subscription without it. TVGC-I offers the following features:

- Flip Bar: The Flip bar appears every time the channel is changed, even during commercials. It displays the currently tuned channel, the name of the program, and its start and end time, and it disappears after 5 seconds.
- Browse: The Browse bar allows the viewers to see program listings for other channels and for other times without missing a minute of the program they are currently watching.
- Listings: Easy-to-read, color-coded program listings are organized by start time, by Channel, and by various genre categories, including movies, sports, and children.
- Program Search: The Program Search feature is an alphabetical organization of all the programs available to the viewer. Programs can be found quickly by entering the first few letters of the desired program's title.
- Program Info: Simply pressing the INFO key on the remote while highlighting a program listing will display a brief description of that program.

- Reminders: Reminders can be set for any program that comes on in the future. The Reminder will appear a few minutes before the program starts, allowing the viewer to tune to that program, or hide the Reminder and continue watching the current program.
- Impulse Pay-Per-View: Viewers can buy a pay-per-view program with just three clicks of the remote.
- Messages: Brief messages can be sent to the viewer both from TV Guide and from the operator. A little, yellow envelope appears on screen whenever a new message is received.
- Favorites: Viewers can designate channels as Favorites, then quickly navigate to those channels while watching TV.
- Parental Control: Objectionable programming can be restricted by setting locks on channels, movie ratings, TV ratings, and program titles. A purchase code can also be set, preventing purchases without parent's approval.

Creating EPGs and navigation devices is a specialized field. Here's how designer Joy Mountford describes the process: "When you're designing an interface, you need also to ask, 'How can I marry the user's desire with a set of good constraints?' These constraints should allow them to explore. They should be consistent. They should be forgiving. They should be interesting. They should be clear. And they should allow the user to always stay in control."[10] She adds that the interface must give users a system or a scheme for coping. It must allow them to exit at any time.

Steve Schlosstein developed a step-by-step approach to the problem:

- Develop the interface concept
- Identify the functionality of each interface model
- Identify the user population
- Identify high-level constraints, such as hardware and software
- Create a storyboard sketch for each interface metaphor
- Identify the usability goals
- Plan the screen layout
- Analyze the input device and design appropriate action sequences
- Develop prototypes of key screens that include the concept, the tasks, and the look and feel
- Test
- Redesign, based on test results
- Conduct final review and usability testing [11]

A distinctly different point of view is advanced by Kai Krause, who creates graphics tools for computer users. He maintains that the one rule of design is to "keep the user happy." Krause refuses to patronize his customers by dumbing down the interface so that it loses all appeal and surprise. Rather, he builds in an evolution of learning so that as the user continues to interact, new messages and functionalities appear.[12]

Creating Web Content

Content for the World Wide Web is being created rapidly, fueled by the growth of the Web. The Gilder Technology Report estimated that in November, 1999, there were about 340 million Web pages, growing by 1,000,000 pages each day. The Excite.com site alone has 49 Terabytes, Amazon.com grew by 42 Terabytes in 6 months, and Mail.com put up 28 Terabytes in just 1.5 months of operation. Today, the total amount of material on the Web runs into the Petabytes and will soon reach Exabytes—that's a lotta data! The number of Internet domains is also increasing quickly, from 14.4 million in 1996 to 43.2 million in 1999.[13]

In the beginning, the Internet carried only the word—text. Audio first appeared online in the form of highly compressed files in 1990—the first hint that musicians would some day be able to place music on the Net and let fans hear it in real-time, bypassing the traditional industry structure. The ability to handle large multimedia files—graphics, audio, video—is a product of the World Wide Web. The Web began in 1992 as an experiment by a laboratory in Switzerland, the European Laboratory Particle Physics (CERN) to access and transfer graphics files more rapidly than using a server-type mechanism. The attractive graphics the Web enabled proved to be a catalyst for the expansion of the Internet. By 1995, 10% of the enormous Internet traffic was WWW activity, and more than 18 million Americans and Canadians used the Web in a three-month period.

Although theoretically video is just another file format, it has taken considerable engineering ingenuity to enable video to run over the narrowband telephone network. The RealAudio player became available in 1995, and 85% of all streaming media enabled Web pages to use Real software, streaming over 145,000 hours of material each week over the WWW. The two other major companies in the webstreaming player space are the Windows Media Player and the Apple QuickTime Player. Real Networks released RealPlayer 7 on November 8, 1999. It was downloaded by 3 million players in 7 days, bringing the total number of RealPlayers on user computers to 88 million, with 61 million registered users. The Windows Media Player, introduced in the summer of 1998, claimed 35 million users less than a year later.

Specialized software is used for Web navigation. The browser is a graphical user interface (GUI) front end to an application that interprets the language of the Web, HTML, or HyperText Markup Language. The three most popular browsers are Microsoft Internet Explorer, Netscape Navigator, and Opera. They package tools for surfing, accessing mail and news groups, and linking to other programs for building Web pages. In addition, the browser will launch additional viewer and applications programs that add functionality to it, as though they were part of the same software package. These additional programs are called plug-ins and helper apps. Plug-ins become part of the browser software and boot up with it, while helper apps are launched only when they're needed. As of late 1999, about 70% of people online used Microsoft's Internet Explorer browser, 24% used Netscape Navigator, with the remaining 6% split among Opera, WebTV, and others.[14]

Three types of online content are created for users: information, entertainment, and applications. All three types draw on the new, sophisticated tools of Web content creation:

- Streamed audio or video
- Three-dimensional worlds created with VRML and VRML+, Virtual Reality Markup Language
- Fast, crisp, colorful animations, produced with Macromedia Flash
- Fast application support with Java

Portals and Destinations

Companies that build large, expensive sites want to have "sticky" content. Stickiness refers to how long surfers spend on the site, on average, because the longer they stay, the more opportunities there are for promotional and advertising messages to appear. Since many Web users seek rich media content, music as well as visual material, sites that hope to become popular have considerable incentive to use Flash and Shockwave animation in addition to audio and video. Moreover, they need to change the content often in order to give visitors a reason to return. The measure of return visits is another important measure, because return visitors offer marketers the opportunity to deliver multiple impressions to the same user.

The category a Web site fits into determines how important stickiness is to them. Portal sites are venues that people can use to start a Web journey. Typically, they aggregate many different services, utilities, and content types. Yahoo, Disney's Go network, Netscape's NetCenter, and NBC's Snap.com are all examples of portal sites. They will offer search engines, information, stock quotes, travel services, and many other conveniences to the Net traveler. A destination site is one that has content that people will want to take time to explore or engage in activities. For example, "Entertaindom" by Warner Bros. is the first destination entertainment site put up by a major Hollywood studio.

To some extent, portal sites are also destination sites, because they want to keep people as long as possible. However, portals recognize that users will eventually jump off the site, possibly after they have located their next destination through the search engine. However, destination sites are not always portal sites. They want people to stay, and they discourage them from leaving, or at least do not encourage it. For example, game sites don't offer stock quotes or Web-side search engines; their users want to play games, and they surf to the site to engage with other players in games they enjoy.

Visiting sites isn't the only way users get content, and users download a lot of material to use offline. Content that can be utilized locally and offline reaches consumers via "push" or "pull," short for "server push" and "user pull." (See Figure 13.3.) Push means that a server sends content to the user automatically, without a specific request from the consumer. Pull means that the material comes after being requested. For example, when visitors surf to a site, they are requesting the material—so it gets to them through user pull. They notice a promotion for them to see a popular movie trailer, and they click on the button. Before the video (which they have requested, or pulled), an advertising message appears on their screen, a pop-up window. They didn't request the ad—it is pushed to them. After they click to close the ad window, the movie trailer starts running—material that was user-pulled. Some material is pull-push, or "opt-in." Examples are serial e-newsletters and e-zines. In these

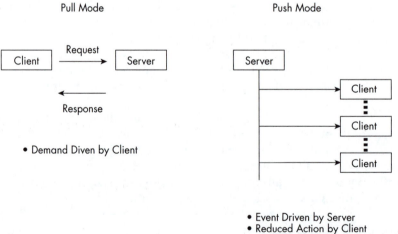

Figure 13.3 Comparison of pull and push modes.

cases, users sign up for a publication—they pull it, requesting it to come to their mailbox on a schedule. After the original request, it is pushed to them regularly.

Entertainment Content

This category represents some of the most popular material on the Web, such as films, videos, animation pieces, music, and games. Indeed, probably the most eagerly-awaited piece of video shown on the Web in its first five years was the *Star Wars Episode I: Phantom Menace* trailer, which reportedly clogged the local area networks of more than one university. Games are universally popular; they will be covered later in this chapter as a genre that cuts across all media platforms.

Users access the work of both little-known artists and highly sophisticated entertainment industry giants, particularly in music, but increasingly in all types of artistic endeavor, from books to photography to digital films. The Web is a relatively inexpensive distribution medium where "garage bands," musicians who have no professional representation, can reach an audience by putting their music on the Internet in MP3 format for download. Finding and downloading music has become so popular that in the summer of 1999, the second most popular search word on the Internet was "MP3"—second only to "sex." There are hundreds, perhaps thousands, of sites that index music, providing clips, cuts, videos, bios, photos, and club and concert dates. The controversy over MP3 and the threat it presents to the protection of intellectual property is a powerful force sweeping through the entertainment and media industries. Most analysts recognize that it is the narrowband harbinger of what will happen with motion pictures, television programs, and other video material as broader band access comes to the home.

The problem of bandwidth constraints in the last mile led to the failure of high-profile Internet "networks," such as Pseudo.com, DEN.net, and others. Pseudo leaned towards live streamed shows. DEN tended to archive episodes of dramatic and comedic fictional programming and edited magazine shows, averaging about 6 minutes in length. AtomFilms and WireBreak offer animated and live-action shorts.

For the most part, there has been much more information about entertainment than actual entertainment on the Internet. The large media companies have been slow to open up shop on the Net, perhaps fearing that it will cannibalize their existing audience. However, this conservative stance is changing. On November 29, Warner Bros. became the first Hollywood studio to put up a destination site devoted primarily to original content and active entertainment—not just marketing and promotional material, chat, and information about entertainment. Called Entertaindom (www.entertaindom.com), the site's content includes animated and live-action shorts, clips from old TV shows and movies owned by the media giant, games, entertainment news, and other features.

On a much smaller scale, News Corp. ran original programming from the creators of the *Blair Witch Project* on their Fox.com site to promote the related program in the 2000 TV season.

Yet at the same time, the company revealed its ignorance of the Net when it made a sadly "bonehead" decision to have its attorneys attack the creators of fan sites for *Buffy the Vampire Slayer*, threatening them with copyright infringement, shutting them down—and alienating their biggest fans and best customers. By contrast, MGM executives were impressed by a 17-year-old Web wizard's pages dedicated to the *Stargate* property and recruited him to contribute to their official site.

Many cable networks have a distinct cyberpresence. Viacom, Time Warner, and Discovery have developed aggressive strategies that give them a high online profile. There is considerable variation among the broadcast television networks, with NBC making the biggest commitment to the Internet, followed by CBS and ABC, while FOX straggles far behind its competitors.

Networked Networks

With the exception of NBC, until 1999, the broadcast networks appeared to inhabit a time-warp that preserved the environment of 1982, enabling them to display a serene indifference to the digital disarray of their business landscape. However, this watershed year showed movement on the part of some media companies to expand their plans to include the proliferation of distribution mechanisms and reception devices and the rapidly changing, increasingly connected audience. Every network except FOX got networked.

Early on, NBC was the most digitally aware of the broadcasters. Back in February of 1995, the network launched NBC Desktop Video. The service allowed PC users to receive live business and financial information from their PCs. Content from NBC News, CNBC, and the Private Financial Network was repurposed for the new service, which also airs live feeds of corporate announcements and significant annual meetings.[15]

In 1999, NBC made its Internet move.[16] The network threw in all its Internet properties and joined with Xoom to create NBCi, and its associated site, nbci.com. The NBC Internet properties made an impressive contribution. NBCi started with a large base of services, including a search engine and directory service, news and media content, user-created content, broadband content, Web community capability, e-commerce, chat rooms, online greeting cards, personalization, E-mail, and digital storage. By offering attractive free services and content and maintaining a large and diverse range of active communities, NBCi will encourage visitors to become members, collecting and repackaging information about them to develop a profitable business.

NBC.com provides information about the shows, stars, and schedule to viewers. It also gives them chat rooms, bulletin board services dedicated to NBC shows, and show-related e-commerce. NBC.com develops specialized content for advertisers. For example, *Jay's Garage* came out of Jay Leno's enthusiasm for cars and driving, which he frequently expresses on-air. The site gives automobile advertisers crosspromotional capability by coordinating television network advertising with sponsorship of *Jay's Garage* on NBC.com.

Another effort, the NBC-in.com Web site, is a local news and information portal developed in conjunction with 104 NBC owned and affiliated TV stations. Television viewers can access news, sports and weather from MSNBC and, in the case of NBC affiliates, the local station itself. They can also access services such as job search, local advertising for real estate and automobiles, restaurant reviews, and telephone directories customized for the relevant geographic area.

Finally, there's VideoSeeker, a video aggregator with a searchable database. The service currently offers users free online access to topical and archival news video from MSNBC, clips from NBC television programming, movie trailers, interviews, and links to music videos and other Web sites with streaming video content. Content includes material from NBC, MSNBC and Dow Jones output deals, Film Scouts, PR Newswire, Universal Studios, Fox, Disney, Sony, and MGM.

NBC starts with 49% ownership, and after a year it will own 54%. NBC's Internet assets include its portal Snap.com and the business units in the multimedia division, including NBC.com, NBC-in.com, and Videoseeker.com. It also threw in 10% of its new CNBC.com site.

The Xoom piece is also important. At the time, Media Metrix measured Xoom as the twelfth-most-visited site on the Internet. The company boasted a sophisticated e-tailing operation, selling computer software, computer accessories and peripherals, consumer electronics, clip art on CD-ROM, gift items, health-related products, long distance services, and personal finance newsletters. It also had a DVD movie club and a travel club, among other products.

But the most important action that Xoom brought to the table was the data the company amassed on 4 million consumers, many of them previous purchasers of Internet goods and services. It also managed a technological powerhouse: a system with 2.5 terabytes of unformatted disk space that supports over 25 million hits per day, has a peak bandwidth of over 90 megabits per second, and transfers 350 megabytes of data each day.

NBCi was the first publicly-traded Internet company to combine portal, content, and community services, all in a strategic relationship with a major TV network. The reach and power of the TV network is used to raise awareness and drive consumers to Snap.com's portal site and navigation services, leveraging Xoom's community and direct e-commerce services. The idea is to make NBCi the foundation for a next-generation media company capable of reaching users through a variety of media including television and the Internet.

It's a powerful and pragmatic combination of high-profile content and back-end e-commerce at work. The Xoom database and the new site-visitors that respond to the cross-media pitch let NBCi provide advertisers and e-commerce partners everything they need to deliver targeted messages and offers. NBC calls the TV-Web crossplatform synergy its "trans-media capability." The NBC TV network is in an excellent position to monetize this relationship because its primetime programming draws a higher percentage of Internet users than any other television service, according to the results of a study commissioned by NBC from Nielsen Media Research in 1999. Measurement during the February sweeps period among viewers with access to the Internet, found that NBC primetime programming attracted 25% more viewers than the primetime programming of its nearest network competitor, and 65% more viewers than the primetime programming of the top ten cable networks combined.

NBCi will advertise heavily on the NBC network, purchasing $380 million in advertising over a four-year period. The TV network will also leverage its Internet and cable advertising connections to its advertisers, offering them unique ways to reach consumers through its Internet connections. Moreover, there is also the content play. For example, when NBC broadcast *Schindler's List* in March, 1999, TV viewers were directed to Snap.com for a comprehensive collection of Holocaust research, information, and links available on the Web. Snap.com will continue to pursue this integrated strategy of leveraging multiple media platforms for marketing, promoting, and content crossover.

Going forward, the company has a broadly-based strategic plan. Three elements are particularly important: developing vertical hubs, creating broadband services, and pursuing international opportunities. Vertical hubs are Web sites that serve an active set of users with similar interests. Aggregated consumers provide marketers with specific sales targets they believe are likely to be interested in their products and services, and they pay premium prices for such customers. This is an area where NBCi will seek acquisitions and partnerships. A broadband play arises from the availability of high-speed cable and telephone service, and the development of digital programming and services for them. NBCi plans to expand the nature and amount of broadband content they make available, both through development and acquisition, particularly in the areas of video, audio, animation, and gaming. The international moves include developing local language sites that offer service to local advertisers and e-commerce companies. NBCi has targeted specific European and Asian countries and wants to find partners there to provide local support and promotion.

Of all the broadcast networks, NBC's NBCi is the most comprehensive and ambitious convergent undertaking. It will be closely watched to see how its "trans-media" capability evolves.

The company also plans to deal with the problems it sees in broadband delivery. NBCi has pulled together a seamless national network with its servers located in major markets to cache content on the edge. Moreover, it has made a deal with a CLEC DSL partner and will negotiate with other telephone companies to place its own set-top box in the home with the NBCi interface on the desktop. Moreover, there will also be a scan converter to pass through broadband signals to the TV screen.

The CBS Internet strategy also became clear in 1999, second only in sweep to the Peacock Network's initiative. CBS New Media is responsible for the company's involvement with evolving technologies, including its Internet holdings and partnerships in Web services, and made a series of Net-related strategic acquisitions and alliances:

- An arrangement with America OnLine, making CBS the only branded broadcast news partner on AOL and Compuserve's news channels
- Extension of an equity-for-promotion agreement with SportsLine USA. SportsLine produces the league Web sites for major league baseball, the PGA tour, and NFL Europe League and is the main sports content provider to AOL, Netscape, and Excite.
- Exchanged promotion and advertising services for 35% of hollywood.com
- Exchanged promotion and advertising services for 50% of storerunner.com
- Acquired 35% of switchboard.com, an online directory for personal and business use
- Acquired 30% of thirdage.com, a site tailored to the interests of people 45–64
- Purchased a 33% equity stake in office.com, a site that provides information for small and medium-sized businesses
- Took a 35% minority interest in MedScape to produce a new consumer health site
- Took a 20% interest in Rx.com, an online pharmacy

In many cases, these were shrewd deals that were not made with just cash. The network swapped promotional time at premium rates in return for hefty ownership percentages. These new ventures are in addition to CBS's 37% ownership in marketwatch.com, which also appears on the broadcaster's newscasts as a source of financial information. In addition, the network owns Country.com and the CBS.com sites directly related to its television programming operations.

Putting these disparate moves together, the CBS Internet moves give it interactive digital content that comprehensively mirrors the material it owns and produces in the broadcasting arena: business and personal finance (marketwatch.com, office.com, storerunner.com); sports (sportsline.com); health (medscape.com; rx.com); entertainment (hollywood.com); lifestyle (thirdage.com); and connection (switchboard.com). Although the company has never laid out its Internet strategy, CBS could roll together quite a complex TV-Internet powerhouse. CBS news material is available on AOL, Netscape, and Excite. In the reverse direction, there is the branded marketwatch.com and sportsline.com elements on CBS News programs.

Christopher Davidson who works in current programs at CBS, says he's trolling the Internet for ideas all the time. "At last count, I think there are 25 or 30 independent chat rooms for our programs alone. We all monitor them as part of our jobs to be aware of what is going on in the world. I check them out, listen to what people have to say, getting an idea of what's going on so we can make a more direct connection with the audience," he said.[17] He believes that network programmers are challenged to develop programs that approach the level of personal and emotional connection that people experience in chat rooms on the Internet. Along those lines, CBS plans to tie some new broadcast programs into the Web by providing chat rooms so that people watching a show can chat about it as they view it.

ABC is still a distant third behind its more aggressive competitors, adopting an "extend and defend" strategy, which means to extend the brand and defend existing operations from encroachment from the new medium. ABC's plans were not coordinated with those of its corporate owner, via the Disney-owned portal, the Go network, or its Infoseek search engine. The network's 1999 Internet effort came out of a ten-month-long strategic study that included senior management and marketing people across the company, and demonstrates beautifully the pitfalls of asking the executives of an existing operation how to grasp the opportunities presented by a new technology.

The best ideas they could generate were limited to using the Net as a rather banal marketing tool. In addition to "webvertising," the company hoped to create events and promotions online, including promotional webcasts of new shows. Copying Showtime, the network also created an E-mail database to push requested information about programs to viewers, and more online advertising, using a rich multimedia format.

FOX is even less forward-looking. It takes real effort to find Internet intelligence within the network's ranks, because the top executives avoid the Net like bad breath. Essentially, FOX has let show producers make their Internet presence. When the producer is *The X-Files'* Chris

Carter, who is right on top of the Net, then there is an imaginative and thoughtful use of the medium; when the producer is a nonuser, there is little initiative.

The least involved broadcasters are the new PAX network, the struggling UPN, and the WB netlet. In this last case, however, the stance makes sense. They are in the throes of establishing a new entity.[18] However, it's impossible not to wonder how the effort might have been helped by judicious use of the Web, given the Gen-Y demographic target of 12-to-24-year-olds, essentially the first Net Generation. It's impossible not to question why these popular shows were not more integrated in the launch of Warner Bros. massive Web site, Entertaindom.

Informational Video and Audio Programming on the Net

Virtually every large news organization from the broadcast and cable networks and many local stations let users see moving video of their stories. Thousands of radio stations have a Web presence that replicates their over-the-air signal by such content aggregators as Broadcast.com (www.broadcast.com). Many large companies like General Motors and wireless-phone-maker Ericsson have Net-based video or audio streaming from their sites. High-quality Flash animations are increasingly popular as well. The site at www.2000enFrance.com showed a VRML (Virtual Reality Markup Language) representation of some of the art pieces created for the millennium celebration in Paris.

Streamed graphic information has become so common that it requires a book, perhaps even an encyclopedia, to detail all the content that is on the Web. Fortunately, the Web itself is the best source of information about the information available on it. Before it closed its doors, FasTV (www.fastv.com) was the first consumer-oriented site that provided a search engine to let users look for the video and audio material from many different sources on the Net. The site used video streaming technology to deliver the selections to them. Users typed in natural language to indicate what they wanted to see, clicked on "find," and FasTV delivered the results in concise, personalized segments to the users' PCs.

FasTV was the most ambitious of the video search engine sites, indexing a vast, ever-changing storehouse of video from across the Web and elsewhere. (See Figure 13.4.) Another video search engine is Virage's VideoLogger, that is used by C-SPAN and ABC News to log, index, search, and retrieve material on their own Web sites.

Online Applications

Applications are service offerings that allow users to process and store information on the Net. They are mounted on the Web by companies

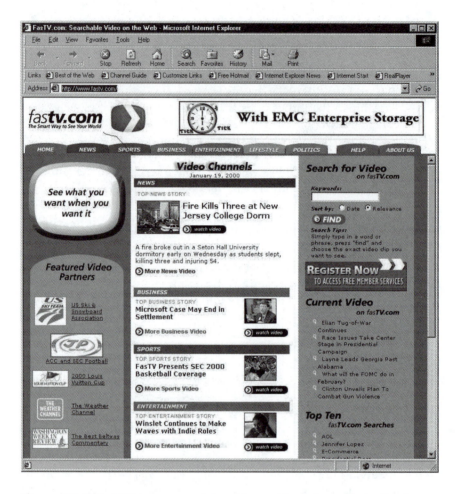

Figure 13.4 FasTV let users search for video and brought up short clips for them to browse. Source: FasTV.

called ASPs, which stands for Application Service Providers. Applications will be a big part of the networked future, prompting some analysts to believe that just as the industrial economic expansion was built on the mass production of goods, so the Internet economic expansion will be built on the mass production of services.

Applications are an emerging form of content that warrant considerable attention because they will be the means by which companies make user-generated material part of their content creation and overall business strategies. The high-dollar business-to-business applications tend to offer access to processing and analytical tools. But video and audio are important in the two consumer areas of collaboration and personal use.

Collaboration applications are E-mail, chat, groupware, and conferencing services. They let people work and play together on a common site even though they are geographically distributed. The experience of The Palace (www.thepalace.com) with young children points to a shift in how people use the Net after they get used to it. Personal applications can be office suites like word processing and spreadsheets to extend productivity, but they might also include software to work with artistic projects that include drawing, digital painting, photographs, and even video.

For example, www.undergroundfilm.com established a community of digital producers, and gives them a way to reach their audience directly. There are many sites that let people post their content, such as www.popcast.com, www.zing.com, www.my-kids.com, www.my-wedding.com, and many more to come. Originally, the Dreamworks/Imagine Entertainment startup, POP.com (www.pop.com) planned to choose user-submitted pieces that would be posted alongside those of the professionals hired by the site managers to provide content. However, it never launched.

Online Marketing, Advertising, and Promotion

The Net extends many benefits to marketing promotion and advertising activities, compared to traditional media. It is cost-effective. The material is current and easily updated. Segmenting potential customers based on their interests is relatively simple. Messages can be personalized and the customer can quickly provide the marketer with detailed information, at little or no cost.

Higher consumer response is making the use of rich media grow in importance for online marketing campaigns. Graduate students at the University of Michigan School of Business Administration, in cooperation with Athenia Associates, found that adding movement to ads through animation or video increased click-through rates from 77% to 228%.[19] Click-through means that users click on an advertisement to learn more about the offered product. Typically, the click-through rate for banner ads is not high, averaging less than 1%. RealNetworks analyzed 52 rich-media ads that use streaming audio or video and the average click-through rate was 5.9%. An interstitial in-stream ad that plays between content pages for General Motors' Buick model garnered a 6.3% click-through, reported the streaming company.[20]

Words from the Web-Wise

When the Web first came into being, there were no Web-site building tools. To create their Interactive World's Fair (IWF) site, one of the first large entertainment destination venues, World, Inc. had to team with Landmark Entertainment, a company that develops amusement centers and location-based entertainment sites. Landmark built Caesar's World in Las Vegas, Nevada, and creative groups from the two companies used Landmark's proprietary construction software to keep track of the millions of software creations it took to build the IWF.

Now the picture has changed considerably, and there is a range of tools available. As a result, attention has turned from sheer mechanics to understanding the parameters of creation for the Web. In *Digital Babylon: How the Geeks, the Suits, and the Ponytails Fought to Bring Hollywood to the Internet*, authors Geirland and Sonesh-Kedar describe how this new medium is so different from traditional media, making program creation an elusive undertaking:[21]

- People on the Web are active, unlike the viewers of movies and TV. They don't necessarily want to be spoon-fed.
- People don't want a simulation of bad TV online.
- Users do not "tune in" once a week or once a day to follow the plot line of an episodic show. Appointment viewing is not Net.
- Users surf with a purpose.
- Creating content for the Net is horrifyingly expensive—and there is no viable business model to support it yet.
- These models are unlikely to emerge soon because of so-far irreconcilable conflicts between the suits (executives and investors), the geeks (techies and engineers), and the ponytails (creatives).
- The book quotes analyst Denise Caruso's comment that "The Web is in fact not a mass medium but a medium for the masses, who are already well along in the process of making it their own."

So much is unresolved. For example, take the oft-repeated goal of many e-execs and creative people to make a site "sticky," to keep people on the site for as long as possible. Advertisers reward it, so site operators will undoubtedly do their best (or worst) to keep people hanging out as long as possible. The problem is that this attitude is profoundly un-Net-like, where hyperlinking is the behavioral norm, enabling geodesic motion along the lines of least resistance.

Costs are always important, no less so for Internet content. There is enormous variation in the amounts spent to build Web sites, so it is difficult to compute an average cost. As a general rule, a professional site with multiple areas, frames, E-mail links, CGI scripts for forms and backend database writing capability, will run between $20,000 and $50,000, according to Joanne Burns. She leads a team that makes Web sites for CBS television shows at Eyemark Entertainment, which inherited a number of domains (*Psi Factor, Caroline in the City, Pensacola: Wings of Gold, Amazon,* and *Dr. Joy Brown*) that were already on the Net. Burns and company redesigned them to maximize their appeal and usability. Speaking at a seminar at the American Film Institute, she demonstrated the ten commandments of Web-content design, which offer useful, specific, user-centric guidelines.[22]

1. Keep people on the site. If you must let them link out, be sure to create a reciprocal link so they can return.
2. Provide the ability to define site sections: Don't use funny and obscure titles.
3. Keep navigation simple and located together, but position the navigation bar so it doesn't constrict design. Keep the navigation bar consistent throughout the site, and provide description for each area, with additional information in "mouseover" text. Make it easy for people to identify where they are within the site.
4. Keep people focused and directed. Design logical placement of all elements. Have the ability to emphasize site areas and show highlights. Emphasize site "special features," and provide a way to post bulletins.
5. Provide an easy way to search.
6. Provide links for users to download any needed plug-ins, such as Flash for animations, media players for audio and video, and Shockwave for games.
7. Personalize and create a community when appropriate.
8. Provide a way for people to use your graphics as wallpaper—it's an inexpensive form of promotion—but be sure to offer instructions on how to download it and use it for wallpaper and screensavers. Provide an example of what the full wallpaper/screensaver looks like, and give options for different size resolutions.
9. Create self-populating content and design form fields to add information to pages.
10. Design a web-based administration; this make updates easy.

Streaming audio and video on the Internet brings some unique creative challenges. First and foremost, except for webcasting live events, producers must realize that the Internet is not about streaming material for an hour. And a key difference in creating a video for streaming, as opposed to preparing it for traditional television, is movement: with streaming, the less movement the better. MTV-style videos do not stream well because there is simply too much information to send. (Recall that compression algorithms depend on frame-to-frame continuity, so every time there is a scene change, the entire frame must be sent. There just isn't enough Net bandwidth for such rapid editing.)

A second consideration is that most people multitask when they are using the Web. Even when the Net is too congested, continuous audio is crucial. As Camille Alcasid, then-director of new technologies and project development at UltimateTV, put it: "You can have some dead video . . . but avoid dead air."[23]

There are 5 steps to creating streamed video (and accompanying video):[24]

- Step 1—Capture material from a digital camera or analog source, a camcorder or videotape. If analog, the video needs to be digitized and stored on disk.
- Step 2—Edit and author video. Editing is putting video sequences together to make up the video, eliminating unneeded material. Authoring weaves the video into the multimedia whole with other elements such as text, graphics, and photographs.
- Step 3—Encode the multimedia package into a streaming file format. The most common format is that of Real Networks, although Windows Media Player .asf format and Apples QuickTime .qt formats are also fairly common, and there are dozens of other less frequently used formats.
- Step 4—Stream material from a media server.
- Step 5—Material plays out on user's PC. The client software receives the stream, buffers it, and plays it out in the browser or its own window. A few "players" function without any browser, providing a VCR-like interface with play, pause, stop, rewind, seek, and fast forward.

SyncTV: Web+TV Is the Art of the Possible

Broadband is coming in a few years, and lots of companies are developing programming for it. But in the interim between 2000 into 2002 and perhaps longer, there isn't a critical mass of viewers to make content profitable. But there are those 44 million co-locators out there, already syncing their TV viewing and PC use. To engage these viewers, a few companies are experimenting with programs designed to reach both appliances.[25]

The pioneer award in the SyncTV space probably goes to ZDTV, which incorporated users' E-mails in to its programming almost from the beginning of the cable channel.

An interesting effort is MTV's *WebRIOT*. In late November, 1999, MTV began a twelve-week minimum run of the first regularly scheduled TV+Web show, that was not, per se, about computers. *WebRIOT* is a stripped (Monday through Friday) interactive game show, hosted by Ahmet Zappa. As many as 25,000 participants whose PCs are synchronized with the cablecast show play simultaneously.

Similarly, ACTV created *CyberBond* with the TBS cable network, a special cablecast plus webcast that kicked off the network's fifteen-movie James Bond movie marathon in November of 1999. Audience members downloaded the HyperTV plug-in from the ACTV site, registered, launched the application, and logged in, clicking to a room, the Las Vegas, Monte Carlo, or Paris, and so forth, and joined a chat with other viewers. The design premise assumed that each viewer co-located the TV in the same room or had a TV tuner card to put the TV picture on their PC screen along with the Web site data. The PC screen had three or four windows: the chat window in the top right, the movie in the top left (if the user was running it over the tuner card), and a Flash window across the bottom half of the screen. Reports were that the servers had a difficult time of it, slowing down the chat experience, and that the Flash animations sometimes shut down as well.

Although the broadcast networks are slow to develop Internet-based programming, they are fast off the mark when it comes to leveraging TV+Web for marketing. NBC's "transmedia" concept describes its efforts to reach consumers and sell them to marketers across media platforms. CBS and News Corp. are pushing what is sometimes called cross-platform advertising; FOX, for example, is emphasizing the advantages of dual platform advertising for *The Simpsons*. CBS offered Fidelity Investments the opportunity to sponsor end-of-show segments of "The American Dream" that are reported by Dan Rather, and to advertise on CBS' radio, billboard, and online properties as well. ABC Sports and ESPN are blending their advertising operations to entice media buyers to target consumers across Disney's entire empire, that includes the ABC network, ESPN cable channels, ESPN magazine, and the various Internet Web sites.[26]

Enhanced TV: Coming Soon to a Small Screen Near You

Television viewers who buy digital TVs, have digital set-top boxes from their cable or satellite service, or subscribe to WebTV will all be able to receive enhanced-TV signals. Although only a few people can get them now, a few shows are already on the air. Producer Larry Namer consulted with Microsoft when that company purchased WebTV to help identify content and services the company could offer.[27] Today, Namer's company ComSpan producers the enhanced version of the popular syndicated show, *Judge Judy*. Originally, the plan called for an enhanced version for only one day out of the five-day schedule, but this proved too difficult to promote among viewers. Enhanced-TV production moved to a five-day per week parallel production schedule with the show. "We had to simplify, simply, simplify," explained Namer. "It is hard work to figure out how to make it simple enough to make it come in on time. We put together our additional information while the TV show is still in production."

This experience provides the first benchmark for how much it costs to produce enhanced TV. According to Namer, it runs about $35,000 for each *Judge Judy* episode. The cost for a minimal channel-level enhancement (meaning that the channel overall is enhanced, generic to every show, rather than specific additions to each show) runs about $1.5 million per year. The staffing required to develop the material includes a producer, designer, writer, programmer, and coordinator.

Other reports indicate that building the enhancements in parallel with the production of the TV program provides greater control, depth, and sensitivity to the work. When they are retrofitted onto an already finished show, the project will take twice as long, therefore doubling the cost. Suzanne Stefanic of RespondTV reports that it takes about three days to create a simple e-commerce scenario along with an original show.[28] She believes the amount of time needed will be reduced as the industry develops coded templates into which enhancements can be dropped, reducing the cycles of testing and optimizing. Enhanced-TV producers can also leverage existing platform functionalities where they are already in place, such as E-mail, chat rooms, and backend e-commerce databases.

Material for Wireless Handheld Devices: Getting Content in Hand

There is a wealth of hype about Web-enabled 3G phones in magazines, trade papers, and even TV commercials from carriers like Sprint. But the reality is that it will be several years before the infrastructure and consumer devices are ready for them, despite the Sprint commercials telling people to come outdoors and get on the Web. The major portal and destination sites have not put in procedures to convert material to WML (Wireless Markup Language) in WAP (Wireless Application Protocol) format for mobile handheld devices.

The main feature governing content creation for handheld information appliances such as telephones is that they are small, in their overall form factor as well as the size of the screen. Today's models display only text, but 3G products will be able to play graphics, video, and audio as well. As with so many of the new communication products, content providers must be prepared to begin with what is currently on the market, migrate to new models as they appear, yet serve both the installed base and new consumers.

Handheld devices are very different from PCs in other ways besides their small size and small display. They do not have the same flexibility of input via mouse or keyboard. They do not have the same amount of processing capacity, nor are they as programmable.

Mobile wireless infrastructure is also a limiting factor, and typically, the bandwidth is quite narrow. In addition, since users are charged for airtime based on the bandwidth they consume, costs to the user must be taken into account. Last, there may be delays in transmission when usage is heavy.

The consumers are different, too. For the most part, they are less sophisticated and experienced than PC users, and have far less tolerance for complexity. Studies of mobile telephone users indicate that every key press that a function requires reduces the use of that feature by about 50%. For example, only about 5% of mobile phone subscribers actually use their voice mail.

Often, the cheerleaders of Web-enabled mobile phones picture a consumer who is difficult to reconcile with actual experience: they envision users with plenty of time to spend surfing the Net on their tiny devices, constantly checking the performance of their stock portfolios, traffic conditions, flight and train information, news and weather—all in spite of limited battery time, input keys, and tiny screens.

On the other hand, it is easy to imagine that users would appreciate intelligent services delivered through their phones. For example, anyone would want to be able to receive a message from an assistant or travel agent that a flight is delayed, a meeting cancelled or changed, or some other important message. All this means that the "push" model of information works better for handheld devices than the "pull" model—that users would welcome the appearance of information they need but they don't have the time or inclination to wade through dozens of sites trying to get information they could just as easily get with a telephone call.

Further, it is by no means clear that video information is a feature that people would pay much to get, although the infrastructure and device requirements would make it rather costly. In short, except for novelty-driven early adopters who want their Dick Tracy watch—or mobile phone—this is likely to be one of the last areas where digital video will appear as a mainstream product.

The 3G mobile phone trial by DoCoMo in Japan will probably provide considerable insight into the services consumers will want and pay for to receive on their wireless phones. In the meantime, narrowband services can be created using WAP emulators provided by pioneer Phone.com. The company offers a free, downloadable multidevice software development kit and a developer's program that includes discussion groups, training, and resources.

PConTV

There's a lot of optimism about this kind of programming. According to Forrester Research, program guides, enhanced broadcasts, and TV-based

Web browsing will generate a cash torrent of $11 billion in advertising and $7 billion in commerce by 2004. However, the differing business arrangements that are being made with interactive systems suggest that most of these dollars will flow to transport operators, rather than to programming networks.[29]

It's not an accident that Forrester combined enhanced broadcasts and TV-based Web browsing: PConTV is rapidly becoming the same thing as enhanced TV. Originally, WebTV planned to provide Internet access from TV sets instead of a PC. Subscribers could surf the Web on their TV and send and receive E-mails via a set top box that connects the TV to the telephone. At that time, 85% of customers wanted easy Internet access, surfing, and E-mail, and 71% wanted long distance telephone calls over the Net.[30]

Now, however, subscribers say they want their Internet TV services to go beyond mere Internet access; they want it to be more like an extension of their regular TV services. Responding to these expressed interests, WebTVPlus is marketed as an enhanced TV experience that enables interactive capabilities such as clicking on a product during a commercial to bring up further information on the product, competing with game show contestants from home via play-along games, accessing data that accompanies sporting events, as well as other information that relates to what is appearing in the television screen. Although it had not yet announced programming as of late 1999, AOL-TV is likely to adopt a similar strategy of enhancing TV programs, although perhaps with more emphasis on the Internet content. The service will probably be available first as part of the DIRECTV satellite service.

CD-ROM and DVD: Creating Disc-Based TV

CD-ROM titles are categorized into entertainment (stories and games), education and "edutainment," and productivity software. Entertainment and education are particularly reliant on rich media to attract buyers, while productivity software is more likely to involve colorful text-based screens. In late 1999, consumers grabbed the *Who Wants to Be a Millionaire* CD-ROM-based trivia game from Disney Interactive when they showed up in retail stores nationwide in November. Within days, the game sold out across the country, and Disney was rushing new boxes to retailers for the holiday shopping season.[31]

This hit continues a trend established in the mid-1990s to repurpose popular TV shows onto CD-ROM games. In previous years, Paramount Pictures licensed *Frazier, Beverly Hills, 90210,* and *Melrose Place.* Another popular game came from HerInteractive.com, which launched *Nancy Drew: Stay Tuned for Danger,* the second in its popular *Nancy Drew*

CD-ROM series. Although games are mostly popular among boys, according to PC Data, sales of girls' CD-ROM titles increased by 38% in 1998, outstripping the 18% overall growth in the games category.[32] (See Figure 13.5.)

In the late 1990s, it became clear that a new market opportunity would revive CD technology—the VCD, or Video CD. Introduced in 1993, VCDs became very popular in Asia, especially China, which supported more than 100 domestic manufacturers that produced around 2 million VCD players per year. The VCD looks the same as a music CD or a CD-ROM, but holds movies that have been compressed in MPEG-1 video. A single disc can only hold about an hour of video, so most movies are packaged into 2-disc sets. The cost of producing VCDs is about the same as for audio CDs, and the technology is well suited for music videos, education, training, movies, and karaoke. The consumer plays back the VCD on a

Figure 13.5 One of the most popular games ever—*Doom*. Source: Atari.

VCD player that costs between $200 and $400, or on a fast PC (Pentium 133) with a CD-ROM drive. Some DVD players will play VCDs as well.

In the nonentertainment categories, the PCData annual survey of 1998 found that reference, finance, and educational software were the big winners of 1998 through retail and mail-order channels. In that year, retail software revenues increased 13% above 1997 to $5.2 billion. At the same time, prices for individual units went down approximately 9%. Growth in reference software was 21.5%, primarily due to a surge in encyclopedia sales. The Learning Company was the top player, commanding 40% of sales revenues, while Microsoft was second with 30%, and IBM placed third with 12%. Encyclopedias accounted for nearly 65% of the revenue. The educational software category grew 17%. The Learning Company was the largest education software publisher in 1998 and accounted for nearly 42% of sales. Havas Interactive came in second with 25% of sales.

CDs are being used in network applications. Media Station, Inc. is partnering with Comcast@Home to trial an interactive service, SelectPlay, in Comcast's cable modem homes around the Detroit metro area.[33] The SelectPlay technology delivers CD-ROM material over broadband networks, enabling users to play titles that are installed on a server at the service provider's facility. The service will give customers instant on-demand access to popular family entertainment and education software, such as dozens of top-rated interactive games, sports, educational, and knowledge-based CD-ROMs on-demand from such producers as Disney, The Learning Company, Havas Interactive, GT Interactive, Humongous Entertainment, and many more.

Methods for creating CDs have been developed and refined over nearly two decades, as presented in Table 13.2. Apple Computers put out the first comprehensive manual. Since then, re-writable CD technology (CD-R) has put CD burners on millions of desktops, and many of them are creating content for them.

DVD Production

By early 2000, there were about 5 million DVD players in U.S. households. Forecasts call for 30 million players by 2003.[34] Prices were dropping rapidly, so that while high-end models sold for about $1,000, there were also products that hit the consumer sweet spot under $300. For the 1999 holiday season, studios hawked multiple motion picture offerings: Twentieth Century Fox marketed the *Aliens* collection; New Line Cinema put out *A Nightmare on Elm Street* films; and MGM went all out with a set of seven James Bond films.

In 2000, there were about 5,000 movies available in the DVD format, compared to the 65,000 movies, TV programs, and video-only releases that are out on videocassette. Studios, careful to protect their assets, plan to

Table 13.2 Creating CDs

Development	Take one or more ideas and flesh them out to make preliminary decisions about platform, format, intended audience, subject matter, characters, plot, interactive structure, purpose, overall look and feel, budget schedules, type of assets (text, graphics, photos, animation, full motion video, film, narration, sound effects, music, dialog).
Design	Select one idea, per assignment. Put together necessary tools, depending on final platform. Assemble information about target consumers. Assign writer. Design navigation and interactive sequences. Storyboard key frames for all scenes, graphics, photos, and visual fields. Create schedules. Hire graphics, programming, music, production, writers, and content experts. Where necessary, get them started working.
Testing 1	Test a prototype version with a focus group of individuals similar in characteristics to intended audience. Execute other tests: User testing—how people use it. Unit testing—how parts of program work. Integration testing—how whole program works together.
Preproduction	Approve script. Make preparations for film, video, and photographic shoots. Cast all performers. Find field and studio locations. Order all equipment. Hire makeup, wardrobe, set design and construction, and prop people. Animation: Approve key frames. Graphics: Select style and text fonts and colors. Music: Approve composition. Sign contracts and firm up schedules.
Production	Approve graphics. Record music, narration, and sound effects. Videotape and/or film scenes with actors. Shoot photographs. Build animated scenes.

Table 13.2 Creating CDs (Continued)

Postproduction	Edit film and video sequences. Mix and equalize final music tracks, sound effects, and narration (separately). Add special effects to film, video, and graphics. Takes about twice as long as conventional postproduction.
Integration	Use authoring program to put all assets together into a single program.
Testing 2	Pre-master a "one-off" version of the program. Conduct alpha testing in-house; beta testing by sending out to potential customers, but give tech support; media testing—how program works on planned medium; stress testing—how platform works; configuration testing—how program works with the platform; content testing—content experts confirm accuracy of assets.
Mastering and duplication	Create a master tape or disc. If a CD-ROM, first copy is nickel-coated, called "father," used to create several "mothers." Mothers are used to make negative masters, or "sons," that are the stampers for positive, mass-duplicated CD-ROMs.
	Mold replicas for distribution.
Assembly	Add liner notes to final medium. Package product manuals, stickers, boxes, or registration cards and shrink wrap or seal.
Marketing, distribution	The product is ready. The execution of the final steps for bringing it to market release falls under the overall business plan.

release only about 150 titles every year. Although most of the high-visibility projects for DVDs have been motion pictures, games are also moving to the format. DVD allows producers to put a large amount of top-quality video on the disc, which is important for games and educational uses. For example, Digital Leisure released *Dragon's Lair II: Time Warp*, promising gamers they could guide Dirk the Daring as he races through time to save the beautiful princess from the clutches of the Evil Wizard Mordroc.

DVD title creation is pretty much limited to the top 200 to 300 postproduction houses and entertainment industry content owners. They have the massive budgets it takes to pay for the expensive systems needed to create DVD titles: authoring and encoding systems run from $30,000 to more than $2 million. In addition, the process involves several workstations connected to a network, because the different parts of the process—video capture, audio capture, menu preparation, authoring, and emulation—are executed on different systems. (See Figure 13.6.) However, the price will drop rapidly in the next few years, and hobbyist users using much lower-cost hardware and software will produce their own DVDs on desktop computer systems.

In early 2000, it costs about $7,000 to $30,000 to produce a professional-looking two-hour DVD movie, depending on complexity. However, special edition DVDs with many menus and supplements cost as much as $100,000 to produce. If you want to do pre-mastering yourself, complete authoring and encoding systems can be purchased from $30,000 to over $2 million. DVD production differs from CD production because of the amount of material and the complexity required to allow users to access it. In the DVD industry, there are basically four stages, as shown in Table 13.3.

Development begins with an idea that is then fleshed out to define the parameters of the project: audience, market, costs, content requirements, level of interactivity, and so forth. Production is shooting and recording the needed audio and video. Pre-mastering costs are the most expensive phase, because this is where all the content is gathered, encoded, and formatted for use on the disc, including authoring of menus and control information, and putting it all together into one multiplexed stream.

Broadband TV

Broadband content is the digital version of interactive television. Many of its proponents believe that its greater flexibility will allow developers to create "addictive content" by combining the interactivity of the Web with the immersiveness of television. Based on the early research, some analysts expect breakthrough interactive experiences sponsored by advertisers will

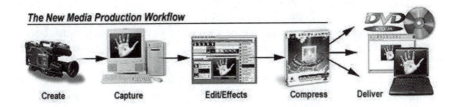

Figure 13.6 Producing DVDs. Source: Panasonic.

Table 13.3 DVD Production

Planning and development	Define scope and structure of project, including resources, equipment, and design specifications. Implement infrastructure for digital asset management.
	Generate and design needed assets (text, graphics, audio, animations, and video).
Production	Create and manage assets.
Pre-mastering	Storyboard—create graphic thumbnails of each section of DVD.
	Layout bit budget—how assets will use disc real estate.
	Asset capture—production.
	Author—create structure of the disc by placing assets on a timeline and creating the interactive paths between them.
	Emulation and test—making sure it all works before replicating the discs.
	Formatting and delivery—convert bitstream to DVD-ready format and deliver for replication.
Mastering	Create glass master—the copy from which all replications will be pressed.

become common by 2002. For example, the broadband ads result in brand-recall rates that average 34% higher than narrowband advertising. One interactive ad that was placed on @Home's portal led more than 50% of the viewers to spend between 30 seconds and 5 minutes interacting with it.[35]

Broadband television requires creators to pay attention to the lessons they have learned from narrowband environments because, unlike traditional television, broadband still has limited bandwidth. It ranges from 128 kbps for dual ISDN and low bit rate DSL service, to 1.5 Mbps on cable systems. The pioneers in this space warn would-be creators for broadband to remember their narrowband lessons: compression is their best friend.

The applications for broadband TV are Web browsing, rich media E-mail, news, chat, shopping, tickets and reservations, and games. The early producers recommend that material be created specifically for the medium or its potential will not be fully realized, and that it be tested over real-world systems. Ongoing bandwidth constraints mean that content for this kind of television calls for video bits that are short and sweet—music clips, news clips, movie trailers, and other snippets. The screen is still not full-size, and at lower bit rates the picture may be jerky. Producers must decide how to allocate bandwidth towards audio versus video. Surprisingly, giving audio priority is preferred, because interruptions in video are usually tolerated easily by audiences, but glitches in audio reception are not.

Like the online environment, consumers are in control of broadband TV: it empowers the edge of the network and the individual. It is not a format where the consumer will humbly accept whatever the entertainment industry chooses to push down to them. This power relationship means that producers need to contextualize their offerings as "touts," propositions, and bids for attention, rather than demands.

Consumer behavior on broadband systems is worth noting. Fifty percent of them use the system on a daily basis, and they use as much bandwidth as is made available to them. For example, they play the highest-quality music and watch long-form video whenever they are able to access these content types.[36]

Creating Broadband Content

The development, production, and delivery of broadband material is inherently the work of large media and entertainment entities. For that reason, making the business case is always the first step in getting a project funded. This activity is sometimes called "scoping the project," or setting the parameters for it, enabling the various business partners to ascertain the desirability of the intended product. It will include the programming strategy based on consumer research, a summary of the advertisers who want to reach that audience, and the e-commerce revenues that can be harvested.

Once the project is greenlighted for development, the producers will create a schematic or flow-chart diagram of the menu structure of the program. It provides a single snapshot of everything viewers will see; a menu breakdown; template of sources; and the paths that link the on-air program to the backend e-commerce database. At the same time, art direction is applied to the schematic model to develop screen representations of all the associated elements. Video streams will be represented by sample frames and inserted in the flow chart.

Basic descriptions of the interactive sequences show how they relate to user behaviors. Creatives will produce graphic representations and tied to the hardware that the consumer uses. Developers must also create guidelines for synchronization of the various materials. A team of designers and art directors will then consolidate the application functions and structure, and coordinate them with appropriate channel and brand extensions of the various partners.

This work requires extensive collaboration between the application designers and the creative people who work on the channel presentation. If it is successful, it will lead to a proof of concept document, often a video-tape showing how the elements play together. Finally, an initial prototype is created and tested before the final product is produced and distributed.

Creating Material for Interactive Television

There is little difference between broadband TV and interactive TV except that IATV is received over the television set, while broadband TV can be directed to either a TV or PC screen, most often the computer. Interactive television (IATV) is a networked environment where consumers access a server at the service provider's headend or central office facility. In the iteration of IATV in the early 1990s, the applications thought likely to be the "killer apps" were video-on-demand (VOD) and near-video-on-demand (NVOD). System operators still look at on-demand services as a potentially rich revenue stream, based on the 1998 video rental market of $9.85 billion, up 9.2% from $9 billion in 1997.[37]

VOD means that requested material arrives on the subscriber's TV screen immediately, while NVOD means people must wait for it to appear. Indeed, exactly when the material appears preoccupied operators for more than a decade. The amount of time between a request and the start time is called "latency." There is always some latency, even if it is a nanosecond. If the latency is five minutes or less, the service is called VOD. If the latency is longer than five minutes, usually 10, 15, 20, or 30 minutes, the service is called NVOD.

To determine what kind of service to offer and how many channels should be set aside, programmers need to know how long a viewer will wait for an ordered program and how much he or she will pay for it. The data is analyzed in terms of "buy rate" or "take-up rate," sometimes cited as a measurement over a baseline of ordering pay-per-view selections. Suppose that 5 out of 100 subscribers will buy a current popular movie, priced at $5.00, with a 2-minute wait. What will the percentage be if it is priced lower but there is a 5-minute wait? A 10-, 15-, or 30-minute rate? At each time increment, how much does the price need to drop for the offered program to attract the same number of viewers?

These questions began with pay-per-view (PPV), deployed on many cable systems in the mid-to late-1980s. Satellite services, with their enormous 300-channel capacity, have the widest PPV choice, and replicate and time-shift premium movie channels like HBO and Showtime as many as six times, essentially providing NVOD viewing of these popular networks. Small cable systems are able to set aside only a few channels for PPV, while larger ones may have 20 or more. Most service providers also have a continuous loop "barker channel" to promote PPV offerings, showing movie trailers and information about price, ordering, and start times.

PPV movies run consecutively, with one or two movies per channel, so the viewer must wait at least 90 minutes for the next showing. By contrast, NVOD takes up many more channels because its start times are 10, 15, 20, or 30 minutes apart. (Sometimes NVOD is called by the apt term, "stagger-casting.") The running time for most movies is about 100 minutes, but some run as short as 90 minutes, and a few run even longer. Thus, a movie that starts every 30 minutes takes up 3 or 4 channels (with 4 being average); one that starts every 20 minutes takes up 5 channels; one that starts every 10 minutes will require 9 or 10 cable channels.

Subscribers order PPV and on-demand material in several ways. Older systems that have no return path may require people to use the telephone to place an order. Satellite services and most cable operators have "impulse PPV," which sends a signal to the set-top box. The satellite services use an STB that descrambles the program and stores the request. It is connected to a telephone line and, in the middle of the night, dials up the service and downloads data about the PPV order.

Cable systems, even when they don't offer two-way service for Internet access, often have a very small return path that they use for system troubleshooting and diagnostics. PPV, VOD, and NVOD require only a low level of interactive capability with negligible return bandwidth because the information from the consumer consists of nothing more than a few pulses sent to a server to request desired programming. At a slightly more sophisticated level, the system could even support transactions, allowing customers to indicate they want to buy a product, and a call center would call them back on telephone to make payment and delivery arrangements.

The Internet has put paid to these limited ideas of interactivity. High-speed access for site-seeing, E-mail services, and chat functionality top current consumers' wish lists. And network operators recognize that there is an additional upside potential in e-commerce transactions of all kinds.

Once service providers move beyond on-demand services, developing interactive services becomes more difficult. Television viewers are, of course, different from PC users. Couch potatoes are not usually interested in technology and will respond only if interactive applications are easy to use and can be accomplished with the remote control. As Josh Bernoff, principal analyst at Forrester Research commented, "To succeed on TV, an interactive application needs to work for viewers who have a remote in one hand and a can of beer in the other."[38]

His point is that interactive applications cannot require too much thought or attention. Forrester suggests that developers create "mini-applications" like purchasing products while programs are on the air, such as CDs and videos along with music shows, quick opinion polls with talk shows, and sports strategies and decisions accompanying athletic events. The Forrester analysts believe that interactive television depends on the buildout of two-way cable networks because that is the only infrastructure truly capable of providing a high-bandwidth two-way connection between subscribers and service providers. The research firm predicts that there will be 6 million digital cable subscribers by 2002 who will use interactive TV services. According to a December, 1999, study, that number will rise to 30 million by 2004, generating $10 billion in revenue.[39] The research firm argued that a new programming and business model would emerge from a Web/television hybrid, forcing entertainment and media companies to reinvent their strategies as they move ahead.

Today, television's business model is based on advertising, while the Internet model is based on transactional revenue earned through e-commerce activities. According to Jupiter, in the future, these companies will need to find a balance between the two kinds of revenue streams. A further consequence of the convergence of the two media will be a new content model. The Jupiter analysts believe that today's players will need to involve the viewer by adopting new tactics:

- Establish and nourish a relationship with the audience members across media platforms. Participate in small-scale commerce ventures on the Internet to develop a direct line of communication with people.
- Develop complementary Internet programming that engages fans.
- Use the Internet to drive traffic to their real entertainment business. Experiment sparingly with entertainment online, but use the Internet as a relatively low-cost experimental platform for interactive TV. Companies should budget 80% of their money on promotion and audience relationships, and only 20% on creating online content.

Cross Platform Genres: Information, Entertainment, and Games

There are three types of programming genres that span every kind of platform, perhaps every form of human communication. These evergreen types are information, entertainment, and games. This next section will look at how each fares in the digital networked environment.

Information

Information lends itself easily to interactive programs and applications. It allows considerable user control over professionally produced material, and leads easily and naturally to transactions. News, much of the commentary and activity surrounding athletic events, how-to programming, magazine segment shows, biographies, and documentaries—all these are enduring formats that have been popular since the first Movietone newsreels appeared in theaters.

Story-Based Entertainment

Stories have proved to be the problem child of the interactive domain. Telling stories, so natural to books, films, and television, seems awkward, contrived, and barely believable as soon as the viewer becomes a user. One particular problem with the narrative form is how to let viewers take control but still make it possible for them to suspend disbelief. This suspension allows people to become involved in a fictional piece as though it were real.

Suzanne Stefanic, an executive at RespondTV, a company that has been working on interactive advertising projects commented on the challenge of creating interactive material: "In general, this format requires intrigue, allure, and respect for the user. Since campfires, narrative has owned your mind, you submit to it. Suddenly we are asked to interact. It is more difficult than any one of us thought it would be."[40]

Why is it so difficult? John Ziffren believes that one reason is that it involves the creation of a language. "It's a remarkable challenge and one of the overriding concerns is not to be overwhelmed by the tools. Ultimately this is storytelling. Ask, Why should this story have interaction? What are the differences in the style of enhancements—autonomous, parallel, or integrated? The challenge, as with any medium of entertainment, is to tap into the emotion and passion of viewers that drives them or pulls them farther into the story."[41]

When asked what they believe interactive storytelling is, many people respond by saying that it lets the audience pick the end of the story. Today's developers have a far more sophisticated view based on the idea that endings are generated from actions. So it is by virtue of taking action that people affect the outcomes of the story. The audience doesn't so much view, read, play, or choose the story—they participate in it.

David Riordan, senior producer/creative director for POV Digital Entertainment, refers to it as the creation of a "variable state environment." Says Riordan: "The model sets up a variety of events which can occur in an environment, and then the environment molds and changes to accommodate what the user is doing."[42] He tells storytellers to begin by setting out their themes, the logic of the created universe, and the traits and attributes of the characters. The second stage, counsels Riordan, is logistical. The writer must lay out what users can do, where they can do it, and how abilities and resources they acquire in one place can be used in another environment. In addition, he or she must specify what players need to do to get from place A to place B, and what will happen if they leave without doing what they need to do.

The storyteller creates a design document that defines the graphic design and any sets that might be involved. The script is written last, and Riordan says most scripts have about 150 pages of dialog. He recommends economy of dialog, with attention given to how it is accessed and how it changes as the story progresses. He warns writers to allow players to bypass it if they choose, since discs are experienced many times.

It is difficult to separate games from interactive stories. Generally, the more expensive the production of a game is, the closer it resembles a movie. The experience with CD-ROM technology is instructive on this point. An interactive story costs about $1.5 million to produce; a game usually costs less than $750,000. But what is *Myst?* This title and others such as *Voyeur* and *7th Guest* are stories that allow a role-playing user to explore a fantasy world or a simulation, leads them into a puzzle and maybe a little adventure, and lets them win (or lose). Are they games or are they stories? Robert Weaver, president of InterWeave Entertainment, Inc., who directed *Voyeur* and *Voyeur II*, believes that all interactive entertainment is a game, because the key to interactivity is mastery, and gaining mastery requires players to know how well they are doing. He thinks the feedback process that is part of interactivity turns a story into a game.[43]

Games

Games are a rich source of revenue on virtually every media platform—the stickiest Web site on the Internet is Gamesville.com. Four video game categories make up the overall market: console games (Sega, Sony, and Nintendo), computer games (PC and Mac), online games, and finally, arcade games. Taken together, they add up to a $17-billion industry, with consumers spending about $12 billion on software for proprietary consoles and $5 billion for PC-game software. In 1999, one in four U.S. households had a console in the living room or play room and the hottest console was the 64-bit Sega DreamCast—until it was shouldered aside by the Sony PlayStation II in 2000. On March 9, Microsoft introduced the X-Box, taking on the two Japanese giants.

In 1999, game software sales surpassed annual U.S. movie box office for the first time. Console games are more popular than computer-based games by a factor of two to one, and online games are a distant third. According to Jupiter Communications, the number of online game players will grow substantially over time, as shown in Table 13.4.[44]

Video games are the fastest growing segment of the entertainment industry. The business shares with other sectors a dependence on hits. Out of the thousands of games released each year, perhaps twenty become viable products, and two or three will become gigantic hits. It's an intensely competitive arena. According to PCData, at any one time, there are on the average 3,500 new games in the retail channel at any given time. In 1998, the best-selling games were *Deer Hunter* and *Quake 2*.[45]

Games are categorized into well-known genres: sports; real-time strategy; first-person shooter; racing simulation; puzzle; adventure; role-playing games (RPGs); flight simulation; strategy; god simulation; racing shooter; action-strategy; and card games. Action and adventure are the most popular, followed by first-person shooter or "twitch" games.[46] Genre games also have identifiable flavors, including modern; cyber; medieval; World War II; alternate reality; post-apocalyptic; sci-fi; fantasy; medieval; and cyberpunk. Finally, they must also be categorized as single-player or multiplayer game formats.

Game developers must take into account the tribal culture that forms around a game, which can become remarkably elaborate. For example, when players join *Ultima*, a combined CD-ROM and online multi-player game (www.ultima.com), they come into that world much as they do this one . . . naked, alone, owning nothing. Over time, they build an identity and acquire possessions, including Ultima gold. Ultima IDs and gold have turned up for sale on eBay, worth real money there!

Producing games involves three types of activities: creative, technical, and marketing. Game developers usually build a game around a licensed game engine that fits one of the formats listed above; for example, a first-person shooter. They then create graphics around the engine that make the game a unique experience. The amount of advance money they get varies from a nominal $30,000 to millions, depending on the track record of the developers. The creators usually get the advance plus 5% of the backend royalties.

Developing and debugging a game is a relatively long process, taking about 18 to 24 months. The marketing cycle for packaged games, however, is very fast, allowing only about 90 days for the game to sell or be trashed. The process begins with a one- or two-page description of the game and a list of the art and sound assets that will be needed. Developers then build virtual models and environments, linking them with elements of programming. Critical decisions must be set out in software statements, such as how much user movement makes an effect, what happens when one piece bumps into another, and so forth. Even small companies have a library of code that allows them to cut and paste, and there are even game developer sites on the Internet that post free software modules.

Table 13.4 Number of Online Game Players, 1997–2000

	1997	1998	1999	2000	2001	2002
(millions of players)	2.0	3.7	7.2	13.0	20.2	26.8

While online games are far behind console and PC-based games, they are nevertheless growing in influence. People like forming communities, so they flock to imaginative, well-designed multiplayer games. And there are also important business issues. The Internet greatly reduces distribution costs, a major problem in the packaged game industry. This translates to a lower risk factor with greater opportunity to tweak the product in fast cycles of innovation. Along these lines, it also lets producers upgrade the game and add new features and characters to keep the game fresh and gamers interested. Some observers believe that online games will soon be "programmed" in episode-like increments so that players have to sign in often to make sure they don't miss important events.

The Internet is also beginning to affect the marketing of packaged games. Some companies are beginning to release a simplified version of the game on the Web. When the game is over, a screen pops up and makes a proposition to up-market the gamer to a more sophisticated game they have to pay for. In 1999, game company Electronic Arts paid AOL $81 million to develop game channels for AOL customers, indicating just how important EA believes the online marketing opportunity is becoming.

Video games are the source of much unease, and they are often linked in news events with adolescent misbehavior. Pushed by public outrage and the likelihood of congressional action, the Recreational Software Advisory Council Association (RSAC) and the Computer Games Work Group have a rating system for video games, requiring game developers to submit storyboards or video of the most extreme scenes before retail distribution. However, this is not done for Net offerings, and it is difficult to see how such a process could be enforced online.

The Creative Worker

The settings for creating content vary considerably, but more and more such work is done on networked computers, as shown in Figure 13.7. Megamergers and alliances between industry giants establish huge creative departments with lavish budgets, while "garage band" creatives work in basements and garages everywhere. Neither situation may provide the most nourishing environment for creative talent, which appears to produce the best work in small but stable boutiques.

Pulitzer-winning journalist Jonathan Freedman describes the plight of the creative person in the midst of continuous corporate restructuring most plaintively. He recounts that his last book was published by Atheneum, a small quality publishing concern. While he was writing the book in 1990, Atheneum merged with Scribners. Macmillan then purchased the joined company. In 1991, Macmillan was sold to Simon and Schuster. Simon and Schuster was a previous acquisition of Paramount. So in 1994,

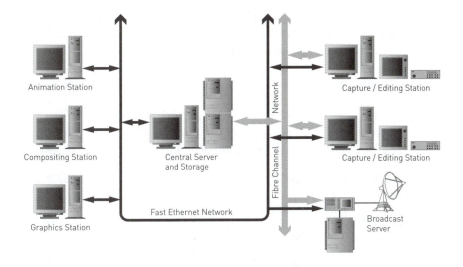

Figure 13.7 A networked environment for the digital production of news.
Source: Matrox.

when Viacom bought Paramount, it got Simon and Schuster along with the deal—and Atheneum was put out of business by its corporate master.

As Freedman wrote in a column entitled "Fried green writers at the Viacom Cafe":

> *Megamergers impoverish American culture and thwart the search for truth. The billions of dollars invested in acquisition, debt service, and technology reduce the money available to introduce new writers and subsidize provocative, if not wildly popular, nonfiction books. The demand for entertainment puts dollars above sense; diversity is sacrificed to the vast bland middle market.*[47]

While many creative people laud the Internet as offering artists freedom from the oppression of agents, labels, publishers, and the like, it is not clear that such direct marketing will actually work to their benefit. In a world where everyone is a creator, a publisher, and a broadcaster, how will individual artists get attention and the revenue they need to live and work? Instead, there could be a nasty downward spiral: the audience splinters into tiny segments, reducing the amount of money any given artistic effort can generate. Budgets decline, quality declines, and fewer people watch, splintering the audience even more.

It is possible that this scenario would in fact strengthen the hand of major media companies who have the marketing clout to break through the clutter. Using the drawing power of stars and hit properties, they may fare very well in the new world of ubiquitous connectivity. In the next chapter, we will consider the ways that content providers and distributors are going about ensuring that they have a place in the e-landscape.

Notes

1. C. Rosendahl. Telephone interview, October 1998.
2. J. Lippman, "Networks push for cheaper shows," *Los Angeles Times* (February 19, 1991):D1
3. M. Schrage, "Humble pie: Japanese food for thought," *Los Angeles Times* (November 4, 1994):C1, 4.
4. V. Grosso. Telephone interview, April 1994.
5. K. Makal, "3 Ring Circus on Interactive Design." *TVindustry.com* (July 29, 1999). Posted on Web site at www.tvindustry.com.
6. J. Niemack, Panelist at a seminar at the American Film Institute's Los Angeles campus, sponsored by the National Association of Television Programming Executives (NATPE) and Entertainment Technology Commerce (ETC), November 1999.
7. C. Rosendahl, op. cit.
8. D. Kohler. Personal interview, Malibu, CA, February 1998.
9. K. Papagan. Telephone interview, May 2000.
10. J. Mountford, "Essential interface design." *Interactivity* (May/June 1995):60–64.
11. S. Schlossstein, "Intelligent user interface design for interactive television applications," *1995 NCTA Technical Papers* (May 1995):165–170.
12. E. Holsinger, "Kai's Power Tools," *Digital Video* (April 1995):37–40.
13. Figures posted by Nua at http://www.nua.com.
14. Media Metrix tracks these ever-changing statistics, posted at http://www.mediametrix.com.
15. M. Berniker, "NBC Desktop Video to deliver news to PCs," *Broadcasting and Cable* (July 8, 1994):26.
16. S. Hardin, executive, NBCi. Personal interview, San Francisco, February 2000.
17. C. Davidson, programming executive, CBS. Personal interview, Pasadena, July 1999.
18. Jamie Kellner, president, WB.
19. Banner Ad Placement Study. Athenia Associates (April 1998). Available online at http://webreference.com/dev/banners/research.html.
20. See latest research at http://www.real.com.
21. J. Geirland and E. Sonesh-Kedar, Digital Babylon: How the geeks, the suits, and the ponytails fought to bring Hollywood to the Internet, New York: Arcade Publishing (1999).
22. J. Burns, Panelist at seminar at the American Film Institute's Los Angeles campus, sponsored by the National Association of Television Programming Executives (NATPE) and Entertainment Technology Commerce (ETC), November 1999.
23. C. Alcasid, former director of technology, UltimateTV.com. Personal interview, Sherman Oaks, CA, July 1998.

24. B. Waggoner, "Making great Web video," *Digital Video* (October 1999). Available online at http://www.dv.com/magazine/1999/1099/webvideo1099.pdf.
25. T. Swerdlow, "Telewebbers on the rise," *Interactive TV Newsletter* (June 14, 2000). Available online at http://www.itvt.com.
26. G. Johnson, "Big media firms are offering advertisers one-stop shopping," *Los Angeles Times* (September 28, 1999):C-4.
27. L. Namer, president, ComSpan. Personal interview, Westwood, CA, November 1999.
28. S. Stefanic. Panelist at seminar at the American Film Institute's Los Angeles campus, sponsored by the National Association of Television Programming Executives (NATPE) and Entertainment Technology Commerce (ETC), November 1999.
29. J. Bernoff, et. al. "Interactive TV Cash Flows." Forrester Research study (August, 1999). Search Forrester Research at http://www.forrester.com.
30. Iacta press release, "Iacta releases two-year landmark study of WebTV and its users" (May 1, 1999). Available online at http://www.iacta.com/news_0399_1.htm.
31. Disney Interactive press release, "Consumers' Final Answer is 'Yes' to *Who Wants to Be a Millionaire* CD-ROM" (December 7, 1999). Available online at http://biz.yahoo.com/bw/991207/ca_disney__1.html.
32. PCData press release, "PCData Releases 1998 U.S. Software Sales Statistics" (January 27, 1999). Available online at http://www.pcdata.com/.
33. MediaStation press release, "Comcast, @Home, and Media Station Team up to Launch Market Trial of SelectPlay Interactive Software Service" (December 6, 1999). Available online at http://biz.yahoo.com/bw/991206/ca_media_s_1.html.
34. "The Market for Video and Multimedia Services 1998–2003." Insight Research Corporation report (October 1998).
35. "Interactive TV to Emerge as Hybrid of Web and TV Models." Jupiter Communications research (December 8, 1999). Jupiter press release available online at http://biz.yahoo.com/bw/991208/ca_jupiter_1.html.
36. P. Kearney, manager, content and services, broadband data services, MediaOne. Telephone interview, November 1998.
37. "Consumer home video rental activity, 1997–1998." Convergent Data report (1999). Available online at http://www.convergentdata.com/trade/tradata.htm.
38. J. Bernhoff, op.cit.
39. Jupiter Communications, op. cit.
40. S. Stefanic, op. cit.
41. J. Ziffren. Panelist at Intel/AFI one-day program to present work-in-progress material from recipients of grants for enhanced-TV programming, Los Angeles, September 1999.
42. Quoted in C. Buchman, "Back to the future: The art of interactive storytelling," *FilmMaker* (Summer 1994):34–39.
43. R. Weaver. Speaker at one-day conference, "Your future in multimedia: Interactive Hollywood, new media, new jobs, new markets," organized by Stuart Fox for the Academy of Television Arts and Sciences and the International Interactive Communications Society, North Hollywood, CA, July 22, 1995.

44. M. Mooradian, "Internet Games: Five Year Outlook—Revenue Models and Technology Development." Jupiter Communications Research Study (March 1998). Contact www.jup.com.

45. For more information from *PCData*, contact James Carey Director, Public Relations at 703-995-5902, or E-mail to jcarey@pcdata.com.

46. D. Hudson. Instructor's lecture at UCLA Extension School, October 1999.

47. J. Freedman, "Fried green writers at the Viacom Café," *Los Angeles Times* (May 14, 1994):F2.

14

Cache Flow: From Bitstreams to Revenue Streams

On the whole, business is business, no matter where it is, what it is, or who runs it. That said, the virtual world does present one anomaly that is worth thinking about: It is an economy of abundance rather than scarcity. Internet "territory" comprises an almost unlimited number of Web sites (the stores), a torrent of available information (the products), and an explosion of bandwidth (the marketplace). No one really knows how to do business in an environment where many key elements are so readily available that they are free or nearly so.

In spite of the uncertainties, many people believe there is money to be made on the Internet, and they are adapting traditional business processes to enter the new arena. Companies that hope to be a real participant in the information economy need to raise money, either by borrowing money or attracting investors. Both these funding mechanisms require companies to design a forward-looking business plan, one of the most basic documents for commercial enterprises.

The plan lays out the specifics of the organization, its management, products, marketing and sales opportunities, and financial arrangements and goals. Part of the business plan is a business model, which shows how the planned company's goals will be achieved if the resources and processes are applied to the model. Ideally, if someone plugged in the numbers from the plan into the model, it would project the future financial position of the company. In casual parlance, the term "business model" has come to be synonymous with business plan, partly because executives cannot discuss actual numbers from the business plan in public. However, they can and do talk about their business model, usually referring to the strategies the company believes will lead to a profitable outcome.

A generic business plan holds true for any enterprise, whether it is making widgets, wagons, or wastebaskets. However, business models reflect the unique customs and practices that define a particular industry or sector. For example, during the difficult transition to digital broadcasting, a station group executive admitted that: "Our business model right now is just staying in business through the transition." More broadly, another industry insider summarized media and entertainment business models succinctly as "Deliver eyeballs, get ads; deliver value, get subscriptions."

The business model for a media and entertainment company includes four components: a content model, a distribution model, a marketing model, and a revenue model. The content model describes the material that will be used to attract and hold an audience, and influence them to act, which usually means to make a purchase. The distribution model lays out how the content will reach consumers. The marketing model clarifies how a potential audience can be turned into an actual audience: how they will learn about a content or service offering and be convinced to try it out. The revenue model details how the company will make money.

This chapter examines the business models that have come to prominence as the marketplace of digital media and entertainment has unfolded in the past few years. Divided into content, distribution, marketing, and revenue models, they can be combined with one another in any number of ways to generate unique overall models tailored to the particulars of content, audience, product, distribution, and transactions.

Content Models

The content model begins with creative people developing ideas for content that they believe will appeal to a particular audience. For example, the traditional broadcast network content model was known as "least common denominator" programming. The audience was everyone, so the content was designed to exclude no one; hence, it appealed to the widest set of interests shared by the population as a whole. Over time, the most universal interests turned out to be sex, violence, news, and music.

Cable, radio, and print produce content that may appeal to a wide audience, like *Time* magazine or all news channels, or to specialized niche audiences that are much more narrowly defined. Thus there are several publications for quilters, and cable channels and radio stations that transmit content for Christians. Although cable and radio create material for the consumer audience; the print medium is divided into consumer and industry trade segments.

Similarly, online content may appeal to a broad audience such as Yahoo! or address a microniche audience, like www.itvt.com, which presents material for people interested in interactive TV. The questions surrounding just how narrow a niche should be are important, as they relate to both creating content and products. Mark Cuban calls it the "feature versus product" problem.

Suppose someone is interested in Porsche automobiles and they want to turn a time- and money-consuming hobby into an online business. Specifically, this person collects the older 911S models. He is aware of the difficulty of getting some parts for the car: shock absorbers and carburetors. Should the site try to target Porsche enthusiasts; classic Porsche fans; 911S collectors; difficult-to-obtain parts for Porsches; classic Porsches, or the 911S; a parts exchange for classic Porsches or 911S cars; shock absorbers and carburetors for 911S cars—or incorporate all of the above with different areas within a single Web site? Which choice will make a viable business, and how wide does the targeting have to be? Which choices are features of a larger overall product?

(This problem permeates the development of many products in the new digital communications space, such as the set-top box. Should an STB have a hard drive in it? Should it provide input/output capabilities for Web access? How much software? How many services? Should it be extensible to cable, satellite, phone, and DSL? Every selection has an implication for the final cost of the product, and manufacturers are wrestling with these decisions now.)

Online efforts are particularly problematic in this regard because the cost implications have less impact and there are fewer guidelines to go by. However, like the print medium, the plans for online content usually call for deciding whether it will be targeted to a consumer or trade audience. Specialized content, even an entire Web site, is created to appeal to that group. Large corporate Web sites can have a public area, supplemented by password-protected private areas devoted to vendors, suppliers, distributors and resellers, institutional buyers, and employees.

When a site serves the general public or consumer, it is engaged in B2C or business-to-consumer activity. When the site serves businesses, then the content reflects a B2B or business-to-business orientation. Depending on whether the activity is B2C or B2B, the content on the site is apt to be quite different.

Broadly, there are four kinds of online content types: information, entertainment, services, and applications. It's difficult to separate one from the other, and they are usually part of a mix of content provided on a site, often in interrelated formats. Nearly all sites provide information, which includes news, facts, anecdotes, and opinions, as well as all the associated text, graphic, and AV elements. Entertainment is such material as fictional stories, games, and music. (Information about entertainment falls somewhere in the middle!) It is particularly difficult to differentiate services and applications. Services do something for people; applications let people do something.

B2C sites rely on information and entertainment. Even when there is no entertainment per se, developers try to present material in a pleasant and visually stimulating way. B2B sites generally don't put much effort into entertainment. They may not invest the material with any overtly attractive elements at all, although there is typically a token effort to make content readable and actionable. B2B efforts are skewed towards product information and sales. They are increasingly likely to offer services and applications, selling them outright or offering them on a per-use or subscription basis.

Services are products or free offerings, like 24/7 availability, price comparisons, online ordering, customer service, gift registries, live personal shopping assistance (voice or text), and so forth. Applications are search, payment calculation, currency conversion, use of an online program to perform a task like photo or video editing, spread sheet manipulation, and so on. Here again, there is some overlapping—a search provides a service to the consumer, but it invokes a program application in order to carry it out. On the other hand, a live personal shopping assistant calls upon a whole array of programs to provide the service.

Applications have led to a new class of online content providers called ASPs, or application service providers. These purveyors offer their customers the ability to use software to accomplish their objectives. An ASP can provide anything from a quick lookup of stock market ticker symbols to extended sessions using sophisticated enterprise resource programs (ERP) or digital video editing. The ASP market is predicted to grow substantially to a multibillion dollar market in the next few years, although estimates vary widely because the online business of providing applications is so new.

Whether they appeal to consumers or businesses, content models fall into three classes: content aggregation, audience aggregation, and audience segmentation models. Content aggregation models begin with a focus on the content itself and pulling together enough material to appeal to any number of different audiences. Audience aggregation and segmentation models start with a focus on an audience and create or acquire content that appeals to a designated audience. Table 14.1 presents the various models that have emerged in the digital media and entertainment space.

Content Aggregation Models

Probably every digital content provider considers several of these models when they make their plans. They can combine them easily enough as well. A company that selects the consumer experience strategy may decide that the interface is the most important part of the process and that the landing page is the most important part of the interface, and they may also decide to syndicate their content or to add such material to their site. What they have in common is that these models call for the content designers to focus on the content first.

Consumer Experience Model   This model holds that all the content, taken together, must provide the consumer with a unique and pleasant

Table 14.1 Content Models in Digital Media and Entertainment

CONTENT MODELS		
Markets: B2B, B2C		
Content types: Information, entertainment, services, and applications		
Content Aggregation Models	*Audience Aggregation Models*	*Audience Segmentation Models*
Consumer experience	Horizontal portals and destinations	Vertical portals and destinations
Bundling and buckets	TV portals	Affinity
Interface control	Free service	Internet Jockeys (IJs)
Screen real estate		Communities
Enhanced TV		
User-created		
Syndication and licensing		

experience, not just information, data, or service. The crux of this idea is that just as goods and services are products, so are experiences, and they are distinct from traditional product categories. In this view, the industrial economy was replaced by the service economy, which is now being moved over by the experience economy.[1] As Joseph Pine, II, and James H. Gilmore say:[2]

> *Make no mistake: Information isn't the foundation of the new economy. Information is not an economic offering. As John Perry Barlow likes to say, information wants to be free. Only when companies package it in a form customers will buy—informational goods, information services, or informing experiences—do they create economic value.*

Shapers of media and entertainment utilizing this approach will want to ensure that the customer will not be confused, angered, frustrated, or otherwise affected negatively by the entire process of finding, retrieving, viewing or playing, sampling, ordering, buying, and ultimately using the product or site. It is easy to think that designing a positive experience can be taken for granted, but research shows that at least 25% of people who have started a sales transaction on the Internet do not complete it. Since it takes many clicks to get through the process, such a result is not surprising. However, the high cost of usability and focus group research means only the largest sites can afford detailed studies.

A good example where concern about the user's experience is foremost in the designers' minds is the Web site devoted to the syndicated show *Caroline in the City,* produced by Joanne Burns, then at Eyemark Entertainment, a subsidiary of the CBS television network, at http://www.carolineinthecity.com/. Burns described some of the considerations that went into the site to make it easy to enjoy for site visitors. On the home page, a graphic of the main cast members and the show title are up front and center—the guest has no doubt that he or she has found the Web site for the show. The typeface and design carry through the theme of lighthearted, romantic whimsicality that pervades the program.[3]

All the main choices are there on a page that requires no scrolling to get it all in view. The choices are clear and highlighted, with special features emphasized. The site has a very personal tone, and there is an invitation to join the "Caroline Club" on the home page. The main selections are clearly stated, and text that more fully explains what a click-through will bring appears when the mouse hovers over a choice—a "mouseover" or "rollover." Community is encouraged by an easy-to-access message board, which adds new, self-populating content to the site all the time.

The navigation bar is positioned at the top of the page so it doesn't constrict the visual design of pages. Site subsections are readily identifiable. Frames at the left hand side always clearly indicate to visitors exactly where they are within the site, and it is easy to move from one area to another. Special promotions are also clickable from every page.

These are just the structural considerations—it does not include a discussion of the subsections or individual pages, each with its own informational, interactional, or transactional requirements. If these were added to the discussion, the requirements to provide a smooth, unified experience for visitors to the Web site would continue for any number of pages!

Bundling and Buckets   Bundling means putting many different kinds of content together into attractive packages. The content can come from different sources and reflect many different types. In the cable, satellite, and PC software industries, consumers usually pay a single price for the bundled content.

Online, content is usually free. But Web sites realize that in order to build traffic, they need to appeal to many people. The way they do it is by placing "buckets" of content that people can sift through to meet their precise individual needs. Material may be produced or commissioned by the site, licensed from copyright holders, and displayed through affiliate or affinity agreements with other Web-site operators.

Interface Control Model   According to this model, what is really important is the consumer interface. It is the specialized content that engages the consumer throughout the session on a Web site, or on a programming lineup for an interactive TV network, channel, or stream. The interface sets the context for the content: it establishes a thematic design, structures the choices a viewser can make, and provides mechanisms for navigation.

Control of the interface gives the ability to manipulate the customers' experience in a number of ways. It defines the number and type of choices customers can make and the procedures they must follow in order to execute them. The interface can add the element of consistency. By jumping out of programming back to the interface, people can tell when they reach a particular environment. An analogous example might be that driving on the highway is uncertain, but checking into a Holiday Inn at least brings the day to a predictable end. The motel will take credit cards, offer a set of known features, and meet certain standards of cleanliness.

Powerful incentives lead companies to provide the user interface.

> *The great faceoff is not over the traditional gatekeeper issues like "do you get my signal from cable, telco, satellite?" It's about the fact that whoever gains control of the last interface has the best shot at maintaining that relationship. Therefore, the interface and its navigation are very strategically important.*[4]

Navigation is essential for customers. From the time they turn on the set and get the "splash" or welcome screen, and throughout the viewing session, they will want to know where they are, where they are going next, and where else they could be. Interactive navigation systems also let people explore for more information if they want it, and allow operators to present new sales opportunities.

Consumers agree that navigation is important. According to a Jupiter/NFO Consumer Survey of online users, 68% of those surveyed rated "ease of navigation" important, and 51% cited it as directly related to their satisfaction.[5] And it's even more important for TV viewers that have skill levels stretching from Junior the Netsurfer to Grandma the Regis Philbin fan.

Important as the interface is to consumers, it is even more important to content providers because it can confer considerable business leverage. The mechanism for content to hook into the interface software is called an API, or application programming interface. Without the specs for the API, a content provider cannot ingrain the material so that it becomes one of the choices available to the consumer. Essentially, control over the interface structures some aspects of consumer behavior—and it may also act as a gatekeeper of which programming is included in the lineup. The interface owner may own a piece of the e-commerce and T-commerce (television commerce) value chain, since it is the vehicle through which the consumer orders products, or gets to the order-taker. Finally, it may also give access to log files of consumer behavior, data that enables marketers to target consumers better.

An example of this strategy is the AT&T Broadband Interactive Service (ABIS), the high-speed digital offering sent to consumers over the company's cable systems. The interface is designed for the TV, not the PC. The services include travel, voice messages, E-mail, business news, local living, education, entertainment, shopping, kids' programming, games, hobbies, financial information and services, and Internet access as a premium service. ABIS controls every aspect of the interface—each element that makes up the look and feel, all the programs, content, application, products, and services that are available, and all the processes subscribers carry out when they are using the service.

From a business standpoint, this means that ABIS stands to make money at every turn. It can charge content providers for reaching customers, and charge customers for accessing content and services. The company also commands a percentage of any T-commerce transactions.

Screen Real Estate Model

This is a variation of the interface model. Many of the advantages conferred upon the company that controls the interface also applies to the control of screen real estate.

This model assumes that of the entire interface, the two most valuable parts are:

- The welcome or landing screen, or the first screen that viewers encounter when they turn their systems on. In some interfaces, the landing page may be called the "home page," and customers return to it over and over again throughout a viewing session.
- The data glove, or the top, bottom, and right-hand or left-hand side of the screen, as separate from the TV screen or the center of a Web page where the consumer-requested information is displayed.

The inventor of the concept of controlling consumer access to content and providers' access to consumers was Bill Gates, whose masterful desktop-real-estate strategy was an important arrow in the Microsoft quiver until the advent of Netscape. When the enormity of the threat to the Windows desktop hegemony posed by the Internet browser became clear, Microsoft moved decisively to package its own Internet Explorer browser with its Windows operating system. This action is at the heart of the U.S. Department of Justice monopoly suit against the software giant, and the proposed division of the company splits the operating system/application units from the Internet browser/application units.

Companies that are well-positioned to execute a screen-real-estate strategy are electronic programming guide services like TV Guide Online. NBCi, the Internet enterprise formed between the interactive divisions of the NBC broadcast network and other partners, plans to execute an interface strategy as part of a deal it is making with the DSL provider Telocity. When a household gets DSL service, it will also get NBCi interface software. But executives in the company are keenly aware of the potential inherent in controlling the PC and TV screen.

As an NBCi executive explained:

> *The key advantage lies in recognizing the selection of components in the network of sites. We will package and present the content information—every choice on the personalized dashboard (as we have dubbed our "navbar") will offer streamed media. And it will take advantage of the "always-on" feature of high-speed Internet access. We'll use the navigation components of Snap!, manage, share, and store data components and handle e-commerce transactions through our partner Xoom's technology, all integrated with the advertising business of Snap!*[6]

Enhanced-TV Model   This method of developing and presenting content allows owners of linear programming, such as TV shows and movies, to make their content interactive. The material is digitized and additional information about it is written and marked up in ATVEF (Advanced Television Enhancement Forum) code. It will call out to the Internet or other storage site to retrieve graphic information that can be overlaid on the video screen. When the viewser clicks on a "hot spot" on the screen, a message is sent upstream via telephone, cable, or wireless device to a headend, central office, or server. More and more, enhanced TV is linked to an Internet site, and the additional interactive material comes from the Web.

User-Created Model   The job of populating a Web site with content is demanding and expensive. And many people want to express themselves and create material that they can share with others. Sites that post material for the public often feature it as the majority of their content.

Content providers following this model find that it does require some administration. The procedures for posting material must be clear. Similarly, rules must be prominently displayed and strongly enforced. Sites for the general public most often guard against pornography and other offensive graphics and language.

User-created material makes the most sense for the Internet. This type of content requires a great deal more tweaking in other digital television venues such as over-the-air TV, cable TV, and even broadband TV. For example, *America's Funniest Videos* entailed an entire staff searching out videos, gathering them, ordering them into a rundown, and editing them. The producers contracted with expensive talent to host the show wraparounds and "tweenies" and assembled a studio audience to give the program the added energy of a live group of viewers.

But on the gazillion channel Internet, user-supplied content is a natural. Sites that have adopted this strategy and made user material the centerpiece of its material include Popcast.com (www.popcast.com), iFilm.com (www.ifilm.com), Internet Photos (www.iphotos.com), and The Palace (www.thepalace.com). iFilm.com puts up user-created videos and animations. iPhotos.com display photographs and allow users to assemble them in albums to show to friends and family. The Palace, one of the earliest graphic sites on the Web, allows users to design their own room in the Palace and invite people to see and experience the room they've created.

Sites that put up user-created content may have quite different objectives. Many, many sites have areas of user-created content—message boards, chat rooms, and content display galleries. Popcast.com and iPhotos.com want to appeal to the general public and generate as much traffic to the sites as possible. As expected, they offer users tools to help them put together their content and post it on the site. By contrast, iFilm.com is geared toward professionals who create content as a full-time occupation. They have a staff that digitizes tapes and posts them, with an eye towards making sure the creator is either part of or wants to be in the professional marketplace.

Naturally, user-created content is the primary offering of personal Web sites, and many thrilled content creators—artists, musicians, writers—have enthusiastically pitched the Internet as the ultimate way for creative people to reach an audience. But there is some disagreement with this position. The counterargument is that there is such an avalanche of material on the Net that no one person can be seen or heard.

In fact, say these people, the ubiquitous Internet gives all the advantage to those with an existing brand name—stars and hit labels. More than ever, a recognized and recognizable brand will be able to generate the trial and purchase of their creations, a trend that can only increase as more and more people access the Net and attempt to aggregate an audience for their creations. It is indeed an irony that the most democratic medium ever invented should serve to reinforce the elite content creators.

However, as covered in the previous chapter, some observers believe that there will never be any real hits on the Net at all. In this view, the ability of people to match their wants and needs with precision will undercut any attempts to aggregate large numbers of consumers for any length of time. The only way a site can build an audience is to aggregate enough content to attract a lot of individuals, a strategy that often involves including user-created material.

Syndication and Licensing   From a content-producer perspective, syndication and licensing are ways of bringing in revenue. From the point of view of a site operator, they are content-aggregation strategies. They are new and exciting developments in the digital content industry because they allow producers to create once and publish in many venues. It's also good news for Web-site owners because they don't have to manage a staff of people to create material. Instead, for reasonable licensing fees, they can populate a site with content that is geared toward their particular audience.

Syndication companies aggregate content from many different providers and send it to Web-site owners and operators for them to post on their sites. Right now, much syndicated material is repurposed from print, so that the columns are wordy blocks of text. More and more, we will see material especially created for the Internet with more graphics, streamed media, and page-formatted for horizontally-biased PC screens, instead of vertically-biased print media.

The technical underpinnings that will allow an automated process to enable syndication are just now being promulgated. The language of choice for syndicated content is XML, an all-purpose markup language that is becoming universal in multisite, multicompany e-commerce applications. This new language has no option, permits no variation, and allows more sophisticated searching than do the hijacked proprietary versions of the HTML language. Another important enabling development for syndication is the emergence of a protocol for exchanging content on a server-to-server level: ICE, the Information and Content Exchange protocol.

In early 2000, the major content syndicators of text, graphics, and small applications are iSyndicate (www.isyndicate.com), Screaming Media (www.screamingmedia.com), and The Content Exchange (www.content-exchange.com). Each of these makes different business arrangements, however, there are several models in play.

- Fee for licensing for a series of individual columns or articles, branded by buyer;
- Fee for licensing packages of content at a reduced rate, co-branded by site and content provider;
- Free to Web site but co-branded by site and content providers, and links take visitors to content provider site.

Syndication companies for streamed audio and video were just entering the scene in early 2000. Indeed, the digital content marketplace is just being born. Whether it will become its own industry or merge with traditional content markets is not yet clear. However, it is certain to grow, because the advent of broadband brings with it higher production values and their associated costs. The days when every Web site creates original material are numbered, because it is too inefficient given the unlimited number of Web sites and the cost of producing material.

Audience Aggregation Models

The content models we just looked at start with aggregating content and then look for audiences for it. Others models reverse that process and begin with strategies for aggregating a particular audience or set of audiences, then assembling content that will appeal to them. Broadcast television is the most successful industry ever to use an audience aggregation model, beside which all other media pale. Cable TV, radio, and print have a very few broadly popular properties, but many more of their products serve niche audiences.

Horizontal Portals and Destinations   The early success of portals in capturing tremendous traffic on the Net set off a race by media and entertainment giants, racing to get their own branded portal on the Net. In rapid order, Disney, NBC, and Warner Bros. bought and built their way to mega-portal status. Portals provide all the needed information services to send visitors on their journeys on the Internet. The horizontal portals like AOL, Yahoo!, Microsoft Network (MSN), and others appeal to as many people as possible. For the most part, they started by offering search engines that let people locate other sites, but as time has gone on, they have also become destination sites in their own right.

A portal is a way station; a destination is a place to spend time, a "sticky" site where people stick around for awhile. The most important feature of a portal is its search engine, which captures as many listings of Web-site content and services as possible and displays the returns in an efficient and readable manner. However, a destination site requires many more services and bits of content than a site designed only to offer search engine services.

What does it take to be a portal? The best way to answer this question is to examine the landing page of Yahoo!. It's a densely populated page composed almost entirely of links to other areas on the site, revealing an extraordinary array of departments. While the entertainment and e-commerce portals are somewhat narrower in focus than the search-based ones, their offerings still attract almost everyone—who doesn't want to have fun, buy, or sell?

Table 14.2 lists the top fifteen U.S. Web sites and the number of visitors they received in April, 2000, as measured by Media Metrix, Inc. Eight of them feature a search engine as the main attractor; three get much of their audience as a result of being an Internet service provider; five can be described as entertainment portals; and two are e-commerce portals.

The difficulties of establishing a horizontal portal today should not be underestimated. The Walt Disney Company, no stranger to aggregating an audience, acquired search engine Infoseek and put it together with the company's other Internet entities to launch Go.com. In 1998, Go.com took in $323 million and lost $991 million for the year; in 1999, the site took in

Table 14.2 Top 15 U.S. Web Sites

	Media Properties	Unique Visitors (1000)	Initial Site Type
1.	AOL Network— WWW and Proprietary	77,883	ISP Portal
2.	Yahoo! Sites	58,592	Search Portal
3.	Microsoft Sites	48,592	ISP Portal
4.	Lycos	47,159	Search Portal
5.	Excite@Home	32,244	Search and ISP Portal
6.	Go Network	30,117	Entertainment Portal
7.	NBC Internet	22,347	Search and Entertainment Portal
8.	About.com Sites	16,267	Search Portal
9.	Amazon	16,084	E-commerce Portal
10.	AltaVista Network	14,174	Search Portal
11.	Time Warner Online	13,319	Entertainment Portal (Entertaindom)
12.	Real.com Network	12,966	Entertainment Portal
13.	Ask Jeeves	12,791	Search Portal
14.	Go2Net Network	12,335	Search Portal
15.	eBay	12,299	E-commerce

Source of ratings data: Media Metrix, Inc.

$348 million and went $1.06 billion into the red. It is now number 6 in the most-visited sites—but that's a lot of money to pay for 30 million visitors.[7]

The "Portal TV" Model   This model transfers the idea of a portal from the Internet on the PC screen to high-speed Internet access on the TV set. Just as AOL and Microsoft Network have become popular portals through their roles as ISPs, so broadband ISPs like Excite@Home hope to become portals through their ability to bring subscribers to the Net through their own interface on the TV set. Probably the only cable companies with a sufficient subscriber base to adopt a "portal TV" model would be AT&T and Time Warner. Now that AOL has taken over Time Warner, these cable subs will become part of the AOL juggernaut.

DSL subscribers typically have high-speed service to their PC. However, the number of telephone company DSL customers is growing rapidly, and with a home gateway, it is possible that customers could order it for their TV sets. Despite their growing ISP business, telephone companies have not shown any creative flare in developing portals, either online or for the TV. No local telephone company is on the Media Metrix Top 50 list of popular Web sites; the only telco that appears on the list at all is AT&T, at number 31 in April, 2000.

Free Service Models   Building an audience by offering free service, where the customer must come to the provider's Web site in order to access the service, has become a popular strategy in 1999 and 2000. (See Figure 14.1 for one of the most successful free offerings on the Net.) Some sites that offer free services are:

- Netzero.com (www.netzero.com) offers free Internet access
- MySpace (www.myspace.com) gives away 300MB of virtual disk space for secure storage
- Mydomains.com (www.mydomains.com)—Reserve a domain name through the service and you can park it here for free
- Blink.com (www.blink.com) lets users upload their bookmarks to the site so they can access them from anyplace on the Net

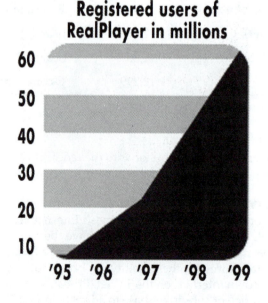

Registered users of RealPlayer in millions

Figure 14.1 Real Networks offers a free version of its player and sells an upgrade. Source: Real.

Audience Segmentation Models

The Internet is the inverse of broadcast television. It has an unmatched ability to reach specialized audiences, and relatively low cash outlays for production and distribution makes it cost-effective to do so. Interested in aviation history, aviary management, or .avi file conversion? There's bound to be a treasure of information on the Net, just waiting for you to discover it. The following models all exploit this ability of Internet to attract people who are united through some common experience or interest.

Vertical Portal and Destination Model How wide is a horizontal portal and when does it become a vertical portal? The term vertical means that it is designed to appeal to a target group. Verticals may be very wide. iVillage (www.ivillage.com) is designed for women. ThirdAge (www.thirdage.com) is a site for middle-aged people. Bolt.com (www.bolt.com) targets adolescents, and Surfmonkey.com (www.surfmonkey.com) is for kids.

There are also portals that are very narrow, serving vertical interests like Bikersites.com (www.bikersites.com/) and Adobe's file conversion info Web site (http://file-conversion.web-design-tools.com). Here's narrow for you: bird lovers who want to know how to build an aviary for budgerigars can always turn to www.budgerigars.co.uk/manage/. Summit Strategies ties the evolution of narrower vertical portals with the growing availability of software services on the Net and predicts that future portals are likely to be virtual workspaces where people go to perform personal and professional tasks.

Affinity Models Affinity models segment and then aggregate as many people as possible who are related to the target segment. An Internet affinity group is composed of people who share an interest in some kind of content, although they may have vastly different reasons for doing so. Nevertheless, based on this interest, it is probable that they will also be interested in related compatible or complementary content, services, and products. To exploit this relatedness commercially, a partnership agreement between two or more Web sites that reach similar demographics establishes URL links between them.

For example, a site that features real estate might partner with lenders. Similarly, a guitar manufacturer could execute agreements with a site that hosts a MIDI community (a digital music interface between an instrument and a computer), a songwriter's site, a musical instrument Web store, repair services, and so forth. It doesn't really matter that one guitar player wants to play funkadelic and another classic flamenco. The House of Blues (HOB) is an interesting example because the company maintains an active affinity program even though it is rather difficult to pinpoint groups that might like blues music.[8] While there may be some slight demographic skew among blues-lovers, they exist among every conceivable grouping.

HOB lets Web-site owners decide if they think HOB content would enrich their site and if their visitors might be interested in it. When they join the HOB affinity program, they download music and a House of Blues clickable logo. When the visitor clicks on the logo, they hear the music and have the option of clicking on a link to HOB where they can listen to more music and purchase it online. If they make a purchase, the originating site gets a percentage of the sale.

Internet Jockeys (IJs) Models   The best example of a site following this model was Pseudo.com (www.pseudo.com) and founder Josh Harris is a proponent of the idea.[9] He argues that that the huge amount of information on and off the Internet means that people will look to others whose opinions they respect for guides to the material. His notion of an IJ is something more than the radio disk jockey. Internet Jockeys, or Ijs, are people who embody the interest group that is targeted. They are not merely knowledgeable or authoritative; they must actually be immersed in the subworld into which they are leading audience members.

Harris notes that Gen X and Gen Y do not have anticommercial attitudes if advertising and promotions are upfront and obvious. So he sees IJs being supported by products they wear, use, and otherwise promote through close association. Thus the IJ doubles as a mannequin, demonstrator, and pitchman for the products that support the content property.

Community Models   In some ways, making community a centerpiece of a Web site or digital service is a particular case of user-generated content. However, not all user-generated content is community, which involves creating services that allow members of the group to participate in communication with one another. The means of communication may be an asynchronous message board or E-mail discussion list, where participants post their comments, and then sign on later to read responses and post again. Or it might be real-time chat that lets participants conduct text-based discussions.

Letting people choose avatars to represent themselves in chat rooms has been around since 1995. Now, several companies are working on ways to let people actually talk when they are chatting. Audio-enabled chat rooms require members to have a microphone hooked up to their computer.

Following the community model doesn't mean there is no other content. Usually operators put additional material believed to be of interest to the members of the group on the site. Information, services and applications, and e-commerce interactives are quite commonly found on community sites.

Communities can form around any sort of interest or concern, and there are thousands, perhaps hundreds of thousands of them on the Internet. In addition to being a nearly essential part of every Web site that is not merely a personal home page, there are Internet news groups and discussion lists. One site that tracks discussion lists is at http://webscout-lists.com/. Another interesting business is eGroups, which lets people start their own E-mail discussion lists. eGroups takes care of adding and deleting members from the lists—and it offers group-buying opportunities through MobShop. An ad on the jump page explaining the group shopping offer says: "eGroups and MobShop have joined forces to bring eGroups members the benefits of group buying. The concept is simple: the more eGroups members that join a buying cycle, the more savings for all. Save up to 30% off of retail prices on tons of products!"[10]

Distribution Models

Once content is developed, a distribution model lays out how it will get to consumers, in terms of media platforms and technologies. In times past, the product itself defined its distribution. Movies were shown in theaters and later rented from video stores; songs were played on the radio, then purchased in stores that stocked vinyl record stores and later CDs; TV shows were shown on TV; and so forth.

In the digital environment, it is no longer clear how a given content vehicle will reach its consumers. A song is played on the radio, on an audio pay service over cable and satellite, on the Internet, and soon over a mobile phone. Three distribution models that address this new reality are currently extant: windowing, cross-media/platform, and the walled garden.

Before considering the models themselves, there is one further wrinkle in distribution—the relationship between providers and consumers in delivering material. The two-way Internet has fostered a new characterization of distribution media and formats as "push," "pull," and "opt-in." Recall from the last chapter that that push means a distribution effort that sends out content to people whether they want it or not, such as broadcast TV and radio, spam E-mail, and direct mail. Pull means a distribution that is requested by the receiver, such as going to a Web site and downloading a file: a request-receive format.

Opt-in is a new category that might be characterized as pull-push. It reflects the need for companies to avoid irritating the consumer. The client makes a request which is then fulfilled at the sender's discretion. Many cable networks have opt-in E-mail programs where subscribers receive on-going notification about up-coming programs they might want to watch. Opt-in distribution usually has some way for the user to "opt out" as well through a request to end service.

Windowing Models

This model marries a distribution and a revenue model together. "Windowing" means to release a property in stages to different distribution mechanisms for a specific length of time—the window. In the motion pic-ture industry, it describes the wholesale level of film distribution. The order of the release windows is based on the revenue brought in per viewer. Domestic and increasingly international theatrical release come first because each viewer pays for a ticket, $4.50 to $12.50 (in Japan). Pay per view is next, with a charge of $4 to $5 per household, followed by home video rental for $3 to $4.50 per night. After these follow premium cable, foreign TV, network TV, syndication packages for TV stations and basic cable channels, each of which brings in a smaller and smaller flat fee or package price.[11]

Revenue per viewer is not the only conceivable way to structure release windows. Strategies for reaching particular audiences might define them—or media platforms. For example, one analyst proposes that film studios create an early release window via satellite HDTV channels. A "First-Run HDTV Channel" would feature pretheatrical releases of movies to the 100,000 people or so around the world who have HDTV sets. Dale Cripps, who developed this idea, believes that there are about that many people who would pay $10 to be the first ones to see new films. The income would be $10 million per month, and it could provide valu-able word-of-mouth to support the theatrical release right afterward. (Except for disappointing films!)[12]

Windows now extend into the digital realm, with new ones added such as DVD, video CDs, and eventually the Internet and other networks. Low-resolution samples could limit the value of previewed material such as trailers. And slicing and dicing can permit the distribution of snippets, portions of text, individual scenes, and single photos. The distribution of music is now in turmoil because of MP3 and Napster, but it is possible that windowing could apply to songs as well, with launch at a concert, pay-per-song over the Internet, then CD-ROM, and finally package prices for subscription audio channels over cable and satellite.

Cross-Media/Platform Models

The cross-media/platform model means distributing content across more than one medium or to more than one type of reception device in order to maximize the reach to a target audience. Consider that the same person may watch TV, listen to the radio, and surf the Internet at different times, in different locations, with varied motivations. They may also combine a broadcast medium with the Internet, the so-called "co-location" phenom-enon. (See Figure 14.2.)

And we have discussed the many new reception devices that are coming on the market in the next few years, such as portable wireless devices that let people download material from the Net and copy the material on a laptop, MP3 player, and so forth. This means that content

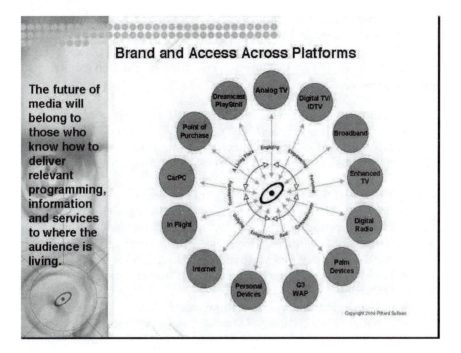

Brand and Access Across Platforms

The future of media will belong to those who know how to deliver relevant programming, information and services to where the audience is living.

Figure 14.2 An elaborate cross-platform model. Courtesy: Pittard Sullivan.

providers and distributors must prepare to reach consumers wherever they happen to be on whatever devices they choose to use. Creating material for different delivery platforms, network access speeds, formats, reception devices, and consumer characteristics is another way of saying that content must be customized, tailored to address many different environments, conditions, and consumers. In the digital world, the most efficient way of thinking about a production is to "create once, publish everywhere," or COPE, a phrase adopted by a coalition of companies including Sun Microsystems and iXL.[13]

Sun calls this strategy "chasing the consumer," delivering content to consumers and tracking their responses across multiple media and platforms. The target audience may be broad or narrow, but the content is tailored to the needs and desires of that group. It is created once, and automated procedures convert it to needed formats for distribution across multiple media. This means that content must be flexible. It must be an exciting print vehicle, a compelling television program, a sticky Internet attraction, and an e-commerce bonanza. The cost of the content is amortized from revenues derived from all sources, across the various media where it appears.

Probably the best example of a company following the cross-media model is Martha Stewart Living Omnimedia (MSLO) LLC, with business ventures divided into four segments: Publishing, Television, Merchandising, and Internet/Direct Commerce. Her content creation empire and media exposure includes:

- A monthly magazine (*Martha Stewart Living*).
- A quarterly magazine (*Martha Stewart Weddings*).
- A syndicated one-hour television show (*Martha Stewart Living*), distributed to 91% of TV households by Eyemark, a division of CBS.
- Weekly *CBS This Morning* appearances and periodic CBS prime-time specials, all in partnership with divisions of CBS Television.
- Half-hour program twice a day, seven days a week on Food Network cable channel that consists primarily of food-related segments from previous Martha Stewart Living television programs.
- Books written by Martha Stewart and the editors of *Martha Stewart Living*.
- A syndicated newspaper column (*askMartha*).
- A national radio show, *askMartha*.
- A mail-order catalog and online merchandising business (Martha By Mail).
- An Internet Web site that features integration of television programs, radio shows, newspaper column, and magazines, as well as seven distinct channels on the site, each devoted to a core content area—Home, Cooking and Entertaining, and so on. Each channel offers live discussion forums from 10:00 A.M. to 6:00 P.M. EST; 24-hour bulletin boards where visitors can post advice, queries, and replies; and weekly live question- and-answer hours with in-house and guest experts.
- Strategic merchandising relationships with Kmart, Sherwin-Williams, Sears, Zellers, and P/Kaufmann. MSLO designs products that are then manufactured by or on behalf of strategic partners. The products designed for Kmart are sold exclusively in Kmart stores in the United States. The Kmart stores carry Martha Stewart Everyday Home products (bed and bath products, kitchen textiles, window treatments, and bath accessories); Martha Stewart Everyday Garden products (currently patio furniture and selected garden tools, growing in early 2000 to a full line of garden tools and accessories, live plants and seeds); the Martha Stewart Everyday Baby line (infant bedding); and, in late 2000, the Martha Stewart Everyday Housewares line (dinnerware, flatware, beverageware, cookware, bakeware, mirrors, picture frames, and lamps). Products designed for Kmart and Zellers are

sold exclusively at their stores; products designed for other strategic partners are sold through a variety of retailers, including Sears, Canadian Tire, specialty paint stores, Jo-Ann Fabrics and Crafts, and Calico Corners stores.

Walled Garden Models

"Walled garden" is a term coined by former Tele-Communications Inc. founder John Malone. It describes a closed network that restricts Internet access and keeps subscribers within a restricted area. Inside the garden, the subscriber can choose from a bouquet of services that are carefully selected, controlled, and often created and operated by the network providing company.

Most cable-operated walled gardens are like the online AOL service, where the content is sectioned into categories. Cable companies such as AT&T's Broadband Interactive Service (BIS) use this model, as does the French satellite program service, TPS (Television par Satellite). If access to other networks such as the Internet is allowed at all, it is available as a premium service, such as the tier arrangement offered by AT&T's BIS.

Marketing Models

Content models define a strategy for how to prepare material that will appeal to some audience, and distribution models lay out how it will get to them. Marketing models establish how customers will find out about the content, products, and services. Part of the process involves developing a value proposition that gives consumers a reason to pay attention to the content, and a communication strategy for delivering the value proposition. In short, a marketing model answers the questions: What appeal does this particular content hold for a given audience segment, and why should these people pay attention to it and consume it? How will they find out about it, and how can the appeal be best communicated?

The core concept underlying most contemporary marketing efforts is branding. It is a set of marketing communication (marcom) activities designed to make people aware of a product, service, or organization in a way that facilitates its business activities. In media and entertainment, branding has achieved dominance in this field because marketers believe that the key to survival in a multichannel environment is to build an identity, a brand name so that the content can stand out in a field of almost unlimited choices.

There are several ways to execute a content branding strategy. Broadcast networks use logos, advertising slogans ("ABC's the One!"), and "signature shows" (for example, "Seinfeld" on NBC) to build their brand. Cable networks title their service to identify their offerings, such as Comedy Central, The Family Channel, The Game Channel, the Cartoon Network, and so forth. No matter the method, successful branding involves:

- Establishing a unique image or position in consumers' minds
- Offering a unique value proposition to the consumer
- Linking the brand with a positive emotion

Writing of ESPN, Michael Wolf describes how a well-positioned brand is an intimidating competitor: "Imitators have discovered that in the entertainment economy, companies have little chance to succeed without a brand that creates the kind of emotional attachment that ESPN has been able to build with its audience. In today's environment, *mind share*—how well the public knows your brand and cares about it—often precedes market share."[14]

Aside from the branding elements embedded in the content, advertising has traditionally been the most important way to establish a brand. Companies may run an image campaign, with branding as the primary purpose. More often, they pay for a product-related campaign that incorporates elements to position the brand as well as to sell or promote specific products.

Beyond advertising, the digital environment offers new ways to establish brands:

- Interactive branding, implemented through the interaction of messages with consumers.
- Service branding, carried out by providing a service (usually free).
- Community branding, originated from discussion among participants who talk amongst themselves, rather than vendor-initiated. As a result, negative branding may occur.[15]

All these opportunities for marketing require even more attention to detail than companies already bring to their communications campaigns. Consistency across the range of materials is crucial, especially in a crowded information arena where consumers receive an avalanche of information. To get the most benefit from the sum total of the branding effort, public and community relations campaigns, image and product advertising and promotion, Web sites, E-mail campaigns, and transaction messages must always be consistent with the overall branding strategy. The next section looks at marketing models that take into account digital media to execute both brand strategies and media and entertainment product promotion and sales: viral models; spiral models; and data mining.

Viral Models

Viral marketing means to create a campaign where one user communicates your marketing message to another, acting as your marketing agent. It's a contagion model of "spreading the word" that resembles how people pass colds and other viruses to each other. The genesis of the term probably came from Douglas Rushkoff's 1994 book, *Media Virus*, which suggested that mass-media programming contained hidden messages intended to spread counterculture attitudes and ideas.[16] However, viral marketing does not consider the content of messages; rather, it is a technique that piggybacks on one-to-one personal communication.

Consider the telephone. It is very difficult to attach a message from a marketer to a telephone conversation. Your mother, clients, boss, significant other—all would most likely hang up if a harangue from an auto dealer or a slogan from the Pillsbury Doughboy came over the telephone before or after your interpersonal exchange. But E-mails are different. Individuals attach their own advertisements at the bottom of their messages through "sig" files.

Similarly, under certain circumstances, advertisers can use materials that are often exchanged between people to get customers to forward marketing messages to their network of acquaintances. The best-known and most successful execution of viral marketing was Microsoft's use of its free Hotmail (www.hotmail.com) E-mail service. It allows users to create a personal account and to receive their E-mail at the Hotmail site. Given the proper mail address, Hotmail will query the user's mail service server, or servers if there are multiple accounts, and aggregate them at Hotmail. Customers with "always on" E-mail service can forward their mail so it goes directly to Hotmail.

Microsoft used Hotmail to promote its Microsoft Network (MSN). After users signed onto their accounts, they could read their E-mail. They can also send replies and original E-mails—and at the bottom of each one the system automatically attaches a message, such as the one used in May, 2000: "Get Your Private, Free E-mail from MSN Hotmail at http://www.hotmail.com." The site home page allows users to choose their language: English, Spanish, German, Italian, or Portuguese. It features an offer for MSN online calendering, a free service that lets users post appointments and due dates on a calendar that they can access anytime they are online. And when the user signs out of Hotmail or MSN Calendar, they automatically land on the MSN home page.

Another instance of content that follows a viral model is Bluemountain.com, which lets users choose from a variety of cards, modify them to add the name of the person it is going to, a personal message, and their own name at the bottom. An E-mail with a clickable link notifies the recipient that a card is waiting for them. When they click on the link, they go directly to the card sent to them on the Bluemountain.com site. At the bottom of the card, they are invited to send a card in reply. The free service is supported by banner ads and related e-commerce opportunities, like ordering flowers.

Spiral Models

You've probably seen all the dot-com advertisements on television, prominent in the 2000 Super Bowl broadcast. Much of the new media advertising on traditional media is driven by the popularity of the spiral model, which says that marketers need to use multiple media to establish a brand. The main idea is that each medium should be used for what it does best, as summarized in Table 14.3.

The model calls for using traditional media first—TV, radio, print—because they reach the most people, and they are also best at raising interest and building an emotional connection between the brand and the viewer. The message should direct consumers to a Web site for several reasons. It is an actionable medium—people can get more information and order conveniently if they want to purchase a product. The Web site should incorporate a service component, like price comparison, customer support, gift registries, custom ordering, online price discounts, and so forth. In the process of serving the customer, the company can get personal information about the customer, including the E-mail address. This leads to the final step of using E-mail to establish a one-to-one relationship with each consumer. With E-mail, the marketer can close the loop with a thank you, notices, and reminders, always sending the person around the spiral again by directing them to a TV or radio broadcast or print story.

It's important to repeat this spiral as quickly and as many times as possible. It also calls for consistency in the content, so consumers stay involved with the brand and recognize it across media. For example, NBC always uses the same three-note audio signature and the peacock no matter where their content appears.

The power of spiral marketing comes from the fact that if it is repeated quickly and often enough, it builds a "virtuous circle" This is a set of events, where one propitious event leads to another, finally returning to the starting event to begin again. An example might be the development of an innovation, which leads to investment, effective marketing, increased productivity, higher economic growth, and greater consumer buying power, which create profits that are used to underwrite efforts to develop more innovations. In this case, the virtuous circle leads to a seamless blending of brand management and customer relationship management (CRM).

The problem with spiral marketing is that it is very expensive. Funded by a rush of capital investment, dot-com spending was more than $3 billion in 1999, many of them hoping to be the first-to-market in a given

Table 14.3 Media as Defined in Spiral Models

	TV/Radio/Print	Online	E-mail
Consumer benefit of medium	Content	Convenience	Communication
Content	:30 commercial, print ads	Searchable, informative Web pages	Invitations, notices, and reminders
Communication structure	One to many	One to many (one at a time)	One to one
Content appeal to viewer/user	Richness	Rewards	Relationship
Medium's most effective marketing use	Selling	Service	Recommendations
Element of influence (All provide information)	Emotion	Actionability	Personal relationship
Viewer/user reception behavior	Passive	Active	Interactive
Consumer cognitive response	Attention	Intention	Retention
Maximal consumer response	Interest	Involvement	Interaction

Source: Based on a presentation by Jesse Berst.[17]

online niche.[18] In some cases, the cost of acquiring customers is more than they can hope to recoup in any reasonable amount of time.

Data Mining

Viral and spiral marketing both lean toward reaching new customers. Steve Milanovich of Merrill, Lynch refers to monetizing customer relations as "hunting or gathering." Reaching out for new customers is hunting; pitching existing customers for more transactions is gathering.

Data mining is a technique for using information about customers to create targeted value propositions to them. If a marketer knows a set of facts about a past buyer, analyzing that data in detail can provide guidance for such efforts. Sometimes additional information, such as zip code characteristics and census data, can be added to the mix, giving powerful additional clues about how to approach a given customer. The more interactions a seller has had with an individual, the more targeted the approach can be. It is the ultimate gathering technique.

Data mining is inherently cost-effective. Snagging a new customer on the Internet costs about $50, while making a sale to an existing customer is substantially less. However, data mining has not yet come of age because marketers may have too little, too much, or just the wrong information. (Too much data is confusing and useless without the knowledge of what is worth looking at.) Moreover, individuals can be unpredictable.

Over time, however, even eccentricity becomes remarkably predictable! Expect data mining techniques to become more sophisticated, increasingly guiding marketers to ask the right questions and make the relevant observations about customers' behaviors.

And as e-commerce expands, techniques that allow further inferences to be made about future buying behavior means that information about consumers will become ever more important. There are limits on the usefulness of information, such as the quality of the data—its freshness and accuracy. And privacy issues loom large. As of mid-2000, there were more than 300 bills in the U.S. congress addressing what is believed by both consumers and lawmakers to be an important concern.

Revenue Models

Revenue models deal with product packaging, pricing, mechanisms for receiving money (or services in the case of barter), and revenue-sharing schemes. Many of the models used by traditional media, as shown in Table 14.4, carry over into the digital marketplace. Not surprisingly, there are usually some unique twists when this analog to digital conversion takes place.

Revenue models in new media markets may have the same name as those used in traditional venues. But they often mean quite different things. For example, pay-per models can cover smaller informational pieces than in the cable industry. And affiliation means something else altogether on the Internet than it does in the TV industry. Table 14.5 shows new media models.

Table 14.4 Traditional Revenue Models by Industry

Industry	*Traditional Revenue Models*
Cable and satellite multichannel services	• *Subscription*: Subscribers pay a flat monthly fee. • *Bundling and tiering*: Content is packaged together in salable bundles and sold in tiers added onto the basic subscription fee. • *Value added*: Pay extra for premium service. • *Pay per view*: Pay to see specific events, like an important boxing match, concert, or other special attraction.
Telcos	• *Flat service fee*: A monthly basic phone bill. • *Access fees*: Extra charges for long distance access. • *Usage fees:* Pay per-minute charges for long distance and local long distance service, based on metered service. • *Network utilization*: Charges incurred when your phone is "off-hook," like an 800 number or cellular phone.
Power utilities	• *Usage fees*: Metered service. • *Flat service fee*: Some companies allow customers to pay the same amount every month, but it's adjusted at the end of the year anyway.
Online (Internet Service Providers, ISPs)	• *Subscription*: Monthly flat fee. • *Usage fees*: Per-minute charges and use of some information services. • *Ad-supported*: Advertisers pay ISP, per user or click-through or by contract.
Film	• *Transaction*: Consumer buys individual ticket at box office. • *Windowing*: At the wholesale level, studios use a distribution and revenue model called windowing, a staged release of motion pictures to different distribution outlets for a specific period of time, based on average per-viewer revenue. Current windows: domestic and international theaters, pay-per-view services, home video rental and purchase, premium pay services, network TV, foreign TV, syndicated TV.

Table 14.4 Traditional Revenue Models by Industry (Continued)

Industry	Traditional Revenue Models
Broadcast TV	• *Advertising-supported* at the consumer level. • *First-run licensing fees:* Substantial payment for exclusive rights to a distributor to run a property. • *Syndication licensing fees:* After first run, shows are licensed to downstream nonexclusive distribution to foreign TV, cable networks, and local TV stations.
Home Video	• *Transactional pay per*: Purchase or rental. You pay, you get.

Table 14.5 New Media Revenue Models

Content- Supported	Ad- Supported	E-Commerce– Supported	Misc.
Tiering	Sponsorship	Longitudinal cohort marketing	Multiple revenue streams
Subscription	Availabilities: Cost-per-thousand impressions	Affiliate	Cybermediary
Pay-per (bit, stream, program)	Availabilities: Cost-per-thousand click-throughs		Consumer data sales
Tentpole			
Big bite			

Content-Supported Revenue Models

People will pay for content they want. Timeliness and time-to-market are always important, because information and entertainment products are highly perishable, like fresh flowers or lettuce. Freshness is especially critical for financial and business information, and companies and investors are sometimes willing to pay considerable sums to get it first. People like to be the first to receive entertainment but it is equally or more important for it to be—well, entertaining.

Tiering Models   Tiering is a way of pricing a range of products and services to get the maximum revenue for them. This is well-developed by multichannel TV providers, but less used in other environments. Tiering by time is charging the first tier of buyers more than subsequent tiers. Products may also be tiered by charging different amounts for content packages or bundles of programming, a preferred technique by the cable, satellite, and computer software industries.

In the cable industry, the first tier is basic cable. The second tier is composed of "extended basic" packages, each one costing an additional monthly fee. The third tier is made up of individual premium channels; the fourth is a digital tier of bundled channels; and the highest tier is pay-per-view, charges for receiving individual programs, usually movies and sporting events. (High-speed Internet access and telephony are not considered a tier; they are services.)

Subscription Models   Tiering is frequently an adjunct to subscriptions, but not always. A subscription is a revenue scheme where the subscriber pays an agreed upon amount for a specified amount of time. In the cable world, system operators charge a monthly fee for the basic tier but per-channel/per-month fees for premium channels. Ultima Online, a persistent world game (PWG), is a subscription site that costs $9.95 per month to play.

An example of subscription combined with tiering is AOL, which charges one subscription fee for a limited number of hours online and another amount for unlimited access. CompuServe is a subsidiary of AOL and will allow unlimited access to both services for still a third price. In addition, on CompuServe, other premium services may carry surcharges.

Pay-Per Models   Long before the term "pay per" came into being, people paid per ticket to attend movies, plays, and performances. Then telephone companies introduced pay-per-minute charges for long distance and local long distance. Cable systems launched "pay-per-view" service, which in the case of the NBC 1984 coverage of the Olympics, was extended to mean "pay per event" or "pay per package." And the advent of videocassettes and players brought about pay- per-night rentals.

Pay-per models are attractive to both buyers and sellers, and there are various proposals for pay-per download, pay-per bit, article, photo, song, and video sequence. The problem with many of these schemes is that of "micropayments," small transactions that are more expensive to meter and collect than the value of the payment. Until some form of "digital wallet" technology is standardized, it is likely to be difficult for content providers to make money on small units of information.

Tentpole Models   This idea is the opposite of "multiple revenue source" plans, where money comes into the cash register from as many directions as possible. Tentpoling is a strategy where one content product or service supports the entire business. In the TV business, it is not unknown for a single successful show to carry a production company for some length of time.

A good example of an online pioneer was the Internet soap opera, *The Spot*. However, after a few months, the site was unable to support itself on the single property, and it stopped production and closed down. This experience has led some analysts to believe that while it may work in the TV industry, tentpoling is not a viable Web-site strategy because no single property can generate sufficient traffic for any length of time.

Big-Bite Models Big-bite models are the opposite of windowing, which monetize content by releasing it in stages to different distribution systems. "Forget windowing. It'll never work in the age of digital re-creation," cry the proponents of the big-bite model. Instead, they argue that content owners must monetize the first bytes out of the box and move on. The motto might be, Eat the apple, then toss the core, because if the product can't make a decent profit when it first rolls out, it will probably incur a loss.[19]

Content providers have to get the money while the getting is good— immediately. As soon as the product is widely available on the Net, its sales price will quickly drop to zero. Downstream money will be found money—and just about as likely to occur. Consider the game business. It takes eighteen months to two years to create the product and a couple of months to sell them out—or bulldoze them under.

Ad-Supported Revenue Models

Advertising is the stalwart of over-the-air television and a major source of revenue for basic cable and satellite channels, through 30-, 20-, 15-, and 10-second commercial spots. It is increasingly important to Internet Web sites as one of several sources of revenue. Marketers can buy a place on a Web page, usually the home or landing page. These availabilities include space for banner advertisements, pop-up windows, streaming video button boxes (V-box), and click-on animations, logos, banners, and buttons for audio clips. Sponsoring a Web site will give an advertiser access to all these forms of placement as well.

In traditional media, audiences are sold to advertisers on the basis of CPM (cost per thousand) impressions. On the largest Internet sites that command millions of visitors, like AOL and Yahoo!, this formulation works well. However, on the majority of Internet sites, advertising rates are less certain. Even though sites deliver customers with specific demographic and psychographic profiles, and their behaviors can be monitored and recorded, measurement issues still cloud the picture.

One controversy is that of impressions versus click-throughs. Site operators want to be paid for the number of targeted consumers they bring to the site. They have no control over the effectiveness of the ad itself, so they believe site traffic is the key. Advertisers want click-throughs—and sell-throughs. They'd rather pay on a per-click-through basis or on sell-through as a percentage of the transaction (an e-commerce-supported model).

The measurement of streamed content is particularly difficult. Edge technologies that position servers on different networks and corporate networks that use proxy servers all make it quite difficult to get accurate data about how many people see the content. Moreover, discussion lists covering webcasting and streaming are filled with acrimonious accusations that content-providers and their infrastructure partners exaggerate the number of people who logged on and minimize the number who failed to receive the stream through a myriad of technical difficulties.

The International Webcasting Association (www.webcasting.org) has established a committee to begin resolving these problems in a formal way. They hope to standardize reporting procedures and data formats by engaging the talents of people from companies engaged in streaming media. Only when such tools are recognized and accepted will the Internet be able to capture a substantial share of advertising budgets.

When high-speed broadband access becomes more widespread, advertising will play an important role in supporting content for it. Video and audio are much more expensive to produce than the mainly text+graphic sites that are now the norm on the Net. Broadband distribution will make content providers more dependent on commercial revenues to produce palatable material—and is likely to reduce many of the irreverent, zany, flippant, obscene, snotty, thoughtful, silly elements of the rich stew that now characterizes online content.

e-Commerce-Supported Revenue Models

One of the advantages that the Internet offers is that it is "actionable," meaning that it can be acted upon immediately. On the Net, a click and a credit card will do it. Contrast this ease of purchase with television and radio, where the viewer or listener has to drive to a store or place a phone call and execute a long series of actions to get a product. For this reason, e-commerce models are not suitable for one-way media; they apply only to interactive platforms such as the Internet and interactive TV. There's also the money; e-commerce brings in cash, as shown in Table 14.6.

One way that designers conceptualize the development of Web sites is "wrapping e-commerce opportunities around content." Sometimes it is relatively seamless, like on Pseudo.com shows when the Internet Jockey wears branded clothing or hawks advertisers' products. Or they may be obnoxiously intrusive pop-up windows, consistently reported by users as the most disliked form of online advertising.

Table 14.6 Estimate of e-Commerce Revenues by Category

Category ($ in millions)	1998	1999 (est.)	2002 (est.)
Books	649.7	1,138.8	3,661.0
Music	134.7	280.6	1,590.6
Entertainment tickets	122.2	274.4	1,809.9
Videos	50.0	104.5	575.2

Source: Jupiter Communications.[20]

On many sites, the relationship between content and e-commerce is taken directly from the company's operations in the bricks-and-mortar world. Information on a Web site may go little beyond scanning in the catalog. Other sites present a bewildering hodge-podge of specs and other data. Only in a few cases is there a carefully thought-through and executed plan to guide the visitor through an extended experience of the content and the e-commerce opportunities. Two models suggest ways e-tailers might think about their site and customers.

Longitudinal Cohort Models   This revenue model goes particularly well with audience segmentation content schemes, which offer material to a specific group of consumers. It tries to anticipate the array of wants and needs members of the group have in common, to provide content that interests and supports them, and to wrap around and embed appropriate e-commerce solutions. Mariana Danilovich, CEO and president of Digital Media Incubator LLC, is an articulate advocate of this perspective. "This whole game is not about the creation of great programming. It is about serving a niche demographic. Now you can't serve a niche demographic with a single property. You can be very creative, but it's the whole experience, designed with the entire lifestyle of the audience in mind, allowing you to fully serve the target demographic," she advises.[21]

A key element in this model is its longitudinal aspect. Marketers must continue to profile their consumers over time so that the content and e-commerce opportunities evolve in sync with the changing customer. The Web is still so young that this aspect of cohort marketing is only now surfacing as people in the Gen X cohort enter their 30s. E-tailers who have focused on that group must now shift their strategies from urban, fashion-conscious, entertainment-driven singles to suburban married parents with children buying their first homes and entering the substantive years of their careers.

DEN, the Digital Entertainment Network, a site aimed at Gen Y, based its operations on this model. Figure 14.3 is a photograph from one

Figure 14.3 Before its demise, Digital Entertainment Network (DEN) put an impressive lineup of shows for Gen Y on the Web. Source: DEN.

of its dramatic webosodics, *Tales from the East Side*. It went belly-up because of the actions of the organizations involved in its IPO, rather than a failure of the business model DEN pioneered.

Another example of this kind of cohort model is MXGOnline, a site aimed at teenage girls. MoXieGirl was originally a catalog that carried fashion merchandise for adolescent females. Now it is a media company with a magazine, *MXG*, a Web site, and a broadband streaming site, MXGtv.com. The site features articles, interviews, entertainment news, reviews, streamed audio and video, and user-supplied content, all free to users. For revenue, MXGOnline markets apparel, shoes, fashion accessories, jewelry, perfume and beauty-related products, home furnishings, and CDs through a partnership with CDNow.

iVillage.com revolves around women, Marthastewart.com around the home, and Bolt.com around Gen Y teenagers. On all these sites, it is more than a matter of just content, or just e-commerce. It is the ongoing symbiotic relationship between them as experienced by a specific cohort that can result in a substantial stream of revenue over time.

Affiliate Networks   In television, an affiliate station carries the programming of a broadcast network, and gives up commercial time that the network can package with stations all over the U.S. to sell to national advertisers. On the Internet, an e-commerce site puts together an affiliate network of other sites, distributing clickable content, logos, banners, and links. When customers click, they are taken to the e-tailer's Web storefront to make the purchase. In return, the seller shares the revenue with the referring site. The use of these kinds of agreements has grown as the cost of placing banner ads on portal sites has risen.

Essentially, it's a way of reaching out to potential customers instead of waiting for them to come to an e-tailer's site. If the affiliates have content compatible with the target audience, then visitors who are looking for the type of products on offer are reasonably likely to click on the link. Compatibility is crucial, just as it is with cohort marketing. Sites that offer information about entertainment are a natural for selling books, videos, music, and tickets. Sports sites can sell event tickets and sports equipment. Personal finance dot-coms might market research on individual stocks and mutual funds.

Here is how a site might analyze the business case for establishing an affiliate network:

- 250 affiliated Web sites, with 1,500 daily page-views per site, to get 11,250,000 impressions per month;
- A 3.5% click-through rate generates 393,750 targeted, self-selected sales prospects;
- A 1% conversion rate of the prospect results in 3,937 products sold per month;
- If the average purchase is $25, total sales = $98,425 per month;
- Less 10–50% commission to referring Web site ($9,842 to $49,212) + cost of goods sold = profit.

Establishing and managing an affiliate network presumes a certain scale and sophistication of infrastructure. The e-tailer must have a product database and the means to process transactions and fulfill orders. Affiliate referrals need to be integrated into the database and tracked so payments can be calculated and distributed.

An example of an affiliate network is Barnes and Noble's agreement with the *New York Times*. Visitors to the *NYT* Web site who are reading book reviews can click and buy it from the *B&N* Web site. On Sportsline.com, the MVP shop sells team merchandise. Its affiliate program will kick back a 10% commission on sales to sites whose visitors click on a link to MVP.sportsline.com and buy their merchandise. The pioneer of affiliate networks is still the biggest. Amazon.com has more than 230,000 affiliated sites, and it is carried on five of the six most visited sites on the Internet.

Amazon.com pays a 15% commission, but there is considerable variation in what different e-tailers pay. Garden.com gives a $10 referral bounty; Lending Tree Branch pays up to $12 a head; V-Store pays up to a 25% commission; and CarPrices will share up to 50%.[22]

Other Revenue Models

There are a few miscellaneous models that do not fit conveniently into the other categories. The multiple revenue stream model brings in money from any and all of its operations. The cybermediary model is the electronic incarnation of "middlemen," performing the functions of traditional intermediaries between product manufacturers and consumers (and a few more) in the online environment. Finally, in a business environment where sales are targeted to specific consumers, there is a robust market for information about them that some sites may also be able to market as a product in itself.

Multiple Revenue Stream Models   The multiple revenue stream contrasts with the tentpole model, where an enterprise is supported by the income generated by a single product or service.[23] The value of multiple sources of revenues in media and entertainment became clear with the rise of the cable industry. Broadcasters who had always relied on money from the sale of commercial time found themselves in competition with TV programming providers that could deliver multiple channels and bring in receipts from monthly subscriptions, advertising, premium channels, pay-per-view, and now high-speed Internet access and telephony. And they are thrilled with the prospect of interactive TV, which would give them a piece of T-commerce as well. Like cable operators, Internet service providers and Web sites almost always adopt a revenue model with multiple sources of income.

Cybermediary Models   In U.S. commerce, there is a huge infrastructure between the makers of products and the buyers of them. A study of the U.S. distribution and sales of high-quality shirts found that the wholesale-retail chain, the intermediaries that facilitate the exchange, account for about 62% of the final price to the buyer. At the same time, many analysts have advanced the notion that networking enables a direct communication between the parties to the transaction that could eliminate these middlemen. They call this flattening of hierarchy "disintermediation."[24]

Keep in mind that in this view, bricks-and-mortar retailers are intermediaries to the final sale, along with the more traditional definitions of middlemen in the distribution process. However, some intermediaries earn their money by performing valuable functions. They include packaging products into attractive, saleable merchandise categories; providing consumer-friendly product information; matching consumers to products; providing transactional economies of scale, and managing risks for both producers (verifying checks and credit cards) and consumers (handling returns).[25]

Indeed, so valuable are some of these services, they have fostered "re-intermediation" by cybermediaries. So online fingers do the walking on Yahoo! and Lycos instead of through the local Yellow Pages. Car buyers can get the Blue Book value of their used car from Kelley's Blue Book online (www.kbb.com/) instead of calling their cousin Pete who works at the General Motors dealership. They can shop in online malls, with multiple storefronts lodged in a single site in place of driving to the store. Internet fan clubs, discussion groups, and chat rooms exert peer influence on the products people choose to buy. eBay.com acts as a cybermediary between millions of buyers and sellers. Business-to-business sites provide real-time spot market and barter networks. And shopping "bots," short for robots, scour the Net, bringing back product and price comparisons.

Some cybermediaries generate revenues through commissions as well as per-query charges, such as the Kelley's Blue Book site. Search and directory sites like Yahoo! and AltaVista bring in money from advertising and contractual arrangements to position some businesses high on the list of returns to users. And, of course, e-commerce through the product sales or service provision directly support many others.

Consumer Data Sale Models   In addition to the dollar amount spent by customers, there is also the data the company owns about the consumers and their behaviors. Indeed, this information about actual purchasers may well be the most valuable asset the company will ever own. It can be used as a bargaining chip to partner with others, or simply sold. Depending on the depth and breadth of the data, it can also be repackaged into segments for use or sale to multiple buyers.

New Markets, New Models—New WWWorld

Many of the traditional principles and ways of doing business will change less than we might anticipate. But they are changing much more than a lot of people in business are comfortable with. And consumers are going through their own painful adjustments, even as they follow up on many new opportunities.

The next chapter will look at how media and entertainment companies are responding to a networked business environment. It will examine the transformations and trends that are appearing as connectivity becomes ubiquitous. And it will summarize some of the problems that pose a challenge to the vision of instant broadband communication anywhere, anytime, for anyone.

Notes

1. B. Joseph Pine, James H. Gilmore, and B. Joseph Pine, II, *The experience economy: Work is theatre and every business a stage,* Cambridge, MA: Harvard Business School Press (1999).
2. J. Pine, II, and J.H, Gilmore, "Are you experienced?" *Industry Standard* (April 9, 2000). Available online at http://thestandard.net/article/article_print/0%2C1153%2C4167%2C00.html.
3. J. Burns. Panelist at seminar at the American Film Institute's Los Angeles campus, sponsored by the National Association of Television Programming Executives (NATPE) and Entertainment Technology Commerce (ETC), November 1999.
4. K. Papagan, Vice President, iXL. Telephone interview, April 25, 2000.
5. For more information, see http://www.jup.com.
6. S. Hardin, NBCi. Personal interview, San Francisco, CA, March 17, 2000.
7. Information taken from Edgar database, Walt Disney Company, available at http://www.sec.gov.
8. Stephen Felisan, Senior Vice President, House of Blues Digital. Telephone interview, March 17, 2000.
9. J. Harris, founder Psedo Networks. Personal interview, Los Angeles, November 1999.
10. eGroups is at http://www.egroups.com/.
11. H.L. Vogel, *Entertainment industry economics*, 4th ed., Cambridge, England: Cambridge University Press (1998):75–77.
12. D. Cripps, consultant and HDTV guru. Telephone interview, November 1999. See his Web site at http://web-star.com/hdtv/history.html.
13. This idea is articulated in a White Paper on the iXL Web site at http://www.ixl.com/whitepapers/index.html.
14. M.J. Wolf, *The entertainment economy*, New York: Random House (1999):223.
15. J. Berst, "Secrets of spiral branding," *ZDNet.com* (November 13, 1998). Online article available at http://www.zdnet.com/anchordesk/story/story_2745.html.

Also, see related Powerpoint presentation at http://www.zdnet.com /anchordesk/story/story_2745.html.

16. D Rushkoff, *Media Virus*, New York: Ballantine (1994).

17. J. Berst, op. cit.

18. G. Johnson, "Investors have dot-qualms on ad spending," *Los Angeles Times* (May 8, 2000):C-1, 5.

19. B. McClellan of PricewaterhouseCoopers articulated this model in "The Future of the Entertainment and Media Industries: 2005." A perspective from PricewaterhouseCoopers (2000).

20. E. Neufeld, "Consumer Internet Economy Portal Landscape, Revenue Strategies, Five-Year Projections." Jupiter Communications Research Study (July 1998). For information, contact Jupiter Communications for information at 627 Broadway, New York, NY 10012; 212-780-6060; www.jup.com.

21. M. Danilovich. Personal interview, Las Vegas, April 2000.

22. These offers came from a Web site that provides information about affiliate and referral programs in March 2000. For a current list, see www.referit.com /main.cfm?screen=info/topten.

23. L. Thurman. Personal interview, Westwood, CA, October 1999.

24. R. Benjamin and R. Wigand, "Electronic markets and virtual value chains on the information highway," *Sloan Management Review* (Winter 1995):62–72.

25. M.B. Sarkar, B. Butler, and C. Steinfield (1995), Intermediaries and cybermediaries: A continuing role for mediating players in the electronic marketplace," *Journal of Computer-Mediated Communication* 1:3 (1995). Online at http://www. ascusc.org/jcmc/vol1/issue3/sarkar.htm.

15

Brave New WWWorld

The fundamental premise of this book is that communication and media are being revolutionized by the transformation of standalone, stranded information processing machines into connected devices and appliances that all talk to one other and exchange data. Ubiquitous connectivity is the brave new networked wwworld, bringing in its wwwake changes to networks, reception devices, content, and human behaviors. It is immersing everyone in a communal media soup of messages that sometimes seems more real to us than the people and objects we encounter in the physical world.

It is difficult for people in the United States to see this "mediasphere" because it originates largely within that milieu. But to the rest of the world, the cultural skew of the Net is quite clear. Make no mistake about it, the business of the United States is information and communication. These industries are a major source of employment, and make up a sizable portion of the nation's exports. According to figures released by the Census Bureau in late 1999, they provided jobs for 3 million people and generated almost $700 billion in 1997, as shown in Table 15.1. Now, we must add to this economic powerhouse the receipts from Internet service and access providers, the production of Internet content, and the jobs and companies created by the e-conomy.

Actually delivering this information and entertainment to consumers is by far the bigger part of the revenue stream. Network providers will continue to partake of a large slice of the income pie as consumers opt for broadband access, as shown in Table 15.2. This last chapter looks at how companies are responding to the networked environment, the trends they must accommodate, and the strategies they employ to meet the opportunities and challenges they face.

Table 15.1 Revenues of the Information and Communication Economic Sectors

Industry segment	Receipts (in billions)
Wired telecommunications carriers	$346.3
Wireless telecommunications	$ 37.9
Satellite telecommunications	$ 5.1
Television and broadcasting	$ 29.8
Radio broadcasting	$ 10.6
Cable and satellite broadcasting	$ 34.9
Publishing	$179.0
Software	$ 61.7
Newspapers	$ 41.6
Magazines	$ 29.8
Books	$ 22.6
Databases and directories	$ 12.3
Greeting cards	$ 5.3
Motion picture, video, and sound recording	$ 55.9
Filmed entertainment	$ 20.1
Video distributors	$ 12.5
Record production and distribution	$ 8.7
Exhibitors	$ 7.6
TOTAL	$699.5

Table 15.2 Global Broadband Access Service Revenues (U.S. $ millions)

	1998	1999	2000	2001	2002	2003	2004	2005	2006	2007
Residential	386	1,068	3,026	5,885	10,081	15,707	23,064	29,833	39,149	49,815
Business	100	481	1,545	3,548	6,758	10,515	14,593	18,710	22,848	27,090

Source: Pioneer Consulting.

The Theory of Diffusion of Innovation

Nothing characterizes the modern age so much as change—technological, economic, political, and behavioral. This chapter begins with a look at the theory of the diffusion of innovation, a communication theory that explains the core of social change by describing how something new spreads throughout a large group of people.[1] The process by which innovations diffuse is pretty much the same whether the "something new" is digital television, data mining, dishes, or democracy, and whether it occurs in Manhattan or Mozambique. This useful and flexible theory is the dominant paradigm guiding much of the market research that underlies business plans. It enables organizations to anticipate the factors that affect the likelihood of adoption, the kind of individuals who will buy and use innovations, and when and why they will do so.

However, it is important to keep in mind that many innovations fail to spread. For example, although the demand for communication bandwidth has grown steadily for the past hundred years, and media are not so much subject to disappearance as redefinition, specific devices may fail at introduction or fall out of favor after a period of popularity. Examples include interactive videodiscs, videotext services, AT&T's PicturePhone, early facsimile machines, vinyl records, and 8-track cassettes and their players, tube radios, and countless other products.

Early diffusion studies began in the field of geography, often looking at how agricultural products were carried from one locale and culture to another. The research retained a spatial bias until the work of communications scholar Everett Rogers. Rogers became familiar with diffusion processes at the University of Iowa, where he studied agricultural sociology. As a student, he learned about the work of Wilbur Schramm and the then-new field of communication research. Rogers realized that the emergence of mass media and electronic communication changed the processes of diffusion and adoption by overcoming the spatial limitations of earlier eras. In 1962, he brought these ideas to the forefront in his compelling work, *The Diffusion of Innovation.*

The theory presents a stage-process model of the spread of innovations and the resulting social change:

- Introduction or invention of a new technology, idea, or practice;
- Adoption and diffusion (or rejection) of that innovation through communication;
- Changes brought by the innovation to individuals, societies, and cultures.

The second stage, adoption, gets particularly close attention because it is composed of a complex set of communication events. Adoption begins with awareness. People may develop an interest in an innovation through its being known. They seek information about it, evaluate it, and decide whether to adopt it. These steps are: awareness, interest, information-seeking behavior, evaluation, and decision to adopt or reject.

The diffusion S-curve in Figure 15.1 illustrates the accumulation of adopters over time. Acceptance begins slowly, growing additively as one adopter after another uses the innovation. At a certain point, the number of adopters escalates sharply as more and more people see how the innovation is useful, resulting in an exponential rise in the number of people accepting the innovation: This point is called "critical mass," which marks the point when the innovation is accepted by enough people to make its adoption throughout the social group likely, at least for a time.

Critical mass is especially important to communication infrastructure because of the "sufficient pair problem," the ability of a person to reach enough people whom they want to reach to make adoption of the communication system worthwhile. For example, few people would pay for a telephone system that only reached people in a city 2,000 miles away. The theory also presents a profile of adopters based on when they accept an innovation:

- Innovators—the first to adopt. They are young, have high social status, and are willing to accept risks. They have close contact with scientific sources and other innovators and use the media to get information. They seek information from many sources, and some act as opinion leaders within their social network.
- Early adopters—role models with high social status. Secure within their local groups, they make independent decisions about adoption. They get information from mass media and contact with change agents, such as marketers and industry outreach efforts.
- Early majority adopters—deliberate decision-makers. They accept an innovation only after peers have tried and liked it. They have middle social status but may be leaders within their local group, and they have considerable contact with change agents.
- Late majority adopters—reluctant skeptics. These followers adopt an innovation only after there is almost no alternative. They have below-average social status and income, and get information from like-minded others and occasionally from adopters in the early majority—as opposed to the mass media.
- Laggards—traditional and past-oriented. They usually have low social status and income, and they are often older members of society. They get most of their information from interpersonal sources who share their values, or they may be social isolates.

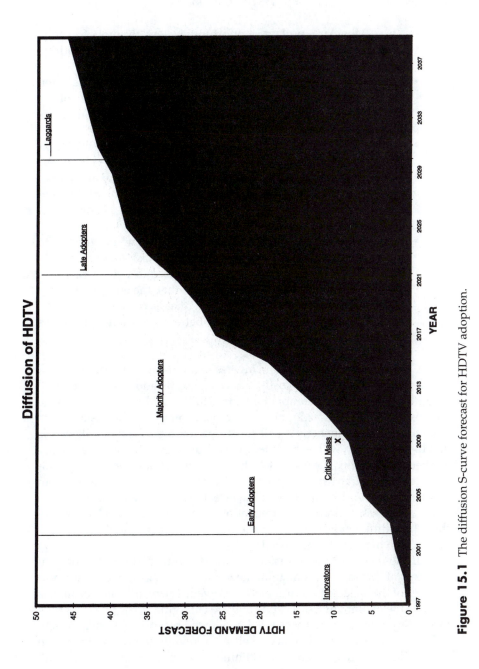

Figure 15.1 The diffusion S-curve forecast for HDTV adoption.

The theory covers attributes of the innovation itself that play prominent roles in whether or not it will be adopted:

- Cost—Monetary outlays, opportunities lost through foregone options, social and personal costs, and other subtle or hidden penalties for accepting an innovation.
- Complexity—The number of different parts that have to be taken into account (from the adopter's point of view) regardless of the underlying technology. For example, automobiles and computers are complex; televisions and staplers are simple.
- Compatibility—How an innovation interacts with existing technologies, cultural and social practices, and individual preferences.
- Trialability—The extent to which a person can try out an innovation before they adopt it, like test-driving a car. One problem in selling computers has been that until people reach a certain level of expertise, they can't really try it.
- Relative advantage—Sometimes called substitutability, it refers to the ability of an innovation to accomplish a task more efficiently than the old product or practice. According to Peter Drucker, a product must be at least ten times better than the one it replaces in order to gain a foothold in the market; that is, more efficient by an "order of magnitude." Five times better won't cut it, observes Drucker.[2]

Cost and complexity have an inverse relationship to adoption. To the extent that an innovation is costly and complex, its adoption will be hindered by those factors. Compatibility, trialability, and relative advantage are positively related to adoption. The more compatible, the easier it is to try, and the more functions it will substitute for, then the more likely it is that the innovation will be accepted. This clear, eminently practical theory has been extraordinarily useful to scholars, national development specialists, technology transfer programs, service providers, and marketers from every industry. A careful study of the research in this tradition will guide those who wish to introduce innovations, and, later in the chapter, will show how different adopter categories require marketers to package their products to meet customers' needs.

If the theory of the diffusion of innovation underlines the constant elements of introducing something new into a social group, there are also realities of the specific time and place involved. The next section examines some of the overarching directions that are emerging in the first decade of the twenty-first century. These will be followed by specific trends in technology and consumer behavior and observations about how companies in the media, information, and entertainment value chain are dealing with them.

Trends—Signs of the Info Times

Responding to change is essential because it is so pervasive. Every industry is affected by the increased coordination made possible by real-time communications across geographical boundaries. And, as we have seen, the ways that many individuals work, handle their personal affairs, participate in social relationships, and spend their time are also undergoing some degree of transformation. We are only at the beginning of the digital revolution. However, we can begin to make out the areas of the business landscape that are affected by the presence of the high-speed, high-capacity backbone networks that already link every continent. We start with a look at the broader trends underlying today's events.

The Dialectics of Transformation: Convergence, Globalization, and Competition

If it is true that we are only at the start of the digital age, then it is probable that the larger changes are yet to come. One indication that we are seeing the emergence of a thorough-going transformation—rather than just a few modifications to business life—lies in the fact that the trends are bimodal; that is, they occur in opposite ways at the same time. In other words, there appears to be a significant reshaping and realignment of factors that are always present in some form: convergence and divergence; globalization and localization; and competition and alliance.

Convergence and Divergence   We have already considered technological convergence, the coming together of digital and analog technologies, such as computer technologies and telephone networks or cable networks. Sometimes the term is used to describe events that go beyond technology to mean the greater blending of entire industries or segments of industries. Examples of this trend are the increased overlapping of filmed entertainment and the music business, or the tangled web of relationships between cable operators, satellite companies, and content providers.

The convergence of previously separate industrial sectors and organizations is a precursor to changes within organizations. "Restructuring" in the 1980s, which resulted from computer processing and enterprise networking, brought internal consolidation, the redefinition of many jobs, altered workflows, and the blurring of previously understood lines of authority. Now that networking is becoming globally ubiquitous, more changes are on the way. It is still unclear how this convergence will play out over the next decade. There will probably be some maximal ways for enterprises to organize themselves to best exploit unified communication networks, but precisely what forms these arrangements will assume remains elusive.

One solution is to create content-conduit (or programming-pipes) powerhouses. However, as we will see, consolidation between companies that own networks and those that provide content to form entertainment conglomerates, presents some thorny conflicts that are not easily resolved. Producers of content want to get the most money they can for their product; but owners of conduits want to buy content as cheaply as possible. When the same company is in both lines of business, they must decide where the greater profit lies.

Convergence at the network level is matched by consolidation at the organizational level, which in turn is driving convergence in content as well. Many companies are adopting cross-platform strategies, publishing their content to several different media, sometimes simultaneously, sometimes serially. In the spring of 1999, a pay-per-view Sarah McLachlan concert was simulcast live on DIRECTV and over the internet at http://www.woodstock.com. The concert, the first Internet/TV pay-per-view simulcast, was then viewable on the Web for one month. Another instance of convergent programming is the increasing inclusion of Web addresses in TV commercials and the use of E-mail to promote daily TV newscasts, strategies employed by KTVT-TV in Dallas, WIFR-TV in Rockford, Illinois, and WMTV in Madison, Wisconsin.[3]

At the same time, divergence is taking place. As Bennett McClellan of PricewaterhouseCoopers puts it: network convergence; device divergence. He notes that networks used to be industry-specific. There were telephone, cable, broadcast, and computer networks, each with its own architecture, standards, functionalities, bandwidths, and uses. Now, networks are networks are networks, as shown in Table 15.3, which details the evolution to Ethernet.

By contrast, reception and user devices are diverging, reconfiguring, and proliferating like mad. They include PCTVs, TVPCs, set-top box +

Table 15.3 Network Infrastructure by Industry

Industry	1980s	1990s	2000s
Telephone	DS-3 access ISDN	Sonet/ATM ADSL	DWDM xDSL Ethernet to home
Cable	Analog coax/ fiber trunks	Analog and digital fiber	HFC, Ethernet to home
Computer	10 Mbps LANs	100 Mbps LANs	Gigabit Ethernet
Power	Fiber for internal Communication	Remote metering via fiber	Electricity + signals over fiber-insulated powerlines

personal video recorder + data receivers, phone+TV, phone+PC, every appliance in the world (refrigerator, microwave oven, cars, etc.) + PCs, MP3 player + digital recorder, CD player + MP3 player, Internet-capable cell phones, media screens and tablets, and portable Internet appliances. And there are probably many more appliances to come.

The networks all deliver broadband, full-screen, full-color moving images. They must accommodate the same functionalities. But the reception devices have specific requirements that content owners and network transporters must address. A golden age of all-purpose end-user devices may well emerge in the future, but it won't come anytime soon. It's a convergent/divergent world—get used to it.

Globalization and Localization   Globalization is a very real phenomenon. One measure of it is to look at how much of the world's products are open to competition in worldwide markets. According to estimates, more than $21 trillion of the world economy's output will be in such markets in 2000, rising from $4 trillion in 1995.[4]

Many factors are driving globalization, but high-capacity communication networks certainly play a large part. A truly global business, as opposed to nationally- or regionally-based subsidiaries that pursue their own objectives, requires the ability to conduct and maintain global coordination. Today's communication network infrastructure provides the means. In just the last decade, the buildout of robust infrastructure has enabled the output from dispersed, distributed work groups to be assembled into a unified effort, with people in different locations and time zones contributing their labor.

On the consumer side, communication networks make it possible to assemble a worldwide audience as well, connecting people together regardless of where they live, allowing them to form communities of interest. The ability to reach a global audience enlarges the sphere of any business to a staggering degree. In the entertainment industry, producers will be able to create niche programming and software that a single national audience could not support. For example, a quilting network or an organic farming network stands a much better chance of achieving financial viability if people from around the globe access the material, because the small number of people who would be interested from any single nation may not provide enough cash flow.

Worldwide networks tilt the balance between local and regional and global entertainment products in unpredictable ways. On the one hand, they give existing giants ever more efficient distribution to anyone, anytime, anywhere. But they also give dancers in Byelorus, singers in Mombassa, and guitarists in Barcelona the opportunity to put their art before the world and to build an audience for their work.

Added to the effects of networks is the way digital technologies serve content up to consumers. Increasingly, product is sliced and diced, customized to the tastes, preferences, and interests of the individual requesting it. And overall, audience preferences tend to be decidedly local.

Simultaneous globalization and localization—producers and media distributors must accommodate both trends if they are to succeed. The new paradigm is: organize globally; productize locally. Producers of content must increasingly identify local talent, foster it regionally or nationally, and then internationally market it outwardly to appropriate cultural, language, and ethnic groups.

Competition and Cooperation   Today, there is considerably more competition than ever before along the entire value chain of all sectors of the communications industries. Whether the product is long distance telephony, wireless PCS, Internet access, multichannel television delivery, or content and service provision, consumers behold a staggering array of choices. One driver towards greater competition is the proliferation of multiple platforms and distribution technologies, many of which exploit the profitable opportunities created by digital efficiency.

When some innovations come on the scene, they lead to a specific form of competition. Clayton Christenson calls them "disruptive technologies."[5] He describes a scenario where a product is based on an existing technology: for example, Detroit cars in 1972, or mainframe computers in 1982. Sales have been moving along a smooth price/performance curve.

Then, along comes a disruptive technology (like Japanese cars or the PC), which enters the market at a price/performance point that is substantially lower. The new entrant may not offer the same quality and reliability as the preexisting product, but it does offer some entirely new features. As a result, the innovation brings in entirely new buyers who finally are able to purchase something they've wanted but that didn't exist at their price point. The new product establishes a market for itself at its own price/performance curve, but lower than the existing product. Over time, the new product becomes profitable.

Up until that point, the existing market leaders haven't considered the disruptive technology as a serious threat. In the case of the PC, it was clear that desktop machines could not perform anywhere near the level of mainframes—they couldn't handle the processing for a human resources department or the payroll. So mainframe manufactures didn't worry about PCs. But they did begin to note that there was a market where they had not anticipated one to exist.

It is at this point, says Christenson, that the traditional companies do exactly the wrong thing: They listen to their customers too much. They ask them: "Would you use this product? What features would you like us to include in our product?" Their customers assure them that they wouldn't use this new, lesser performing version of the technology. Furthermore, they'd like a boatload of bells and whistles.

So the traditional market leaders add features, adding cost and raising prices. In the meantime, the companies making products based on disruptive technology are making money, and have plowed profits into improving their products. They are now more reliable. The new features and advantages they offer become clear. In the example of the PC, the advantages of local capabilities emerge, like word processing and spreadsheets, functions it is now possible to perform without timesharing on a mainframe computer.

The combination of the increased reliability and greater functionality of the products that were built on the disruptive technology, plus the higher price of the new features added to those based on the existing technology, encourage traditional customers to at least try the new, improved new products. Now the older technology is in a race for its life—one that it may well lose. In the cases of the examples used above, Detroit cars did not become competitive again until they adopted the new disruptive Japanese automotive technologies. IBM entered the PC market but underwent many years of organizational difficulties that forced the company to restructure its operations to address the huge PC market.

This entire story is summarized in an elegant set of curves designed by Christenson, as shown in Figure 15.2.

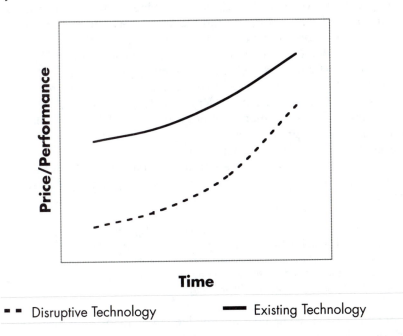

Figure 15.2 Price-performance curves of existing and disruptive technologies.

Disruptive technologies are just one source of competition. Another is the entrance of companies into new markets. As one observer put it: "Everyone wants to eat everyone else's lunch."

No matter the cause, consumers like competitive markets because in many cases, prices decline. An example taken from the arena of multichannel TV program delivery is Glasgow, Kentucky, a market where the local electric company also provides cable services.

Not surprisingly, prices are significantly lower than nearby comparable noncompetitive areas. This pattern prevails in other two-provider markets, as well.[6] The situation is similar when the local cable operator offers high-speed Internet access and the telephone company provides DSL service. Competition will drive the cable offering to between $20 and $40 per month, and the telephone company product to less than $50.

The competition between content products is somewhat more difficult to characterize because each information service, finished program (film, TV show, book), and performance is unique, appealing to a core group of users or fans. In the digital economy, the scarce element is time, even more rare than disposable income. Thus, before content competes for money, it competes for attention, a fact of life now encapsulated in the description, "the attention economy."[7]

At the same time, the global marketplace brings with it a heterogeneous audience composed of multiple language groups, customs and mores, and tastes. When globalization is combined with the proliferation of platforms and content formats and the need to operate internationally on a 24/7 basis, it becomes clear that only the largest multinational companies can carry out their operations in the face of such complexity. The solution for most enterprises is to acquire, merge, partner, and ally with other entities to serve these diverse markets. In other cases, there may be alliances between two divisions or departments of companies, yet all the while competition rages among the noncooperating business units. The phenomenon of the presence of these seemingly contradictory elements has even been given a name, "coopetition."

David Waterman, a media economist at the University of Indiana, points out that limited competition is the norm, despite the public image of competition:

> There is a vision of huge giants with many arms slugging it out in the ring until one succumbs and the other stands up on top, grunting and snorting. In reality, only the different particular segments of these big companies (e.g., record distribution, movie distribution, cable networking) really compete with each other directly, and this competition is essentially unchanged by conglomeration.[8]

In many cases, companies are involved in many or all of the various kinds of partnering arrangements, and there is an ever-changing mosaic of

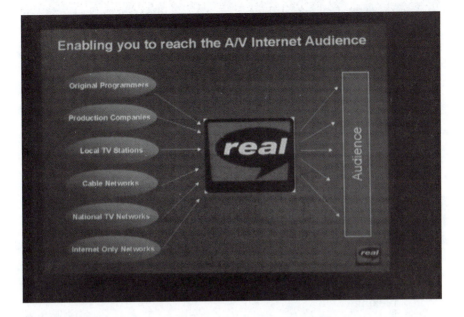

Figure 15.3 The television industry's disruptive technology—video over the Internet.

relationships in the telecom, media, and entertainment industries. The multifaceted business landscape is also a major factor driving consolidation across sectors and companies, as well as within companies. This chapter will look at some of the problems of consolidation in a later section.

Within the framework of the overarching trends of convergence/ divergence, globalization/localization, and competition/cooperation, it is no surprise that the media and entertainment industries are struggling to understand and accommodate the new business landscape they face. And in addition to the larger transformations that are occurring, there are also some specific changes they must address. Technological innovations continue and fuel intense competition between network providers. On the other side of the network, an active, connected consumer is a much different customer than members of a passive, fragmented audience. And it takes a new kind of producer and packager to create content that appeals to him or her. The next section looks at trends in technology and networks, consumers, and content.

Technology Trends

Earlier chapters in this book explored the rise of digital media and devices, the development of wired and wireless networks, and the growth of global bandwidth, so I will not re-plow those fields. Table 15.4 pulls all these issues together to consider the various stages of computing.

Networking offers extraordinary economic efficiency when compared to earlier forms of communication. Table 15.5 compares the cost of commonly used modalities to deliver a 42-page document from New York to locations in NYC, Los Angeles, and Tokyo. (These figures were published in the mid 1990s, so of course today's speeds are much faster, the cost of building networks is lower, and networks are much more pervasive, lowering the costs even more.)

Moore's Law, the rapid growth of processing power, and Metcalf's Law, the increasing value of networks as more people join them, continue to have an impact on communication and are likely to do so for some time into the future. The acceleration of technology will make broadband networking a reality around the world at some point in the twenty-first century. The signposts that this goal is coming closer include (but are not limited to):

- Continuing advances in dense wave-division (DWDM) multiplexing
- Advances in switching that extend IP across optical networks and petabit routing technologies
- Lower-cost, higher-capacity storage that will lower the costs of edge technologies
- Commoditized, end-to-end distribution over broadband networks
- Redundant, ubiquitous multicasting of audio/video streams across public networks
- Less expensive "last-mile" broadband connectivity, such as fixed wireless, cheaper copper wire solutions, or other technologies

Not All Networks Were Created Equal   Even if all the technical challenges were overcome tomorrow, it still might take a surprising amount of time for ubiquitous broadband networking to reach end-users. While the newer networks are likely to be quite similar in design and architecture, most industries have legacy networks in place. And, as we have seen, these were deployed for very different purposes, resulting in different equipment, standards, functionalities, and reception devices.

To the extent that industries are saddled with existing networks, one must consider the advantages and limitations of each, and the time and effort it takes to integrate new technologies into them. Given the cost of building new networks and upgrading existing ones, it takes considerable pressure for a company to revamp its infrastructure. The traditional rule of thumb to estimate the cost of building two-way hybrid fiber-coax (HFC) broadband networks is "$1 billion per million"; that is, costs would

Table 15.4 The Evolution of Computing

	STAGE 1 *300BC* *Abacus–Babbage–PC*	STAGE 2 *1961* *DARPA network*	STAGE 3 *1994* *WWW*	STAGE 4 *2015?* *N-Dimensional Networking*
Form of usage	Thinking: Standalone computation	Interpersonal communication	Combines interpersonal and mass communication	Integrated communication
"Killer apps": Popular implementations	Data processing and storage	Text E-mail	Web surfing, webcasting: push and rich A/V media	Sun's concept of Java; Motorola's LON networks
Sender/ receiver format	User/programmer	Point-to-point: one-to-one, one-to-few	Point-to-point or multipoint: one-to one, or one-to-many	Combines all: adds multipoint-to-multipoint or many-to-many
Interface	OS and application software	Browser	Interim: plug-in, media player Final: TBA	TBA
Hardware and infrastructure	A computational device	Computer, modem, telephone line	Computer, modem, high-speed network	Ubiquitous information appliances and richly connected high-speed networks
Image	Extension of individual brain and nervous system	Extension of a personal network: wheel and spokes	Extension of a social network: clustered network	Unknown: A geodesic sphere?

Table 15.5 Relative Costs of Delivering a Document

From New York to:	New York	Los Angles	Tokyo
E-mail	9.6 kbps/2.8 cents	9.6 kbps/2.8 cents	9.6 kbps/2.8 cents
U.S. Air Mail	2 days/$3.00	2 days/$3.00	3 days/$7.40
Overnight service	1 day/$15.50	1 day/$15.50	1 day/$26.25
Fax	31 min./44 cents	31 min./$9.85	31 min./$28.83

Source: Prepared by UUNet/MFS.

run about $1 billion to install an interactive broadband system to 1 million homes. That $1,000 per household includes $750 to $800 or so for the network and $200 to $250 for the set-top box.

At a billion per million, it will cost the cable industry nearly $100 billion to wire the 92 million U.S. TV households. The price tag for fiber to the curb (FTTC) is even higher, estimated to be about $3,500 per residence, so this architecture would cost providers in the United States about $350 billion. On the whole, wireless networks cost less to build than wired systems because they do not require the same rights-of-way, digging, and beneath-the-earth maintenance. However, they are bandwidth constrained and, except for broadcasters, they have to pay for spectrum rights. MMDS systems cost between $400 and $600 per household, and LMDS systems run between $250 and $600, depending on the assumptions made about the population density. Thus, it would cost between $23 and $55 billion to provide advanced wireless service to all 92 million households.

These costs are hardly stable. In 1992, installing hybrid-fiber/coax (HFC) cost $400 per home passed; in 1995, that amount was $800. By 1999, with considerably more experience under their belts, experts say costs range between $500 and $2,200 per household, depending on the geography, the bandwidth, the number of homes per node, and many other factors. Of course, as more broadband networks are deployed, Moore's law and economies of scale come into play. It is estimated that Singapore, which is wiring every home and business in the city-state, will pay about $400 to connect each location because of the lower costs associated with doing all the work at once.

DSL installation is about the same as HFC, running between $800 and $1,500 per port (household). Telephone companies enjoy an advantage though, because their networks allow them to upgrade one central office at a time and to hook up each customer as requested. Other last-mile bandwidth providers, cable operators and wireless providers, must upgrade their entire systems regardless of how many people actually subscribe.

One of the reasons cable operators, broadcasters, and telcos have only reluctantly (and slowly) embraced digital infrastructure is that they cannot figure out how to recoup the considerable expenditures it requires to do so. By contrast, satellite companies spent billions to start up because their business model showed them how they could make substantial returns on their investment by going digital, using it both to gain market share and to save money over traditional distribution channels.

The costs of running a wireless system, as well as initial building costs, is also lower than it is for wired systems. MMDS operating costs are about 25% of gross revenue, while wired cable operating expenses run 36% to 39%. With wireless systems, there are no broken cables to fix, no rights-of-way to lease or franchise fees to pay, and they have a larger functional range, about a 35 mile radius, depending on the specific technology.[9] The few trials of LMDS indicate that the cost per customer for LMDS service falls between $250 and $600, depending on how many subscribers live within a 2- to 3-mile cell area.

The above discussion provides a sense of the real costs to conduit providers. To conceptualize their cost/income projections and develop a business model, they begin with a brute force approach. Suppose an organization planning to deliver interactive service over a wired system estimates it must spend on average about $1,000 per home passed. An optimistic projection of a 25% signup rate would make the actual cost of wiring each subscriber home $4,000. If the average cable bill is currently about $33 per month, the yearly income from that home is about $400 a year. Assume that the new services will add $12 to the bill; on the average, the operator will take in $45/month or $336/year. At that rate, allowing for $17 per month (about 37.5%) in expenses for product and maintenance, it will take more than 12 years to recoup the investment.

Twelve years would tax the patience of most investors. Unfortunately, the business case is probably somewhat worse than the one outlined above. In the past few years, experience shows that 25% is at the high end of the signup rate. In many areas, penetration actually hovers between 6% and 10%. In 1999, Excite@Home brought an expensive demonstration to neighborhood malls and was able to persuade 25% of those who saw the demo to buy the service, which rated headlines in the trade press. If the turnover rate is kept low, signups are additive, so that over time, the percentage of two-way subscribers will move ever higher. Nevertheless, for service providers to amortize their costs over anything less than 30 years, they must find ways to lower the cost of building the network, sell more services, or sign up more customers.

Besides weaknesses in the access-provider business model, another incentive to delay implementation is the rapid decline in processing costs as time passes, the phenomenon predicted by Moore's Law. Everything related to processing—switching and routing, modulation, multiplexing, servers—becomes cheaper to the company that waits. The cost of storage has gone down even more dramatically. So the longer a company waits to build or upgrade its networks, the lower the costs will be.

A final consideration in how networks differ is the cost of providing content. Some networks lend themselves to user-supplied content, like phone calls over telephone networks. In the United States, telephone companies are considered common carriers, which means they cannot control the information that flows over their network. Although this prevents telcos from making money on added-value content, it also means they don't have to shell out any money to create or acquire programming. The Internet by its very nature offers an uncontrolled and nearly uncontrollable environment for content. By contrast, broadcasters and cable and satellite operators jealously guard access to their systems, permitting only material they believe will bring them a profit to reside on the network.

A Tale of Six Industries   Six industries are actual or potential network providers to deliver digital television to consumers: cable, telephone, utility power, computer, satellite, and terrestrial broadcasting. For the past few years they have enacted the most intricate corporate choreography, jockeying for competitive advantage and joining for cooperative benefit. How do they stack up against one another?

When it comes to the financial ability to pay for the expensive infrastructure advanced television systems, the power companies have the most capital with $200 billion in annual revenues.[10] They are followed by the telephone companies that bring in $100 billion annually, and the cable companies, with $50 billion yearly income.[11] For the most part, the computer industry doesn't build infrastructure. Computer networks are either provided by the telephone companies (and perhaps by cable and power companies in the future), or by organizations that communicate over LANs (local area networks) and WANs (wide area networks), using some combination of proprietary infrastructure and leased lines.

Except for the satellite industry, which built its network to take advantage of digital technologies, all the others own a substantial installed base of legacy equipment that requires a tremendous investment to position that industry in the forefront. Table 15.6 shows the advantages and limitations each industry faces.

It may be surprising, but decisions about network design and architecture are not just technological in nature. Organizational cultures are inextricably bound to the networks that individual companies and entire industries build. These cultures are nearly invisible, yet they exercise considerable power over the way industries conduct business, shaping their assumptions, strategies, and investments.

Table 15.6 Industry Advantages, Limits, and Challenges

Industry	Network Attributes	Network Limits	Industry Challenge
Telephone	Ubiquitous Two-way	Narrow bandwidth Expensive switches Load coals and bridge taps limit DSL Standards conflicts	Increase bandwidth Overbuild IP data networks
Cable	Large capacity	Geographically fragmented One-way shared local loop: security problems limit number of users	Geographic consolidation Upgrade to two-way
Computer	Fast Inexpensive	Exist only in company LANs, reliant on telcos IP limits quality of service Voice service difficult	Build IP networks around the world
Power utilities	Ubiquitous Good customer relations	Not designed for Communication Need to shield bits from electricity No measurement, management or billing for communication services	Get going, despite a slow-moving culture Install network management for communication services
Satellite	Ubiquitous Large capacity	One-way	Partner with others to get backhaul capacity Build LEO systems with two-way capacity
Local broadcast	Ubiquitous Large capacity Free bandwidth	One-way Expensive upgrade underway	Change business model/image from programmer to spectrum manager

For example, the copper wire plant of the telephone companies has taken 100 years to build at a cost of $1,500 per person![12] Naturally, telcos have been interested in technologies that would extend the use of this extensive, expensive infrastructure.

From the beginning, the national telephone company, AT&T was committed to dumb terminals (telephones) and smart networks (switches and SS7 signaling). The business model involved keeping the cost of universal service low, a loss leader, combined with marketing smart services that the consumer could obtain only from the telephone company's network.

This thinking pervaded the company, long after the breakup of Ma Bell into regional operating companies and the introduction of the PC and computer LANs. The commitment to the idea of placing intelligence in the network was so strong that when AT&T employee David Isenberg wrote his seminal paper, "The Rise of Stupid Networks," he faced corporate banishment. (He was not actually fired, but he was made quite uncomfortable. He is now flourishing as a consultant and sought-after analyst and speaker.)

If telcos can't shrug off the legacy of networks past, cable operators seem unable to ease their grip on control of the content over their networks. They risk alienating their own subscribers with frequent rate hikes brought on by network upgrades and the rising cost of programming. In 1995, the industry spent nearly $5 billion for programming; in 1999, costs were $8 billion.

Yet competition from satellite makes it more important than ever for cablers to exploit their two-way networks fully by embracing telephony and the free-for-all environment of the Internet. In an attempt to manage interactivity and to harness the Net, major MSOs (multiple system operators) like AT&T are trying to develop a limited interactive capability over their networks. They are driven by the vision of a new revenue stream from interactive television (IATV) that they call T-commerce, or television commerce. Forrester Research believes IATV will reach 7 million viewers by 2001, and Jupiter Communications predicts it will be a $10-billion market by 2004. A host of companies are involved in making the conversion possible, such as WorldGate, Commerce.TV, Interlink, and many others.

Since most of their programming is linear TV, they plan to "interactivate" it through ATVEF (Advanced Television Enhancement Forum) compliant enhancements that allow viewers to pause the video, get more information (usually marketing-related material), and buy products related to or featured in the content. Some programming will be originally produced for the cable digital tier and DSL systems. Finally, wholly interactive content will be ported directly from the Web, such as Quokka's broadband deals with cable and satellite companies, which will package it as part of their interactive TV offerings.

Cablers have been quick to see the advantages of linking individual cable systems within regional areas. Every major market now has interconnects that link nearly all of the cable systems in the area and inserts advertising that runs on the same channels at the same time. So, if a viewer is watching CNN at 8:32 P.M. in Glendale, California, they see the same Ford dealership commercial as the viewers in Malibu, Cerritos, and Altadena, even though they may all subscribe to different local cable systems. This geographic consolidation has been furthered by an overall industry consolidation, as major MSOs acquired one another and swapped systems among themselves.

Power utility companies are potentially major players in the telecommunications delivery business, but despite considerable testing, they are slow-moving behemoths that have not been able to commit to the fast-moving world of telecommunications. The industry enjoys a number of advantages, particularly right-of-way. Like railroads, power utilities are granted governmental right-of-way, which allows them to put up poles, dig ditches, and string cable wherever they need to without paying fees like cable companies must.

Since the 1970s, the power companies have been laying fiber optic cable to meet their own internal communication needs. The reason they were such early adopters is that fiber allowed them to send communication along with power—electricity interferes with the information passing along copper wire (or any other metal). However, a silicon core acts as an insulator, so it can carry data right next to a power cable, with no interference, consuming no more than 5% of the bandwidth for energy management.[13] Power companies are planting an enormous amount of fiber optic cable. In the mid-1990s, the industry had put in about 18,000 miles.[14] But by the end of the decade, plans called for considerably more. For example, in September of 1999, New Century Energies, a holding company that owns several state power utilities and new communication subsidiaries, partnered with Touch America to build an 18,000-mile fiber-optic network—a single enterprise that equals the fiber optic mileage of the entire industry just a few years earlier![15]

Similarly, the broadcasting industry is wrestling with its traditional notions of how it does business. Executives of companies are trying to understand how to make money from digital spectrum in addition to what they take in from local programs and single-sourced network feeds. In the old days, the spectrum was just a vehicle they took for granted—now it is a commodity that has value independent of the content that flows across it. But conceptualizing themselves as "bandwidth managers and sellers" is so far from the TV industry's traditional view of itself that the current generation of managers seems unable to re-orient themselves to take advantage of the new opportunities within their reach.

In the end, these are all powerful industries that can potentially influence the evolution of the global infrastructure. But none of them can entirely predict or control it. Starting with their unique technological configuration, each of them faces different network design and construction challenges, yet must offer essentially the same capabilities—a network capable of delivering and receiving switched analog and digital signals for voice, data, and video from multiple sources to anyone, anywhere, anytime.

We move now from network technologies to the people who use them: the audience, fans, end-users, customers, and consumers. The digital age has changed them, too, forcing businesses to rethink how they conduct and manage their relationships with the people whom they want to buy or use their products and services.

The New Empowered Consumer—Dancing Click to Click

On the Net, all news travels fast. Day trading, the battle in Seattle over World Trade Organization policies, the Intel math coprocessor PR disaster—each of these events points to the fast living and powerful impact that connectivity brings. And although the characteristics of connected consumers are becoming more mainstream, there is little chance that they will ever resemble the TV mass audience of old. When the World Wide Web came to the Internet and people began going online in droves, the early adopters tended to be young, highly educated, technologically savvy, upper-income males, with females making up only 20% of regular online users. By the end of 1999, females composed 48% of WWW subscribers. Although there are some gender-specific behaviors, they are no greater online than they are in other media usage patterns, such as television viewing—or offline behavior for that matter.

It's not surprising that as going online goes mainstream, the demographics of Web users more and more resembles the distribution of age, gender, socioeconomic, and ethnic characteristics of the general population. However, there is quite a difference in the psychological stance of active users versus relatively passive viewers and listeners. Perhaps nothing sums up this difference as well as the old saying, "Information is power." And a corollary might be: The faster and easier it is to get the information, the greater the power.

In a way, this access to information brings to fruition the classical free market economic theory, in which the prediction of ever-lowering of prices due to competition was based on the assumption that consumers had "perfect" information about the market. Before the Net, this condition was never close to being met and couldn't be met. Professional sellers almost always knew much more about market dynamics than mainstream consumers: trends in wholesale prices, overall retail prices, sales costs, product inventory and supply, and raw materials supply.

Widespread Net access begins to approximate the availability of perfect information. People can send out software bots to check prices on any given item. They can access industry information, talk to other consumers, and conduct their own in-depth research about nearly any product, and they are becoming increasingly sophisticated about finding the path of least expense. Information empowers the consumer so significantly that some analysts believe that it tilts the balance of power towards consumers and away from sellers. Certainly no one disputes that the relationship between the two parties is undergoing vast change.

In the past, producers and programmers targeted consumers and aggregated an audience. In the audience-empowered world, consumers target and aggregate the material they want and display it on the media platform of their choice. A vast number of people now have access to the information that interests them, anytime, anywhere, in any format. Kiran Rao, Vice President of technical development at Steeplechase Media calls it a change from the WYSIWYG (what you see is what you get) to the WYWWYWI (what you want, when you want it) model.[16]

In the past, programmers created images to appeal to a targeted demographic group, the products were inviolable, protected against change and redistribution. That model is history. In the new digital environment, they are challenged to create material that allows individuals to recreate it in their own image. And if they don't figure out what consumers want and give it to them, users will bypass them altogether and create and exchange images of their own making.[17]

An augur of how this workaround might occur is apparent with programs like Napster and Gnutella. Napster lets consumers go to the company site, search for MP3 files that reside on the hard drives of other users, and download the files they want. Gnutella takes this model even further by making files of many different kinds available, not only MP3 music, and does not require the user to go to a central site. It is a true peer-to-peer (P2P) arrangement at the very edge of the network, where each user connects directly to the computer of another.

In the near future, there will be many more instances of applications that allow consumers to bypass traditional businesses and their offerings altogether—the ultimately empowered consumer. In a sense, eBay users are already shopping in a virtual mall of their own creation. In this environment, companies need to take to heart some important business lessons.

- Make it easy for people to do business with you.
- Offer buyers real value.
- Make money by gathering information about buyers and reduce profit from goods and services accordingly . . . but let consumers know what you are doing.

- Consider the value of sales to the customer over the life of the relationship, not just the immediate sale.

Content Trends

Content providers and the packagers of information and entertainment material find themselves in the infancy of a one-creation-fits-all entertainment landscape. In the near future, the most efficient and profitable operations will manage their content and automate the generation of multiple versions of a single program or concept. Content will have to reach people wherever they are, whatever device they are using, whenever it is requested. Providers will transmit material to consumers who want to access content on the run on cell phones with video screens, on the road over the satellite to car computers, at the office over the LAN to their desktop PC, and at home over the air or the wire to TV screens, screen phones, and wireless e-tablets.

From a production perspective, this is a challenging task. The traditional four-step content creation process—development, preproduction, production, and postproduction (followed by marketing and distribution)—are well-understood by industry professionals. But creating material for the emergent entertainment landscape requires profound reengineering to adapt legacy workflows to the new requirements.

David Wertheimer, chairman and CEO of Wirebreak.com describes how the creative process takes place in that company: "One of our greatest assets is our knowledge of how to develop and create shows made for the Internet—121 steps from the pitch to the time it is an interactive, digital experience on the Internet."[18] According to Wertheimer, only about 20 of the steps involve the making of the video, so that traditional skills play less than a 20% role in putting together an interactive product. The other 80% requires putting together all of the pieces that make the product right for the Internet medium, focused on answering one fundamental question: Why should this product exist on the Internet as opposed to television, feature film, or video? "On most sites, you don't get a compelling reason for the material to be on a computer, so can you expect a visitor to come back day after day, week after week? Why would you put something on the Internet? Because there are things about the medium that make the content more fun, engaging, and entertaining—that is the central focus," says Wertheimer.

The question he poses of why material should be on the Internet may well shift to every other platform as well. Why is this material on the TV set? Why is it on a cell phone, a wireless tablet, a car computer? It is not easy to see how entertainment (except perhaps games) will shift easily to the power- and bandwidth-constrained mobile environment, at least until a number of years in the future when these problems have been solved for portable devices. In the meantime, information products will probably migrate across platforms more easily.

On the Internet, producers make compromises for entertainment content. Audio and animation survive transport over dialup modems. To the extent that they use video, producers usually break it up into short segments. AtomFilms.com, iFilm.com, breaktv.com, wirebreak.com, and aentv.com all feature shorts and break long-form material into 3- to 5-minute episodes for the Internet.

Media Asset Management (MAM)   MAM is one solution to the complexity of creating, marketing, and distributing content across multiple platforms to consumers who request different versions, matching it to their own wants and needs. MAM is a set of technologies that allow smooth, timely, largely automated responses to requests for viewing or using assets. The term "asset" refers to digital creations—text, graphics, audio, and video.

In addition to customizing content, the networked landscape requires efficiency. To be profitable and successful, companies have to manage resources carefully, operate cost-effectively, and transport their product-to-market quickly. For any company even thinking about delivering content, MAM won't be merely useful—it will be necessary for survival.

Media asset management involves the acquisition, indexing, storing, retrieving, tracking, and versioning of material. The performance of all these functionalities requires a complex system with four components: a hardware platform, a network, software, and a set of procedures. The hardware includes a central server or a distributed system of multiple, connected servers. The network can be a local area network (LAN), an organizational wide area network (WAN), a within-enterprise Intranet that uses HTTP, a virtual private network (VPN) provided by the telephone company, or the public Internet. Many organizations use a combination of networks to link a wide variety of users—employees, suppliers, vendors, and customers. Depending on how large the organization is, there may be any number of work stations attached to the network, accommodating upwards of several thousand simultaneous users.

Multiple layers of software run the network and provide security, communication, and user interfaces. There is also a shared application layer for generating, altering, repurposing, and managing digital assets. Finally, local machines are loaded with the specific software that users need to do their particular jobs.

The procedures and practices of the organization that is implementing media asset management technology define how and how much the system must be customized. Few big companies can adopt a completely turnkey system and expect a smooth transition to an all-digital operation. Automated radio stations, and to some extent television stations, may be exceptions, however, since they tend to have similar operational procedures even before bringing in a media asset management system.

There are at least a dozen reasons why diverse industries and enterprises are now focusing on media asset management systems. But they can all be summarized into two categories: MAM will either make money or save money. As noted earlier, MAM systems enable information and entertainment companies to meet the needs and wants of individual customers through mass customization and personalization.

Traditionally, customization and low cost have been mutually exclusive. Mass production made goods available at low prices but were uniform. Customization required some kind of special handling that greatly raised costs. And personalized products came from designers and craftsman, reserved for the wealthy.

Today, customers can register unique requests for entertainment and informational products through an interface. A MAM system slices and dices stored material to meet those demands through an entirely automated process. It gives customers what they want. And it keeps costs low. Such efficiency means that customization of existing products for everyone—mass customization—is now possible and, within limits, may even allow for mass personalization in the sense that unique combinations of existing material can be generated. (Of course not even MAM systems can create new products from scratch. So far, that requires the creativity of human beings.)

MAM lets copyright holders market and exploit the material they own more effectively. They have immediate access to their legal rights to the content and its availability for sale and use. Most assets have contractual rights and obligations attached to them. Digital rights management (DRM) facilitates rapid determination of what the company owns, how it is allowed to use it, how much stakeholders across the value chain should be paid for it, and the terms and conditions that apply to buyers' rights and end-users of the assets.

MAM is especially important to media and entertainment companies that own the rights to large libraries of legacy content like movies and TV shows. Content owners can repurpose and distribute material quickly. They can market component elements of the product in addition to the whole finished product, providing other incremental revenue streams. It allows them to "webify their material" to make it interactively searchable and retrievable to customers by putting a simple user interface at the front end of a complex database.

All these functions reduce the "cycle time to market," a particularly important capability where highly perishable information products such as films, television programs, magazines, and books are involved. MAM brings these kinds of business efficiencies wherever it is implemented. It lets companies save considerable money by reducing the number of times material is digitized and the considerable costs of transporting negatives, tapes, and reels from one location to another over the course of a project.

Many people send images and audio clips, sometimes even video clips, across the Internet or another computer network.

MAM systems are not installed all at once across a company's creation, distribution, marketing, and fulfillment activities. Organizations usually begin by installing a MAM system in one area of their business. For example, TV stations often start with the newsroom. A film studio might begin by putting the video, still images, drawings, and text needed for marketing, advertising, and publicity. Table 15.7 shows the benefits that accrue when MAM is implemented within specific activity centers.

A MAM system offers benefits beyond meeting consumer demands and efficiency. It enables entirely new functions, such as remote collaboration among employees, suppliers, vendors, and customers. More than one worker can access the same asset to perform work, allowing people to work in parallel on the same project or to complete multiple projects at the same time without waiting for someone else to finish his or her work and free up the original material. Collaboration and multiple access to digital assets also permit staff to accomplish many tasks in less time than they did before the system was available. MAM technology can also automate the approval process of finished material by routing material to the appropriate executives simultaneously.

Organizations that have implemented this new streamlining technology throughout the value chain create an effective barrier to entry in their markets. New entrants face lean, mean, fully networked competitors that are able to respond to the wants and needs of individual customers. That's a difficult challenge for companies without MAM systems to beat.

Strategies for Survival

Facing the challenges of constant change, global operations, and a fast-moving marketing environment, companies have to be ready to transform themselves and their own operations, and merge or partner with other companies. The next section looks at organizational and financial strategies that enterprises are adopting to be adaptable and profitable in the face of rapid change. It must be said that by themselves, organizing down to the lowliest paper clip and executing a perfectly rationalized financial scheme are not sufficient for the long-term and rarely successful in the short-term. They are part of an overall business model, but success in the media and entertainment industries also requires well thought out approaches to content, revenue, and marketing.

Organizational Strategies

One response enterprises must make in periods of transformation is to look to their own structure and their domains of operation. Digital technologies,

Table 15.7 Benefits of Media Asset Management to Activity Areas within Media Organizations

Area/Activity	Operations	Content Creation	Postproduction	Distribution
Overall Business	Greater overall efficiency throughout organization	Asset tracking Improved workflow Personnel and scheduling efficiencies Collaboration both in and out of organization Approval tracking	Collaboration Versioning Approval tracking	Asset exploitation and marketing Rights management Royalty payments Availability and scheduling
Content Production Floor	Workflow Tracking data	Collaboration Simultaneous multiple access to same material Asset tracking Versioning	Collaboration Simultaneous multiple access to same material Asset tracking Versioning Repurposing existing assets for new uses and markets	Tracking Exploitation of elements as well as whole finished products
Postproduction Facility or Department	Workflow Tracking data	Distribute work across multiple organizations, maintaining tight coordination	Automated output of products to multiple formats	Simultaneous real-time distribution to clients and distribution servers

both internal and external, affect how companies organize themselves. The "restructuring" of the 1980s and 1990s came at least partly in response to internal uses of computers, while mergers and acquisitions tend to reflect changes in the market. Future reorganization is likely to come from external market conditions as well. In this section, we will cover consolidation through merger and acquisition in the media and entertainment industries, across sectors and companies, as well as consolidation within companies. We will also look at partnerships and alliances—and a special kind of this activity, "camp-building."

Organizational Strategy: Synergy and Leverage through Alliance and Merger A rapid pace of consolidation in telecommunications, media, and entertainment has been ongoing for the past two decades. By August 1999, there were $172 billion in consolidation deals in telecommunications and $126 billion in the radio/TV industry.[19] Leverage and synergy both pay homage to the notion that $2 + 2 = 5$—that the whole is greater than the sum of the parts. More prosaically, if a merger or acquisition adds to an organization's horizontal or vertical control of a market, the deal is said to leverage the company's efforts or to provide synergy across previously separate business units.

In the past decade, all four of the major television networks have been taken over by multinational conglomerates that operate many different divisions to address specific markets. With the exception of NBC, all these entertainment giants also own a major Hollywood film studio. The film studios have production units that create TV programming, so the acquisitions stem from the belief that there is a synergy (efficiency) in creating properties that flow across multiple media platforms, and that properties can be leveraged (i.e., increased in value) across them.

The advantages are thought to be even greater when the conglomerate owns both content and conduits. Consider that there are two major categories of activity in the media and entertainment industry, production and distribution. Companies produce and distribute content, sometimes performing only one of these functions, sometimes both.

Media economist David Waterman of Indiana University provided a thoughtful response to the Disney acquisition of ABC:

> *Vertical integration creates strategic and cost advantages, but in my view they are much more limited than people seem to imagine. ABC Network has approximately the same incentive to carry Disney programs that they did in the past, and from an organizational standpoint, the risk is that a decision to carry a Disney program gets made without the discipline of the market really putting the program's value in the ABC schedule to the test. And ABC network is hardly the only distribution outlet for Disney programs.[20]*

Figure 15.4 Virage VideoLogger is the leading software for ingesting, indexing, and retrieving video. Source: Virage.

Another high-profile series of mergers and acquisitions involve Time Warner. The company was formed through a content-conduit deal that married Time's publishing empire to the Warner cable systems and movie studio. Time Warner's aggressive acquisition strategy included a buyout of Turner Broadcasting, adding the popular CNN brands to its roster of content offerings.

Now Time Warner will itself be part of the AOL empire (if the merger is approved by regulating authorities), the execution of a content-conduit strategy by AOL. Why did AOL make this purchase? Because AOL was concerned that it might be banished from cable systems, unable to compete with MSO-owned Internet access services, RoadRunner and Excite@Home. Presumably AOL determined that failure to gain carriage on cable's two-way broadband networks could have damaged the long-term prospects for the giant service provider.

Two models describe the relationship of conduits to content. The oldest is the "common carrier" model, developed out of the regulation of telephone companies. In this view, the provider of the conduit is prohibited from providing content and must interconnect its conduit with others in an international infrastructure. The other model is what Gilder terms the "Malone model," derived from John Malone's control over the Tele-

Communications, Inc. (TCI) conduit and its content as well.[21] In the United States, there has traditionally been some suspicion towards allowing the owners of the means of content distribution to also own the content itself. For example, in the film industry, the federal government's anti-trust policy dictated that film studios could not own theaters, so creation and distribution were mandated to be separate. In television, regulation dictated a partial separation of over-the-air broadcasting, so that networks could own only a limited number of television stations. Over time, suspicion has waned and no such limitation was imposed upon cable companies. Many multiple system operators (MSOs) also own cable networks and were able to leverage their ownership of both content and conduit to their financial advantage.

Consolidation doesn't just occur between huge companies in different sectors, like AOL and Time Warner and Disney and ABC, or even between companies in the same sector, like MCI and WorldCom and now Sprint. It also occurs within companies. A newly conglomerated entity commonly eliminates divisions, departments, and business units that existed when there were two separate companies, made redundant by the unified organization. For instance, in late 1999, Disney brought its Buena Vista Television Production group under the ABC umbrella, taking it out of Disney Studios' filmed entertainment company.[22] The results are a mixed bag: The move will cut costs, and probably improve the unit's batting average at supplying shows to ABC, but it could also undercut its ability to provide shows for other networks.

Internal consolidation may be driven by other considerations. In May, 2000, NewsCorp announced that it would combine some of its most valuable companies into a single entity. Included in the deal will be satellite services, BSkyB and Asia Star TV (jointly owned by Cable and Wireless HKT Ltd.), News Digital Systems, and NewsCorp's 50% interest in Stream. The two reasons advanced for the reorganization were providing an integrated platform to ensure deeper and stickier relationships with customers, and to monetize the valuation of these business units at a higher level than can be obtained if they are under the NewsCorp corporate umbrella.[23]

Rollups: Consolidating Small Companies   Content providers often start as very small boutique operations, with product development coming from the minds, imaginations, and hands of a small number of people. Growth may entail otherwise creative individuals spending time on the business, away from their core activities. If one of these companies is profitable, it often adds people through an acquisition strategy. When the purchases involve several boutiques, the resulting bigger fish is called a "rollup."

Probably one of the most impressive of all the rollups among digital companies is iXL, the Atlanta-based interactive marketing services and consulting agency. It has acquired creative agencies in cities around the world to build its twenty offices worldwide. The Los Angeles office was built from three local companies purchased by iXL.

Organizational Strategy: Partnerships and Alliances Many factors are driving companies to work together, and both partners can derive any number of benefits. Joint activities may allow them to take advantage of economies of scope and scale, and to share the costs of global marketing and distribution and capital-intensive advanced technologies. They may deprive competitors of access to markets or platforms. And they can open up new lines of business or markets around the world.

Alliances and partnerships come in many different degrees of complexity and duration. Internet ventures seem to encourage relatively permanent arrangements, and some analysts point to Japanese models of intercompany agreements, such as the keiretsu and the zaibatsu. A keiretsu is a group of aligned companies, usually with a bank or other funding mechanism at the core. There is often cross-ownership of the key execution units, but all the connected companies work together to provide complete execution of a many-faceted business model. A zaibatsu is similar but is more tightly linked, acting as a single entity.

Net partnerships operate much more closely than alliances between traditional organizations. They are more likely to focus on their core business, outsourcing or partnering to provide necessary services. For example, a company might package top-flight technical and marketing talent, while leaving field sales, customer support, and fulfillment organizations to third parties or partners. When dot-coms enter agreements, they are open to sharing essential resources like fulfillment, cooperative financial systems, customer data, and even human resources, which many traditional businesses would scrupulously avoid.

Partnering is also a way for an organization to learn how to address a new market. When Fox Sports wanted to produce digital sports programming that could be used as a linear program on NewsCorp.'s broadcasting network and as interactive content for the Web, they turned to Broadband Interactive Group. BIG will work with Fox producers to tailor the product, Gotcha.tv for the network, replacing the existing show, *Rush Hour.*[24]

A by-product of partnering is the "co-branding" agreement, which can refer to a myriad of agreements covering how content, a product, or service offered by two or more companies must be identified. With co-branded syndicated online content, for example, it can mean that the Web site operator controls the top, bottom, and left frame of the co-branded Web page. The central portion carries the partner's logo to indicate they provided the content. In some agreements, the site can change the background; in others, the supplied content is untouchable.

The trend towards partnerships and alliances does not change the reality that while companies are cooperating in one arena, they may still compete in others. In the film industry, studios are getting together to pay for expensive movies more frequently, and sharing in the revenues. *Titanic* is a good example, where 20th Century Fox (Newscorp.) and Paramount Pictures (Viacom) jointly financed the picture. Meanwhile, the two companies also pushed their own films that year.

Organizational Strategy: Camp-Building   Some partnerships that purport to be made for developing product are actually exercises in camp-building. Companies engage in building camps when they need to share technical and marketing information to make their products available and compatible with one another.[25] It most commonly occurs in the process of building a market during periods when it is essential to set standards.

For several years, the Internet has been the stage for a complex interaction between two giant camps: the Microsoft-Intel "Wintel" camp and the Sun-AOL (including Netscape) camp. Each is guided by a philosophy. The Wintel group believes that processing power should be at the edge in the end-user's machine. Microsoft's long-time advertising slogan, "Where do you want to go today?" locates the action in the end-user. The Sun-AOL camp believes that processing power should be in the network, expressed in Sun's advertising slogan, "The network *is* the computer." This group sees the Internet as a giant server-client network with requests from somewhat intelligent end-users getting information from and storing their applications on servers.

Thousands of smaller companies are attached to each camp, adopting the standards advanced by their leaders. The Wintel companies generally support Windows NT servers, Microsoft standards, and the Windows Media Player for streaming. The Sun-AOL companies are more Apple oriented, and tend to support Sun Sparc servers, Java, and the Real Networks streaming media player.

On a much more limited scale, there are also two camps that have formed around the standards for digital television. On the one side are TV station groups that favor the COFDM (Coded Orthogonal Frequency Division Multiplexing) transmission standard, led by Sinclair Broadcasting. They are facing off against the ATSC (Advanced Television Standards Committee) camp that includes General Instrument, Zenith, Philips, and the other HDTV developers, the TV networks, and the FCC. It seems like an uneven battle, but this ongoing struggle has shown how an influential group whose support is essential to actually implementing a standard can stave off even a very powerful, entrenched, and determined array of forces.

Chris Halliwell studied the role of standards in camp-building, and she found that standards may play a different role, depending on the stage of the market for a given technology. She used the categories of the theory of diffusion of innovation to describe how camp-building takes place. Table 15.8, adapted from Halliwell's model, shows why each type

Table 15.8 Diffusion and the Standards-Setting Process

	Innovators	Early Adopters	Early Majority	Late Majority	Laggards
Attitude towards tech.	Tech for its own sake	Compet. Advantage or personal advantage	Practical use	Use, not tech-oriented	Tech averse
Attitude towards standards	Don't care	Don't care	Want stable, accepted stds.	Demand end-to-end solution	Cling to old stds.
Role of stds.	Verify technology	Deploy stds.	Partially accept stds.	Industry-wide stds.	Extend prev. stds.
Develop market-	Mfgrs agree to grow mkt.	Cultivate early users	Get partners for pkg. Solutions	Develop stds. and products	Invest in educ.
Eval. of stds.	Meets needs of users	Meets needs of users	If no de facto stds. Need them	Multiple stds. required	Use stds. to grow mkt.
		All: Address cost of adoption to users.			

of adopter buys (or does not buy) a new product, the role that standards play in the buying decision, what producers need to do to increase the size of the market, and how to evaluate standards.

This fascinating analysis suggests that standards are important throughout the life of a product, not just at its beginning. In the early stages, producers need to get together to develop the standards needed for interoperability so there can be a market. As the Early Majority begins to buy, the market enters a mainstream stage and shared standards are essential if it is to grow. The Late Majoritarians demand end-to-end solutions and applications—specific products that require multiple products that absolutely comply with the shared standards.

Camp-building efforts are likely to be led by companies that are the "gorilla" in their market.[26] Geoffrey Moore describes a new high-tech product category as bursting upon the marketplace like a tornado; examples are the Internet, PCs, CD-ROM, DVD, and MP3 players. Inside the tornado is a gorilla, a company that grows at an extraordinary rate and quickly dominates the market.

As Moore writes: "All of a sudden, you go from 'We're not doing it,' to 'We do it all at once.' What it does is create phenomenal market demand that wildly outstrips supply. As it turns out, the tornado will always have a gorilla. In fact it must have a gorilla because the gorilla becomes the basis for de facto standards. Technological deployment can't happen without standards." In this way, a new company becomes the center of a larger camp-building exercise when it is joined by existing companies that move in to develop the new market created in the tornado.

Organizational Strategy: Bricks to Clicks, Clicks to Bricks   In today's lexicon, traditional businesses are called bricks-and-mortar operations, what Web entrepreneurs call "the dirt world" or the Real World. Internet businesses are called clicks-and-mortar operations, existing in the virtual world of cyberspace. In the future, it will probably be necessary to have a foot in both the "dirtual" and the virtual worlds.

The implication of this assessment is that bricks-and-mortar businesses need to add clicks, while e-tailers should add bricks and go to ground. But why? Why can't a company do one or the other and not both?

The answers depend to some extent on the nature of the product and the business model. But here are some considerations. Virtual operation is global. It provides customers with a 24/7 storefront. It shifts a great deal of work from the sales side to the buyer side, such as looking for products, filling out forms, and tallying up the final bill. But unless the product is entirely digital, it will not place a single molecule in the hand of any consumer anywhere.

To the extent that physical goods are involved, some advantage is conferred upon bricks-and-mortar operations. Potential purchasers can touch, poke, fondle, sniff, compare, sit on, and lift items they might buy.[27] They can take them home now, this minute. If they have problems, they can return them and deal with a person face-to-face.

For suppliers and sellers, thousands of years of business experience suggest that relying on others to carry out important parts of your supply chain is risky. So is letting others peek at your customer data. The ownership of consumer information is tenuous among online partners, and there have been some high-profile disagreements about who owns what and how the data can be used. And so far, we have not even considered privacy concerns.

Only time will reveal how these wrinkles will be ironed out. In the meantime, companies will have to struggle to find the right mix of bricks and clicks that best suits their particular operations. Expect experimentation.

Organizational Strategy: Strategic Prototyping   In large organizations, there is always a group that is responsible for strategic planning, which puts together a plan for the direction of the company over some period of time. Ideas are developed and thoroughly explored before they are handed off to operational units for implementation. Some people believe that the speed of change in the marketplace of digital information— music, news, chat rooms, and other forms—outstrips conventional strategic planning.

Robert Levittan, cofounder of iVillage, calls the new way of making plans for the future, especially in the creation and distribution of content, "strategic prototyping."[28] Using the Internet as an experimental petri dish to grow new strains of products, Levittan says companies should put product on the market as quickly as possible, get feedback, and revise it in light of what customers say: "Launch, listen, learn," advised Levittan. "Both advertising targets and audience behavior around content are not just tonnage—you have data based on exact measurement. The Web is a 24/7 focus group. In the old days, you created a beautiful product and put it out there and that was it. Today, you keep your fingers on the pulse and react quickly to make any needed changes."

Levittan cites the experience of iVillage. Launched in 1995, the company initially put advertising around content commissioned by the company. But advertisers weren't getting click-throughs. So creatives redesigned the site, putting community first, then content. This proved to be the key factor. "People didn't want to hear what we were writing. They wanted to hear from others about their personal experience," explains Levittan.

The ever-changing Internet holds some appeal as a proving ground. As the speed of all commerce accelerates, Internet entrepreneurs believe they are forging the new ways of working that will be needed in a period of continuous transformation. Perhaps many businesses involved in communication over the Net will find that they need to adopt some form of strategic prototyping to replace or, more likely, accompany strategic planning in the bricks-and-mortar arena.

Financial Strategies

Companies that are playing in the new media space develop a financial approach to raise money, or shore up their revenue model, or both. Financial schemes are no substitute for an appealing product, a robust revenue model, and a clever marketing idea. However, even given some or all of these advantages, startups have to figure out how to fund their operation.

Financial Strategy: Go IPO   IPOs aren't just for unknown companies. Drawn by the huge amounts of money that startups attract, some of the biggest names in media and entertainment are spinning off their Internet operations into standalone IPO companies, even when they are taking in little or no revenue. Noting the turnaround in attitude on the part of TV networks towards the Internet and its potential, Rick Ducey, then Senior Vice President of the Research and Information Group for the National Association of Broadcasters, said: "Broadcasters are more receptive to the Internet because a lot has happened. Looking at the market caps of Internet companies—broadcast.com, eBay—it makes you sit up and notice. Broadcasters use one business model but they see people in similar businesses taking their companies to such heights in the equity market . . . well, it has gotten their attention."[29]

To Sandra Kresch, strategy partner in PricewaterhouseCoopers' entertainment and media practice, spinning off its Internet units may lead conglomerates down the wrong path.[30] "There will be much more integration of the media," she believes. "So all the splitting out, making them standalone, split-off operations, while attractive from a financial and training standpoint (building different skill sets), could actually undermine their ability to get value out of the brands and content in a cross-media world." An alternative Internet strategy would be to look at the entire picture, particularly as it relates to content and the consumer, and understanding how to these elements will play out in the digital environment.

Kresch advises media and entertainment giants to pay attention to brand-building, to engage in more experimentation, and to emphasize their relationships with customers. Putting distance between their Internet initiatives and their other businesses could impede their ability to develop a learning curve about how to benefit from the synergy between the media platforms.

A risky form of an IPO strategy is called "flipping." The term is taken from the real estate business and means to buy a property with no intention of ever using it (or even holding it for any length of time), instead selling it as soon as a new buyer can be found who will pay a higher price.

Sometimes startup companies will act as if they want to go IPO, but are really aiming for takeover by a larger entity, either before or after they go public. This scenario occurred any number of times during the Wall Street love affair with Internet stocks in the late 1990s, with startups forming, raising money, generating hype, then leaping into the arms of a corporate sugar daddy that needed to round out its product or service offerings.

Financial Strategy: Customer Valuation   How do you place a value on an enterprise that has few physical assets, does not produce any physical products, owns no real estate or leases, maintains no office space, and outsources most of its operations? Consider that the cost of an Internet customer is about $50. In other words, dot-coms must spend $50 in marketing and promotion, on average, to attract each customer. If an organization has thousands of customers who spend, on average, more than the cost of acquiring them, then that customer base is itself an asset subject to valuation, and some companies are using this method to establish their worth.

Some customers are more valuable than others, depending on their LTV score—their Life Time Value. The ability to track purchases and to analyze other data about people allows marketers to calculate the LTV. Essentially, it is the sum of all future purchases that a customer is likely to make from a company, less the marketing and customer-service costs paid out to acquire and maintain the relationship. Author Jeremy Rifkin views it all with a jaundiced eye:

> . . . these large-scale efforts to create a surrogate social sphere tucked inside a commercial wrap are, for the most part, going unnoticed and uncritiqued, despite the broad and far-reaching potential consequences for society. When virtually every aspect of our being becomes a paid-for activity, human life itself becomes the ultimate commercial product, and the commercial sphere becomes the final arbiter of our personal and collective existence.[31]

Charting the Challenges

Sometimes opportunity knocks, and sometimes it knocks you out. Located as they are at the epicenter of the digital revolution, media and entertainment companies face a turbulent and uncertain future. They are fighting on two fronts, simultaneously struggling to grasp radically new ways of doing business, while holding fast to any advantages they currently enjoy. The challenges are real, and some of them pose serious problems that could threaten their very existence.

The difficulties of the next phase of the transformation in communication have little to do with technology. Five years ago, this was not the case. The interactive TV test trials revealed a host of technical obstacles that were simply too expensive to solve in ways that would support a rational business case. As we have seen, two essential pieces of the puzzle are largely solved: increased processing capacity brings fast, inexpensive switching and routing, and fiber-optic technology expands bandwidth dramatically.

For businesses, the challenges are moving sharply away from hardware concerns to fundamental questions about how to do business in a networked environment. Consolidation itself engenders unmanageable webs of relationships. And there is little evidence that bigger is smarter—but it is definitely less innovative and agile in a market that rewards innovation and agility. The ease of digital content creation, re-creation, and distribution could be catastrophic for companies that profit from their control of these activities. The empowered audience exhibits a worrisome independence. And competition is fierce at every level.

The headlines say it all: "Networks face a bumpy ride," "Hollywood Studios: The hits keep on comin' and so do the misses," and "ISPs face a choice: Evolution or extinction." The next section looks at the coming wave of these challenges to media and entertainment companies in conducting business. It will then turn to consumer and social concerns.

Conflicts of Interest Created by Content-Conduit Conglomeration

It's a natural—bandwidth owner buys content-production capability, or giant content producer buys bandwidth provider. And as covered earlier in the section on organizational consolidation strategies, most large entities have already made a conduit-content play. Broadcast networks always had a foot in both camps—they can own up to twelve local stations (previously five). They round out their network by contracting with affiliates, for which they provide nationally distributed content. Like them, AOL quickly moved from merely offering Internet access to delivering a wealth of content and services to their subscribers.

In the cable industry, multiple system owners (MSOs) own all or part of cable programming services. Satellite services providers are mainly infrastructure providers—except for NewsCorp, which began as a content company and founded its own global satellite service, bringing content to Australia and New Zealand, the United Kingdom, Asia, the Middle East, and Latin America. Telephone companies, historically prohibited from owning content, have stayed on the bandwidth-provision side, and film studios, long-prevented from outright ownership of theaters where their material was shown, have focused on producing content.

In the past decade, however, the stakes have become much larger. The worldwide buildout of entertainment infrastructure, rising income, privatization, and increasing urbanization have made media and entertainment global businesses. Consolidation has increased accordingly. All the major film studios are part of the portfolio of a corporate owner, as are all the broadcast networks.

Now some problems with this kind of organization are coming to the fore, chief among them the conflict involving some of the most highly sought-after creative people in the content-producing industry. Most successful creatives form their own companies to better control their working conditions and assets. But their basic business goals may conflict with the media giants who commission, license, and distribute their properties.

Distributors, the conduits, want exclusive content. They make money by using it to force consumers to come to their conduit in order to get the material they want. By contrast, exclusivity may not be the best strategy for content producers. They want to place their products wherever they can get the most attractive terms. And they can maximize their returns if they place it in more than one venue, either through simultaneous or windowed release.

In the past few years, several conflicts have hit the courts. In late 1999, producer Steven Bochco filed suit against 20th Century Fox, alleging that the company sold reruns of *NYPD Blue* to its own cable network, without looking for other possible buyers.[32] According to the filing, Bochco maintained that he lost $15 million on the deal. Other high-profile instances are a dispute between *X-Files* star David Duchovny, who sued Fox over a similar complaint—that the studio had cheated him by selling reruns to its own FX channel. And creators of *Home Improvement* sued the Walt Disney Company, saying that WDC let its ABC broadcasting subsidiary renew the show for less than its market value.

The suit filed by the producers of *Homicide* against NBC has an interesting twist. The TV network aired the show and owns the downstream rights as well. In this case, the creative team accuses NBC of cheating them of $20 million by packaging and selling the crime show with another NBC series that had flopped.[33]

A growing number of conflicts have arisen in the music business as well, as it transitions to a digital environment. Metallica, the Offspring, Aerosmith, Rage Against the Machine, and the Goo Goo Dolls have all fought with their labels to control their own Web site names. The Goo Goo Dolls band members were forced to resort to cyber-war guerrilla tactics to wrest http://www.googoodolls.com from their label, Warner Bros. Records. The Buffalo music group objected when WBR set up the domain name and, in retaliation, registered http://www.googoodolls.org. After the fracas became public, the label backed down and both sites are registered to the band.[34]

Both sides want to own the rights to the Web site because they can intercept the traffic, allowing them to maintain the relationship with fans, promote related goods and services, and collect data on visitors. Record companies anticipate that popular artists will become the online center around which e-commerce is built. Beyond controlling the cash register, a label can use the site to cross-promote its other properties—a step that the Goo Goo Dolls and other bands with their own franchises may not think is in their own best interest.

Changing Business Models

The rich variety of business models in the previous chapter shows how much experimentation digital media and the Internet is fostering. For startup companies, this variety brings with it the opportunity to innovate in business as well as in product development. But for established companies, it mostly brings trouble. They have sunk costs in equipment, facilities and salary overhead, and tried-and-true procedures that are embedded in the people and machines in the organization. And it is difficult for them to estimate just how much of a threat a new technology, trend, or competitor will ultimately pose.

It's hardly surprising that the first strategy is usually one of holding the line against change, maintaining business models just the way the company likes them—the way they are. The media and entertainment giants' empires rest on the control of content creation and its distribution. The music industry's defense of their business model against destruction by the MP3 compression format and the Napster file-sharing program provides a good example of the set of dilemmas that will eventually face all media and entertainment companies. The complicating factor for the music business (as it will be for others providing information and entertainment content) is that the technology is in the hands of its consumers and its suppliers, rather than a competitive company. Its enemies are the very people it is dependent upon at both ends of its value chain: the creative artists who provide the content, and the audience that consumes it.

The artists need to be financed in the early stages of their careers. Music labels perform this function, laying out the cash for that first, second, and perhaps third CD. Between the recording costs and the marketing, hundreds of thousands of dollars might be involved, and everyone is aware that there are many more misses than hits. One wonders how the hit-producing artists feel about paying for the mistaken judgment of the label's talent scouts.

Mark Cuban's fascinating analysis of this process is worth attention.[35] He points out that most artists produce one CD per year that sells for about $15. They earn about the same on a CD as the label does, $1 or $2 per CD. This amounts to 16 to 20 cents per month per fan. "Not very much," concludes Cuban, continuing:

> Now does anyone out there think there might be a better way for an artist to make more than 16 to 20 cents per month from its fans? I certainly do. Remove all the physical aspects of music and video, and make it so that rather than buying a single

or a CD, artists deliver music in a manner that maximizes their revenues and prof-its and optimizes the amount of content that consumers receive.

No downloaded singles. No downloaded ten-song groups called CDs. Its going to be all that you can eat. Instead of paying $15 bucks for a CD from your favorite band, you pay $15 per year to your favorite band, and they allow you to download 100% of the music they create now, from their libraries, and to new music for as long as you subscribe. The artists make incredibly more amounts of money. They can release music they want, when they want, for whatever reason they want.

The labels wouldn't be in such a difficult situation if they actually marketed music. But they don't. They are only middlemen, arranging deals with radio stations who actually play the music to reach the consumer. This means that artists can cut side deals with the radio stations themselves. Or hire their own marketers on their own terms, offering stations a deal with no ASCAP, BMI, or RIAA (Recording Industry Association of America) fees when they play the music if they mention the artist's Web site URL as part of the broadcast. Such an agreement would allow musicians to sell tickets, merchandise, or whatever to their subscribers to generate even more revenue.

And this just the artists' side of the equation. Take a look at the consumers. The labels' relationship with their customers is the stuff of a public relations disaster, heaped on top of what may already be an insoluble business problem.

The prospect of the FBI arresting 16-year-olds for keeping music from their favorite band on their hard drives and E-mailing songs to friends puts the labels in a particularly ugly light. But what are these teenagers really doing? True, they are stealing intellectual property from the creators and the labels. But the real crime in the eyes of the recording companies is that fans are challenging the control of the music and its distribution, the physical fulfillment that labels reserve for their cut of the proceeds—and that's how labels make money, the heart of their business model.

Stand in the way of your suppliers making money and offend your customers. Or stand by and watch your business model blown up by a disruptive technology. Or scramble as fast as you can for an alternative. It's not an enviable set of choices—and every content provider and distributor in the entertainment, information, and media industries will find itself in this situation at some point during the digital revolution.

Protection of Intellectual Property: The Impossible Dream?   Inexorably tied to the ability to control is the right to own intellectual property (IP) and to enforce those rights. Technologist George Gilder once noted: "If you can't protect something, then it isn't property."[36] But the inherent flexibility of digital data endangers the protection of IP.

The commoditization of information, that is, the concept of information as a mass-produced, marketable product, is a twentieth-century phenomenon. It has led to an entirely new industry as well, one that creates information products, and markets, distributes, and sells them. As long as the product was tied to a physical medium, a film, video, record, CD, book, and so forth, much progress was made towards the establishment of a content marketplace. In the analog days, copying was a problem that license-holders came to terms with by allowing casual, informal copying and fair use, while prosecuting professional copiers and sellers. Now such practices are in the hands of consumers themselves who can distribute a perfectly reproducible content product to anyone who wants it in real time.

The issue is much greater than any particular technology or copying software such as MP3, Napster, or Gnutella. And there may be no antidote, simply because the material must be translated into a form usable, readable, visible, or audible to the end user. The implication is that once it is decoded, it can be redigitized with little difficulty. Digital watermarking embeds information in material that identifies its owner, and accompanies the material from user to user. However, this is a salvo from the producer side that can effectively be only temporary. Expect to learn about the "scrubber," a program that removes the current generation of digital watermarks from content. Then producers will build a better mousetrap and users will create smarter mice. Ultimately the problem remains: The material must eventually be decoded so it can be consumed by purchasers, and that point will always be the Achilles heel for content owners.

Certainly technology can impede the casual user for some period of time by making copying difficult, time-consuming, and expensive. Such a strategy will stave off the immediate distribution of intellectual property for awhile. Other solutions will surely emerge. For example, the "big bite" revenue model covered in Chapter 14, "Cache Flow: From Bitstreams to Revenue Streams," proposes to go around the problem by controlling distribution and monetizing the content's value for the limited time it is possible. In time there are sure to be other ideas.

Audience Fragmentation   A gazillion channels generate a gazillion audiences. This scenario may not be as disastrous as some analysts predict, although it may sink a large number of the vessels currently carrying content cargo. The reason is that while a person may have some interests that are quite individualized, there are also common life experiences that almost everyone wants to learn about and share. For example, there are few people in the world who are not interested in family, jobs, food, health, money, relationships, religion and spiritual matters, and psychological well-being. A very high percentage of them also follow ethnic, regional, and national issues. Many want to further themselves or their children through education. The desire to travel is certainly widespread; so is an attraction to art, culture, and entertainment. In short, no matter who people are and where they are, there is a mix of interests that span the spectrum from broad to narrow, extending perhaps to ideas and activities indulged in by only a very few individuals.

So while the future will indeed bring a fragmented audience, it will continue to support audiences of all sizes. Moreover, as the Internet becomes more mainstream, it is beginning to reflect the patterns observed in other media. Michael Lambert, CEO and founder of iBlast, a company that will datacast Internet content over local TV stations' digital television signal, observed: "Just as most people want to see the same five movies, watch the same TV shows, listen to the same Top 10 songs, read the same books . . . they also go to the same Web sites. In mid-2000, over 99% of Internet traffic goes to fewer than 1% of sites. The Yahoo! home page is downloaded 40 million times per day."[37] So while it is still uncertain whether there will ever be hits shows, programs, or content on the Net, hit sites and hit applications already exist in cyberspace.

Taken together, it appears that well into the future, media that can aggregate audiences—television, for example—can fare quite well if they make the effort to understand how to reach them. It will certainly be a far more competitive marketplace and the consumer will more or less control their own consumption. However, it is unlikely that traditional media channels will be extinguished, and certainly not anytime soon.

Free versus Fee   The empowered networked consumer, armed with instant price and product comparisons and detailed information, tends to move profit margins toward zero—the effect predicted by classical economics. It creates a competitive free-for-all, great for people as consumers, tough on them as producers, since most people participate in activities on both sides of the commercial equation. On the Internet, there are many cases of marketers accepting zero profits, even loss leaders, to attract buyers for other revenue-generating schemes. And a good number of folks give away information and entertainment that companies would package and sell, just for fun!

It's a situation that could become more gnarly as time moves on, forcing companies to figure out how to deal with competitors who give away the product upon which an important revenue stream depends. Microsoft faces this dilemma on several fronts. In the operating system marketplace, they have to compete with the free (or relatively inexpensive commercial versions that provide customer support) Linux OS. Microsoft's server software is not as popular as the free Apache software. And in 1999, Sun Microsystems released Star Office, an office suite that performs many of the same functions that Microsoft Office does. Indeed, there are at least five free, or nearly free, such programs available on the Internet by download. In the hardware arena, stores advertise rebates that bring the price of a PC down to free—$99, $399, or $499—with ISPs footing the bill. Or buy the PC from a manufacturer and get two years of Internet access for free.

Companies have to have flexible business models and make fast, agile adaptations to new market conditions. Multiple revenue streams provide some protection by giving them time to come up with a creative response. The secure world of decade-long business models, enforced by FUD-mongering (fear, uncertainty, and doubt) is probably over.

Consumer Concerns

Mainstream consumers are interested in content, not technology. They pay little attention to the newest gadgets, and do not swoon over boxes, wires, and processors. They are not nerds.

The Internet has entered a new phase. There is now so much material on the Internet that, for consumers, the technology is taking a back seat to the pleasures of instant gratification. Mark Cuban sums up the situation eloquently:

> *Media on the Net has nothing to do with the technology. It's about access to what people want. If the Net is the only place for you to get it, you will watch it at 28.8 kbps. I remember being a kid watching cartoons on a small black and white TV while holding the antennas to try to get a picture, constantly adjusting it to keep it from rolling every few minutes . . . . just to watch Under Dog. Or driving around in my car while in Indiana to listen to sporting events fade in and out. We all have our programs that we can only get on the Net, and that is why we watch and listen on the Net. The greater the desire, the less important the quality. Quality is not the driving factor—love for the programming is.[38]*

The thrill of engrossing entertainment or of receiving the exact information the person has long sought does not mean that consumers do not worry about a ubiquitous network. For more than a quarter of a century, since the earliest experiments in interactive technologies, they have steadfastly pointed to concerns about privacy and the security of their personal information. "Privacy in a networked world? It won't happen—get over it," Sun Microsoft's Scott McNealy once said. But there's no evidence that people will get over it: Every survey reflects pervasive preoccupation with these two issues.

Privacy and Security   It's probably possible to have a secure network that will protect users' privacy but it won't be cheap or fast—at least not as cheap and fast as a network that is relatively insecure. "Relatively" is the operative word. The Internet is far more secure than wireless phones. Dialup and DSL are more secure than cable modems. Encrypted data is more secure than unencrypted. But in the end, nothing is 100% secure.

The best bet for people who really care about their personal data is to work on their computers, employ permanent data erase programs, encrypt the information using state-of-the-art programs, save the material to a floppy or recordable CD-ROM, shut down the computer to clear the RAM cache, remove the disc or diskette from the drive, and store it in a safe. As far as giving private information such as credit card numbers to Net businesses, it's safer than handing it to an unknown waiter and cashier in a busy restaurant. And the same limits on liability for fraudulent purchases apply to the Internet as anywhere else.

McNealy is almost certainly right. Digital data and technology elude attempts to dam them, hold them in place, or rope them off. Privacy on a network is as uncertain as the protection of intellectual property. But like intellectual property, technological solutions can work for a time, so the avid privacy seeker can probably protect their personal information if they are careful and stay on top of state-of-the-art methods.

The Digital Divide The great issue in the nineteenth century in Western culture was social inequality in the ownership of the means of production. In the twentieth century, social concerns centered on unequal access to consumption. In the twenty-first century, the challenge will be to provide equitable access to the means of communication by making high-speed networks globally ubiquitous and affordable.

Country by country, region by region, ubiquity will precede affordability. If technologies such as the Teledesic LEO (low earth orbiting) satellite fleet come into being, and high bandwidth fiber miles continue to grow, penetrating even moderately populated areas, then perhaps some day most people on Earth will be connected to each other through communication technologies. For telephone penetration, rural Native American households (76.4%) rank far below the national average (94.1%).[39] Similar disparities exist among other minority groups in the United States, a situation referred to as the "Digital Divide."

The Clinton-Gore administration will have created an enduring legacy with its visionary policies towards communications technologies. The support of Internet 2 and the call to action to transform the digital divide into digital opportunity for all, were both important turning points if all boats are to rise in the networked world. A heartening list of companies, institutions, and organizations joined together to support the National Call to Action declared on April 4, 2000, stated in encouraging words:

> *Access to information technology and the Internet and the ability to use this technology effectively are becoming increasingly important to full participation in America's economic, political, and social life. While computer and Internet access has exploded in recent years, America faces a "digital divide"—a gap between those who have access to Information Age tools and the skills to use them and those who don't.*
>
> *America has an important choice to make: we can allow unequal access to deepen existing divisions along the lines of race, income, education level, geography, and disability—or we can use technology as a powerful tool to help make the American dream a reality for more people. To help create digital opportunity for more Ameri-*

cans, we must create strong partnerships between government, industry, and the rich mosaic of America's civil society—including educators, labor unions, librarians, civil rights leaders, faith-based organizations, foundations, volunteers, and community-based organizations.[40]

It will take this kind of commitment if we are to have a future where every citizen will have access to the network and the means to communicate. And work towards these goals must be a high priority for anyone interested in the continued expansion of economic success and commercial prowess along with the preservation of democracy. We may disagree vociferously as to the means of accomplishing these ends, but there is widespread agreement that if we fail, we fail separately. If we succeed, we network together.

The Net is our *doppelwelt*, our doubled world—a digital mirror of our analog existence. We are just beginning the immense project of populating the network until it becomes a virtual replica of the Real World: All that *can* go on the Net, *will* go on the Net.

Notes

1. E.M. Rogers, *Diffusion of innovations*, New York: Free Press of Glencoe (1962). Also see, E.M. Rogers, *Diffusion of innovations*, 4th ed., New York: Free Press (1995).
2. P. Drucker, *Innovation and entrepreneurship*, New York: Harper and Row (1985).
3. "TV and the Internet," Electronic Media Special Report (June 28, 1999).
4. "What's new about globalization," The *McKinsey Quarterly* 2 (1997):179.
5. C.M. Christenson. *The innovator's dilemma: When new technologies cause great firms to fail.* Cambridge: Harvard Business School Press (1997).
6. G. Lawton, "The utilities' role: Building the ubiquinetwork," Part 5, *Communications Technology* (December 1994):80–85.
7. R. Adler, *The future of advertising: New approaches to the attention economy*, Aspen, CO: Aspen Institute of Human Studies (June 1997).
8. D. Waterman, professor, Indiana University. Telephone interview, August 1995.
9. D. Tobenkin, "The wireless system that could," *Broadcasting and Cable* (May 1, 1995):20.
10. P. Lambert, "Utility networking percolates," *On Demand* (December 1994 /January 1995):16.
11. T. Coyle, "1994 local telco revenue per average switched line," *America's Network* (June 15, 1995):14.
12. Sanjay Kapoor, speaking at Digital World, Los Angeles, April 1995.
13. S.R. Rivkin, "Positioning the electric utility to build information infrastructure," *New Telecom Quarterly* 3 (1995):30–33.
14. H. Jessell, "Infohighway power play," *Telemedia Week* (December 1994):26–28.
15. L. Wirbel, "Fiber-Optic Network," *EE Times* (September 1, 1999). Available online at http://internetwk.com/story/INW19990901S0005.
16. K. Rao, Vice President, Technical Development, Steeplechase Media. Telephone interview, March 2000.

17. Thanks again to Bennett McClellan for explaining this point to me.

18. D. Wertheimer, founder Wirebreak.com. Telephone interview, May 2000.

19. T. Mulligan, "The new oligopoly boom," *Los Angeles Times* (August 22, 1999):C-1.

20. D. Waterman, op. cit.

21. G. Gilder, "Washington's bogeyman," *Forbes ASAP* (June 6, 1994):115.

22. S. Hofmeister, "Disney plans to consolidate two of its television groups," *Los Angeles Times* (July 8, 1999):C-1.

23. Reuters, "News Corp. planning to bundle its 'crown jewels' in a new company," *Los Angeles Times* (May 17, 2000):C-5.

24. L. Earnest, "New-media firm strikes deal with Fox on extreme sports," *Los Angeles Times* (October 12, 1999):C9, 19.

25. C. Halliwell, "Camp development: The art of building a market through standards," *IEEE Micro* (December 1993):10–18.

26. G.A. Moore, *Inside the tornado,* New York: Harper Collins (1995).

27. Again, Bennett McClellan is so right!

28. R. Levittan. Panelist at NATPE/ETC one-day seminar at the American Film Institute, Los Angeles, November 1999.

29. R. Ducey, former Senior Vice President of research at the National Association of Broadcasters. Telephone interview, November 1999.

30. S. Kresch, consultant, PricewaterhouseCoopers. Telephone interview, August 1999.

31. J. Rifkin, "Cradle to grave, you're a customer first," *Los Angeles Times* (April 24, 2000):B-7.

32. S. Hofmeister, "Bochco sues 20th Century Fox over 'NYPD Blue,'" *Los Angeles Times* (September 14, 1999):C-9.

33. B. Lowry, "The fittest? Make that survival of the biggest," *Los Angeles Times* (April 11, 2000):F-1.

34. G. Boucher, "Feud between band and its record label over ownership of Web site reflects growing battlefront in the industry," *Los Angeles Times* (November 3, 1999):F-1.

35. M. Cuban. Personal interview, Manhattan Beach, CA, May 1999.

36. G. Gilder. Personal interview, Squaw Valley, CA, September 1998.

37. M. Lambert, CEO and founder, iBlast. Telephone interview, June 2000.

38. M. Cuban, op. cit.

39. U.S. Department of Commerce, National Telecommunications and Information Administration (NTIA). Fact Sheet: Native Americans Lacking Information Resources. Available online at http://www.ntia.doc.gov/ntiahome /digitaldivide/factsheets/native-americans.htm.

40. Posted on a government Web site: http://www.digitaldivide.com.

Bibliography

"40-Gb/s soliton transmission travels 70,000 km." *Photonics Technology News* (March 1999). Online. Available: http://www.laurin.com/Content/Mar99/techNippon.html.

Abel, M., Bell, R., Perey, C., and Zakowski, W. "The MMCF transport services interface." *New Telecom Quarterly* 4 (1994):38–44.

Acker, S. "Designing communication systems for human systems: Values and assumptions of 'socially open architecture.'" *Communication Yearbook* 12, Newbury Park, CA: Sage Publications, 1989:498–532.

"Activmedia: Web improves relationships." *Activmedia Research*. Available: http://www.glreach.com/eng/ed/it/150998.html.

Adler, R. *The future of advertising: New approaches to the attention economy.* Aspen: Aspen Institute of Human Studies, June 1997.

Akhavan-Majid, R. "Public service broadcasting and the challenge of new technology: A case study of Japan's NHK." Paper presented at International Communication Association, Miami, FL, May 21–29, 1992.

Akwule, Raymond. *Global telecommunications: The technology, administration, and policies.* Boston, MA: Focal Press, 1992.

Allied Business Intelligence. "Wireless Systems Outlook 2000: Markets, Systems, and Technologies, Report Code:WSO00." Available: http://www.alliedworld.com/.

Altman, I., and Taylor, D.A. *Social penetration.* New York: Holt, 1973.

Aoki, K. "Virtual Communities in Japan." Paper presented at Pacific Telecommunications Council, University of Hawaii, Honolulu, HA, 1994.

Argyle, M. *Social interaction.* Chicago: Aldine, 1969.

Arlen, G., and Krasilovsky, P. "Got a minute? And the need for speed?" *Convergence* 10 (June 1995).

Banks, Mark J. "Low power television." In A.E. Grant (ed.), *Communication technology update*, 3rd ed. Boston, MA: Butterworth–Heinemann, 1994:107–115.

Banks, Mark J., and Havice, M. "Low power television 1990 industry survey." Unpublished report of the Community Broadcasters Association (December 14, 1990).

Banner Ad Placement Study. Athenia Associates, April, 1998. Online. Available: http://webreference.com/dev/banners/research.html.

Barish, Charles. "Superman's now super digital." *Videography* (Oct., 1993):30–32, 101–102.

Barlow, J.P. "Crime and puzzlement: Desperadoes of the datasphere." Online. Available: http://www.eff.org/pub/Legal/Cases/SJG /crime_and_puzzlement.

Barlow, J.P. "Is there a there in cyberspace?" *Utne Reader* 68 (March-April 1995):54.

Bateson, Gregory. *Steps to an ecology of mind*. New York: Ballantine Books, 1972.

Baylin, Frank. *Miniature satellite dishes: The new digital television*. Boulder, CO: Baylin Publications, 1994.

Becker, L. "A decade of research on interactive cable." In *Wired cities: Shaping the future of communications*, eds. William Dutton, Jay Blumer, and Kenneth Kraemer. Boston: G.K. Hall, 1987:75–101.

Benjamin, R., and Wigand, R. "Electronic markets and virtual value chains on the information highway." *Sloan Management Review* (Winter, 1995):62–72.

Berniker, Mark. "Microware creates de facto operating system for interactive TV." *Broadcasting & Cable* 124, no. 30 (July 25, 1995):30.

Berniker, Mark. "NBC Desktop Video to deliver news to PCs." *Broadcasting & Cable* (July 8, 1994):26.

Berniker, Mark. "PacTel joins wireless migration." *Broadcasting & Cable* (April 10, 1995):35.

Bernoff, J., et. al. "Interactive TV Cash Flows." Forrester Research study (August, 1999). Available: http://www.forrester.com.

Berst, J. "Secrets of Spiral Branding." *ZDNet.com* (November 13, 1998). Online. Available: http://www.zdnet.com/anchordesk/story /story_2745.html.

Besen, S.M., and Farrell, J. "The role of the ITU in standardization." *Telecommunications Policy* (April 1991):311–321.

Blankenhorn, Dana. "Wireless cable operators form alliance." *Newsbytes* (June 22, 1994).

Blaschke, C.L. "CAI effectiveness and advancing technologies: An update." Paper presented at International Communication Association, 1987.

Boeke, C., and Fernandez, R. "Via satellite's global satellite survey satellite industry trends and statistics." *Satellite Today* (July 1999). Online. Available: http://www.satellitetoday.com/viaonline/backissues /1999/0799cov.htm.

Borgman, C.L. "Theoretical approaches to the study of human interaction with computers." Unpublished paper, Institute for Communication Research, Stanford University (1982).

Borland, J. "Industry squabble thwarts DSL progress." *CNET News.com* (September 9, 1999).

Boucher, G. "Feud between band and its record label over ownership of Web site reflects growing battlefront in the industry." *Los Angeles Times* (November 3, 1999):F-1.

Bowling, Tom. "A new utility." *New Telecom Quarterly* (Fourth Quarter, 1994):14–17.

Briggs, J., and Peat, F.D. *Turbulent Mirror.* New York: Harper & Row, 1989.

Brinkley, J. *Defining vision.* New York: Harcourt Brace, 1997.

Brinkley, J. "Disk vs. disk: The fight for the ears of America." *NY Times on the Web* (August 8, 1999). Available: http://www.nytimes.com /library/tech/99/08/biztech/articles/08disk.html.

British Telecom. "BT World Communications Report 1998/9." Online. Available: http://www.bt.com/global_reports/1998–99/regional.htm.

Brown, R. "The return path: Open for business?" *CED Communications Engineering & Design* (December 1994):40–43.

"BT World Communications Report 1998/9." May 1998. Online. Available: http://www.bt.com/global_reports/1998–99/index.htm.

Buchman, Caitlin. "Back to the future: The art of interactive storytelling." *FilmMaker* (Summer 1994):34–39.

Burgess, John. "U.S. withdraws support for studio HDTV standard: Japanese suffer setback in global effort." *The Washington Post* (May 6, 1989):D-12.

Burgoon, J., and Hale, J.L. "The fundamental topoi of relational communication." *Communication Monographs* 51 (1984):193–214.

Burgoon, Judee K. "Nonverbal Signals." In *Handbook of interpersonal communication*, eds. M.L. Knapp and G.R. Miller. Beverly Hills: Sage, 1985:344–390.

Burns, J. Panelist at seminar sponsored by the National Association of Television Programming Executives (NATPE) and Entertainment Technology Commerce (ETC), November, 1999, at the American Film Institute's Los Angeles campus.

"Buy rates skyrocket for DBS pay-per-view." *Interactive Video News* (April 3, 1995):1–2.

"Cable television developments." Washington, DC: National Cable Television Association (Spring/Summer 2000):1.

Cahners In-Stat. "Wireless Cable report." *Cable Datacom News* (April 30, 1999).

"Canadian Internet Survey Fall '97." A.C. Nielsen Company. Online. Available: http://www.acnielsen.ca/sect_studentcor/studtcor_en.htm.

Capella, J. "The structure of speech-silence sequences." *Human Communication Research* (1982).

Careless, J. "Pumping up sales: VSATs and the automotive industry." *Satellite Today* (July 1998).

Carter, N.M., and Cullen, J.B. *Computerization of newspaper organizations: The impact of technology on organizational structuring.* University Press of America, 1983.

Casson, H.N. *The History of the Telephone.* Chicago: AC McClurg, 1910:224–5.

Casti, John. *Complexification.* New York: Harper Collins, 1994.

Christenson, C.M. *The innovator's dilemma: When new technologies cause great firms to fail.* Cambridge: Harvard Business School Press, 1997.

"Cisco Drives Industry Standards for Broadband Wireless Internet Services." Cisco press release (October 26, 1999). Available: http://www.cisco.com/warp/public/146/october99/26.html.

"Cisco finances FirstCom in Chile, Peru." *ATM Report* (November 22, 1998).

Clampett, E. "Qualcomm, U.S. West test high speed wireless solution." *Internetnews.com* (February, 1999). Online. Available: http://www.internetnews.com/bus-news/print/0%2C1089%2C3_65391%2C00.html.

"Commission background note on digital TV." *Reuters News Service* (December 6, 1993), archived on NEXIS.

Commission of the European Communities. "Wide-screen television lifts off." *RAPID* (press release IP:94–21, January 14, 1994), archived on NEXIS.

Consumer Electronics Manufacturers' Association, press release, "Video product sales picture perfect for the month of July." August 11, 1999. Available: http://www.ce.org/.

"Consumer home video rental activity, 1997–1998." Convergent Data report. (1999). Online. Available: http://www.convergentdata.com/trade/tradata.htm.

Coran, S.E. "Low power subscription television." *Wireless Broadcasting Magazine* (April 1995):14–16.

Cox, B. "Study: Web no threat to traditional media." *Internetnews.com* (October 6, 1999). Online. Available: http://dev-www.internetnews.com/IAR/print/0%2C1089%2C12_212391%2C00.html.

Coyle, Tom. "1994 local telco revenue per average switched line." *America's Network* (June 15, 1995):14.

Cripps, D. "An open letter on Sinclair's challenge to VSB." Available: http://web-star.com/hdtv/cofdmvs8vsb.html.

Cripps, D. "DIRECTV goes HD." *HDTV News* (Jan. 22, 1998).

Crishna, V., Baqai, N., Pandey, B., and Rahman, F. "Telecommunications Infrastructure: A long way to go." *South Asia Networks Organization.* Available: http://www.sasianet.org/telecominfrastr.htm#hdg.2.

Crossman, M. "Dollars and sense: Cyclical growth ahead for satellite manufacturing." *Satellite Today* (June 1999).

Crutchfield, E.B., ed. *Engineering handbook*, 7th ed. Washington, DC: National Association of Broadcasters, 1985.

Curtin, Michael. "Beyond the vast wasteland: The policy discourse of global television and the politics of American empire." *Journal of Broadcasting and Electronic Media* (Spring 1993):127.

Dallas, J. "Digital DTH Duels: Sharpening the Aim in Latin America." *Multichannel News International* (1998):18, 20, 36–7.

D'Amico, M.L. "Internet has become a necessity, U.S. poll shows." *CNN.com* (December 7, 1998). Online. Available: http://www.cnn.com/TECH/computing/9812/07/neednet.idg/.

"Data CLECs: xDSL Markets and Opportunities for Small and Medium-sized Businesses." Study by Pioneer Consulting (1999). Available: http://www.pioneerconsulting.com.

Davis, D.M. "Illusions and ambiguities in the telemedia environment: An Exploration of the transformation of social roles." *Journal of Broadcasting & Electronic Media* 39 (1995):517–554.

Dawson, Fred. "The state of the display: Flat-panel screens coming soon to a PDA or computer near you." *Digital Media* 3, no. 9/10 (February 1994):11.

Depp, Steven W., and Howard, Webster E. "Flat-panel displays." *Scientific American* (March 1993):90–97.

Development of a U.S.-based ATV industry. Washington, DC: American Electronics Association, May 9, 1989.

Dickinson, John. "Financial crash on the digital highway: Lack of home banking standards." *Computer Shopper* 14, no.4 (April 1994):68.

"Direct Broadcast Satellite: 10 Million and Still Growing." Yankee Group study, August 1999.

Disney Interactive, press release. "Consumers' final answer is 'Yes' to 'Who Wants to be a Millionaire'." CD-ROM, December 7, 1999. Online. Available: http://biz.yahoo.com/bw/991207/ca_disney__1.html.

"DMX Music Service first DBS user of AC-3 audio." *PR Newswire* (January 17, 1994).

Dorgan, M. "Malaysia: High tech utopia?" *San Jose Mercury News* (July 7, 1997).

Douglass, E. "Wireless cable may prove golden alternative to copper wires." *Los Angeles Times* (October 11, 1999):C4.

Drucker, Peter. *Innovation and entrepreneurship*. New York: Harper & Row, 1985.

Dupagne, M. "High-definition television: A policy framework to revive U.S. leadership in consumer electronics." *Information Society* 7, no.1 (1990):53–76.

Earnest, L. "New-media firm strikes deal with Fox on extreme sports." *Los Angeles Times* (October 12, 1999):C9, 19.

Eastman, Susan Tyler. *Broadcast/cable programming: Strategies and practices*, 4th ed. Belmont, CA: Wadsworth, 1993:29–30.

Edgar database, Walt Disney Company. Available: http://www.sec.gov.

Elmer, S. "Internet usage and commerce in Western Europe 1997–2002." International Data Corporation (IDC) research, February, 2000. Press release available: http://www.idc.com/Press/default.htm.

Emberg, J., and Rose, J. "Draft AISI/HITD Internet Connectivity Sub-Programme Framework." (April 1997). Online. Available: http://www.bellanet.org/partners/aisi/proj/itufram.htm.

Engineering Report. Washington, DC: National Association of Broadcasters, September 4, 1989:1.

English-Lueck, J.A., Darrah, C., and Freeman, J.M. "Technology and social change: The effects on family and community." Unpublished paper. June 19, 1998.

"Ericsson, Nextlink launch wireless broadband trial." *IDG News* (September 22, 1999). Online. Available: http://www.idg.net/crd__85607.html.

"European cable and satellite economics." *Screen Digest*, research report, April 1999. Available: http://www.screendigest.com.

Fantel, Hans. "HDTV faces its future." *New York Times* (February 2, 1992):H17.

Farrell, J., and Shapiro, C. *Brookings papers: Microelectronics 1992.* Washington, D.C.: Brookings Press, 1992.

"FCC. MM Docket No. 99–292 Establishment of a Class A (RM-9260 Television Service) NOTICE OF PROPOSED RULE MAKING." Washington, DC: Federal Communications Commission, September 29, 1999.

Featherstone, Mike. An Introduction. In *Global culture: Nationalism, globalization, and modernity*, ed. M. Featherstone. London: SAGE Publications, 1990:6.

Flanigan, James. "TV networks evolve from dinosaurs to darlings." *Los Angeles Times* (October 5, 1994):D1, 2.

Foley, Theresa. "Mega operators: Big fish in a very big pond." *Satellite Today* (November 1999). Available: http://www.satellitetoday.com/viaonline/issue/1099cov.htm.

Forrester, C. "Europe's New programming paradigm: The Continent's channel explosion is ushering in a host of new programming concepts." *Multichannel News International* (October 1999).

Freedman, Jonathan. "Fried green writers at the Viacom Café." *Los Angeles Times* (May 14, 1994):F2.

Fujio, T. "High-Definition television systems." *Proceedings of the IEEE* 73 (April 1, 1985):646–655.

"G7 countries agree to 11 'GII testbed networks.'" *NextNet* 4, no. 5 (March 13, 1995):13.

"G-7 nations hop on info superhighway." *Los Angeles Times* (February 27, 1995):D5.

Gagnon, D. "Toward an Open Architecture and User-Centered Approach to Media Design." *Communication Yearbook* 12, Newbury Park, CA: Sage, 1988:547–555.

Gardiner, D., et. al. "Third generation mobile: Market strategies." *Ovum* (October 1999).

Geirland, J., and Sonesh-Kedar, E. *Digital Babylon: How the geeks, the suits, and the ponytails fought to bring Hollywood to the Internet.* New York: Arcade Publishing, 1999.

Gergen, Kenneth. *The saturated self.* New York: Harper Collins, 1991.

Gilder, George. "The new rule of the wireless." *Forbes ASAP* (April 11, 1994):99–110.

Gilder, George. "Washington's bogeyman." *Forbes ASAP* (June 6, 1994):115.

Gilder, George. *Life after television.* Knoxville, TN: Whittle Direct Books, 1990.

Golding, Peter. "The communications paradox: Inequality at the national and international levels." *Media Development* 41 (April 1994).

Graf, James E. "Global information infrastructure: First principles." *Telecommunications* (May 1994):72–73.

Green, Richard. Speaker, National Association of Broadcasters' convention, April 1998, Las Vegas, NV.

Halliwell, Chris. "Camp development: The art of building a market through standards." *IEEE Micro* (December 1993):10–18.

Halonen, Doug. "FCC: Who pays for advanced TV?" *Electronic Media* (March 1995):1, 75.

"Hamamatsu City designated HiVision City by Ministry of Posts." *COM-Line Daily News Telecommunication* (September 6, 1994), archived on NEXIS.

Hanson, Jarice. *Connections: Technologies of communication.* New York: Harper Collins College Publishers, 1994:100.

Harmon, A. "Sad, lonely world discovered in cyberspace." *New York Times* (October 30, 1998).

Harmon, Amy. "A digital visionary scans the info horizon." *Los Angeles Times* (June 1, 1994):D6, D8.

Hayes, Mary. "Working online, or wasting time?" *Information Week* 525 (May 1, 1995):38.

"HDTV cooperation asked." *Television Digest* 29 (May 22, 1989):9.

"HDTV developments in Japan." *Financial Times* (May 9, 1989), archived on NEXIS.

"HDTV live broadcasts." *Japan Economic Newswire* (February 23, 1988), archived on NEXIS.

"HDTV production standard debated at NTIA." *Broadcasting* 116 (March 13, 1989):67.

"HDTV transmission tests set to begin next April." *Broadcasting* 119 (November 19, 1990):52–53.

"HDTV: Broadcasters look before they leapfrog." *Broadcasting* 117, no. 11 (September 11, 1989):24.

Henig, P.D. "Qualcomm, Ericcson unite over new wireless standard: Red herring online." (March 29, 1999). Online. Available: http://www.redherring.com/insider/1999/0329/news-qual-comm.html.

Hofmeister, S. "Bochco sues 20th Century Fox over 'NYPD Blue.'" *Los Angeles Times* (September 14, 1999):C-9.

Hofmeister, S. "Disney plans to consolidate two of its television groups." *Los Angeles Times* (July 8, 1999):C-1.

Holsendorph, Ernest L. "CBS cable bid cleared by FCC." *New York Times* (August 5, 1981):D-1.

Holsinger, Erik. "Kai's power tools." *Digital Video (DV)* (April 1995):37–40.

"Home Telephone selects Newbridge for LMDS trial." Available: http://www.mainstreetexpress.com/doctypes/customer-story/hometel.jhtml.

Hontz, Jenny. "Infohighway bill passes Senate panel." *Electronic Media* (August 15, 1994):1.

Horton, Bob. "Standardization and the challenge of global consensus." *Pacific Telecommunications Review* (September 1993):16–22.

Hotz, R.L. "String of missteps doomed orbiter." *Los Angeles Times* (November 11, 1999):A1, 17.

House., K. "To Connect or Not to Connect: Home-Networking Market Review. IDC research Study and Forecast, 1998–2002. Report #W18220." March 1999.

Iacta, press release. "Iacta Releases Two-Year Landmark Study of WebTV and Its Users." May 1, 1999. Online. Available: http://www.iacta.com/news_0399_1.htm.

"I must go down to the seas again." *Public Network Europe* (December 1998/January 1999).

"Interactive TV to emerge as hybrid of Web and TV models." Jupiter Communications, December 8, 1999.

"Internet surfing not impacting TV viewing." Burke study at *CyberAtlas.* Available: http://cyberatlas.internet.com/big_picture/traffic_patterns /print/0,1323,5931_214791,00.html.

"Internet usage threatens TV viewing." *Inside Cable* (August 16, 1998). Available: http://www.inside-cable.co.uk/n98q3alt.htm.

"Iridium files chapter 11." *CNNfn Online* (August 13, 1999). Online. Available: http://cnnfn.com/1999/08/13/companies/iridium/.

Isenberg, D. "The rise of the stupid network." (1998). Available: http://www.isen.com.

Jacobi, Fritz. "High-definition television: At the starting gate or still an expensive dream?" *Television Quarterly* 16 (Winter 1993):5–16.

Jaggi, P., and Steinhorst, L. "IP over photonics: Making it work." *Lightwave* 16, no. 1 (January 1999).

James, M.L., Wotring, C.E., and Forrest, E.J. "An exploratory study of the perceived benefits of electronic bulletin board use and their impact on other communication activities." *Journal of Broadcasting & Electronic Media* 39, no. 1 (1995):30–50.

Jensen, J.F. "The concept of interactivity." *Interactive Television*. Aalborg, Denmark: Aalborg University Press, 1999:25–66.

Jensen, M. "Where is Africa on the Information Highway? The status of Internet connectivity in Africa." Paper presented at RINAF Day/CARI 98, Dakar, Senegal October 16, 1998. Online. Available: http://www.unesco.int/web-world/build_info/rinaf/docs/cari98.html.

Jessell, Harry. "Infohighway power play." *Telemedia Week* (December 1994):26–28.

Johnson, G. "Big media firms are offering advertisers one-stop shopping." *Los Angeles Times* (September 28, 1999):C-4.

Johnson, G. "Investors have dot-qualms on ad spending." *Los Angeles Times* (May 8, 2000):C-1, 5.

Jones, K. "Jolly old Cambridge targets high-tech success." *Inter@ctive Week* 3, no. 47 (November 30, 1998.).

Kao, K.C., and Hockham, G.A. "Dielectric-fibre surface waveguide for optical frequencies." *Proceedings of the IEEE* 133, no. 7 (July 1966):1151–1158.

Kapoor, Sanjay. Speech at Digital World, April 1995, Los Angeles.

Katz, E., and Blumler, J.G. "Uses of mass communications: Current perspectives on gratifications research." Beverly Hills, CA: Sage, 1974.

Keller, J.K. "AT&T secret multimedia trials offer clues to capturing interactive audiences." *Los Angeles Times* (October 6, 1993):C1.

Kelly, Lindsey. "Group delaying interactive goal." *Electronic Media* (March 20, 1995):22, 36.

Kennard, W.E. "Unleashing the totential: Telecommunications development in Southern Africa." Keynote speech before the Annual General Meeting Telecommunications Regulators Association of Southern Africa (TRASA), August 11, 1999, Gaborone, Botswana.

Kennard, W.E. "Vision to mission: A blueprint for architects of the global information infrastructure." Speech before the World Economic Development Congress, September 23, 1999, Washington, D.C.

Kiesler, S., Siegel, J., and McGuire, T.W. "Social psychological aspects of computer-mediated communication." *American Psychologist* 39, no. 10 (1984):1123–1134.

Kim, B., Chung, Y., and Kim, S. "Dynamic analysis of widely tunable laser diodes integrated with sampled- and chirped-grating distributed Bragg reflectors and an electroabsorption modulator." *Institute of Electronics, Information, and Communication Engineers* E81-C: 8 (August 1998):1342–1349.

Krauss, Jeffrey. "NGSO satellites—Creating new spectrum capacity." *CED* (March 1999).

Kupfer, Andrew. "The U.S. wins one in high tech TV." *Fortune* 60, no. 4 (April 8, 1991):123.

Kwong, K.C. Speech at Asia Pacific Smart Card Forum and the Electronic-Business Symposium, November 1998, Melbourne, Australia.

Kwong, K.F. "American Peacock on Chinese Soil: The Challenge Facing NBC Asia in Greater China." A Project Report Submitted in Partial Fulfillment of the Requirements for the Degree of Master of Arts in Communication, School of Communication, Hong Kong Baptist University, August 1998.

Lambert, Peter. "Abel says multichannel options could pay for HDTV." *Broadcasting* 122 (October 26, 1992):44.

Lambert, Peter. "Utility networking percolates." *On Demand* (December 1994/January 1995):16.

Lambert, Peter. "ACATS orders issue of broadcast multichannels." *Multichannel News* 15, no. 9 (February 28, 1994):3.

Lambert, Peter. "First ever HDTV transmission." *Broadcasting* 122 (March 2, 1992):8.

Lane, Earl. "The Next Generation of TV." *Newsday* (April 5, 1988):Discovery section, 6.

Lawton, George. "Deploying VSATs for specialized applications." *Telecommunications* 28:6 (June 1994):27–30.

Lawton, George. "The utilities' role: Building the ubiquinetwork—Part 5." *Communications Technology* (December 1994):80–85.

Lazarus, D. "Asia's racing toward wiredness." *Wired.com* (July 30, 1997).

Levine, Marty. "Critical time nears for setting HDTV standard." *Multichannel News* (June 12, 1995):12A.

Levittan, R. Panelist at NATPE/ETC one-day seminar at the American Film Institute, November 1999, Los Angeles.

Lewyn, Mark, Thierren, Lois, and Coy, Peter. "Sweating out the HDTV contest." *Business Week* 33, no. 6 (February 22, 1993):92–93.

"Licensing framework for new pay and specialty services in a digital world: The need for new rules and immediate action for more consumer choice, addressing Public Notice CRTC 1999–19." Online. Available online at: http://www.crtc.gc.ca/ENG/PROC_BR /NOTICES/1999/1999–19e/co0322(submission).doc.

Lippman, John. "Networks push for cheaper shows." *Los Angeles Times* (February 19, 1991):D1.

Lippmann, Andrew. "HDTV sparks a digital revolution." *BYTE* (December 1990):297–305.

"LMDS and the Buildout of Broadband Wireless Networks: Worldwide Market Opportunities for LMDS, MMDS, and MVDS Technologies." Pioneer Consulting proprietary research, August, 1998.

Lowndes, Jay C. "14 seek direct broadcast rights." *Aviation Week and Space Technology* (August 10, 1981):60.

Lowry, B. "The fittest? Make that survival of the biggest." *Los Angeles Times* (April 11, 2000):F-1.

Luff, Bob. "Why take an interest in DAVIC?" *Communications Technology* (April 1995):18–19.

Luther, Arch C. *Digital Video in the PC Environment*. New York: McGraw-Hill, 1991.

Madhavan, N. "Asia lags in bumpy Internet revolution." *Reuters* (September 9, 1998).

Magel, Mark. "The box that will open up interactive TV." *Multimedia Producer* (April 1995):30–36.

Makal, K. "Three-ring circus on interactive design." *TVindustry.com* (July 29, 1999). Available: www.tvindustry.com.

Marshall, Tyler. "E.U. panel urges tighter TV import quotas." *Los Angeles Times* (February 9, 1995):C-1.

McClellan, B. "The future of the entertainment and media industries: 2005." PricewaterhouseCoopers, 2000.

McConnell, K. "The future of the Earth station industry: A billion dollar bet." *Satellite Today* (May 1999). Available: http://www.satellitetoday.com/viaonline/backissues/1999/0599cov.htm.

McCullagh, D. "Report: Mideast misses the net." *Wired Online* (July 8, 1999).

McElvogue, Louise. "Cannes now a prime-time player." *Los Angeles Times* (October 13, 1995):D4, D5.

McLaughlin, Laurianne. "Pentium flaw: A wake-up call?" *PC World* 13, no. 3 (March 1995):50–51.

McQuillan, J.M. "Technology analysis: What's next for the Net?" *Broadband World* 1, no.1 (1999), special advertising supplement.

MediaStation, press release. "Comcast, @Home and Media Station team up to launch market trial of SelectPlay Interactive Software Service." December 6, 1999. Online. Available: http://biz.yahoo.com/bw/991206/ca_media_s_1.html.

Meherabian, A. *Silent messages*. Belmont, CA: Wadsworth, 1971.

"Member states ready to impose E.U. norm on digital TV." *European Insight* (June 10, 1994):615.

Menn, J. "Pioneer paints portrait of graphics' future." *Los Angeles Times* (August 9, 1999):C1, 5.

"Microsoft and British telecom join forces against symbian alliance." *EPOC Times* (February 10, 1999). Online. Available: http://pdacentral. flashnet.it/5alive/Archives99/Feb10_613.htm.

Middleton, G. "University unplugs last electronic bottleneck." *TechWeb* (September 27, 1999). Online. Available: http://www.techweb. com/wire/story/TWB19990927S0003.

Mitchell, William. "When is seeing believing." *Scientific American* (February 1994):68–73.

Mooradian, M. "Internet games: Five-year outlook—revenue models and technology development." Jupiter Communications Research Study, March 1998.

Moore, Geoffrey A. *Inside the tornado.* New York: Harper Collins, 1995.

Motorola corporate press release, September 6, 1999. Available: http://www.apspg.com/press/090199/8vsb.html.

Mountford, Joy. "Essential interface design." *Interactivity* (May/June 1995):60–64.

Mowlana, H., and Wilson, L.. *The passing of modernity: Communication and the transformation of society.* New York: Longman, 1990:43–75.

Mulligan, T. "The new oligopoly boom." *Los Angeles Times* (August 22, 1999):C-1.

"Multimedia Communications Forum establishes workgroup to establish MIB." *OSINetter Newsletter* 9 (January 1994).

"Nazcasaatchi. Internet Survey." 1999. Online. Available: http://www. nazcasaatchi.com/

Ness, Susan. DTV in the Desert Symposium Consumer Electronics Show, January 8, 1998, Las Vegas, Nevada. Available: http://www.fcc.gov /Speeches/Ness/spsn808.html.

"Net consultants identify six types of surfers." April 20, 2000. Online. Available: http://canadacomputes.com/CCP/Print/1,1040,3310,00.html.

"Net usage versus offline media." *eMarketer* (November 27, 1999). Online. Available: http://e-land.com/estats/usage_net_vs.html.

Neufeld, E., et al. "Consumer Internet economy portal landscape, revenue strategies, five-year projections." Jupiter Communications Research Study (July 1998). Jupiter Communications, 627 Broadway, New York, NY 10012, phone: 212–780–6060, www.jup.com.

Neufeld, E., and Beauvillain, O. "Online landscape: Vision report— France, Vol. 2." *Jupiter Communications* (December 19, 1999).

Newton, Harry, and Horak, R. *Newton's Telecom Dictionary,* 16th ed. Telecom Books/Miller-Freeman, 2000.

"Next Generation Broadband Satellite Networks." A proprietary study by Pioneer Consulting, October 1999.

"Next Generation Broadband Satellite Networks: Executive Summary." *Pioneer Consulting* (September 1999). Online. Available: http://www.pioneerconsulting.com.

"NHK and nine private stations for trials of EDTV." *COMLine Daily News Telecommunication* (September 1, 1994), archived on NEXIS.

Nickell, Joe. "Tune in, turn on . . . to what?" *Wired News* (September 10, 1998). Available: http://www.wired.com/news/news/culture/ story/14781.html.

Nickelson, Richard I. "The evolution of HDTV in the work of the CCIR." *IEEE Transactions on Broadcasting* 35, no. 3 (September 1989):250–258.

Nicolis, G., and Prigogine, I. *Self-organization in non-equilibrium systems.* New York: Wiley, 1977.

Nielsen Media press release, November 11, 1998. Available: http://www. nielsenmedia.com/newsreleases/releases/1998/HHtuning.html.

Niemack, J. Panelist at a seminar sponsored by the National Association of Television Programming Executives (NATPE) and Entertainment Technology Commerce (ETC), November 1999, at the American Film Institute's Los Angeles campus.

Nolle, T. "Take a look at the dark side of fiber." *LANTimes* (October 1997). Online. Available: http://www.lantimes.com/97/97oct/710a063b. html.

Nortel Networks, press release. "Liberty Cellular will be using Nortel Networks equipment." Available: http://www.nortelnetworks.com/.

"Online radio listeners are the new breed of Internet consumers." Online. Available: http://www.newmediamusic.com/ps/arbitron_net _services.html.

Owen, Bruce M., and Wildman, Steven. *Video economics.* Cambridge: Harvard University Press, 1992:270–271.

Patton, Carl. "Digital HDTV: on-the-air!" *ATM* 1 (Advanced Television Markets) no. 5 (April 1992):1–2.

PBS, release. "Intel and PBS collaborate on digital broadcast programming." April 6, 1998. Available: http://www.pbs.org/insidepbs /news/intelpbs.html.

PCData, press release. "PC data releases 1998 U.S. software sales statistics." January 27, 1999. Online. Available: http://www.pcdata.com/.

Pelton, Joseph N. "Geosynchronous satellites at fourteen miles altitude?" *New Telecom Quarterly* (Second Quarter, 1995):11.

Peters, P. "Going places: Emerging markets roundup." *Wireless Week International* (June 22, 1998).

Pierce, John. *An introduction to information theory: Symbols, signals, and noise,* 2nd ed. New York: Dover Publications, 1980.

Pine II, B. Joseph, and Gilmore, James H. "Are you experienced?" *Industry Standard* (April 9, 2000). Online. Available: http://thestandard.net/article/article_print/0%2C1153%2C4167%2C00.html.

Pine, B. Joseph, Gilmore, James H., and Pine II, B. Joseph. *The experience economy: Work is theatre and every business a stage.* Cambridge MA: Harvard Business School Press, 1999.

Pine, B.J., Gilmore, James H., and Pine II, B.J. *The experience economy.* Boston: Harvard Business School Press. 1999.

Platt, Charles. "Satellite pirates." *WiReD* (August 1994):8, 122.

Podlesny, Carl. "Hybrid fiber coax: A solution for broadband information services." *New Telecom Quarterly* 1 (1995):16–25.

Press release. *Canada News Wire.* Available: http://www.newswire.ca/releases/August1998/19/c3402.html.

Price, Derek, et al. "Collaboration in an invisible college." *American Psychologist* 21 (1966):1011–1018.

PricewaterhouseCoopers research. Reuters report, October 1, 1999. Available at Nua Surveys: http://www.nua.ie/surveys/index.cgi?f=VS&art_id =905354444&rel=true.

PricewaterhouseCoopers. "Digital Television '99: Navigating the transition in the U.S." 1999, PricewaterhouseCoopers Entertainment & Media Marketing Group, 212–596–3737.

PricewaterhouseCoopers. "Digital Television '99: Navigating the transition in the U.S." Position paper from the *PwC Entertainment/Media Group* (1999):35–6.

Quittner, Joshua. "500 TV channels? Make it 500 million." *Los Angeles Times* (June 29, 1995):D2, D12.

Ramnarayan, S. "Telewebbers on the rise." Gartner Group, research publication, June 5, 2000.

Rao, M. "Indian ISPs form alliances to tackle market growth challenges." *Indialine.com* (March 4, 1999). Online. Available: http://www.india-line.com/net.columns/column66.html.

Rast, Robert. *Statement of Robert M. Rast, Vice President, HDTV Business Development, General Instrument Corporation, Communications Division.* Washington, D.C., FDCH Congressional Testimony, (March 17, 1994), archived on NEXIS.

Reece, K. "Modeling the Propagation and Latency Effects of the Integration of ATM in a Seamless Terrestrial and Satellite-Based Wireless Network." Project at the Center for Wireless Telecommunications, Virginia Tech.

"Refined HDTV cost estimates less daunting." *Broadcasting* 118 (April 9, 1990):40–41.

"Report No. Mm 97–8 Mass Media Action April 3, 1997, Commission Adopts Rules for Digital Television Service, (Mm Docket No. 87–268)." Available: http://www.fcc.gov/Bureaus/Mass_Media /News_Releases/1997/nrmm7005.html.

Reuters News Service Report 3, June 17, 1994, archived on NEXIS.

Reuters. "India on threshold of Internet age." *San Jose Mercury News* (January 23, 1999). Online. Available: http://spyglass1.sjmercury.com /breaking/docs/005829.htm.

Reuters. "News Corp. planning to bundle its 'crown jewels' in a new company." *Los Angeles Times* (May 17, 2000):C-5.

Reuters. "AOL, TiVo hook up on interactive TV service." *Los Angeles Times* (June 15, 2000):C3.

Rheingold, H. *The virtual community: Homesteading on the electronic frontier.* Reading, MA: Addison-Wesley, 1993.

Rifkin, J. "Cradle to grave, you're a customer first." *Los Angeles Times* (April 24, 2000):B-7.

Rivkin, S.R. "Positioning the electric utility to build information infrastructure." *New Telecom Quarterly* (Third Quarter, 1995):30–33.

Robinson, B. "HDTV grand alliance faces tough road." *Electronic Engineering Times* (May 31, 1993):1, 8.

Rogers, Everett M. *Diffusion of innovations,* 4th ed. New York: Free Press, 1995.

Rogers, Everett M. *Diffusion of innovations.* New York: Free Press of Glencoe, 1962.

Rosenthal, Edmond. "FBC studies multiplexing strategies." *Electronic Media* (February 28, 1994):26.

Rubin, D.K., and Angelo, W.J. "Level 3 grows on high fiber diet." *Telecommunications* (November 1, 1999). Online. Available: http://enr.com /new/C1101.asp.

Rutter, Derek. *Looking and Seeing: The role of visual communication in social interaction.* New York: Wiley, 1984.

Sablatash, Mike. "Transmission of all-digital advanced television: State of the art and future directions." *IEEE Transactions on Broadcasting* 40 (June 1994):2.

Samuelson, Pamela. "Digital media and the law." *Communication of the ACM* 34, no. 10 (October 1991):23–29.

Sanchez, Jesus. "Two phone giants open cable TV's door." *Los Angeles Times* (February 12, 1995):D1, 2.

Santo, Brian, and Yoshida, Junko. "Grand alliance near?" *Electronic Engineering Times* (May 31, 1993):1, 8.

Sarkar, M.B., Butler, B., and Steinfield, C. "Intermediaries and cybermediaries: A continuing role for mediating players in the electronic marketplace." *Journal of Computer-Mediated Communication* 1, no. 3 (1995). Online. Available: http://www.ascusc.org/jcmc/vol1/issue3 /sarkar.html.

"Satellite communications: The birth of a global footprint." *Voice and Data* (February 1997). Online. Available: http://www.voicendata.com /feb97/4ib0081101.html.

Schlossstein, Steven. "Intelligent user interface design for interactive television applications." *1995 NCTA Technical Papers* (May 1995):165–170.

Schmitz, J., Rogers, E., Phillips, K., and Paschal, D. "The Public Electronic Network (PEN) and the homeless in Santa Monica." *Journal of Applied Communication Research* 23, no. 1 (February 1995):26–43.

Schoechle, T. "Toward a theory of standards." Online. Available: http://www.standardsresearch.org/presentations.html.

Schoenherr, S. "Recording technology history." Online. Available: http://history.acusd.edu/gen/recording/digital.html (revised February 24, 2000).

Schrage, M. *Shared minds*. New York: Random House, 1988.

Schrage, Michael. "Humble pie: Japanese food for thought." *Los Angeles Times* (November 4, 1994):C1, 4.

Schwartz, Frank. "Set-top standards." *Electronic Design* 42, no. 9 (September 19, 1994):151.

SeniorNet and Charles Schwab Inc., study, October 1998. Available: http://www.nua.ie/surveys/
index.cgi?f=VS&art_id=905354444&rel=true.

Shannon, Claude, and Weaver, Warren. *A mathematical theory of information*. Urbana: University of Illinois Press, 1949.

Sheer, D.I. "Cost-effective DTV: Transmitter companies offer digital transition strategies." *Digitaltelevision.com* (December 1998). Available: http://www.digitaltelevision.com/business1298bp.shtml.

Short, J., Williams, E., and Christie, B. *The social psychology of telecommunications*. New York: Wiley, 1976.

Skoros, J. "LMDS: Broadband wireless access." *Scientific American* (October 1999).

"SMPTE: Seeking a universal, digital language." *Broadcasting* 121 (November 4, 1991):62.

"Soliton waves double fiber-optic capacity." *New Telecom Quarterly* (Second quarter, 1993):6.

Somheil, T. "Real life in a box." *Appliance* 508, no. 8 (August 1993):41.

Stallings, William. *Data and computer communications*, 4th ed. New York: Macmillan, 1992:803.

Stallings, William. *Data and computer communications*, 4th ed. New York: Macmillan, 1992:803.

"Star Media Annual Report, 10-K405." (March 30, 2000). Available on company Web site at: http://www.starmedia.com, or through the EDGAR database at: http://www.sec.gov.

Steckley, E. "Broadband wireless access: Dawn of a new era." *Telecommunications Online* (February 1998).

Stefanic, S. Panelist at seminar sponsored by the National Association of Television Programming Executives (NATPE) and Entertainment Technology Commerce (ETC), November 1999, American Film Institute's Los Angeles campus.

Stein, Mark. "Satellites: Companies, nations fight for spots in space." *Los Angeles Times* (September 20, 1993):A1, A16.

Stern, Christopher. "FCC moves to strengthen wireless cable." *Broadcasting & Cable* (June 13, 1994):11.

Stern, Christopher. "Telcos hedge bets with wireless wagers." *Broadcasting & Cable* (May 1, 1995):14.

Stilson, J., and Pagano, P. "May the best HDTV system win [an interview with R. Wiley]." *Channels 10* (August 13, 1990):54–55.

Stone, A.R. "Will the real body please stand up?: Boundary stories about virtual cultures." In *Cyberspace: First steps*, Michael Benedikt, ed. Cambridge: The MIT Press, 1991:81–118.

Swedlow, T. "Enhanced television: A historical and critical perspective." Commissioned by the AFI-Intel Television Workshop, July 1999.

Swerdlow, F. "Growth abroad requires change by U.S. sites wishing to compete." Jupiter Communications, October 25, 1999.

Swerdlow, T. "Telewebbers on the rise." *Interactive TV newsletter.* (June 14, 2000). Online. Available: http://www.itvt.com.

System Subcommittee Working Party 2. Document SS/WP2–1354. Washington: Advisory Committee on Advanced Television Service, September 1994.

Tadjer, Rivka. "Low-orbit satellites to fill the skies by 1997." *Computer Shopper* 15, no. 3 (March 1995):49.

Takashi, F. "High-definition television systems." *Proceedings of the IEEE.* 74, no.4 (April 1985):646–655.

Taylor, D., and Brentnall, T. "Display Technologies for Advanced Television Overview." 1998 Society of Motion Picture and Television Engineers Seminar on Production and Display Technologies for HDTV, May 16,1998, University of Southern California, Los Angeles.

Teledesic news release. Available: http://www.teledesic.com/newsroom/articles/2000–05–17%20ico.htm.

"The CommerceNet/Nielsen Internet Demographic Survey." Research by CommerceNet/Nielsen, June 1998. Press release available: http://www.commerce.net/news/press/19980824b.html.

"The Internet in the Middle East and North Africa: Free expression and censorship." Compiled by *Human Rights Watch* (July 1999). Online. Available: http://www.hrw.org/reports98/publctns.htm.

"The Market for Video and Multimedia Services 1998–2003." Insight Research Corporation report, October 1998.

Tobenkin, David. "The wireless system that could." *Broadcasting & Cable* (May 1, 1995):20.

Travers, Jeffrey, and Milgram, Stanley. "An experimental study of the 'Small World Problem.'" *Sociometry* 32 (1969):425–443.

Truxal, John.G. *The age of electronic messages.* New York: McGraw-Hill, 1990:309.

Turing, A. "Computing machinery and intelligence." *Mind LIX* (1950):11–35.

"TV and the Internet." *Electronic Media Special Report* (June 28, 1999).

Twain, Mark. *Pudd'nHead Wilson and those extraordinary twins.* New York: Harper, 1922.

Tyrer, Daniel. "The high definition television programme in Europe." *European Trends* (Fourth quarter, 1991):77–81.

U.S. Congress, Office of Technology Assessment. *Global standards: building blocks for the future, TCT-512.* Washington, DC: U.S. Government Printing Office, March 1992:8.

U.S. Department of Commerce, "National Telecommunications and Information Administration (NTIA). Fact Sheet: Native Americans Lacking Information Resources." Online. Available: http://www.ntia. doc.gov/ntiahome/digitaldivide/factsheets/native-americans. htm.

U.S. Federal Communications Commission. "In the Matter of Advanced Television Systems and Their Impact Upon the Existing Television Broadcast Service. FR, MM Docket No. 87–268." (Adopted: December 24, 1996, released: December 27, 1996). Available: http:// www.fcc.gov/Bureaus/Mass_Media/Orders/1996/ fcc96493. txt.

"U.S. Industry and Trade Outlook 1998." *National Technical Information Service.* Washington, DC: U.S. Department of Commerce (1998). Online. Available: http://www.ntis.gov/.

"U.S. staging comeback in technology." *Los Angeles Times* (September 19, 1994):C3.

U.S. Wireless Broadband: LMDS, MMDS and Unlicensed Spectrum. Strategis Group proprietary report, December 6, 1999.

Van Tassel, Joan, and Rose, S. "The evolution of the interactive broadband server." *New Telecom Quarterly* 1, 2 (1996).

Van Tassel, Joan. "World domination." *The Hollywood Reporter: Anniversary Issue* (December 1997):S-1, 5, 7.

Van Tassel, Joan. "Yakkety-yak, do talk back: Santa Monica's PEN System." *WiReD* (January 1994.)

Van Tassel, Joan. "Act two: The curtain rises on the 1996 telecommunications reform act." *The Hollywood Reporter, Special Issue on the New Regulatory Environment* (April 30, 1996):S1–3, 4.

Van Tassel, Joan. "The computer industry re-boots the U.S. regulatory system." *New Telecom Quarterly* 1 (1997):21–40.

Vogel, H.L. *Entertainment industry economics,* 4th ed. Cambridge: Cambridge University Press, 1998:75–77.

Vonder Haar, S. "Portal player cultivates south-of-border market." *Inter@ctive Week* 5, no. 47 (November 30, 1998):48.

Waggoner, B. "Making great Web video." *DV Magazine* (October, 1999). Online. Available: http://www.dv.com/magazine/1999/1099 /webvideo1099.pdf.

Warner, R.M. "Speaker, partner, and observer evaluations of affect during social interaction as a function of interaction tempo." *Journal of Language and Social Psychology* 11, no. 4 (1992):253–266.

Weaver, R. Speaker, "Your future in multimedia: Interactive Hollywood, new media, new jobs, new markets." Academy of Television Arts and Sciences and the International Interactive Communications Society, July 22, 1995, North Hollywood, CA.

"Webaholics Anonymous." Online. Available: http://www.concentric.net/~Astorm/iad.html.

Weinman, Richard. "Anytime, anywhere communication." *New Telecom Quarterly* (Fourth Quarter, 1994):18–22.

Weissman, R.X. "Connecting with digital kids." *American Demographics* (April 1999). Online. Available: http://www.demographics.com/publications/ad/99_ad/9904_ad/ad990405d.htm.

Weld, D. "Africa and the global information wave." *The Journal of Public and International Affairs* (1997). Online. Available: http://www.wws.princeton.edu/~jpia/1997/chap2.html.

Wharton, Dennis. "HDTV org threatened by flexibility." *Daily Variety* (August 8, 1994):32.

"What's new about globalization." *The McKinsey Quarterly* 2 (1997):179.

Wiener, N. *Cybernetics: Control and communication in the animal and the machine*. New York: Wiley, 1948.

Wilcox, J. "Who will win the DVD standards fight?" *CNET News* (August 10, 1999).

Wiley, Richard E. "High tech and the law." *American Lawyer* (July 26, 1994):6.

Wiley, Richard. "Entertainment for Tonight (and tomorrow too)." *The Recorder* (July 26, 1994):6–10.

"Will shift to digital HDTV, Japan firm says." *Los Angeles Times* (June 9, 1994): C3.

Wilson, C. "Network architectures of the future." *Interactive Week White Paper* (May 4, 1998).

Wilson, Carol. "Telco networks take the fast lane." *Inter@ctive* 2, no. 9 (May 8, 1995):35.

Wilson, Peter R. "Standards: Past tense and future perfect?" *IEEE Computer Graphics & Applications* (January 1991):47.

Winston, B. "HDTV in Hollywood." *Gannet Center Journal* 3, no. 3 (1989):123–137.

Wirbel, L. "Fiber-Optic Network." *EE Times* (September 1, 1999). Online. Available: http://internetwk.com/story/INW19990901S0005.

Wolf, M.J. *The entertainment economy*. New York: Random House, 1999:223.

Yoshida, J. "DVB digital TV format gains on ATSC standard." *EE Times* (June 5, 2000).

Ziffren, J. Panelist at Intel/AFI one-day program to present work-in-progress material from recipients of grants for enhanced TV programming, September 1999, Los Angeles.

Zukav, Gary. *The dancing Wu Li masters*. New York: William Morrow, 1979:12–13.

Personal Interviews

Julius Adams (Consultant)

Gray Ainsworth (MGM)

Camille Alcasid (UltimateTV.com)

Robert Alexander

Gary Arlen (Arlen Communications)

Ken Aupperle (Hauppauge Digital)

Robert Bakish (Viacom)

Dave Banks (U.S. West)

Mary Barnsdale (AT&T)

Richard Bauarschi (Synctrix)

Michael Bebel (Universal)

Avi Bender (IBM)

Josh Bernhoff (Forrester Research)

Jim Bloom (Sony)

Jim Boyle (YCTV)

Flory Bramnick (Belo Television Group)

Mike Brand (Ameritech)

Alan Brightman (Apple)

James Bromley (Engineer)

Jerry Brown (U.S. West)

Cynthia Brumfield (Broadband Intelligence)

Jan Brzeski (Sonic Foundry)

Bill Buhro (EDS)

Mark Bunzel (Intel)

Brian Burke (AutoFilms)

Kevin Burmeister (Brilliant Digital Entertainment)

Nancy Buskin (Viacom)

Joseph Butt (Forrester Research)

Ed Cafasso (eMedia)

Ken Calvert (Georgia Institute of Technology)

Tom Campo (GeoCast)

James Carlson (Jones Intercable)

Frank Casanova (Apple)

Gordon Castle (CNN)

M. Cevis (Babelsberg Studios, Berlin)

Gareth Chang (Hughes International)

Jim Chiddix (Time Warner)

Adam Clampitt (VidNet.com)

Adam Cohen (Digital Island)

Scott Cooper (Vela Research)

Anita Corona (IN)

Bill Correll (Sun Microsystems)

Jim Cracraft (Dreamworks)

Allen Crawford (StationX)

Dale Cripps (HDTV Analyst)

Mark Cuban (Founder, broadcast.com)

Drew Cummings (AENTV)

Jason Danielson (Silicon Graphics)

Mariana Danilovich (Digital Media X)

C. Davidson (CBS)

Jonathan Davis (Screamingly Diffferent)

Craig Decker (Spike & Mike)

Dr. Sadie Decker (TCI)

Joe DeMauro (NYNEX)

Marcia DeSonne (National Association of Broadcasters)

Richard Doherty (Envisioneering)

Patrick Donoghue (formerly of Big Band Media)

Bob Doyle (Digital Video Group)

Kevin Doyle (BellSouth)

Rick Ducey (Former S.V.P., Research, National Association of Broadcasters)

Ellen East (Cox Cable)

Scott Evans (Interaxx)

James Fancher (Pacific Ocean Post)

Peter Fannon (Advanced Television Testing Center)

Tom Feige (Time Warner)

Stefan Felisan (House of Blues)

Bob Ferguson (Southwestern Bell)

Ginger Fisk (Bell Atlantic)

Dr. Joseph Flaherty (CBS)

Fred Fletcher (Burbank Public Service Department)

Linda Frazier (formerly of Discovery Communications)

Mark Friedmann (Prisa Networks)

Jack Galmiche (InTouch)

Neil Gaydon (Pace Technologies)

Phyllis Geller (WETA-TV)

Bill Geppert (Cox Communications)

Massoud Ghaemi (Los Angeles Department of Water and Power)

George Gilder (author)

Jonathan Gill (former White House staffer)

Jon Gluck (formerly of YCTV)

Howard Gordon (formerly Xing)

Irwin Gottlieb (TeleVest)

August Grant (2Wire.com)

Virginia Gray (Southern New England Telephone Multimedia)

Dr. Richard Green (CableLabs)

Vincent Grosso (AT&T)

Sam Gustman (Shoah Project)

Mike Gwartney (Family Channel)

Mark Haefeli (CenterSeat)

Tom Hagopian (ESPN)

Allen Haines (New Wave)

Austin Hamson (Mediatrip.com)

Scott Hardin (NBCi)

David Harrah (then at IBM)

Bill Harris (formerly of Cinebase)

Josh Harris (Pseudo Networks)

Lew Harris (E! Online)

Neil Harris (SohoNet Limited)

Gregory Harrison (Massive Media Group)

Eddy Hartenstein (DIRECTV)

Hamid Hashemi (Muvico)

Bill Hausch (Sony)

Jerry Heller (formerly of General Instruments)

Robert Hersov (Telepiu)

Ray Hodges (Technology Futures, Inc.)

Ed Horowitz (Viacom)

Rob Howe (CompUSA)

Mark Jeffrey (The Palace)

Bruce Johansen (National Association of Television Programming Executives)

Stacy Jolna (TiVo)

Dick Jones (GTE)

Andrea Kalas (Dreamworks)

Lorin Kalisky (Journalist)

Larry Kasanoff (Threshold Entertainment)

Gerry Kaufhold (Cahners In-Stat)

Patrick Kearney (MediaOne)

Jamie Kellner (WB)

Robert Kietzman (GTE)

Ed Knudson (Intellocity)

Dylan Kohler (Digital Asset Management Guru)

Tara Kolla (K Media Relations)

Sandra Kresch (PricewaterhouseCoopers)

Marty Lafferty (EON)

Michael Lambert (iBlast)

Steve Lange (U.S. West)

Carolyn Layne (Informix)

Sylvan LeClerc (Groupe Videotron)

Bruce Leek (WebTV)

Bruce Leichtman (Yankee Group)

Michael Liebhold (then at Apple)

Kenneth Locker (Worlds, Inc.)

Barrie Loeks (Loeks-Star Entertainment)

Jack Loftus (Nielsen Media Research)

Ray Lopez (FORE Systems)

Steve Mack (Real Networks)

Doug McCormick (formerly Lifetime Network)

John McCormick (Frost & Sullivan)

Leigh Meredith (Microsoft)

Peggy Miles (Intervox and International Webcasting Association)

Avram Miller (Intel)

Dan Miller (On2)

P. Mitchel (Novocom)

Jim Mitchell (TCI)

Carlos Montalvo (Virage)

Michael Moon (Gistics)

M. Moritz (National Theater Owners, California and Nevada)

Thom Mount (CastNet)

William Mutual (Popcast.com)

Bob Myers (Viacom)

Russ Myerson (WeB)

Larry Namer (ComSpan)

Ron Nutt (LPTV station owner, KCTU)

Woo Paik (formerly of General Instruments)

Paul Palumbo (DFC Intelligence)

Dave Pangrac (Pangrac & Associates)

Ken Papagan (iXL)

Chuck Pennock (BreakTV.com)

Ken Phillips (Founder, Public Electronic Network, City of Santa Monica)

Larry Plumb (Bell Atlantic)

Susan Portwood (U.S. West)

Greg Pulier (Digital Evolution)

Jim Ramo (TVN)

Kiran Rao (Steeplechase Media)

Robert Rast (formerly General Instruments)

Mitch Ratcliff (Digital Media Analyst)

Steve Rose (Viaduct Corporation)

Daryl Rosen (Oracle)

Lee Rosenberg (William Morris Agency)

Carl Rosendahl (Pacific Data Images)

Mark Rosenthal (MTV)

Lawrence Rowe (Berkeley Multimedia Research Center, U.C. Berkeley)

William Samuels (ACTV)

Peter Samuelson (Starbright)

Edmond Sanctis (NBCi)

Steve Saville (USAi Networks)

Alison Savitch (Threshold Entertainment)

Joseph Schmitz (University of Tulsa)

Mark Schubin (HDTV Analyst and consultant)

Doug Schultz (Loudeye)

Mariah Scott (Intel)

Andy Setos (FOX Broadcasting)

Jonathan Seybold

Rob Shambro (StreamSearch)

Steve Shannon (Replay)

Garrett Smith (Paramount)

Chris Speer (New Media Hollywood)

Steven Spielberg

Christopher Strachan (Bulldog)

Paul Sturiale (EON)

Bill Sullivan (Prevue)

Rob Swartz (MXG-TV)

Jonathan Taplin (Intertainer)

John Taylor (Zenith)

Gary Teegarden (U.S. West)

Tamiko Theil (Worlds, Inc.)

Linda Thurman

Richard Titus (Razorfish)

Scott Tolleson

Ed Trainor (Paramount)

Antoon Uyttendaele (ABC)

Ellen Van Buskirk (Sega)

Charlene Steele Vaughn (Click2Send)

Morgan Warstler (formerly at LoadTV)

David Waterman (Indiana University)

Winnie Wechsler (Lightspan)

Jake Weinbaum (Checkout.com)

Marc Weiss (POV Interactive)

Kevin Wendle (iFilm.com)

David Wertheimer (Wirebreak.com.)

Joseph Widoff (Advanced Television Test Center)

Russell Wintner (Cinecomm Digital Cinema)

Patty Zebrowski (PacTel)

Index